T0331136

Depressariidae

Microlepidoptera of Europe

Series Editors

Ole Karsholt (Zoological Museum, Copenhagen, Denmark)
Marko Mutanen (University of Oulu, Finland)

Editorial Board

Zdeněk Laštůvka (Mendel University of Agriculture and Forestry,
Brno, Czech Republic)
Sergey Yu. Sinev (Zoological Institute RAS, Saint-Petersburg, Russia)

Language Editor

Martin Corley (Faringdon, United Kingdom)

VOLUME 10

Depressariidae

By

Peter Buchner
Martin Corley

BRILL

LEIDEN | BOSTON

Cover illustration: *Agonopterix cervariella*, Austria, Gumpoldskirchen, larva 25.v.2013 from *Peucedanum cervaria*, moth emerged 20.vii.2013.

The Library of Congress Cataloging-in-Publication Data is available online at https://catalog.loc.gov
LC record available at https://lccn.loc.gov/2024948107

Typeface for the Latin, Greek, and Cyrillic scripts: "Brill". See and download: brill.com/brill-typeface.

ISSN 1395-9506
ISBN 978-90-04-41272-9 (hardback)
ISBN 978-90-04-71311-6 (e-book)
DOI 10.1163/9789004713116

*Dedicated to everyone interested in microlepidoptera,
in the hope that a better knowledge of the species will also contribute
to their protection and that of their habitats.*

∵

Contents

Abstract   IX
Introduction: Depressariidae through the Years   XI
Acknowledgements   XIII
Abbreviations of Museums and Private Research Collections   XVI

1   Classification of Depressariidae   1

2   Morphology of Depressariidae   3

3   Descriptions of Genitalia   6

4   Bionomics of Depressariidae   16

5   Methods   20
     Field Collecting Methods   20
     Dissecting and Preparing Male Genitalia   21
     Dissecting and Preparing Female Genitalia   26
     Photographic Methods   31
     DNA Barcoding   31
     Barcoding as a Taxonomic Aid   32

6   Notes on the Illustrations   37

7   Key to European Genera of Depressariidae   38

8   Checklist of European and Macaronesian Depressariidae   39

9   Systematic Treatment of the Genera and Species of the European
     Depressariidae   50

10   Excluded Species   361

11   Distribution Table   362

12   Colour Plates   376

13   Male Genitalia Illustrations   436

14   Female Genitalia Illustrations   499

References   577
Index to Host-Plants   596
Index to Entomological Genus Names   600
Index to Entomological Species-Group Names   601

Abstract

This volume reviews the European genera and species of the gelechioid family Depressariidae.[1] 192 species are described (including two with no European records) and figures of whole moths, male and female genitalia are given. Information is provided on life histories and distribution of the species. Distribution data are summarised in a table showing the records for each European country and the larger islands.

Six new species are described: *Agonopterix paracervariella* sp. n. (Switzerland, Italy, Slovenia); *Agonopterix richteri* sp. n. (Bulgaria, Greece); *Agonopterix galicicensis* sp. n. (North Macedonia); *Agonopterix langmaidi* sp. n. (Sicily and Crete); *Agonopterix uralensis* sp. n. (Russia); *Depressaria hansjoachimi* sp. n. (Sicily, Croatia, Greece, Turkey, Armenia). In addition one new subspecies is described: *Agonopterix putridella* subsp. *scandinaviensis* subsp. n. (Norway, Sweden, Finland).

The following nomenclatural innovations are introduced: *Luquetia osthelderi* (Rebel, 1936) comb. n. for *Epigraphia osthelderi* Rebel, 1936, a lectotype is designated for this species; *Levipalpus* Hannemann, 1953 syn. n. of *Exaeretia* Stainton, 1849; *Exaeretia nigromaculata* Hannemann, 1989 syn. n. of *Exaeretia thurneri* (Rebel, 1941); *Depressaria exquisitella* Caradja, 1920 syn. n. and *Exaeretia amurella* Lvovsky, 1990 syn. n. of *Exaeretia mongolicella* (Christoph, 1882); *Agonopterix crassiventrella* (Rebel, 1891) syn. n. of *Agonopterix adspersella* (Kollar, 1832); *Agonopterix inoxiella* Hannemann, 1959 syn. n. of *Agonopterix nodiflorella* (Millière, 1866); *Agonopterix banatica* Georgesco, 1965 syn. n. of *Agonopterix purpurea* (Haworth, 1811); *Agonopterix kotalella* Amsel, 1972 syn. n. of *Agonopterix curvipunctosa* (Haworth, 1811); *Agonopterix dictamnephaga* Rymarczyk, Dutheil & Nel, 2012 syn. n. of *Agonopterix pupillana* (Wocke, 1887); *Depressaria echinopella* Chrétien, 1907 syn. n and *Agonopterix mendesi* Corley, 2002 syn. n. of *Agonopterix straminella* (Staudinger, 1859); *Agonopterix budashkini* Lvovsky, 1998 syn. n. of *Agonopterix squamosa* (Mann, 1864); *Agonopterix seneciovora* Fujisawa, 1985 syn. n. of *Agonopterix cotoneastri* (Nickerl, 1864), *Depressaria cyrniella* Rebel, 1929 syn. n. and *Depressaria genistella* Walsingham, 1903 syn. n. of *Agonopterix scopariella* (Heinemann, 1970); *Depressaria perstrigella* Chrétien, 1925 syn. n. of *Agonopterix nervosa* (Haworth, 1811); *Depressaria duplicatella* Chrétien, 1915 syn. n., *Depressaria adustatella* Turati, 1927 syn. n., *Depressaria delphinias* Meyrick, 1936 syn. n. and *Depressaria subtenebricosa* Hannemann, 1953 syn. n. of *Depressaria radiosquamella* Walsingham, 1903; *Depressaria venustella* Hannemann, 1990 syn. n. of *Depressaria manglisiella* Lvovsky,

---

1 Depressariidae as defined in this book contains just five European genera. Some recent classifications of Lepidoptera include other groups in the family. Our reasons for excluding these are discussed in the section on Classification of Depressariidae.

1981; *Depressaria corticinella* Zeller, 1854 syn. n. of *Depressaria badiella* (Hübner, 1796); *Depressaria fusconigerella* Hannemann, 1990 syn. n. of *Depressaria subnervosa* Oberthür, 1888. *Depressaria pavida* Meyrick, 1913 is reduced to subspecies of *Agonopterix adspersella*; *Depressaria tenerifae* Walsingham, 1908 and *Depressaria rungsiella* Hannemann, 1953 are restored to species status. A neotype has been chosen for *Depressaria beckmanni* Heinemann, 1870.

Introduction: Depressariidae through the Years

Some depressariids have been known since the earliest years of scientific interest in Microlepidoptera. *Agonopterix heracliana* and *A. alstromeriana* were described respectively by Linnaeus in 1758 and Clerck in 1759. These are species with larvae that are easily found and reared in early summer; their adults hibernate and in winter are frequently disturbed by human activity such as taking an armful of firewood from the woodshed. By the end of the 18th century 21 European species had been described, many by Denis & Schiffermüller (1775). In the 19th century an additional 88 species were described, initially from the north and central parts of the continent with a number of important contributors including Haworth (1811), Treitschke (1835) followed by Zeller and Stainton and others in the middle part of the century. At this time some species were also described from South Europe notably by Zeller (1847) and Staudinger (1859) but knowledge of the South European fauna always lagged behind central and northern areas. By the end of the century and into the early years of the 20th century increasing numbers of species were being described from the Balkan countries and the Canary Islands, particularly by Rebel over a long period of time. North African species were being described by Chrétien.

The first comprehensive study of genitalia in the family started in 1953 by Hannemann followed by numerous papers including some North African and Asian species, up to 1990. He was able to bring some much needed order to the family with the recognition of numerous synonyms. Towards the end of the century Lvovsky added several species mainly from Russia. In northern Europe there were regional publications covering the family in the Netherlands (Van Laar, 1961, 1964) and Fennoscandia (Palm, 1989). Hannemann (1995) treated the species of Germany, actually covering a much wider area of Central Europe. By the end of the century 160 of the species recognised in the present work were known.

The present century has seen a substantial number of new species described, bringing the European total to 190, many as a result of work for this book, but also by several other authors including Šumpich & Skyva (2012), Šumpich (2013), Rymarczyk, Dutheil & Nel (2013) and Vives & Gastón (2017).

The project to write this book was initiated around 20 years ago when Ole Karsholt asked MC if he would consider the idea. MC was fully aware that he would have limited time and that he lacked the required skills in photography and in making quality genitalia preparations. Nevertheless he accepted the challenge. Initially Ian Thirlwell volunteered to provide the required photographs. Specimens were borrowed from Ian's neighbour, John Langmaid who also had a keen interest in Depressariidae and from the Natural History Museum, London for description.

Before long, Ian's circumstances changed and he was obliged to withdraw from the project. It gradually became clear that the project was likely to fail. Fortunately in 2010 Peter Huemer had the idea that PB should collaborate with MC as he had all the skills that MC lacked.

PB's work on the project has involved the examination of thousands of specimens, genitalia preparations of many hundreds and visits to most of the major natural history museums in Europe. He has also examined many specimens from outside Europe, allowing a better understanding of the European species. He has contributed all the photographs and the distribution table. MC is responsible for most of the descriptive text, but other parts are a collaborative effort.

From the outset we expected to discover previously unknown species from southern Europe, but it was more surprising to find that the well-worked fauna of central Europe also contained undescribed and misidentified species.

While we have been able to solve many problems in the European depressariid fauna, we are aware that there are still unresolved questions and probably some species complexes that need to be untangled. The *Depressaria douglasella* group is likely to include further species and there are probably additional species related to *D. libanotidella*. Resolution of these problem groups will entail the study of specimens from many geographical areas, establishing which males and females are conspecific and more information on host-plants.

Acknowledgements

We are indebted to Peter Huemer for bringing us together as a team and to Ole Karsholt for initiating the project. They have been the most important contacts for questions that arose again and again during the preparation of the manuscript. Likewise we thank Marko Mutanen who joined the editorial team and has provided invaluable insight into matters relating to DNA barcoding.

Special thanks to the Canadian Centre for DNA Barcoding (Guelph, Canada), whose sequencing work performed by Genome Canada was funded by the Government of Canada through the Ontario Genomics Institute.

We particularly thank Jan Správce and Zdenek Laštůvka, who reviewed the completed manuscript.

We are most grateful to Cornelia Schlup-Sonderegger for permission to use the wing venation drawings made by her late father, Peter Sonderegger. Peter was working on a book on Depressariidae of Switzerland for which he had excellent illustrations and much bionomic information on which we have drawn heavily, but his untimely death left that project unfinished.

Frédéric Rymarczyk and Monique Dutheil have very generously allowed us to use their extensive data on host-plants of Depressariidae resulting from their fieldwork in France.

Mark Shaw, NMSE, Edinburgh, United Kingdom obligingly wrote the paragraph on parasitoids in the chapter on Bionomics.

Only the help of many colleagues has made it possible to complete the present volume, whether as curators of public collections or as private collectors. They enabled access to specimens of the European species under their care, for photography, preparation of genitalia slides and genetic studies, and helped with finding all the scientific literature needed. Only museum collections could provide specimens of species which have declined or not been found in Europe for decades. On the other hand, for genetic examination relatively fresh specimens are necessary, these were easier to find in private collections. Studies of the range of variation in each species, both externally and in genital characteristics made it necessary to examine many individuals of a species, preferably spread over its entire distribution area. This provided data on host-plants, phenology and overall distribution. In order to understand the relationships of some groups of species, it was also necessary to include non-European species in the studies.

We are indebted to all the following who assisted us in one way or another.

Leif Aarvik, NHM, Oslo, Norway; David Agassiz, Weston-super-Mare, United Kingdom; Hazumu Arashima, Kyushu University, Fukuoka, Japan; Günter Baisch, Biberach, Germany; Giorgio Baldizzone, Asti, Italy; Zsolt Bálint, HNHM, Budapest, Hungary; Patrizio Barberis, Italy; Hannes Baur, NMBE, Bern, Switzerland; Stella Beavan, Zeal Monachorum, United Kingdom; Knud Bech, Ølsted, Denmark;

Bengt Å. Bengtsson, Färjestaden, Sweden; Kai Berggren, Kristiansand, Norway; Jan-Olov Björklund, Herräng, Sweden; Hans Blackstein, Rathenow, Germany; Stella Brecknell, OUMNH, Oxford, United Kingdom; Rudolf Bryner, Biel, Switzerland; Uwe Büchner, ZfBS, Reden, Germany; Ulf Buchsbaum, ZSM, Munich, Germany; Danielle Czerkaszyn, OUMNH, Oxford, United Kingdom; Michael Dale, Talke, United Kingdom; Georg Derra, Reckendorf, Germany; Hans-Peter Deuring, Blumberg, Germany; Helmut Deutsch, Bannberg, Austria; Monique Dutheil, Nice, France; Marek Dvořák, Smrčná, Czechia; Gyulay Fábián, Budapest, Hungary; Per Falck, Nexø, Denmark; Michael Falkenberg, SMNK, Karlsruhe, Germany; Imre Fazekas, Pécs, Hungary; Gabriele Fiumi, Forli, Italy; Sabine Gaal-Haszler, NHMW, Vienna, Austria; Jose Manuel Gaona, Cadiz, Spain; Javier Gaston, Getxo, Spain; Christian Gibeaux, Avon, France; Paolo Glerean, MFSN, Udine, Italy; Stanislav Gomboc, Kranj, Slovenia; Friedmar Graf, Bautzen, Germany; Keld Gregersen, Sorø, Denmark; Theo Grünewald, Landhut, Germany; Thomas Guggemoos, Ohlstadt, Germany; Danijela Gumhalter, Stuttgart, Germany; Bert Gustafsson, Stockholm, Sweden; Heinz Habeler (+), Graz, Austria; Penny Hale (+), Casares, Spain; Alfred Haselberger, Traunstein, Germany; Robert Heckford, Plympton, United Kingdom; Marcel Hellers, Bissen, Luxemburg; Martin Honey, NHMUK, London, United Kingdom; Peter Huemer, TLMF, Hall, Austria; Catrin Hühne, NLMB, Braunschweig, Germany; Fernando de Juana, Vitoria-Gasteiz, Spain; Jari Junnilainen, Vantaa, Finland; Urmas Jürivete, Tallinn, Estonia; Lauri Kaila, ZMUH, Helsinki, Finland; Christian Kaiser, Rötha, Germany; Jari-Pekka Kaitila, Vantaa, Finland; Claudia Kamcke, NLMB, Braunschweig, Germany; Rudolf Keller, Dachau, Germany; Muhabbet Kemal, Yüzüncü Yıl University, Van, Turkey; Sibel Kızıldağ, Yüzüncü Yıl University, Van, Turkey; Ahmet Koçak (+), Yüzüncü Yıl University, Van, Turkey; Andreas Kopp, Simach, Switzerland; Stanislav Korb, Nizhny Novgorod, Russia; Zoltán Kovács, Miercurea Ciuc, Romania; Jaakko Kullberg, ZMUH, Helsinki, Finland; Bernard Landry, MHNG, Genève, Switzerland; Jean-François Landry, AAFC/AAC, Ottawa, Canada; John Langmaid (+), Southsea, United Kingdom; Knud Larsen, Dyssegård, Denmark; Alejandro A. Lázaro, Galicia, Spain; David Lees, NHMUK, London, United Kingdom; Patrice Leraut, MNHN, Paris, France; Martin Lödl, NHMW, Vienna, Austria; Alexander Lvovsky, ZIN, St. Petersburg, Russia; Geoff Martin, NHMUK, London, United Kingdom; Anton Mayr, Feldkirch, Austria; Ruben Meert, Lebbeke, Belgium; Heidrun Melzer, Leipzig, Germany; Wolfram Mey, ZMHB, Berlin, Germany; Joël Minet, MNHN, Paris, France; Carlo Morandini, MFSN, Udine, Italy; Lucio Morin, Ronchi dei Legionari, Italy; Rolf Mörtter, Kronau, Germany; Marko Mutanen, Zoological Museum, Oulu, Finland; Steve Nash, St Mellion, United Kingdom; Wolfgang Nässig, Senckenberg, Gemany; Jacques Nel, La Ciotat, France; Erik J. van Nieukerken, RMNH, Leiden, The Netherlands; João Nunes, Valongo, Portugal; Kari Nupponen (+), Espoo, Finland; Timo Nupponen, Espoo, Finland; Matthias Nuß, MTD, Dresden, Germany; Eivind Palm, Højer, Denmark; Gabriel Pastoralis, Komárno, Slovakia; Charles Perez, Gibraltar; Pedro Pires, Southampton,

United Kingdom; Colin Plant, Bishops Stortford, United Kingdom; Luis O. Popa, MGAB, Bucharest, Romania; Hossein Rajaei, SMNK, Karlsruhe, Germany; Emili Requena, Igualada, Spain; Hans Retzlaff, Lage, Germany; Ignác Richter, Malá Čausa, Slovakia; Ivan Richter, Prievidza, Slovakia; Oliver Rist, Graz, Austria; Jürgen Rodeland, Mainz, Germany; László Ronkay, HNHM, Budapest, Hungary; Jorge Rosete, Louriçal, Portugal; Hartmut Roweck, Kiel, Germany; Teet Ruben, Harjumaa, Estonia; Walter Ruckdeschel, Übersee, Germany; Frédéric Rymarczyk, Nice, France; Paul Sammut, Rabat, Malta; Klaus Sattler, NHMUK, London, United Kingdom; Nikolay Savenkov, LDM, Riga, Latvia; Benjamin Schattanek-Wiesmair, TLMF, Hall, Austria; Jürg Schmid, Illanz, Switzerland; Willibald Schmitz (+), Bergisch Gladbach, Germany; Tina Schulz, Rodenberg, Germany; Andreas Segerer, ZSM, Munich, Germany; Rudi Seliger, Schwalmtal, Germany; Christian Siegel, Fussach, Austria; Sergey Sinev, ZIN, St. Petersburg, Russia; Peder Skou, Stenstrup, Denmark; Jan Skyva, Prague, Czechia; Peter Sonderegger (+), Biel, Switzerland; Jan Správce, Ľubomír Srnka, Lehota pod Vtáčnikom, Slovakia; Mihai Stănescu, MGAB, Bucharest, Romania; Wolfgang Stark, Trübensee, Austria; Hartmuth Strutzberg, Weimar, Germany; Jan Šumpich, NMPC, Prague, Czechia; Czaba Szabóky, Budapest, Hungary; Jukka Tabell, Hartola, Finland; Tom Tams, United Kingdom; Gerhard Tarmann, TLMF, Hall, Austria; Franz Theimer, Berlin, Germany; Ian Thirlwell, Southsea, United Kingdom; Giovanni Timossi, Preganzio, Italy; Zdenko Tokár, Šaľa, Slovakia; Robert Trusch, SMNK, Karlsruhe, Germany; Kevin Tuck, NHMUK, London, United Kingdom; Thierry Varenne, Nice, France; Francesca Vegliante, Langebrück, Germany; Ebbe Vesterhede, Denmark; Antonio Vives Moreno, Ciudad Universitaria, Madrid, Spain; Andreas Werno, Nunkirchen, Germany; Christian Wieser, KLM, Klagenfurt, Austria; Wolfgang Wittland, Wegberg, Germany; Hans-Peter Wymann, NMBE, Bern, Switzerland; Josep Ylla, Barcelona, Spain; Xiaoju Zhu, SDAU, Tai'an, China; Alberto Zilli, NHMUK, London, United Kingdom; Boyan Zlatkov, BAS-IBER, Sofia, Bulgaria.

Abbreviations of Museums and Private Research Collections

Note: the mentioned private collections were correct at the time the specimens were checked (years 2013–2021). Such specimens may have been transferred into a public collection since then or will be transferred in the future.

AAFC/AAC	Agriculture and Agri-Food, Ottawa, Canada
BAS-IBER	Institute of Biodiversity and Ecosystem Research, Bulgarian Academy of Sciences, Sofia, Bulgaria
ECKU	Collection of Ecology-Centre, Kiel University, Kiel, Germany
HNHM	Hungarian Natural History Museum, Budapest, Hungary
KLM	Kärntner Landesmuseum, Klagenfurt, Austria
LDM	Latvijas Dabas Muzejs, Riga, Latvia
MFSN	Museo Friulano di Storia Naturale, Udine, Italy
MGAB	"Grigore Antipa" National Museum of Natural History, Bucharest, Romania
MHNG	Naturkunde Museum Genf, Switzerland
MNCN	Museo Nacional de Ciencias Naturales, Madrid
MNHN	Muséum National d'Histoire Naturelle, Paris, France
MTD	Museum für Tierkunde, Dresden, Germany
NHMUK	Natural History Museum, London, United Kingdom
NHMW	Naturhistorisches Museum, Wien, Austria
NLMB	Staatliches Naturhistorisches Museum, 3Landesmuseen Braunschweig, Germany
NMBE	Naturhistorisches Museum, Bern, Switzerland
NMPC	National Museum, Prague, Czechia
NMSE	National Museums of Scotland, Edinburgh, United Kingdom
OUMNH	Oxford University Natural History Museum, Oxford, United Kingdom
RCAM	Research Collection of Anton Mayr, Feldkirch, Austria
RCAS	Research Collection of Andreas Stübner, Germany
RCAW	Research Collection of Andreas Werno, Nunkirchen, Germany
RCBB	Research Collection of Bengt Å. Bengtsson
RCBZ	Research Collection of Boyan Zlatkov, Sofia, Bulgaria
RCCM	Research Collection of Carlo Morandini, Italy
RCCP	Research Collection of Colin Plant, Bishops Stortford, United Kingdom
RCCS	Research Collection of Csaba Szabóky, Budapest, Hungary
RCEP	Research Collection of Eivind Palm, Højer, Denmark
RCEV	Research Collection of Ebbe Vesterhede, Denmark
RCFG	Research Collection of Friedmar Graf, Bautzen, Germany

RCFT	Research Collection of Franz Theimer, Berlin, Germany
RCFV	Research Collection of Francesca Vegliante, Langebrück, Germany
RCGB	Research Collection of Günter Baisch, Biberach, Germany
RCGD	Research Collection of Georg Derra, Reckendorf, Germany
RCGF	Research Collection of Gyulay Fábián, Budapest, Hungary
RCGFi	Research Collection of Gabriele Fiumi, Forli, Italy
RCHB	Research Collection of Hans Blackstein, Rathenow, Germany
RCHD	Research Collection of Helmut Deutsch, Bannberg, Austria
RCHR	Research Collection of Hans Retzlaff, Lage, Germany
RCIR	Research Collection of Ignac Richter, Malá Čausa, Slovakia
RCIvR	Research Collection of Ivan Richter, Prievidza, Slovakia
RCJB	Research Collection of Jan-Olov Björklund, Sweden
RCJJ	Research Collection of Jari Junnilainen, Vantaa, Finland
RCJK	Research Collection of Jari-Pekka Kaitila, Finland
RCJN	Research Collection of Jacques Nel, La Ciotat, France
RCJT	Research Collection of Jukka Tabell, Hartola, Finland
RCKB	Research Collection of Knud Bech, Denmark
RCKL	Research Collection of Knud Larsen, Dyssegård, Denmark
RCLM	Research Collection of Lucio Morin, Monfalcone, Italy
RCLS	Research Collection of Ľubomír Srnka, Lehota pod Vtáčnikom, Slovakia
RCMC	Research Collection of Martin Corley, Faringdon, United Kingdom
RCMM	Research Collection of Marko Mutanen, Oulu, Finland
RCOR	Research Collection of Oliver Rist, Graz, Austria
RCPB	Research Collection of Peter Buchner, Schwarzau am Steinfeld, Austria
RCRD	Research Collection of Frédéric Rymarczyk & Monique Dutheil, Nice, France
RCRH	Research Collection of Robert Heckford, Plympton, United Kingdom
RCRK	Research Collection of Rudolf Keller, Dachau, Germany
RCRM	Research Collection of Rolf Mörtter, Kronau, Germany
RCRS	Research Collection of Rudi Seliger, Schwalmtal, Germany
RCSG	Research Collection of Stanislav Gomboc, Kranj, Slovenia
RCTG	Research Collection of Theo Grünewald, Landhut, Germany
RCTN	Research Collection of Timo Nupponen, Espoo, Finland
RCWS	Research Collection of Wolfgang Stark, Trübensee bei Tulln, Austria
RCWSc	Research Collection of Willibald Schmitz, Bergisch Gladbach, Germany
RCWW	Research Collection of Wolfgang Wittland, Wegberg, Germany
RCZK	Research Collection of Zoltan Kovacs, Miercurea Ciuc, Romania
RCZT	Research Collection of Zdenko Tokár, Šaľa, Slovakia
RMNH	Naturalis Biodiversity Center, Leiden, The Netherlands
SDAU	Shandong Agricultural University, Tai'an, China
SMNK	Staatliches Museum für Naturkunde, Karlsruhe, Germany

TLMF	Tiroler Landesmuseum Ferdinandeum, Innsbruck, Austria
TUEE	Tartu University, Estonia
ZfBS	Zentrum für Biodokumentation Saarland, Reden, Germany
ZIN	Zoological Institute, Russian Academy of Sciences, St. Petersburg, Russia
ZMHB	Museum für Naturkunde der Humboldt-Universität, Berlin, Germany
ZMUC	Zoological Museum, University of Copenhagen, Copenhagen, Denmark
ZMUH	Zoology Museum, University of Helsinki, Helsinki, Finland
ZSM	Zoologische Staatssammlung, München, Germany

Classification of Depressariidae

The genera treated in this book as Depressariidae have long been recognised as closely related, but understanding of their relationships to other groups of Microlepidoptera has even now not achieved a complete consensus. Over the last two centuries the concept of the family-group has progressively narrowed resulting in the proliferation of ever smaller families.

The earliest classifications of Microlepidoptera were at such a broad scale that they cannot easily be compared with modern treatment at family level. Moving on, Stainton (1854) used family names that are still recognisable today. His Tineina included 13 families. Among these were Nepticulidae, Tineidae, Hyponomeutidae, Elachistidae, Coleophoridae and Gelechiidae. The last three families are equivalent to the modern Gelechioidea. The *Depressaria* group were included in Gelechiidae.

Meyrick (1906) removed the Oecophoridae (including the *Depressaria* group) from Gelechiidae. Clarke (1941) examined genitalia of the North American Oecophoridae and suggested separation into two groups with gnathos bearing spines or not. This separation was not adopted at family level by Hodges (1974) and the *Depressaria* group remained in Oecophoridae in some works (e.g. Harper *et al.*, 2002) for many more years. Hodges' Oecophoridae included four subfamilies: Depressariinae, Peleopodinae, Oecophorinae and Chimabachinae.

Minet (1985, 1991) recognised the importance of the spined gnathos and correlated this with larval characters. He proposed two gelechioid families with spined gnathos, Coleophoridae and Elachistidae. Elachistidae included subfamilies Agonoxeninae, Elachistinae, Cryptolechiinae, Hypercalliinae, Ethmiinae and Depressariinae.

In the present century, DNA studies have allowed greater insight into relationships within Gelechioidea. Nevertheless, the results from different studies have not always produced the same conclusions. Kaila *et al.* (2011) maintained an "Elachistid *sensu lato*" assemblage, which did not include Elachistidae, but did include Depressariinae, Oditinae, Hypercalliinae, Cryptolechiinae, Ethmiinae and Peleopodidae. Heikkilä *et al.* (2014) reached very similar conclusions, placing these as subfamilies of Depressariidae. Their Depressariidae was exceptionally diverse, including 10 subfamilies. This was partly because they were unable to find clear morphological features to segregate some of the groups.

Sohn *et al.* (2016) recognised a "core group" of depressariid subfamilies which consisted of Depressariinae, Aeolanthinae and Hypertrophinae, the latter two subfamilies being unknown in Europe.

Wang & Li (2020) were unable to fully resolve the Depressariidae group. Ethmiidae was raised to family status and Oditinae placed in Peleopodidae. From their data the remnant Depressariidae are clearly paraphyletic. Cryptolechiinae are evidently distant from the core Depressariidae while Hypercalliinae is also well separated. Nevertheless, they leave these groups in Depressariidae in a wide sense.

Kaila *et al.* (2011) were unable to find a satisfactory subfamily in which to include *Telechrysis* Toll. As a result it was left in Depressariinae. This situation was not resolved by Heikkilä *et al.* (2014). The genus is a misfit in Depressariidae with strongly developed uncus and larval pabulum more akin to Oecophoridae. We do not attempt to resolve its position, but exclude it from Depressariidae.

There has been a consistent failure to find a satisfactory placement for these problem groups: *Telechrysis*, Cryptolechiinae (which may be quite unrelated, in which case Cacochroinae is the valid subfamily name for *Orophia*, *Zizyphia*, *Cacochroa* and *Rosetea*) and Hypercalliinae including *Hypercallia* and *Anchinia*. If they are included in Depressariidae it becomes a "dustbin" family containing morphologically dissimilar groups. This cannot be justified and we exclude them from Depressariidae. No doubt in future studies these problem groups will find better placement. The core depressariids (*Depressaria*, *Exaeretia*, *Agonopterix*, *Luquetia* and *Semioscopis* in Europe) are closely related according to morphology and DNA studies. In our opinion these are the true Depressariidae and they are the only genera treated in this book. It is worth noting that in Karsholt & Razowski (1996) the Depressariidae were essentially the same as in this book, apart from minor differences in generic limits.

CHAPTER 2

Morphology of Depressariidae

Depressariidae is a family in the superfamily Gelechioidea. According to Minet (1991) Gelechioidea share four apomorphies: presence of dense scales on the basal part of the proboscis, strongly recurved labial palps, larval abdominal segments I–VIII with L1 and L2 approximated or on the same pinaculum and pupa with an apical or subapical invagination of the mesothoracic legs.

The Depressariidae as delimited in this book are described below. Several recent classifications of Lepidoptera based substantially on genomic data have included various anomalous groups in Depressariidae which morphologically show no close relationship to the genera included here. For further information on this see the previous section on Classification of Depressariidae.

The European Depressariidae are medium-sized to moderately large Microlepidoptera with wingspan 11–30 mm (to 37 mm in one species), frons smooth-scaled, crown with raised scales usually curved forwards, ocelli present except in *Luquetia*, antenna shorter than forewing, scape with pecten (sometimes fugaceous, absent in *Semioscopis* and the non-European *Agonopterix* subgenus *Subagonopterix*), labial palp segment 2 often thickened with scales or tufted below, longer than segment 3; forewing broad, termen rounded or weakly sinuous, strongly oblique in *Semioscopis*, hindwing broad, often with slightly expanded lobe at anal angle; abdomen often flattened dorso-ventrally, tergites without groups of spines; forewing veins R_4 and R_5 long-stalked, Cu_1 and Cu_2 separate or short-stalked, hindwing Cu_1 and M_3 connate or short-stalked. Male genitalia with well-developed socii, uncus vestigial, gnathos with ranks of fine spines, usually simple, occasionally divided into two parts, valva with costal margin often simple, ventral margin with sacculus often ending in a process (cuiller) directed across valva, other processes sometimes present, a terminal process at end of sacculus in *Exaeretia* and a process (clavus) from base of sacculus in some *Depressaria* species, transtilla more or less overlapped by a pair of transtilla lobes, anellus usually bifid at apex, overlapped by a pair of anellus lobes, saccus usually short, aedeagus[1] more or less curved, ductus ejaculatorius attached basally, not laterally so there is no caecum, although attached at one side

---

1   In this book we have used the term *aedeagus* for the intromittent male organ in preference to *phallus* which has been increasingly used in recent years. *Aedeagus* was used by lepidopterists throughout the 20th century including in several major book series and the early volumes of this series. Kristensen (2003) introduced the term *phallus* as a replacement name for *aedeagus* because of incomplete homology with the *aedeagus* in some other insect orders. However the details of structure and homology of this organ in the most primitive families of Lepidoptera are varied and in some cases not fully understood. In one family (Agathiphagidae) the organ is not a *phallus* in Kristensen's sense but an *aedeagus sensu stricto*. We argue that for the sake of simplicity and

© PETER BUCHNER AND MARTIN CORLEY, 2025 | DOI:10.1163/9789004713116_003

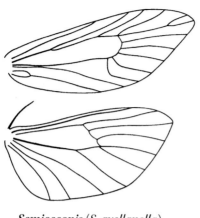

Semioscopis (*S. avellanella*)

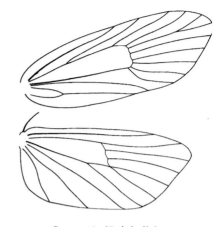

Luquetia (*L. lobella*)

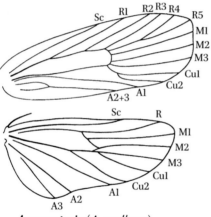

Agonopterix (*A. ocellana*)

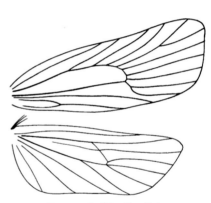

Exaeretia (*E. allisella*)

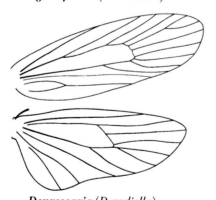

Depressaria (*D. radiella*)

FIGURE 1   Wing venation in Depressariidae

of base in *Semioscopis* and *Luquetia,* cornuti often present or vesica with a dense mass of micro-cornuti. Female genitalia with short oviscapt, apophyses of moderate length, ostium opening on segment VIII, ductus bursae mainly membranous, corpus bursae membranous, signum usually present as a plate with thorn-like teeth on its surface and margins.

---

terminological stability the term *aedeagus* should continue to be used for all Lepidoptera (as it is in other insects) regardless of the finer details of structure and homology.

Kristensen chose the term *phallus* because it already existed in combination for some of the constituent parts (e.g. *phallobase, phallomere*). We consider that it was an inappropriate choice since in the English language *phallus* refers to the engorged male organ of mammals, usually humans. While this is analogous to the organ in Lepidoptera it cannot be considered homologous.

In our opinion it was unnecessary to replace *aedeagus* but if it were to be replaced a more suitable term should have been found.

CHAPTER 3

Descriptions of Genitalia

In the following paragraphs we indicate the ideal position of the various parts of the genitalia in permanent preparations and explain how measurements have been taken in the genitalia descriptions; we also discuss which characters are more reliable and which should be used with particular caution. These remarks apply particularly to *Agonopterix* in which variation in the genitalia and the large number of species present significant problems, but to some extent they apply to the other genera also.

It is important to understand that the descriptions of male genitalia are based on near perfect preparations, which we refer to as standard preparation. Techniques used to obtain such preparations are explained under Methods (p. 21–p. 31). Such preparations have the socii flattened, gnathos extended caudad and aedeagus extracted. Even perfect preparations are subject to variation, particularly relating to amount of pressure placed on cover glass, which in particular affects the apparent position of the cuiller and how far it crosses the valva.

In the descriptions of male genitalia, consideration is mainly given to characters that show some differentiation between species, so may not be described for every species. The way in which structures have been described and measurements have been taken is as follows.

The socii are rarely of shapes that can be readily described in a few words, so it is better to examine the figures.

The uncus is generally very poorly developed and usually hidden in standard preparation.

The gnathos shape is important, width to length proportions are based on the part of the gnathos that is covered with minute spinules.

The tegumen is liable to be distorted in preparation. There are potential characters relating to its length, width and the shape of the basal sinus, but we have not used these because they are not sufficiently reliable.

The shape of the valva is not at great risk of distortion in preparation, but does show substantial variation within species. It is described in the species accounts and there are tendencies towards particular shapes that are of some taxonomic value, but not too much weight should be given to them. The costal margin shows little variation within species, but the ventral margin and apex vary more. The ventral margin usually has an indentation at the end of the sacculus, beyond which the margin curves, usually becoming straight or even slightly concave towards the apex.

The valva usually has one or more processes of which the most prevalent in the family is the cuiller, arising from the end of the sacculus and directed across the valva. In *Exaeretia* the end of the sacculus has a terminal process approximately

© PETER BUCHNER AND MARTIN CORLEY, 2025 | DOI:10.1163/9789004713116_004

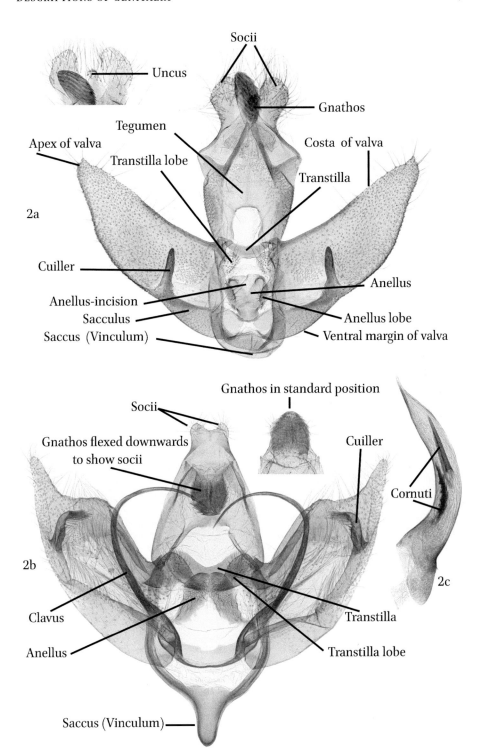

FIGURE 2 Male genitalia embedded in standard position with aedeagus removed; 2a: *Agonopterix graecella*, 2b: *Depressaria tenebricosa*, 2c: *Depressaria fuscovirgatella*, aedeagus

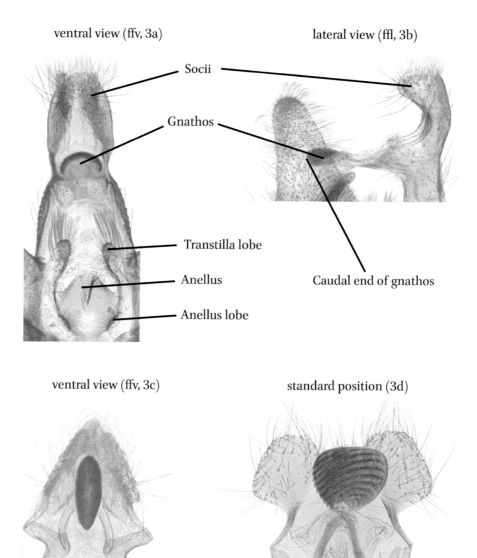

ventral view (ffv, 3a)

lateral view (ffl, 3b)

Socii

Gnathos

Transtilla lobe

Anellus

Anellus lobe

Caudal end of gnathos

ventral view (ffv, 3c)

standard position (3d)

FIGURE 3   Male genitalia in different views
Semioscopis avellanella male, free floating in natural position, 3a: ventral view, 3b:
lateral view (in ventral view, gnathos is seen from caudal end!)
Agonopterix lessini, gnathos, free floating from ventral (3c) and standard position (3d)
compared. If gnathos is a flat disc it turns aside under compression of cover glass in
standard preparation, falsely appearing to be globose

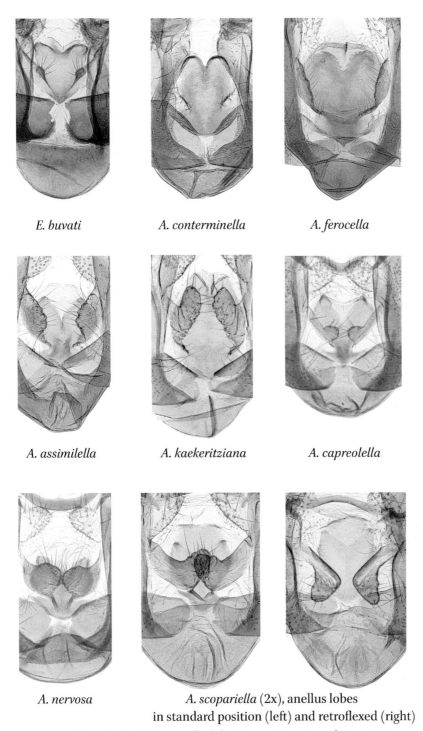

E. buvati   *A. conterminella*   *A. ferocella*

A. assimilella   *A. kaekeritziana*   *A. capreolella*

A. nervosa   *A. scopariella* (2x), anellus lobes
in standard position (left) and retroflexed (right)

FIGURE 4   Examples for different anellus lobes in genus *Exaeretia* and *Agonopterix*

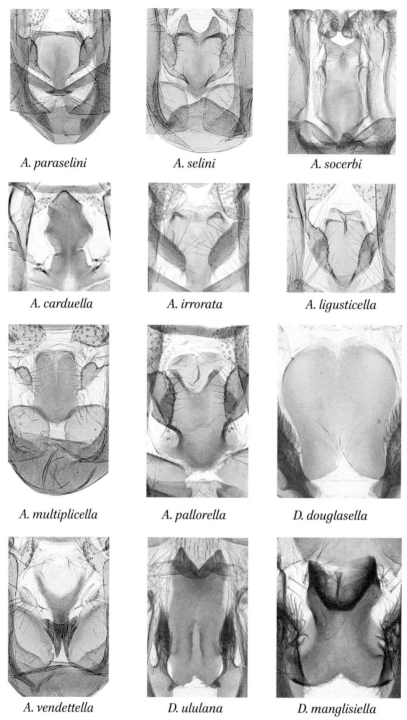

A. paraselini   *A. selini*   *A. socerbi*

A. carduella   *A. irrorata*   *A. ligusticella*

A. multiplicella   *A. pallorella*   *D. douglasella*

A. vendettella   *D. ululana*   *D. manglisiella*

FIGURE 5   Examples for different distal margin of anellus in genus *Agonopterix*
            and *Depressaria*

at right angles to the cuiller. Some *Depressaria* species have a clavus, a process aris-ing from the base of the sacculus and directed caudad.

The length and shape of the cuiller are valuable characters for differentiating spe-cies, but must always be used with caution as the cuiller is affected by preparation, which can alter its curvature, direction and apparent length relative to the valva width. The last character is given as a percentage of the valva width, which is meas-ured perpendicular to the ventral margin through the apex of cuiller (not usually from the base of the cuiller), regardless of the direction or curvature of the cuiller.

The transtilla may be narrow or widened, at least in the central part.

The transtilla lobes, which may show no connection to the transtilla, are not described unless they show useful features.

The anellus is a complex structure consisting of a short-stalked plate-like struc-ture which is lightly sclerotised, often bifid with a posterior incision or sinus. In descriptions of this the stalk is not mentioned. Dorsal to this is an unsclerotised membrane which usually extends beyond the sclerotised part, but is often incon-spicuous. If the aedeagus is left in position this membrane is hidden.

The anellus lobes do not always show any clear connection to the anellus, but overlap its margins to a greater or lesser extent and can show useful characters, but are generally not described in detail as the shapes can rarely be described in a few words, so it is better to examine the figures.

The saccus is normally short and rounded and does not provide useful charac-ters, apart from the *Depressaria veneficella* group, in which it can be longer and varies considerably between species.

The aedeagus is always at least slightly curved. In preparation it therefore lies in a lateral position, showing the curve. Length/width ratio of aedeagus may be a useful feature, however if aedeagus is compressed, width increases, but not length, therefore it is a problematic feature if slides are used where the extent of distor-tion is unknown. We only depict aedeagus uncompressed. The basal half or more is cylindrical or tapering, the distal part is obliquely truncated, tapering to a narrow point; in *Agonopterix* and some *Depressaria* species the base has a sclerotised band, from which a process arises which is of variable length and shape. The shape is not easily seen in standard preparation and in many cases does not provide useful characters, but may do so in a few species, we therefore illustrate the aedeagus from ventral view on the plates. In species where the vesica contains a very large number of minute cornuti, these rarely provide useful characters. The length of the aedea-gus (measured directly from the lowest part of the base to the apex) is compared with the length of the valva (measured from the extreme base on the ventral side directly to the apex).

Female genitalia do not always provide characters that allow a safe determina-tion, but can do so in many cases. However many of the characters must be used with caution as they show variation, both natural and as a result of small differences

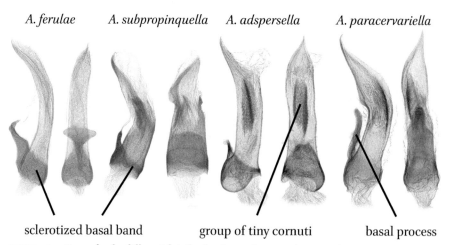

A. ferulae *A. subpropinquella* *A. adspersella* *A. paracervariella*

sclerotized basal band   group of tiny cornuti   basal process

FIGURE 6   Examples for different details of aedeagus in genus *Agonopterix*

in preparation. Absolute measurements have not been used in the descriptions as these can vary within species due to the size of individuals.

In the species descriptions papillae anales are not described. At least in some cases there are differences between species, but rarely sufficiently clear to be of use in identification, and subject to differences in preparation.

The length of the anterior apophysis is given as a proportion of the length of the posterior apophysis. There is often some variation within species, although usually this is quite small.

Segment VIII has useful characters on the ventral side (sternite) but rarely on the posterior side (tergite) which is nearly always more or less concave on its anterior margin. The ratio of width to length of the sternite is given, with width of the segment measured at its widest point and length measured at the middle, thus including any anterior extension or posterior indentation. There is some variation in this character, mainly due to differences in preparation.

The position of the ostium on the sternite is given, based on the midpoint of the circular part of the ostium. There is often a narrower posterior extension. Usually the antrum is not described: when clearly visible this normally consists of a pair of sclerites tapering to narrow points.

The ductus bursae normally begins as a narrow tube, often gradually expanding towards the corpus bursae, but in many species it is more complex with expanded sections and often with complicated surface patterns of folds. However after mating most such structure is lost. Some *Depressaria* species have sclerotised sections mainly in the posterior part of the ductus, sometimes on one side only. The relative length of corpus bursae to length of the ductus can be useful, but especially after mating it is not always easy to ascertain where one ends and the other begins.

The ductus seminalis is attached close to the posterior end of the ductus bursae, the number of spirals at its distal end is sometimes different between species.

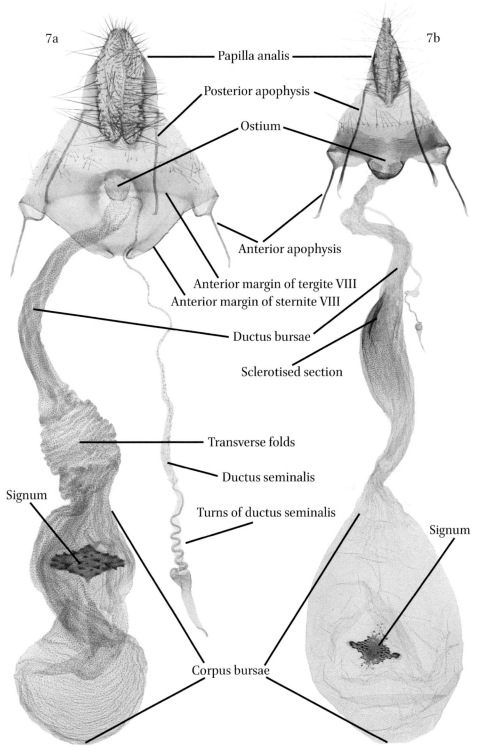

7a

7b

Papilla analis

Posterior apophysis

Ostium

Anterior apophysis

Anterior margin of tergite VIII
Anterior margin of sternite VIII

Ductus bursae

Sclerotised section

Transverse folds

Ductus seminalis

Signum

Turns of ductus seminalis

Signum

Corpus bursae

FIGURE 7   Female genitalia embedded in standard position. 7a: *Agonopterix adspersella*,
7b: *Depressaria heydenii*

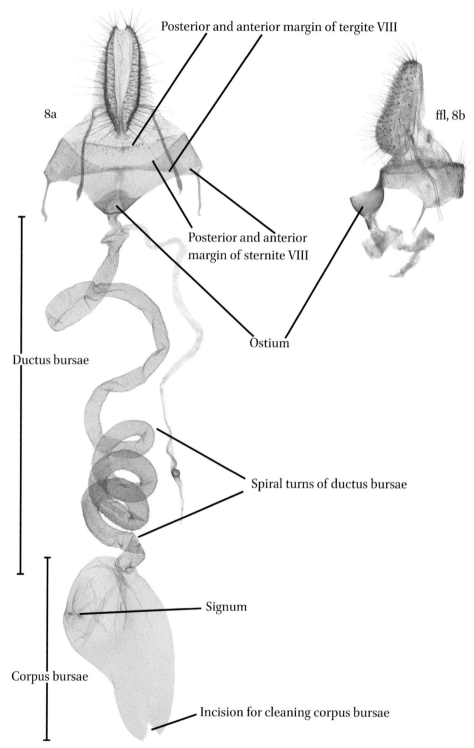

Posterior and anterior margin of tergite VIII

8a

ffl, 8b

Posterior and anterior
margin of sternite VIII

Ostium

Ductus bursae

Spiral turns of ductus bursae

Signum

Corpus bursae

Incision for cleaning corpus bursae

FIGURE 8   *Semioscopis steinkellneriana*, female genitalia: embedded in standard position (8a), papillae
anales and segment VIII with ostium, free floating, lateral (8b)

The signum is most often elliptical or more or less rhombiform, often with shorter or longer anterior and posterior extensions. It can sometimes be useful in distinguishing species, but there can also be significant variation within species. The same is true for the thorn-like teeth on the margins and surface of the signum plate, which sometimes vary considerably in size and number.

CHAPTER 4

Bionomics of Depressariidae

Detailed description of the early stages of depressariids is outside the scope of this book, since the aim is to enable identification of adult Depressariidae. Larvae of many species are illustrated on Lepiforum e.V. (2006–2023). Patočka & Turčáni (2005) provide details of pupae of most Central European species.

Almost all European depressariids appear to be univoltine with the possible exception of two species on the Canary Islands and probably *Agonopterix furvella*. In addition we have found evidence in two species (*A. ciliella* and *Depressaria depressana*) for a partial second generation. *Semioscopis* and *Luquetia* species overwinter as pupae with adults emerging in spring or early summer respectively. In the remainder of the family a slight majority of species are known to hibernate as adults, laying eggs in spring which give rise to the new generation of adults sometime during late spring or summer. These adults may aestivate for a time, usually flying again in autumn before finding a suitable site for hibernation. The flat abdomen characteristic of many species in the family probably facilitates hibernation in crevices such as under loose bark. In Mediterranean climates, aestivation is likely to be long and hibernation brief. The alternative strategy adopted by many species is to lay eggs in late summer or autumn, with larvae feeding in spring and early summer. Adults resulting from these may aestivate for a time in the hottest part of summer. For most species with this strategy, exactly when larvae hatch is not known. *Agonopterix assimilella* larvae can be found feeding during winter months before completing their development in spring. It is quite possible that some Mediterranean species such as *A. thapsiella* also behave like this.

Larvae feed on living dicotyledonous plants. A few species are known to be leaf miners initially, before feeding externally on leaves, flowers or developing fruits, although some feed within stems. Most species make individual spinnings, often forming tubes from leaves or leaflets, moving on to make new tubes as they grow. A few species (e.g. *Agonopterix ferulae*, *A. angelicella* and *Depressaria radiella*) are gregarious. Woody plants are used by *Semioscopis*, *Luquetia* and a few *Agonopterix* including the species feeding on such Fabaceae as *Genista*, *Cytisus* and *Ulex*. Several plant families are used by European depressariids, particularly Apiaceae, Asteraceae and Fabaceae, but also Rutaceae, Clusiaceae, Lamiaceae, Salicaceae, Rosaceae, Betulaceae and Malvaceae (*Tilia*). For a species such as *A. heracliana* there appears to be a gradation of favoured host-plants: we list 24 under this species, but some such as *Anthriscus sylvestris* and *Oenanthe crocata* are utilised predominantly (in different parts of the moth's distribution range), while others are less used and some only very occasionally. However the majority of species are more restricted in choice to one or a few genera of plants. Two *Agonopterix* species that can feed on a fairly wide selection of Asteraceae have also been recorded on plants from other families, namely *A. kaekeritziana* on *Anchusa* (Boraginaceae), *Knautia* and *Scabiosa* (Dipsacaceae) and *A. arenella* on *Knautia* but some of these reports may relate to misidentification of plants before flowers appear. Alternatively they may be examples of xenophagia (feeding on the wrong plant) which can have various consequences: larvae may fail to complete their development, or moths may be of reduced size, or development may be entirely successful which in evolutionary terms could eventually lead to separation of populations on the different host-plants. Probably most such instances are unusual events and have no significance, but without extensive fieldwork it is rare that we have a clear picture of what is happening. We do have examples of normally monophagous species on other host-plants, see for example *Agonopterix rotundella* and *Depressaria daucella*, but we do not know if these can form sustainable populations. *A. squamosa* is remarkable in that it has been found feeding on two species of Asteraceae and one species of Apiaceae. With very few observations of larvae of this species, we cannot know if the larvae on Apiaceae represent xenophagia or if this is a regular occurrence.

Identification of full grown larvae is possible in some species which have characteristic markings, aided by knowledge of host-plants (although larvae in earlier instars may look quite different). A few, such as *Depressaria daucella*, *D. chaerophylli* and *D. ululana*, are particularly distinctive.

Larvae of Depressariidae rarely cause severe damage to their host-plants, although MC has seen all plants of *Seseli tortuosum* with leaves stripped down to the main veins over an area of several hectares inland of sand dunes on the coast of Portugal by larvae of *Depressaria velox*. A few species are known as pests of particular crops, including *Depressaria marcella* damaging carrot seed crops in Italy, *D. daucella* on caraway (*Carum carvi*) in Finland and some Central European countries, *D. bantiella* on anise (*Pimpinella anisum*) in Italy and *D. erinaceella* on artichoke (*Cynara scolymus*).

The host-plants of 51 species of European Depressariidae are unknown or doubtful. As our classification of the family is partly based on host-plants this leaves the relationships of some species in a state of uncertainty. It also throws up the possibility that the failure to find larvae of some species (notably MC's efforts with *D. krasnowodskella* and *D. radiosquamella*) could be because searching was concentrated on the wrong plant family.

We have treated host-plant records from the older literature with caution. They may relate to misidentifications of either moths or plants and also to nomenclatural problems where a moth name has been applied to different species over the years. We have only accepted host-plant records which we have reason to consider reliable, although less convincing records are mentioned. In the Index to host-plants only accepted records are included.

The pupal stage is quite short, often no more than about three weeks, except in *Semioscopis* and *Luquetia* in which it is the overwintering stage. Pupation mostly takes place among detritus on the ground, but in a few species it may be in the larval spinning and some species pupate inside the hollow stems of the host-plant (*D. ultimella* and *D. radiella* when on *Heracleum* but not on *Pastinaca*).

Parasitism of Depressariidae by Hymenoptera is frequent, but there are no groups of parasitoids at genus-level or above that largely or exclusively specialise on Depressariidae as hosts. Several species do so, but they fall into genera with wider host repertoires overall, for example *Agrypon*, *Diadegma*, *Campoplex*, *Enytus* and *Microgaster*, and there are more species in these genera that use this host group only sporadically than do so as specialists. Different groups of Hymenoptera parasitoids can attack eggs, larvae and pupae. Either the host does not progress after attack (by idiobiont parasitoids) or it continues to develop until being killed by the parasitoid larva (koinobiont parasitoids). Most idiobionts are rather unspecialised, and often ectoparasitoids depending on development in concealment, while koinobionts are usually endoparasitoids and more host-specialised. However, some *Pimpla* species, developing as idiobiont endoparasitoids of the host pupal stage, do have regular associations with the easily discoverable *Depressaria* pupae in hollow Apiaceae stems. The same hosts and stage are also the target of a few specialist Ichneumoninae. Koinobiont species usually oviposit into the larval stage, and kill the host as a larva (often prepupal) or pupa; in the latter cases the adult parasitoid emerges from that. One genus of tiny wasps, *Copidosoma*, includes species that regularly parasitise *Depressaria* species; oviposition is singly into the host egg but the parasitoid embryo divides many times as it develops so that a brood of many hundreds eventually results in an aggregation of parasitoid pupae in a distorted host prepupa. Often the kind of parasitoid that a particular host will have is influenced by its feeding biology. Many koinobiont parasitoids attack the host very early in its life, and depressariids that start life as leaf miners tend to be parasitised by groups that have radiated in relation to that feeding habit. Similarly with parasitoids attacking the host larva late: some species of *Exochus* and *Triclistus*, with almost concealed ovipositors, rounded faces and strong legs as adaptations for entering silken tubular structures and grasping the host during oviposition, specialise on certain depressariids, in these cases ovipositing into the final instar host larva but delaying their development to emerge (as adults) from the host pupa. Others, such as *Macrocentrus bicolor*, use their relatively long ovipositors to reach their host larvae. Some tachinid flies have been recorded from Depressariidae but this is quite unusual. Belshaw (1993) mentions five species that have been reared from *Agonopterix*, but all have additional hosts belonging to other Lepidoptera families. The great majority of reared parasitoids are Hymenoptera and, by overwintering in the adult (or possibly egg) stage, many depressariids deny their parasitoids the opportunity to overwinter as larvae in an overwintering larva, which is a frequent way in which koinobiont parasitoids exploit their hosts.

Adult depressariids are almost entirely nocturnal, although some are easily disturbed from hibernation on winter days when conditions are not too cold. It is probable that most depressariid species feed in the adult stage but there is remarkably little information on this. MC has observed *A. heracliana* and *A. arenella* feeding on flowers of ivy (*Hedera*) in the autumn and *A. scopariella* and *A. nervosa* on wine ropes. Sugar bait is known to attract some species of *Agonopterix* and *Depressaria*; PB has recorded *A. heracliana* and *D. albipunctella* at sugar bait.

CHAPTER 5

Methods

Field Collecting Methods

While some Depressariidae can be readily identified in the field or from photographs, reliable identification for most species requires retention of specimens which can then usually be identified by genitalia examination or by DNA barcoding, although neither method is guaranteed to provide a correct identification in every instance.

Specimens can be collected at light, but the brightest lights tend to repel the moths rather than attract them. Actinic or LED lights of low wattage can be more effective than powerful mercury vapour lights. Because many Depressariidae are quite long-lived, light-trapped specimens are often in rather poor condition. Furthermore many species are very rarely recorded at light.

Searching for larvae and rearing through to the adult stage is an effective method of collecting many species, particularly of *Agonopterix* and *Depressaria* and has the advantage of producing specimens in perfect condition. As an example of the effectiveness of searching for larvae, of the 53 species of Depressariidae recorded from Portugal, 15 were first discovered as larvae. Many species are easily found on their known host-plants, often rolling leaves or leaflets into distinctive tubes, or feeding in a spinning among flowers or developing fruits. Larvae in later instars soon pupate and the pupal stage only lasts for a few weeks thus needing relatively little effort from the collector. The majority of species feed on plants in the families Apiaceae, Asteraceae and Fabaceae. Examining plants of these families can be productive, although identification of Apiaceae, often before flowers or fruits develop, can be challenging.

When rearing Depressariidae it is important to keep the larval containers clean and to remove condensation. Fresh host-plant material should be provided regularly if possible. If suitable plant material is not going to be available in the days after the larvae are collected, additional material can be collected and stored in a plastic bag in a refrigerator. Provision of dry sand for pupation is also recommended as this protects the pupae from contact with mouldy plant material.

When attempting to rear adults from larvae found in the wild, it is recommended to collect a few more larvae than actually required to allow for the probability that some will be parasitised. Sometimes all larvae produce adult moths, but usually there are at least a few parasitoids, usually Ichneumonidae. These should be retained and sent to a specialist on hymenopterous parasitoids together with the remains of the dead larva and the cocoon of the parasitoid. The record becomes much more valuable if the host can be identified through successfully reared adult moths.

Not all species are easy to find as larvae because some, including several *Exaeretia* species, feed in stems. There are 51 European Depressariidae species for which the host-plant remains unknown, mainly species of southern and eastern Europe.

Historically Depressariidae were often collected from thatch. Before the invention of the combine harvester cereal crops were stored as sheaves gathered into stacks which were then thatched for protection from the weather until such time as threshing (initially manual work, later done by a threshing machine) was feasible to separate the grain from the straw. Depressariidae species commonly entered the thatch in order to hibernate, whence they could be collected by disturbing the thatch or by smoking them out with a bee-smoker.

Dissecting and Preparing Male Genitalia

The main focus of this section is the description of how to make permanent slides fit for good photos to be published. This does not mean that genitalia preparation must always be a time-consuming and difficult procedure. In cases where the purpose of a dissection is simply to provide or confirm an identity, less complicated methods should bring satisfactory results. Nevertheless, the knowledge of how to bring the parts into standard position is also necessary for precise determination, because otherwise the result of comparisons may be incorrect.

All genitalia preparations photographed are by the first author, if not otherwise specified.[1] Standard techniques essentially followed (Robinson 1976) but modifications were made to obtain the best possible preparations to provide the photographs in this book. These are given in detail below.

First step after maceration is to remove abdomen skin from genitalia. It is recommended not to completely remove intersegmental skin which connects abdomen skin with genitalia at this stage from genitalia, because this brings risk of damage to valva complex. For controlled removal it is often necessary to stain genitalia with remaining parts of intersegmental skin to see all details, because it is nearly invisible otherwise. Our experience is that best results are obtained with a withdrawing direction as shown in fig. 9a with an acute angle between skin-free valva complex and skin. Right angle (fig. 9b) is also possible but obtuse angle (fig. 9c) brings high risk that the dividing line turns into the valva complex. By changing withdrawing direction the course of the dividing line can be somewhat controlled, but full control is not possible, experience is necessary for good results. Ideally the dividing line runs exactly along the border of the sclerotised part, but a reliable prediction cannot be made. Therefore, this preparation step must be carried out slowly and with close observation of the dividing line. If the edge of the valva complex threatens

---

1   The exceptions are males of *Agonopterix rigidella, A. adspersella* ssp. *pavida, A. lidiae, Depressaria pyrenaella, D. fuscipedella* and *D. erzurumella* and females of *A. tschorbadjiewi* and *D. pentheri*.

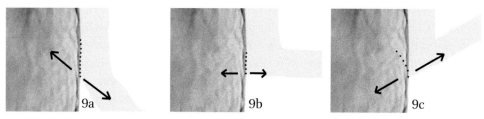

FIGURES 9A–9C   Darker area on left side of each figure = sclerotised parts; pale grey area on right
side = skin; arrows show the direction of tearing apart sclerotised part and skin, each
held by forceps; line of dots shows where the dividing line is likely to be

to tear (the dividing line turns into it), one must stop immediately. An attempt to
separate the skin from the other side is possible. But it is better to leave parts of the
skin attached than to damage the valva complex. It will be difficult or impossible
to spread valva symmetrically if there are remaining parts of skin or, even worse,
damage to valva complex, therefore this first step is essential for further preparation success.

In standard preparations, the abdominal skin is cleaned, stained and added to
the permanent slide. In some genera, e.g. *Coleophora* and *Eupithecia*, it contains
important features for determination, but this is not the case in Depressariidae.
Since cleaning of the skin is very time consuming, it is recommended to check with
the curators of specimens to be dissected, if it is acceptable only to retain the skin
with the specimen e.g. in a glycerol vial.

Before spreading valvae, aedeagus must be removed carefully, taking care not
to compress it. Aedeagus should immediately be dessicated with isopropanol
on a microscope slide without covering with a cover glass, otherwise it would be
ompressed and show an incorrect length/width ratio, this ratio is an important feature in some species groups. As soon as it is dessicated it can be picked up by the
ductus ejaculatorius and put into the staining liquid for as long as necessary.

Preparing valva complex: the genitalia should be laid with ventral side downwards on a microscope slide. Some may prefer one with hollow on one side, but
fixing valva complex is only possible if there is no hollow. Also spreading valva complex is easier if quantity of the liquid is at the lowest level possible, because this
helps to keep the parts spread; in too much liquid they tend to bend back into the
natural position. Good experience was made with preparation liquid of about 1/3
glycerol and 2/3 water with a trace of anionic surfactant. If the preparation work
takes very long, glycerol is important to prevent drying out and the anionic surfactant helps to leave all parts covered with the liquid.

After valvae are spread and put into correct position for the next step (vinculum directed towards preparator), a square cover glass is to be put on, but not over
the whole part at this step, but only over the lower part, leaving about half of tegumen with gnathos and socii uncovered. For putting cover glass on, first it must
be brought into position: on left side (for right handed-workers, this is the edge

opposite to the side where it is held with forceps) it is fixed with a fingertip on the microscope slide, on the right side it is kept closely above the genitalia parts to be covered, held by lower point of forceps. Then there are two different methods to put on the cover glass, neither one is "better", sometimes one and sometimes the other brings best results: forceps can be removed very quickly, and pressure of the finger lets cover glass move down extremely fast, often allowing the genitalia parts no time to turn into an asymmetrical position. Alternatively forceps are moved down very slowly, and as cover glass starts to touch tegumen, it may start moving into an asymmetrical position, and if so, the cover glass must be moved in the other direction, hoping cover glass takes the tegumen with it, and continue to move cover glass down very slowly until it fixes the parts. If covered in this way, it is necessary to check if all parts are flattened symmetrically. If not, cover glass can be removed, the parts set symmetrically and cover glass is put on again. If the same asymmetry is present, this may be caused by remaining parts of skin, which should be checked. Other reasons for persistence of asymmetry may be, especially in old specimens, that the genitalia dried out in asymmetrical position and remained in this position for many years, or there can be malformation. Anyway, it is better to accept some asymmetry than to try too often to obtain perfection, because this brings risk of further damage.

If the parts which are covered now are in a position which is good enough for final slide, the uncovered remainder must be set. The microscope slide should be turned into a position most convenient for the next step. Best tool is a 0.15 mm headless pin, fixed in a suitable carrier. Tip of pin may be completely straight or slightly bent. If bent, it can be easier to set the parts, but the risk of damage is higher. It depends on skill of preparator which is better, it is worth trying both. Gnathos can now be set upright by putting tip of pin under free parts of genitalia to turn it upright, this is essential to calculate exact gnathos position in relation to socii, an important feature in some species groups. In *Agonopterix*, uncus is usually indistinct and without value for determination, but in a few species its length is above average. In these species it may be better to turn gnathos a little aside, just enough to leave uncus visible. Finally socii are spread, this usually works well, but in a few *Agonopterix* species it is very difficult (e.g. *A. heracliana*, *A. alstromeriana*), in such cases it is recommended to check shape of socii in unspread genitalia in lateral position to preserve this information.

As soon as the uncovered parts are in the correct position, they must be covered with an additional cover glass. This is a critical step, because even a little lateral movement of one of the cover glasses can cause the genitalia to turn into an unsuitable position or even destroy some parts. Best to use low magnification, which leaves whole length of cover glass visible. Microscope slide must be turned into a position where the edge with uncovered parts is horizontal, uncovered parts directed towards preparator. Technique is in general the same as for slow putting on first cover glass. Differences are, no fingertip is necessary to fix any of the cover glasses and top left edge of second cover glass must be put very close to bottom left edge of first cover glass, but without touching it, because this unavoidably leads to movement and therefore distortion of covered parts of genitalia. Second cover glass is now put down by slowly removing lower point of forceps, watching the edges of the two cover glasses, the second one must be parallel and very close to the first, leaving almost no parts of the genitalia uncovered. An important detail: no part of any cover glass must overlap edges of microscope slide, because if so, it will drift into centre of it if fixing liquid is dropped on, causing movement and distortion of genitalia.

Once both cover glasses are put on and genitalia are found to be in good position, water-free isopropanol is dropped on to the gap between the two cover glasses. It is essential to use water-free isopropanol because only with this it is possible to go directly into euparal. Water causes euparal to turn milky-white, and ethanol does not mix with euparal without problems. Isopropanol takes water and glycerol out of the genitalia and they become stiff enough not to return to a bent position, but one drop is not nearly enough. Usually isopropanol must be dropped on 5 to 10 times and after a short time (about 30 seconds, but this depends on size of genitalia and may be longer) it must be taken away from microscope slide near edges of cover glasses with filter paper before next drops are put on. Better too many steps than too few as any remaining glycerol can cause problems in euparal, because it is not soluble in it and forms ugly inclusions looking like fat drops. Finally both cover glasses are removed and the result of all former steps is checked. If genitalia tend to turn back a little bit, usually it is result of some water remaining, in this case it is enough to put on cover glass again (one over whole genitalia) and perform some more desiccation steps. If parts of genitalia are found not to be in perfect position, sometimes it is possible to correct this even in the desiccated genitalia covered with isopropanol, e.g. if an anellus lobe is turned outwards instead of inwards. For this, one cover glass is put on in a way such that genitalia are held in place, but part which needs to be corrected is free. Correction can be made e.g. with a 0.15 mm pin with hooked tip in a carrier, but here individual experience is necessary, no description can guarantee success. If genitalia are found to be so far from perfect position that it is impossible to put them right at this stage and there are reasons to bring them into better position, this is possible by putting whole genitalia back into preparation liquid (water with glycerol), where they take on water and become soft (usually within a few minutes), all steps can then be repeated.

Staining: there are different ways, usually every preparator has a favourite method. For this book, all male slides have been stained after fixing in a staining liquid consisting of isopropanol with mercurochrome, which was found in tests to be very good. Advantages are that it can be washed out in cases of over-staining and mercurochrome stains exactly the structures which are essential in male genitalia. Chlorazole black is preferred instead of mercurochrome for females but is not

FIGURE 10   Details in text

recommended for males, because it especially stains components which are not wanted to be visible in males, e.g. thin elastic skins.

Before embedding in euparal, in all male slides used for this book, scales and most bristles were removed. Scales, which are usually found on dorsal side and edges, should be removed completely. Bristles are found especially on ventral side of valva. Although they may bring some information helpful for determination, they are an elusive feature and if relied upon too much, old specimens may appear to belong to a different species from fresh ones. Additionally these bristles often cover important parts of valva complex, e.g. anellus lobes and transtilla lobes. For removing, fully stained genitalia are put on a microscope slide with isopropanol (which must be added again and again to prevent drying out of genitalia), covered by a cover glass in part, while from the free part the bristles are removed using a piece of setting paper cut into a wedge with sharp point, which is held with forceps (fig. 10). To remove scales from abdominal skin often a feather is used, but this is not suitable for cleaning genitalia.

This cleaning risks damaging genitalia and is rather time consuming, therefore it is recommended to try at first with specimens of lower value, and additionally to decide for every specimen if the effort is worth while.

Preparing permanent slides: especially in *Agonopterix* it is recommended to take a photo of aedeagus in ventral view, because the aedeagus is in lateral view in the permanent slide in standard position. Outline in ventral view and details of basal sheath show important features for determination in some groups. For the photos which were made for the book, the equipment shown in fig. 11 was used: microscope slide and cover glass with euparal in between, on one side the cover glass is propped with 2 or 3 superimposed splinters of cover glass to leave enough space for the aedeagus in ventral view without distorting it. With a fine needle it can be put into the correct position.

If the aedeagus is taken out of isopropanol and put into the gap filled with euparal, this should be done rather quickly, otherwise isopropanol evaporates and air can penetrate the aedeagus. If this happens, it must be brought back into isopropanol, here the air slowly dissolves, usually it takes one or two days, but removing the air by compression must be avoided. Once brought into euparal, it again takes

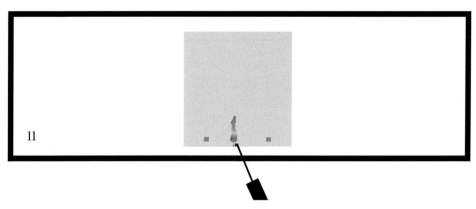

FIGURE 11   Details in text

time till euparal replaces isopropanol completely, this is necessary to avoid artefacts
⋅caused by light refraction. For the photo the aedeagus can be turned into the needed
position with a pin, and when it is done, again it is recommended to wait some time
before taking the photos, because aedeagus tends to move somewhat immediately
after this procedure. For good photo result stacking technique is necessary. As soon
as the photos are taken, the aedeagus is removed from this equipment and placed
together with valva complex into permanent slide. Here again splinters of cover
glass (usually not superimposed, but this depends on diameter of aedeagus) are
used to prop cover glass leaving sufficient space not to compress aedeagus in lateral
position. The valva complex preferably is put near the opposite edge, where cover
glass is not kept away from microscope slide, but in species where it is important
to avoid compression and therefore distortion of gnathos, such spacers are helpful
here also.

   An additional detail on the equipment shown in fig. 11: it can be reused, it is not
necessary to remove cover glass and euparal. Best to store in a little box for genitalia
slides if not in use, where it is protected from dust and from drying out too fast.
If dried somewhat, euparal can be refreshed by a drop of isopropanol on the side
where the aedeagus will be put under the cover glass. It is difficult to give the time it
takes to soften the euparal, it must be checked and maybe addition of isopropanol
must be repeated.

Dissecting and Preparing Female Genitalia

Many general details are already described in the previous section "Dissecting and
preparing male genitalia". Here only additional or different steps are discussed.

   Removing abdominal skin needs even more care in females than in males, especially not to damage proximal edge of segment VIII and ductus seminalis. Therefore
it is essential to stain the genitalia with chlorazole black after a rather cursory

cleaning, but if a photo of papillae anales in lateral view is wanted, this should be done before staining. For this book such photos were made in preparation liquid in the hollow of a microscope slide, covered with cover glass, using stacking technique. Bringing papillae anales into lateral, stable position is not easy, it is helpful if papillae anales are as close to the border of the hollow that they are held in position by the cover glass, but far enough away not to be compressed and distorted.

Because chlorazole black is carcinogenic any contact with human skin must be avoided, therefore a special procedure has been used which both brings good results and minimises the risk of contact with human skin: base is the same mixture of water, glycerol and a trace of anionic surfactant as used for dissections, into which is added a very small amount of chlorazole black which is dissolved in water and stored safely for further use. The concentration in the staining equipment is just enough to be visible as pale blue colour. It is used on a microscope slide with hollow, which must not be filled completely full to prevent the liquid exceeding the border of the hollow. This slide is stored strictly horizontally in a small slide box and can be stored there indefinitely. The liquid only dries out very slowly and can be used many times (approximately 20–50 times) without adding chlorazole black, only one drop of water must be added from time to time, because if air humidity is low, this liquid tends to lose some water resulting in an increase in concentration of both glycerol and chlorazole black. Glycerol lowers the staining speed markedly, this effect exceeds the increase of chlorazole black concentration, therefore we have the paradoxical situation that staining time increases with greater concentration. In general the staining time in this equipment can be rather long (hours to days) but adding a drop of water increases staining speed. Tests are necessary to find out all details. Once stained, the genitalia can be stored on a microscope slide in a drop of dissecting-liquid without losing stain, if there is no time to finish cleaning, which can take rather long. (Chlorazole black cannot be washed out, usually an advantage, but a disadvantage in case of overstaining. If overstaining has happened, a way to reduce or remove it is bleaching with Sodium hypochlorite).

Cleaning: for removing rest of intersegmental skin attached on proximal edge of segment VIII see "male genitalia". Here only one additional remark: if dividing line reaches origin of one anterior apophysis, it can happen that it turns and continues to run along apophysis. This must be watched carefully and if it occurs skin removal must be stopped immediately and parts of the skin must be left in the area around origin of apophysis. If ductus bursae shows loops, these should be preserved. If this seems impossible, at least a photo of the unchanged situation is recommended, because number and other details of such loops are often important features for species determination. Before removing unwanted parts attached to papillae anales and distal area of ductus bursae, a careful check is recommended where ductus seminalis is attached on ductus bursae and where ductus seminalis including the terminal coils runs. Distally, ductus seminalis is attached twice, once it is connected to distal part of ductus bursae and, forming a loop here, a second time to papillae anales. The connection with papillae anales should be removed, the one with ductus bursae preserved. For removing parts which are connected with papillae anales, it is necessary to hold papillae anales with forceps to prevent the tips from invaginating. Papillae anales must be very carefully kept from lateral position, otherwise damage is likely. Also taking parts to be removed with other forceps needs care, e.g. not to damage apophyses. This needs experience and must be practiced at first with specimens of lesser value. The microscope slide where genitalia have been cleaned will be the one for the permanent slide, therefore the genitalia should not be removed from this microscope slide any more, because this could damage delicate parts. Last step prior to desiccation is to set genitalia in the way they should appear in the permanent slide. Papillae anales should be moved somewhat distally, just enough that intersegmental skin between them and segment VIII is fairly stretched, and all should lie with ventral side up, strictly symmetrical if possible, apophyses posteriores should lie on either side of ostium without overlapping it (fig. 12). Cleaning may take rather long, but as long as genitalia are in preparation liquid, work can be interrupted and continued later. As soon as desiccation starts, the work should be continued without interruption until euparal and first cover glass is put on.

Desiccation: the usual method is to cover whole genitalia for desiccation, for this book a different method has been used, leaving papillae anales uncovered at this stage to avoid serious distortion. For this, a cover glass is used which is broken at about its half. Our experience is that cheap cover glasses bring better results here, but anyway not every attempt leads to a satisfying result. The broken edge of the cover glass should be rather straight (dividing line about as in fig. 13a) and the broken edge about perpendicular to upper and lower surface of cover glass (fig. 13c, seen parallel to broken edge), while such results as in figs. 13b and 13d are unsuitable. This broken edge is to be located between distal edge of ostium and distal edge of segment VIII (fig. 14a), at least in the permanent slide, because here there usually are no important features. Photos of such permanent slides show this edge as a fine dark line which must be removed on the computer. This is possible without loss of information only if no significant features are located in this area. But for the desiccation step it is not as important, the edge also may lie upon parts of the ostium at this stage, because cover glass will be removed before final embedding (but preferably it should lie as in fig. 14a).

Before final embedding all preparation liquid must be washed away and replaced by isopropanol. Therefore after several desiccation steps as described for males, the cover glass should be removed, followed by some more steps without cover glass. In some cases, ductus bursae or ductus seminalis adheres either to microscope slide (less problematic) or to cover glass (more problematic) during first desiccation steps. Therefore removing cover glass must be done very carefully, at first only lifting 1 mm or less at a time at the side which lies over segment VIII while it is fixed on opposite side with fingertip to prevent lateral movement, watching if any part is attached to any glass. If so, usually it is enough to move cover glass up and

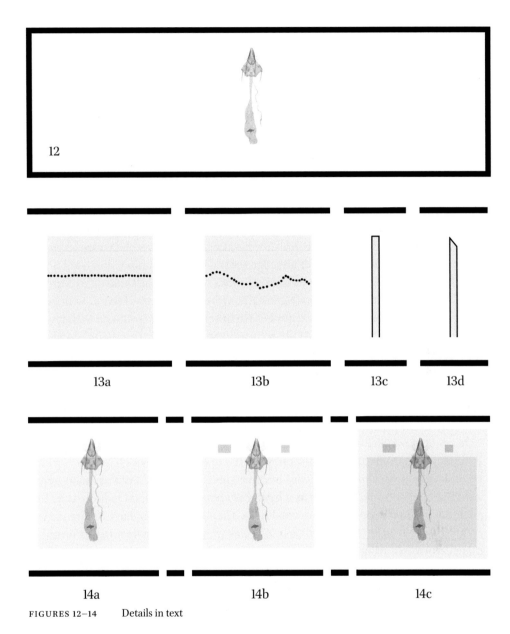

12

13a 13b 13c 13d

14a 14b 14c

FIGURES 12–14 Details in text

down several times, no more than 1 mm each time, this movement of isopropanol gradually separates the glued parts from glass, this rarely takes very long (but it can). As soon as cover glass is removed, several more desiccation steps are made, at this stage it can be helpful to gently squeeze papillae anales laterally with forceps several times when they are covered by isopropanol to expel any remaining preparation liquid. When it is estimated to be clear of preparation liquid, euparal is put on whole genitalia. Along the outer edge of this drop it mixes with isopropanol and it is recommended to clean away this mixture with filter paper and also clean away euparal if there is more than necessary, otherwise the first cover glass of the final slide tends to drift away and therefore comes into an unsuitable position.

Before putting on first cover glass of permanent slide, position of whole genitalia (in centre of microscopic slide, but papillae anales a little closer to long edge of it than corpus bursae on opposite side) and position of all parts should be checked and corrected if necessary. The cover glass used for desiccation can be used also as first cover glass of permanent slide, but usually it is not really clean after desiccation procedure and it may be easier to break another glass for this purpose. Before putting on, its quality must be checked and also if it is large enough to cover whole section needed. Best results were obtained with this procedure (for right-handed workers): microscope slide turned so that papillae anales are on right side, broken cover glass held with forceps with broken edge at right side, cover glass once brought into position touching down with left edge (outside corpus bursae!) and fixed here with fingertip (with or without protecting gloves). Now is the last chance to correct position of cover glass in relation to genitalia without problems. If found to fit perfectly its right side is slowly moved down with lower tip of forceps. To do this slowly is recommended because this helps to avoid air-inclusions. Removing overflow of euparal on the right side with filter paper helps to let cover glass stay in place, additionally it is recommended to wait some time before removing fingertip on left side to prevent any further movement of it. Enough euparal must remain to include papillae anales and also glue the two (for large specimens maybe three) superimposed splinters of cover glass to microscope slide which are subsequently put on either side of papillae anales (fig. 14b). Then this microscope slide has to be stored horizontally in a safe and dust free place until euparal is dry enough for putting on second cover glass. Time for this may be one to several months, if not so much time is available it is also possible to use a heater at about 40 ° C, which reduces time to a few days. When dry enough, a second cover glass is put on (fig. 14c). Technique is about the same as for first cover glass. A check if everything is dust free is recommended before further steps. Left edge of second cover glass also can be placed by closed forceps or needle here, which cannot stop lateral movement as effectively as fingertip, but this is not as essential as for first cover glass. Advantage is that it helps human skin keep clean of euparal. As soon as put on it is recommended to press down the second cover glass very gently with a clean tool at its middle, so that finally it touches glass below it at three areas: the two superimposed splinters of cover glass and the edge of first cover glass on the opposite side. If too much euparal appears just outside the edges of second cover glass, it can be removed by filter paper. The space between first and second cover glass is now filled with exactly the amount of euparal needed.

Subsequently the slide must be stored horizontally for several months (preferably for ever). Taking photos is not recommended immediately after putting on second cover glass, because there is euparal of different structure (more and less dried) around papillae anales, which causes artefacts by light refraction. This effect takes a few days to disappear.

When drying, euparal sometimes contracts so much that in the space between the superimposed splinters of cover glass and the second cover glass the border euparal/air moves towards papillae anales. Therefore the slides should be checked about one month after finishing and if necessary, a drop of euparal can be added, a second check may be useful some months later.

Photographic Methods

Photos of whole specimens were taken with Canon EOS 5D Mark III and Canon lens EF 100mm 2.8 L IS USM at 1:1. Specimens were illuminated with two diffused flashes, using a third flash for setting the background whiteness. Photos of specimen details were taken with Canon lens MP-E 65 at 2:1, using ring flash. Genitalia photos were taken with microscope (Wild Heerbrugg) using a 10× objective and a 2.5× ocular. All photos were taken by the first author and edited using the software Helicon Focus 4.80 and Adobe Photoshop 6.0. For creating the black and white photos, the G alpha channel of the RGB originals was used in males and the Y alpha channel of the CMYK originals in females, due to the different stains.

DNA Barcoding

Full-length lepidopteran DNA barcode sequences are a 658 basepair long segment of the 5' terminus of the mitochondrial COI gene (cytochrome c oxidase 1). DNA samples (dried leg) were prepared according to the accepted standards and were processed at the Canadian Centre for DNA Barcoding (CCDB, Biodiversity Institute of Ontario, University of Guelph) to obtain DNA barcodes using the standard high throughput protocol described in deWaard *et al.* (2008). Further details including complete voucher data and images can be accessed in the the the public dataset DS-DEEUR400 (http://www.boldsystems.org/index.php/Public_Search Terms?query=DS-DEEUR400) in the Barcode of Life Data Systems (BOLD; Ratnasingham & Hebert, 2007). For molecular-based estimation of relatedness, neighbour-joining trees of DNA barcode data were usually used, constructed using MEGA (Kumar *et al.*, 2018) under the Kimura 2 parameter model for nucleotide substitutions. Alternatively, maximum likelihood computation method was used including 500 or 1000 bootstrap replications for each tree.

As this is the first volume of the Microlepidoptera of Europe series to make significant use of DNA barcoding as a taxonomic aid, a guide to barcoding follows including discussion of its uses and limitations.

Barcoding as a Taxonomic Aid

It is not the intention of this section to present comprehensive background information. Only the most important details are addressed to give an overview for those less familiar with the subject. There is plenty of specialist literature and web-based information available for in-depth studies.

The DNA barcode is the sequence of the first 658 base pairs (bp) of the gene for cytochrome oxidase 1 (CO 1 or COX 1; 5' region of the mitochondrial cytochrome c oxidase subunit 1 gene), an important enzyme in respiratory metabolism. Since it occurs throughout the animal kingdom, it is ideal for genetically based comparisons of even very distantly related species.

At first a quick look at genetics is recommended here. The structure and function of each protein, that includes also enzymes, is determined by the sequence of the amino acids. Changing the sequence of amino acids (usually) also changes the function of the protein. The building instructions for these proteins are in the DNA. Four different bases, adenine (A), cytosine (C), guanine (G) and thymine (T) are available, three consecutive bases (a triplet) form the code for the incorporation of an amino acid. With four bases and three consecutive positions, there are four to the power of three, i.e. 64 different triplets. Three have non-coding roles, leaving 61 triplets for amino acid coding, but only 20 different amino acids used to build the proteins are coded by DNA. Accordingly, there are several different triplets that encode the same amino acid. Depending on the amino acid, one to six variants are possible.

The exact number of possible variants of a DNA sequence that result in exactly the same protein can therefore only be calculated if the individual amino acids used in a protein are taken into account. For our consideration, however, a simplified estimate is sufficient, which assumes that there are three different triplet variants for each amino acid. For the barcode region, this means that it consists of approximately 220 triplets, and based on the simplified assumption, 3 to the power of 220 variants result in exactly the same protein. 3 to the power of 220 is about 10 to the power of 105, which is a "1" followed by 105 zeros – an unimaginably large number, more than there are atoms in our galaxy. In addition, some amino acids can be replaced by others without changing the function of the protein. The number of possible genetic variants of the barcode region which do not affect the function of the protein is therefore even higher, although this is irrelevant for further consideration. Consequently all barcode sequences found produce the same or at least functionally identical protein, they are not the result of adaptations to different circumstances, but random products within the framework of the molecular principles of protein synthesis outlined above. Furthermore the number of theoretically possible barcode sequences astronomically exceeds the number of existing variants.

Mutations, i.e. the random exchange of bases, occur again and again when copying DNA. If this leads to a malfunction of the encoded protein, the affected cell may die sooner or later, so such mutations are subject to negative selection. If the functional change leads to new, usable properties, they are subject to positive selection, which is also the genetic basis for the development of new species. Mutations that do not change the function of the protein undergo no selection, they are retained, and as explained above, there is an almost infinite number of such variants available. They always occur as random events. So if you compare the base sequence of the coding gene of a functional protein that occurs equally in two organisms or species, the differences are definitely a useful measure of their relatedness. Because "relatedness" means that two organisms had a common ancestor at some point in the past. "Less closely related" means that the most recent common ancestor is further back in the past. Therefore, the differences in the base sequence of a functionally identical gene found in two organisms are a measure of their relatedness. However, that is only true in general. To mathematically equate the relationship exactly to the base differences would lead to wrong results, because not only the type of mutation, but also its frequency is determined by chance.

The differences are usually given as a percentage of all base pairs considered. Therefore one different bp. in the 658 bp. long barcode sequence means a distance of approximately 0.15%, while 13 different bp. gives a distance of approximately 2%.

A special feature of the barcode region as a mitochondrial gene should be discussed in more detailed. The mitochondria are the organelles of respiratory metabolism in a cell. They were originally free-living bacteria, which were the first and only organisms in evolution to develop energy production through respiration. The eukaryotes (organisms with a cell nucleus, i.e. plants, fungi, animals) did not develop respiration independently again, but the eukaryotes' parent ancestral form brought the bacteria capable of respiration into their plasma as symbionts. This symbiosis is so close that the descendants of these bacteria, i.e. the mitochondria, would no longer be able to survive independently, but they have retained their individuality by having their own DNA and reproduce independently from their host cell. The main consequence of this is that the mitochondrial DNA is not subject to recombination, as is typical for the DNA of the cell nucleus of eukaryotes, and that the mitochondria, because they are in the plasma, are only transferred via the egg cell, i.e. from the mother to the offspring, while the male parent does not provide mitochondria. It is therefore not possible, for example, to use the barcode to identify whether the individual in question is a hybrid.

The choice of this mitochondrial gene for the development of a genetic fingerprint and a genetic basis for comparison of species throughout the animal kingdom has two reasons: on one hand it is the code for a universally present enzyme, and on the other hand there are significantly more copies of mitochondrial genes than of nuclear genes, which greatly facilitates successful sequencing. Despite all the success and the new knowledge gained through barcoding, it must not be forgotten that it has its limits, especially because the barcode region is only a tiny part of the entire DNA of a cell.

As of May 2022, there are already so many barcode results (over 14 million in the BOLD database alone) that the following statement can be made reliably: members of a species sometimes do not differ at all in the barcode sequence, or usually no more than about 2–3%, while the barcode sequence of different species tends to show larger difference and tends to do so more the less closely related they are.

At this point it is necessary to consider the question of what a "species" actually is. There is no simple answer, but the most widely accepted and also the most plausible definition is that a species is a potential reproductive community, i.e. members of one species can produce fertile offspring together, while members of different species can not do so because of a genetic barrier – either they can produce no offspring at all, or if they can, offspring is infertile because the organisation of the genes in the cell nucleus of the parents of such hybrids is so different that the hybrids are no longer able to produce germ cells with balanced genes to form a functional genome.

Since the presence or absence of a genetic reproductive barrier can hardly ever be observed directly when the question arises if two individuals belong to the same species or not, we tend to define or identify species on the basis of their features. But this poses a trap, because it is not the differences in features that determine specific difference, but the existence of a genetic barrier to reproduction.

However, if fundamentally independent features always occur in the same combination, this is an indication of the existence of a reproductive barrier. Without a reproductive barrier, all possible combinations of features would be found through genetic exchange and subsequent recombination. The barcode sequence can also serve as such an independent feature.

Barcodes have a role in the discovery of previously unrecognised species, i.e. the recognition of cryptic diversity. Morphological variation within a species can hide the fact that two or more species may be unrecognised within this variation. In such cases barcodes falling into two or more clearly separate groups give an indication that more than one taxon is involved. With this additional knowledge, clear differences in morphology can often be found which had been previously assumed to be intraspecific variation. There are however also instances where barcodes indicate cryptic diversity, but as yet, no correlated differences have been found.

It is also important to say that, if fundamentally independent features always occur in the same combination, this is only an indication of a genetic reproductive barrier if the populations are sympatric. In allopatric populations, there is a geographical barrier to reproduction that has the same effect on feature correlation. In order to enable a practical handling of the species concept in such cases, it is advisable to follow the suggestion by Mutanen *et al.* (2012): allopatric populations are then classified as different species if the differences are of a similar size or greater than is usual for sympatric species. The barcodes also play an important role here as an independent individual feature.

Frequently the question is asked, beyond what level of barcode difference can we speak of two different species. In this form, however, the question is unanswerable because it was posed incorrectly. The barcodes available so far show that differences of up to 1% within a species are common and differences of up to 3% are not uncommon, while between members of different species differences of less than 1% are rare and differences of more than 3% are common. But the reverse conclusion is not permissible. Identical barcodes can also be found in different species, known as "barcode sharing". On the other hand, very large differences, even over 10%, can also occur within a species. This can occur in species that evolved a very long time ago, but the same species may also have big differences from their nearest relatives. Such large intraspecific divergence can also result from introgression.

In Depressariidae there are instances of barcode sharing in *Exaeretia, Agonopterix* and *Depressaria*. This phenomenon can result from rapid evolution of a subpopulation into a new species by adapting to a newly emerging ecological niche. Such a development can happen very quickly through the selection of genes that are already present in the population, or at least much faster than the emergence of new mutations, including new barcode variants.

Barcode sharing by two or more different species can, for example, also be the result of introgression, i.e. the introduction of "foreign" mitochondria through the extremely rare, but not impossible occurrence of a fertile hybrid with subsequent backcrossing: a female hybrid is created from a female of species A and a male of species B, therefore it has mitochondria of species A. This hybrid female successfully has offspring with a male of species B and their female offspring again mate with species B. If this continues the nuclear genes of type A decrease with each generation and soon they disappear. But the mitochondria with sequence of species A remain because they do not participate in recombination and are passed on only on the mother's side. After this event, either the whole population or a subpopulation of species B has the DNA barcode sequence of species A due to introgression.

Another reason for barcode sharing is rapid evolution of a subpopulation into a new species by adapting to a new ecological niche. By selecting genes that are already present in the population, such a development can happen very quickly, at least much faster than the emergence of new mutations, including new barcode variants.

Species determination is possible solely on the basis of barcodes, excepting known cases of barcode sharing. The most important prerequisite is that a barcode which matches the searched sequence exactly or nearly exactly must be available. If that is the case, there are definitely pitfalls to be aware of, e.g. unrecognised barcode sharing, misidentification of the specimen that provided the matching sequence, contamination of the examined sample with foreign DNA or laboratory error in sequencing. Conclusion: every determination obtained only via barcoding should always be checked for plausibility. To align a barcode sequence to a species is possible via the publicly accessible site http://www.boldsystems.org/ (subsite "Identification") of BOLD (Barcode Of Life Data) system in the University of Guelph, Canada.

Estimating relatedness is an essential area where barcodes are extremely helpful, but the distances between the sequences alone only give a very rough picture, the positions of the deviating bases must also be taken into account. This simple example should show that the distance alone (linear consideration) brings insufficient information, and a multidimensional calculation is required: the starting point is species or sequence K. In species L, six adenines are replaced by thymines, so the sequence differs in six positions or almost 1%. In species M, exactly the same six adenines are replaced by guanines. In species N, also six adenines – but in different positions – are replaced by guanines. The distance of the species K to all three compared species (L, M and N) is six bases, almost 1%. The comparison of L and M also results in a distance of almost 1%, but the comparison of N and M results in a double distance, i.e. 12 positions or almost 2%. The usual construction, which is relatively easy to calculate and also includes the positions of the bases, is the NJ tree (neighbour-joining tree). An optimised calculation is the "maximum likelihood analysis", a computer simulation of many (up to 1000) evolutionary processes to determine which prerequisite is the most likely to lead to the current genetic situation.

Finally, a brief consideration of the conditions that affect successful DNA analysis of a sample. After the death of a cell, DNA is subject to progressive degradation, which takes place relatively slowly in a dry environment but is significantly accelerated in the presence of water. Specimens that have been dried quickly and are subsequently kept in a dry environment still bring good results after 10–15 years. Today there are also sequencing methods (protocols) available that bring good results if DNA is more degraded, up to NGS (New Generation Sequencing), where six subunits of the barcode region are sequenced separately and that still delivers good results with samples that are more than 50 years old. However, storage in a dry place is always a prerequisite. Storage in high humidity accelerates degradation, and softening of dried material destroys the DNA in a short time. The success of the sequencing of older specimens is therefore not easy to assess. Storing samples in pure ethanol preserves the DNA by removing water, but there is as yet no experience as to whether storage for decades is better than storage in dry air.

In work for the preparation of the present volume, barcodes have been obtained for nearly all species and these have been extremely useful in some cases in helping to clarify some difficult taxonomic issues. In the species accounts, barcodes are not routinely mentioned, but where they have taxonomic significance, such as barcode sharing or possible cryptic species, they are discussed under Remarks.

Notes on the Illustrations

For details of photographic methods see p. 31.

Specimens are figured at 2.5 times magnification, if not otherwise specified. If details are shown at larger magnification, this is given in brackets behind the figure number and refers to the basic magnification, e.g. (3×) means the detail is shown 3 times larger than the figure of the moth, therefore at 7.5 times magnification. Specimens are selected predominantly to show intraspecific variability, but it is not possible to show the full range of variation for every species. The label data are given in a standardised form as sex / country / province (if available) / locality / date / collector(s). If a specimen was bred from larva, ex l. is added, but if the date given is the collecting date of larva or date of emergence of the moth is often not specified on the label and therefore remains unknown, but if known the date of emergence is given. Additionally in brackets DEEUR-number and collection is listed. The DEEUR (Depressariidae of Europe) number is a unique number which enables recognition of a particular specimen in a collection, it is pinned to every specimen studied by PB for this volume, in total more than 10,000. Abbreviations for collections are either the usual abbreviations of the museum or for private collections in the form "RCFS" from "Research Collection Forename Surname".

The large majority of genitalia figures is based on slides dissected and photographed by PB, and if not, the preparator and/or photographer is mentioned. The scale of basic magnification is different in males and females and is given on the first plate of each. If details are shown in larger magnification, this is given in the same way as for specimen details. Enlarged details from a basic magnification figure share the same figure number. Structure of segment VIII of abdominal skin is shown only for a few *Semioscopis* and *Exaeretia* male species, in general it is without value for determination of the species presented in this volume. Figures of free floating genitalia have the additional information "ffl" for "free floating, lateral view" or "ffv" for "free floating, ventral view". Minimal label data are provided, further details are accessible on the web, where a complete list of the DEEUR numbers with full data will be published together with this volume.

Key to European Genera of Depressariidae

1a	Antennal scape without pecten; termen strongly inclined, forewing with apex acute	*Semioscopis*
1b	Scape with pecten, termen not or weakly inclined, apex rounded or weakly falcate	2
2a (1b)	Forewing with scale tufts	*Luquetia*
2b	Forewing without scale tufts	3
3a (2b)	Veins Cu_1 and Cu_2 of forewing distant	*Depressaria*
3b	Veins Cu_1 and Cu_2 of forewing stalked	4
4a (3b)	Sacculus ending with a single process directed across valva	*Agonopterix*
4b	Sacculus ending with two processes, one directed across valva, the second at right-angles to the first	*Exaeretia*

© PETER BUCHNER AND MARTIN CORLEY, 2025 | DOI:10.1163/9789004713116_008

CHAPTER 8

Checklist of European and Macaronesian Depressariidae

Semioscopis Hübner, 1825[1]
    Epigraphia Stephens, 1829
1.   avellanella (Hübner, 1793)
    fumicella Amsel, 1930
2.   oculella (Thunberg, 1794)
    anella (Hübner, 1796)
    alienella (Treitschke, 1832)
3.   steinkellneriana ([Denis & Schiffermüller], 1775)
    characterella (Hübner, 1793)
    characterosa (Haworth, 1811)
4.   strigulana (Fabricius, 1787)
    atomella (Hübner, 1796)
    consimilella (Hübner, 1825)

Luquetia Leraut, 1991
    Enicostoma Fletcher, 1929, nec Stephens, 1829
5.   lobella ([Denis & Schiffermüller], 1775)
    thunbergiana (Fabricius, 1781)
    lugubrella (Duponchel, 1838)
6.   orientella (Rebel, 1893)
    abchasiella Lvovsky, 1995
7.   osthelderi (Rebel, 1936)

Exaeretia Stainton, 1849
    Depressariodes Turati, 1924
    Martyrhilda Clarke, 1941
    Levipalpus Hannemann, 1953, **syn. n.**
8.   hepatariella (Lienig & Zeller, 1846)
9.   allisella Stainton, 1849
    lechriosema (Meyrick, 1928)
10. ciniflonella (Lienig & Zeller, 1846)
    klamathiana (Walsingham, 1881)

---

1 The oldest available family name is the rarely used Epigraphiidae Guenée, 1845. An application (Case 3841) has been made to ICZN for suppression of this in favour of Depressariidae Meyrick, 1883 (Van Nieukerken *et al.*, 2022).

smolandiae (Palm, 1943)

isa (Clarke, 1947)

11.   preisseckeri (Rebel, 1937)

gozmanyi (Balogh, 1951)

12.   lutosella (Herrich-Schäffer, 1854)

13.   thurneri (Rebel, 1941)

nigromaculata Hannemann, 1989, **syn. n.**

14.   ledereri (Zeller, 1854)

homochroella (Erschoff, 1874)

xyleuta (Meyrick, 1913)

leviella (Amsel, 1934)

15.   praeustella (Rebel, 1917)

16.   nebulosella (Caradja, 1920)

17.   lepidella (Christoph, 1872)

18.   buvati Nel & Grange, 2014

19.   mongolicella (Christoph, 1882)

leucostictella (Rebel, 1917)

exquisitella (Caradja, 1920), **syn. n.**

amurella Lvovsky 1990, **syn. n.**

20.   lvovskyi Buchner, Junnilainen & Nupponen, 2019

21.   stramentella (Eversmann, 1844)

culcitella (Herrich-Schäffer, 1854)

22.   niviferella (Christoph, 1872)

23.   indubitatella (Hannemann, 1971)

Agonopterix Hübner, [1825] 1816

Epeleustia Hübner, [1825] 1816

Pinaris Hübner, [1825] 1816

Tichonia Hübner, [1825] 1816

Haemylis Treitschke, 1832

Agonopteryx Stephens, 1834

Syllochitis Meyrick, 1910

Ctenioxena Meyrick, 1923

24.   impurella (Treitschke, 1835)

hypericella auct. nec (Hübner, 1817)

25.   liturosa (Haworth, 1811)

liturella (Hübner, 1796), nec ([Denis & Schiffermüller], 1775)

hypericella (Hübner, 1817)

huebneri Bradley, 1966

26.   conterminella (Zeller, 1839)

27.   arctica (Strand, 1902)

nordlandica (Strand, 1920)

28. ocellana (Fabricius, 1775)
     characterella ([Denis & Schiffermüller], 1775)
     signella (Hübner, 1797)
     signosa (Haworth, 1811)
29. fruticosella (Walsingham, 1903)
30. rigidella (Chrétien, 1907)
     rebeli (M. Hering, 1936)
31. olusatri Corley & Buchner, 2019
     chironiella auct. nec (Constant, 1893)
32. leucadensis (Rebel, 1932)
33. adspersella (Kollar, 1832)
     amanthicella (Heinemann, 1870)
     feruliphila (Millière, 1866)
     crassiventrella (Rebel, 1891)
     sabulatella (Turati, 1921)
     rubripunctella (Amsel, 1935a)
     karmeliella (Amsel, 1935b)
33.1. adspersella subsp. pavida (Meyrick, 2013) **stat. rev.**
34. thapsiella (Zeller, 1847)
     linolotella (Chrétien, 1929)
35. chironiella (Constant, 1893)
36. cervariella (Constant, 1884)
37. paracervariella **sp. n.**
38. cadurciella (Chrétien, 1914)
     adspersella sensu Rymarczyk *et al.*, 2013a, *nec* (Kollar, 1832)
39. nodiflorella (Millière, 1866)
     inoxiella Hannemann, 1959, **syn. n.**
40. rotundella (Douglas, 1846)
     peloritanella (Zeller, 1847)
41. purpurea (Haworth, 1811)
     vaccinella (Hübner, 1813)
     banatica Georgesco, 1965, **syn. n.**
42. curvipunctosa (Haworth, 1811)
     zephyrella (Hübner, 1813)
     granulosella (Stainton, 1854)
     amasina (Mann, 1861)
     turbulentella (Glitz, 1863)
     kotalella Amsel, 1972
43. vendettella (Chrétien, 1908)
     iliensis (Rebel, 1936), **syn. n.**
44. alpigena (Frey, 1870)
     sileris (Pfaffenzeller, 1870)

         salevensis (Frey, 1880)

         ragonoti (Rebel, 1889)

         seraphimella (Chrétien in Lhomme, 1929)

45. richteri **sp. n.**
46. coenosella (Zerny, 1940)
47. kayseriensis Buchner, 2020
48. cachritis (Staudinger, 1859)
49. ferulae (Zeller, 1847)
50. galicicensis **sp. n.**
51. langmaidi **sp. n.**
52. lessini Buchner, 2017
53. selini (Heinemann, 1870)
54. ordubadensis Hannemann, 1959
55. socerbi Šumpich, 2012
56. angelicella (Hübner, 1813)
57. paraselini Buchner, 2017
58. parilella (Treitschke, 1835)

         humerella (Duponchel, 1838)

59. ciliella (Stainton, 1849)

         annexella (Zeller, 1868)

60. orophilella Rymarczyk, Dutheil & Nel, 2013
61. heracliana (Linnaeus, 1758)

         punctata (Clerck, 1759)

         applana (Fabricius, 1777)

         cerefolii (Retzius, 1783)

         cicutella (Hübner, 1796)

62. perezi (Walsingham, 1908)
63. putridella ([Denis & Schiffermüller], 1775)
63.1. putridella subsp. scandinaviensis **subsp. n.**
64. quadripunctata (Wocke, 1857)
65. hippomarathri (Nickerl, 1864)
66. astrantiae (Heinemann, 1870)

         isabellina (Klemensiewicz, 1898)

67. cnicella (Treitschke, 1832)
68. melancholica (Rebel, 1917)

         funebrella (Caradja, 1920)

69. alstromeriana (Clerck, 1759)

         monilella ([Denis & Schiffermüller], 1775)

         puella (Hübner, 1796)

         albidella (Eversmann, 1844)

70. capreolella (Zeller, 1839)

         caprella (Stainton, 1849)

71. yeatiana (Fabricius, 1781)
    albidana (Donovan, 1806)
    putrida (Haworth, 1811)
    ventosella (Stephens, 1834)
    atricornella (Mann, 1855)
    oglatella (Chrétien, 1915)
    fuscovenella (Rebel, 1917)
72. silerella (Stainton, 1865)
73. ligusticella (Chrétien, 1908)
    tripunctaria Buchner, 2015
74. medelichensis Buchner, 2015
75. irrorata (Staudinger, 1871)
    anthriscella (R. Brown, 1886)
76. graecella Hannemann, 1976
77. pseudoferulae Buchner & Junnilainen, 2017
78. subtakamukui Lvovsky, 1998
    cluniana Huemer & Lvovsky, 2000
79. guanchella Buchner, 2022
80. furvella (Treitschke, 1832)
81. pupillana (Wocke, 1887)
    dictamnephaga Rymarczyk, Dutheil & Nel, 2012, **syn. n.**
82. rutana (Fabricius, 1794)
    retiferella (Zeller, 1850)
83. subumbellana Hannemann, 1959
84. pallorella (Zeller, 1839)
    subpallorella (Staudinger, 1871)
    divergella (Caradja, 1920)
    mikomoensis Fujisawa, 1985
85. straminella (Staudinger, 1859)
    echinopella (Chrétien, 1907), **syn. n.**
    liodryas (Meyrick, 1921a)
    cryptipsila (Meyrick, 1923a)
    mendesi Corley, 2002, **syn. n.**
86. carduncelli Corley, 2017
87. bipunctosa (Curtis, 1850)
    perpallorella (Morris, 1870)
88. kyzyltashensis Buchner & Šumpich 2020
89. squamosa (Mann, 1864)
    budashkini Lvovsky, 1998, **syn. n.**
90. tschorbadjiewi Rebel, 1916
91. volgensis Lvovsky, 2018
92. kaekeritziana (Linnaeus, 1767)

liturella ([Denis & Schiffermüller], 1775)
sparrmanniana (Swederus, 1787)
flavella (Hübner, 1796)
corichroella (Turati, 1924)
93. broennoeensis Strand, 1920
roseoflavella Benander, 1955
94. uralensis **sp.n.**
95. latipennella (Zerny, 1934)
96. invenustella Hannemann, 1953
97. abditella Hannemann, 1959
98. lidiae Buchner, 2020
99. subpropinquella (Stainton, 1849)
intermediella (Stainton, 1849)
rhodochrella (Herrich-Schäffer, 1854)
himmighofenella (Herrich-Schäffer, 1854)
thoracica (Lederer, 1855)
sublutella (Staudinger, 1859)
variabilis (Heinemann, 1870)
occaecata (Meyrick, 1921)
remota (Meyrick, 1921)
crispella (Chrétien, 1929)
keltella (Amsel, 1935)
amilcarella (Lucas, 1951)
100. nanatella (Stainton, 1849)
aridella (Mann, 1869)
101. kuznetzovi Lvovsky, 1983
102. propinquella (Treitschke, 1835)
103. carduella (Hübner, 1817)
104. ivinskisi Lvovsky, 1992
flurii Sonderegger, 2013
centaureivora Rymarczyk, Dutheil & Nel, 2013
ivinskisi subsp. *daghestanica* Lvovsky, 2018
105. ferocella (Chrétien, 1910)
106. laterella ([Denis & Schiffermüller], 1775)
heraclella (Hübner, 1813)
incarnatella (Zeller, 1854)
107. xeranthemella Buchner, 2018
108. cinerariae Walsingham, 1908
109. arenella ([Denis & Schiffermüller], 1775)
gilvella (Hübner, 1796)
immaculana (Stephens, 1834)
110. petasitis (Standfuss, 1851)

111. cotoneastri (Nickerl, 1864)
     senecionis (Nickerl, 1864)
     sarracenella (Rössler, 1866)
     marmotella (Frey, 1868)
     seneciovora Fujisawa, 1985, **syn. n.**
112. multiplicella (Erschoff, 1877)
     sutschanella (Caradja, 1926)
     klimeschi Hannemann, 1953
113. doronicella (Wocke in Gravenhorst, 1849)
     schmidtella (Zeller, 1851)
     laetella (Herrich-Schäffer, 1853)
114. assimilella (Treitschke, 1832)
     irrorella (Stephens, 1834)
115. atomella ([Denis & Schiffermüller], 1775)
     pulverella (Hübner, 1825)
     respersella (Treitschke, 1833)
116. scopariella (Heinemann, 1870)
     rubescens (Heinemann, 1870)
     genistella (Walsingham, 1903), **syn. n.**
     cyrniella (Rebel, 1929), **syn. n.**
     calycotomella (Amsel, 1958), **syn. n.**
117. conciliatella (Rebel, 1892)
     mutatella Hannemann, 1989
118. comitella (Lederer, 1855)
119. nervosa (Haworth, 1811)
     costosa (Haworth, 1811)
     depunctella (Hübner, 1813)
     boicella (Freyer, 1836)
     dryadoxena (Meyrick, 1920)
     blackmori (Busck, 1922)
     perstrigella (Chrétien, 1925), **syn. n.**
     obscurana (Weber, 1945)
120. umbellana (Fabricius, 1794)
     ulicetella (Stainton, 1849)
     lennigiella (Fuchs, 1880)
     prostratella (Constant, 1884)
     knitschkei (Predota, 1934)
121. oinochroa (Turati, 1879)
122. aspersella (Constant, 1888)
     novaspersella (Spuler, 1910)
     autocnista (Meyrick, 1921)

Depressaria Haworth, 1811
 Piesta Billberg, 1820
 Volucrum Berthold, 1827
 Volucra Latreille 1829
 Siganorosis Wallengren, 1881
 Schistodepressaria Spuler, 1910
 Horridopalpus Hannemann, 1953
 Hasenfussia Fetz, 1994
123. pulcherrimella Stainton, 1849
 semenovi Krulikovsky, 1903
124. sordidatella Tengström, 1848
 weirella Stainton, 1849
 gudmanni Rebel, 1927
 larseniana Strand, 1927
125. floridella Mann, 1864
126. douglasella Stainton, 1849
 miserella Herrich-Schäffer, 1854
127. beckmanni Heinemann, 1870
128. incognitella Hannemann, 1990
129. nemolella Svensson, 1982
130. cinderella Corley, 2002
131. infernella Corley & Buchner, 2019
132. indecorella Rebel, 1917
133. lacticapitella Klimesch, 1942
134. hofmanni Stainton, 1861
135. albipunctella ([Denis & Schiffermüller], 1775)
 albipuncta Haworth, 1811
 aegopodiella Hübner, 1825
136. subalbipunctella Lvovsky, 1981
137. krasnowodskella Hannemann, 1953
138. tenebricosa Zeller, 1854
 albiocellata Staudinger, 1870
 amblyopa Meyrick, 1921
139. radiosquamella Walsingham, 1898
 duplicatella Chrétien, 1915, **syn. n.**
 adustatella Turati, 1927, **syn. n.**
 delphinias Meyrick, 1936
 subtenebricosa Hannemann, 1953
140. emeritella Stainton, 1849
141. olerella Zeller, 1854
142. leucocephala Snellen, 1884
 thomanniella Rebel, 1917

143. depressana (Fabricius, 1775)
     depressella (Fabricius, 1798)
     bluntii Curtis, 1828
     colarella (Zetterstedt, 1839)
     prangosella Walsingham, 1903
     rhodochlora Meyrick, 1923
144. absynthiella Herrich-Schäffer, 1865
     anchusella Nowicki, 1860
     absinthivora Frey, 1880
145. tenerifae Walsingham, 1908, **stat. rev.**
146. artemisiae Nickerl, 1864
     dracunculi Clarke, 1933
147. fuscovirgatella Hannemann, 1967
     pagmanella Amsel, 1972
148. atrostrigella Clarke, 1941
149. silesiaca Heinemann, 1870
     millefoliella Chrétien, 1908
     freyi (M. Hering, 1924)
150. zelleri Staudinger, 1879
151. pyrenaella Šumpich, 2013
152. marcella Rebel, 1901
     cuprinella Walsingham, 1907
     cruenta (Meyrick, 1920)
     chneouriella Lucas, 1940
153. peregrinella Hannemann, 1967
154. heydenii Zeller, 1854
155. fuscipedella Chrétien, 1915
156. ululana Rössler, 1866
157. manglisiella Lvovsky, 1981
     venustella Hannemann, 1990, **syn. n.**
158. chaerophylli Zeller, 1839
159. longipennella Lvovsky, 1981
160. daucella ([Denis & Schiffermüller], 1775)
     rubricella ([Denis & Schiffermüller], 1775)
     apiella (Hübner, 1796)
     nervosa Stephens, 1834, *nec* Haworth, 1811
161. ultimella Stainton, 1849
162. halophilella Chrétien, 1908
163. radiella (Goeze, 1783)
     heracliana auct., nec (Linnaeus, 1758)
     heraclei auct., nec (Retzius, 1783)
     radiata (Geoffroy *in* Fourcroy, 1785)

pastinacella (Duponchel, 1838)

sphondiliella Bruand, 1851

ontariella Bethune, 1870

caucasica Christoph, 1877

164. libanotidella Schläger, 1849

laserpitii Nickerl, 1864

daucivorella Ragonot, 1889

mesopotamica Amsel, 1949

165. bantiella (Rocci, 1934)

166. velox Staudinger, 1859

tortuosella Chrétien, 1908

167. pimpinellae Zeller, 1839

pulverella (Eversmann, 1844)

reichlini Heinemann, 1870

reichlini subsp. *hungarica* Szent-Ivány, 1943

168. villosae Corley & Buchner, 2018

169. bupleurella Heinemann, 1870

170. sarahae Gastón & Vives, 2017

170.1. sarahae subsp. tabelli Buchner, 2017

171. badiella (Hübner, 1796)

corticinella Zeller, 1854, **syn. n.**

brunneella Ragonot, 1874

aurantiella Tutt, 1893

uhrykella Fuchs, 1903

frigidella Turati, 1921

frustratella Rebel, 1936

172. pseudobadiella Nel, 2011

173. subnervosa Oberthür, 1888

fusconigerella Hannemann, 1990, **syn. n.**

174. cervicella Herrich-Schäffer, 1854

175. gallicella Chrétien, 1908

quintana Weber, 1945

176. altaica Zeller, 1854

177. albarracinella Corley, 2017

178. eryngiella Millière, 1881

campestrella Chrétien, 1896

deliciosella Turati, 1924

obolucha Meyrick, 1936

179. veneficella Zeller, 1847

180. discipunctella Herrich-Schäffer, 1854

pastinacella Stainton, 1849, *nec* Duponchel, 1838

181. hansjoachimi **sp. n.**
182. junnilaineni Buchner, 2017
183. pentheri Rebel, 1904
184. hannemanniana Lvovsky, 1990
185. erzurumella Lvovsky, 1996
186. dictamnella (Treitschke, 1835)
187. moranella Chrétien, 1907
      arabica Amsel, 1972
188. hystricella Möschler, 1860
189. hirtipalpis Zeller, 1854
190. erinaceella Staudinger, 1870
      sardoniella Rebel, 1936
191. peniculatella Turati, 1922
      rungsiella sensu Hannemann, 1976, *nec* Hannemann, 1953
192. rungsiella Hannemann, 1953, **stat. rest.**

Uncertain status
Agonopterix dumitrescui Georgesco, 1965.

Systematic Treatment of the Genera and Species of the European Depressariidae

Semioscopis Hübner, 1825

Semioscopis Hübner, [1825] 1816: 402.
    Type species: *Tortrix steinkellneriana* [Denis & Schiffermüller], 1775.
Epigraphia Stephens, 1829: 49.
    Type species: *Phalaena avellanella* Hübner, 1793.

DESCRIPTION. Head with ocelli present. Labial palp porrect or slightly recurved, segment 2 long, about twice diameter of eye, weakly curved, slightly thickened with scales, not furrowed beneath, segment 3 one-quarter to three-quarters length of segment 2, slender, straight. Antenna simple or finely ciliate in outer half, scape without pecten. Thorax with or without posterior crests. Forewing long, costa convex, with strongly oblique termen, dorsum two-thirds length of costa, apex acute; with scale tufts, distinct in *S. oculella*, indistinct in the other three species; vein CuA_2 originates close to CuA_1 (but stalked in *S. steinkellneriana*), at a wide angle, then strongly bent towards tornus. Hindwing without lobe at anal angle; veins CuA_1 and M_3 connate. Abdomen cylindrical.

MALE GENITALIA. Socii well-developed, uncus obsolete; gnathos round to elliptic, with an area free of spines on lower surface; valva slightly curved, tapering to rounded apex, ventral margin with process beyond end of sacculus, sacculus strongly sclerotised, with or without cuiller; aedeagus more or less curved, with cornuti of varied types. Tergite VIII with characteristic sclerotisation, showing slight differences between species (figures on plates 1 and 2).

FEMALE GENITALIA. Papilla analis elongate, as long as posterior apophysis, with or without broad, hooked bristles; anterior apophysis one-quarter to one-half length of posterior apophysis; posterior margin of sternite VIII with wide deep V-shaped excavation, anterior margin protruding around ostium, a field of microtrichia posterior to ostium; ductus bursae as long as corpus bursae and nearly straight or longer and spiral, corpus bursae ovate or elliptical with signum.

DISTRIBUTION. *Semioscopis* has a Holarctic distribution with seven species in the Palaearctic region and six in the Nearctic. Four species are known in Europe.

BIONOMICS. Known host-plants are deciduous trees. Adults fly in spring.

© PETER BUCHNER AND MARTIN CORLEY, 2025 | DOI:10.1163/9789004713116_010

REMARKS. In three species vein CuA_2 originates close to CuA_1 but in *S. steinkell-neriana* they are stalked. In consequence Meyrick (1927) placed *steinkellneriana* in *Semioscopis* but *avellanella* (and by implication the other species) in *Epigraphia*.

To show how problematic a division of *Semioscopis* into subgroups (independent of their taxonomic level) is, here five character states are selected, which at least in part are used successfully to form groups in other genera or families. Each of these five characters produces another grouping:

Forewing venation: *S. avellanella, oculella, strigulana* (CuA_2 originates close to CuA_1) vs. *S. steinkellneriana* (CuA_2 and CuA_1 stalked); cuiller: *S. avellanella, oculella* (present) vs. *S. steinkellneriana, strigulana* (absent); cornuti: *S. avellanella, steinkellneriana, strigulana* (one stout, bilobed cornutus present) vs. *S. oculella* (stout, bilobed cornutus absent); ductus: *S. avellanella* (straight) vs. *S. oculella, steinkellneriana, strigulana* (coiled); papilla analis: *S. avellanella, steinkellneriana* (stout hooked bristles absent) vs. *S. oculella, strigulana* (stout hooked bristles present).

Based on general appearance, wing shape and phenology, the genus shows uniformity. DNA barcodes support the unity of the genus. In contrast to other Depressariidae it appears that wing venation does not provide stable characters at the genus level. We therefore follow Hodges (1974) who treated the North American species in one genus.

1   *Semioscopis avellanella* (Hübner, 1793)

Phalaena (Tinea) avellanella Hübner, 1793: 12, pl. 65.
  Semioscopis fumicella Amsel, 1930: 119.

DESCRIPTION. Wingspan 18–26 mm, male slightly larger than female. Head grey, greyish buff or light brown. Labial palp short, porrect, with segment 3 one-quarter to two-fifths length of segment 2; segment 2 dark brown on outer side, greyish buff on inner side, segment 3 greyish buff with some brown scales on outer side beyond middle, not usually forming distinct band. Antenna fuscous, narrowly ringed grey, finely ciliate in distal half in male. Forewing light fuscous or light grey, with scattered fuscous scales along dorsum and termen and in costal area, not evident in fuscous specimens; blackish markings consisting of a median streak from base to two-fifths, joined before its apex to a curved streak which merges with a median spot on its costal edge, and projects a little beyond; a V-shaped mark at end of cell; some more or less erect whitish grey scales along dorsal edge of streaks and inside V-shaped mark, scarcely visible in grey specimens; fringe whitish grey, with two fringe lines, the basal grey-brown, the distal paler. Hindwing whitish grey, slightly darker towards termen; fringe whitish grey with up to three weakly marked fringe lines. Abdomen greyish buff.

VARIATION. The main variation consists of forms with light grey or light fuscous ground colour. There is some variation in the forewing markings, which may be interrupted, or lack a developed median spot, or be reduced to a weak basal streak. Some specimens have a subapical dark brown dash, and often there are some dark brown spots on termen.

SIMILAR SPECIES. The dark forewing markings are reminiscent of those of *S. steinkellneriana*, but in that species they do not reach the wing base. Weakly marked specimens resemble *S. strigulana* but usually have a trace of a basal streak. Labial palp segment 3 one-quarter of segment 2, but more than one-half length in the compared species.

MALE GENITALIA. Socii in standard preparation triangular, base (connection with tegumen) about as long as height; gnathos strongly compressed dorsoventrally, spinose on upper side only, in standard slides usually directed upwards or downwards (appearing spinose throughout in both cases, because it is translucent), showing broad elliptic outline in this view; valva long, tapering from three-fifths, sacculus with straight, digitate, acute terminal process just projecting from margin of valva, cuiller slightly curved, reaching or just crossing costal margin of valva; transtilla lobes large, broad, rounded, directed posteriorly; anellus lobes well-developed, anellus almost circular, distal margin incised nearly to middle; aedeagus tapering from base, curved about 70 ° near its middle in lateral view, straight in ventral view, acute, with a single cornutus with long and short limbs.

FEMALE GENITALIA. Papilla analis without broad, hooked bristles; anterior apophysis one-third to two-fifths length of posterior apophysis; segment VIII short and wide, tergite with posterior margin straight, anterior margin shallowly excavate, sternite a belt narrowing towards centre, sclerotised over its full length, laterally this sclerotisation gradually tapering toward centre, where anterior margin is produced in form of a narrow loop around ostium, a small field of microtrichia posterior to ostium; antrum slightly sclerotised, ductus bursae expanded beyond antrum with small round sclerite at one-third of length; corpus bursae longer than ductus bursae, elliptic; signum a weakly sclerotised, blunt angled triangle with a few small teeth.

DISTRIBUTION. Middle and northern latitudes of Europe from British Isles and Fennoscandia south to France, northern Italy, Romania and east into Ukraine and Russia. The distribution area extends well beyond the Arctic Circle.

BIONOMICS. Larvae on *Betula* L., *Tilia cordata* Miller and *Carpinus betulus* L. (Hannemann, 1995), *T. platyphyllos* Scop. (Sonderegger, *in litt.*), in a long cylinder at the leaf margin, from May to autumn. Also on *Corylus avellana* L. according to Robinson *et al.* (2010) but we cannot discover the source of this record. Pupa hibernates in a cocoon among detritus. The adult flies in March and April, in North

Europe until mid-May. Males continue flying in very cool temperatures. Both males and females are found by day resting on tree trunks. Males come to light more often than females.

2   *Semioscopis oculella* (Thunberg, 1794)

Tinea oculella Thunberg, 1794: 92.
  Tinea anella Hübner, 1796: pl. 4, fig. 28.
  Lemmatophila alienella Treitschke, 1832: 40.

DESCRIPTION. Wingspan 23–30 mm, male slightly larger than female. Head grey-brown mixed with dark brown scales. Labial palp reddish brown with admixture of grey-brown scales, porrect, segment 3 paler at base and apex, three-quarters length of segment 2. Antenna barred grey-brown and blackish above. Thorax concolorous with head. Forewing pale grey-buff, with dark chestnut-brown scales in central area from base to end of cell extending to costa beyond mid-wing, at base of dorsum and also forming small costal strigulae and scattered singly elsewhere; an indistinct pale buff dot at one-third in cell and another larger and more distinct at two-thirds, forming scale tufts, the latter sharply ringed with blackish scales, which are also present as a streak between the pale dots in cell and forming a spot in the fold at two-fifths; a patch of orange-brown to orange-pink scales distal to outer pale dot, and often an indistinct darker fascia beyond this at five-sixths; fringe very pale grey-brown, darker at base and apex. Hindwing light grey-brown; fringe as on forewing.

VARIATION. There is some variation in the extent of the dark brown central area and the amount of blackish scales within this area; sometimes there is a row of dark grey-brown dots along termen. The thoracic crests may have dark brown scales only.

SIMILAR SPECIES. The two pale dots in the cell in a chestnut-brown area are characteristic.

MALE GENITALIA. Socii in standard preparation triangular, base (connection with tegumen) distinctly longer than height; gnathos compressed laterally, spinose nearly throughout except a small area on lower side, in standard slides usually turned aside, this lateral outline broadly elliptic; ventral margin of valva excavated in middle, sacculus with outward-projecting thorn-like terminal process, cuiller nearly straight but directed inwards, not reaching costal margin of valva; transtilla lobes not conspicuously large: anellus lobes in two parts, a narrow basal fold and a strongly developed distal part with long bristles; anellus wider than long, broadly notched in distal margin, this notch (the guiding area for the aedeagus) in natural (uncompressed) shape reminiscent of the pouring spout of a jug, with its tip on ventral side somewhat below the lowest part of the notch; when compressed in standard slides, this tip becomes distorted and may appear as small irregular triangle; aedeagus tapering from base, curved through about 100–130 ° in lateral view, with one shallow spiral turn in ventral view, acute, with abundant minute cornuti.

FEMALE GENITALIA. Papilla analis with broad, hooked bristles; anterior apophysis one-third length of posterior apophysis; segment VIII short and wide, tergite with posterior margin straight, anterior margin shallowly excavate, sternite narrowing towards centre, where anterior margin is slightly produced around ostium, a small field of microtrichia posterior to ostium; antrum not sclerotised, ductus bursae long, narrow, initially straight then forming a spiral with about four turns, a wire-like sclerite through length of spiral part; corpus bursae narrowly elliptic; signum strongly sclerotised, rhomboidal with numerous small teeth.

DISTRIBUTION. Middle and northern latitudes of Europe from Fennoscandia south to France, northern Italy, Romania and east into Ukraine and Russia. Absent from British Isles and rare in northern Fennoscandia.

BIONOMICS. Larvae form spinnings on *Betula* L. (Hannemann, 1995) from late April to autumn. Pupa hibernates. Adults fly from February to May depending on the timing of spring, earlier in the year than other *Semioscopis* spp. Maximum activity usually soon after sunrise, often even near freezing point. Both males and females can be found by day resting on birch trunks. Males come to light more often than females.

3   *Semioscopis steinkellneriana* ([Denis & Schiffermüller], 1775)

Tortrix steinkellneriana [Denis & Schiffermüller], 1775: 130.
   Phalaena characterella Hübner, 1793: 6, pl. 9.
   Depressaria characterosa Haworth, 1811: 511.

DESCRIPTION. Wingspan 20–27 mm, male slightly larger than female. Head greyish brown. Labial palp, grey-brown, segment 3 three-fifths length of segment 2; segment 2 slender, curved, greyish brown, inner side pale buff, segment 3 pale buff, deep brown beyond middle, tipped buff. Antenna greyish brown, finely ciliate. Thorax concolorous with head, with dorsal crests. Forewing light greyish brown, with scattered fuscous scales, mainly on costa and dorsum where they are organized into short transverse lines, and along veins; deep brown markings consisting of a broad streak from one-quarter to one-half in mid-wing, extended at basal end obliquely half-way towards base of costa, a wide V-shaped mark at end of cell, these marks finely and irregularly margined with whitish buff scales, whitish scales on proximal margin of these two markings slightly erect; smaller deep brown markings include a fine short streak from base along subcostal vein, a short streak below costa, level with V-shaped mark, a plical dot and a series of terminal spots; a weak fuscous cloud between costa and V-shaped mark; fringe light grey-brown, with two fringe lines. Hindwing light grey-brown; fringe as on forewing. Abdomen light grey-ish brown.

VARIATION. The species shows only minor variation in depth of ground colour and development of dark markings, especially in costal area.

SIMILAR SPECIES. *S. avellanella* (*q.v.*).

MALE GENITALIA. Socii in standard preparation roundish, projecting beyond end of tegumen; gnathos slightly compressed dorsoventrally, spinose on upper side only, in standard slides usually directed upwards or downwards (appearing spinose throughout in both cases, because it is translucent), showing roundish outline in this view; sacculus with projecting thorn-like terminal process nearly parallel with costal margin of valva and not much short of apex of valva, cuiller vestigial; transtilla lobes narrow, directed posteriorly, expanding at apex and long-bristled; anellus about twice as long as wide, slightly tapering from base to apex, distally bifid with V-shaped sinus; whole anellus in natural (uncompressed) shape chan-nelled on dorsal side, serving as a guide for the aedeagus, this guide terminating in the distal notch of upper margin and here it is best compared with the pouring spout of a jug, with its tip on ventral side somewhat below the lowest part of the notch; when compressed in standard slides, this complex three-dimensional func-tional structure is damaged and usually some irregular folds appear; aedeagus in form of an open circle, tapering in basal one-third, then parallel-sided, straighter in distal one-quarter, with a single cornutus with long and short limbs.

FEMALE GENITALIA. Papilla analis without broad, hooked bristles; anterior apo-physis nearly one-half length of posterior apophysis; segment VIII short and wide, tergite with posterior margin slightly concave, anterior margin straight, anterior margin of sternite with a large semiglobose "cup" projecting ventrally with ostium at its bottom (only visible uncompressed and in lateral view), in standard slides (compressed dorsoventrally) this cup changes to a large triangular projection cov-ering ostium, an extensive field of microtrichia around and posterior to ostium; antrum short, sclerotised, ductus bursae long, narrow, initially straight including a small sclerite, then forming a loose spiral with about four turns; corpus bursae ellip-tic; signum strongly sclerotised, small, almost divided in two parts, with a few teeth.

DISTRIBUTION. Middle and northern latitudes of Europe from British Isles and central Fennoscandia (north to the Arctic Circle) south to northern Spain, Bosnia, Romania and east to Russia; Sicily.

BIONOMICS. Larvae from May to autumn between spun leaves. The main hostplant is *Prunus spinosa* L., but reported also from *Crataegus* L., *Sorbus* L. (Hannemann, 1995), *Cotoneaster* Medik. (Palm, 1989), *Amelanchier ovalis* Medik. (Sonderegger, *in litt.*). A report from *Fraxinus* L. (Schütze, 1931) appears doubtful. Pupa hibernates in a cocoon among detritus. The adult flies from February to May depending on the timing of spring, slightly later than *Semioscopis oculella*. It flies at dawn, before sunrise. Unlike the other species of the genus, females come to light more often than males.

4   *Semioscopis strigulana* (Fabricius, 1787)

Pyralis strigulana Fabricius, 1787: 233.
   Tinea atomella Hübner, 1796: pl. 2, f. 13.
   Chimabache consimilella Hübner, 1825: 402.

DESCRIPTION. Wingspan 25–30 mm. Head grey. Labial palp grey with scattered dark brown scales, segment 2 twice as long as diameter of eye, curved, segment 3 two-thirds length of segment 2, slender, straight, dark on inner side, with weak band at two-thirds. Antenna barred grey and mid-brown above, darker below; scape grey. Thorax grey with admixture of dark brown scales. Forewing light grey with abundant brownish grey scales all over wing, more or less organised into irregular transverse bands, darker at wing base, with darker grey-brown costal strigulae and series of dots around termen, a pair of dark grey-brown dots formed by erect scales placed transversely at one-third, another pair at end of cell fringe beyond light grey-brown fringe line, whitish with light grey-brown tips. Hindwing pale brownish grey; fringe as on forewing. Abdomen brownish grey.
   VARIATION. There is considerable variation in the intensity of coloration of the brownish grey scales scattered over the forewing.
   SIMILAR SPECIES. The least well marked of the *Semioscopis* species, but some forms of *S. avellanella* (*q.v.*) approach it.
   MALE GENITALIA. Socii in standard preparation four-sided, projecting beyond end of tegumen; gnathos strongly compressed dorsoventrally, spinose on upper side only, in standard slides usually directed upwards or downwards (appearing spinose throughout in both cases, because it is translucent), showing nearly circular outline in this view; distal part of valva rather narrow, sacculus with stout projecting thorn-like terminal process nearly parallel with basal part of sacculus, cuiller absent; transtilla lobes not particularly large, with long hairs; anellus lobes divergent from middle of anellus base; anellus parallel-sided, distally shallowly bifid; aedeagus stout, curved predominantly in its middle, 90–110 ° in lateral view, with one shallow spiral turn in ventral view, tapering throughout; a single cornutus with two unequal limbs.
   FEMALE GENITALIA. Papilla analis with broad, hooked bristles; anterior apophysis one-third length of posterior apophysis; segment VIII short and wide, tergite with posterior margin slightly concave, anterior margin straight, anterior margin of sternite with truncated triangular projection enclosing ostium, a field of microtrichia posterior to ostium; antrum not sclerotised, ductus bursae long, narrow, forming a loose spiral with about three turns, part of one turn with conspicuous sclerite; corpus bursae ovoid; signum strongly sclerotised, small, irregular-shaped, with a few teeth.
   DISTRIBUTION. Middle and northern latitudes of Europe from central Fennoscandia south to France, northern Italy, Bosnia, Romania and east to Russia. Absent from British Isles, Belgium and Netherlands.

BIONOMICS. Larvae from April to July (shorter larval period than other *Semioscopis*) on *Populus tremula* L. (Hannemann, 1995 *et al.*), no further host-plant known. Pupa hibernates. Adult flies predominantly in March and April. Both males and females are found by day resting on aspen trunks. Comes to light, males more often than females.

Luquetia Leraut, 1991

Luquetia Leraut, 1991: 232.
   Type species: *Tinea lobella* [Denis & Schiffermüller], 1775
Enicostoma Fletcher, 1929: 77, nec Stephens, 1829
   Type species: *Tinea lobella* [Denis & Schiffermüller], 1775; incorrect subsequent designation.

DESCRIPTION. Head without ocelli, with raised scales on neck and vertex, the latter falling over the face. Labial palp segment 2 slightly curved, smooth to slightly rough-scaled, not furrowed beneath. Antennal scape with pecten (fugacious, often completely absent in worn specimens), flagellum weakly serrate (particularly *L. lobella* males) from projecting scales in distal half, densely covered with cilia, these very short but up to one half diameter of flagellum in *L. orientella* males. Thorax with posterior crest. Forewing with convex costa, apex less acute and termen less oblique than in *Semioscopis*; with scale tufts; vein CuA_2 originating close to CuA_1 (*L. lobella*) or connate (*L. orientella*), at a narrow angle, then slightly bent towards tornus. Hindwing without anal lobe; veins CuA_1 and M_3 connate. Abdomen cylindrical.
   MALE GENITALIA. Gnathos round; sacculus with cuiller; aedeagus sheath and lateral process absent, cornuti numerous and small. Tergite VIII with weak median sclerotisation.
   FEMALE GENITALIA. Papilla analis triangular in lateral view, segment VIII short and wide, sternite with anterior bulge, a field of microtrichia posterior to ostium; ductus bursae with somewhat thickened wall in posterior section, with one to six spiral turns, nearly with the same diameter throughout; signum a sclerotised plate with thorn-like teeth.
   DISTRIBUTION. Four species in the Palaearctic region, two in Europe.
   BIONOMICS. Larva of *L. lobella* spins leaves of *Prunus spinosa* L.
   REMARKS. Apart from forewing venation the genus shows uniformity, especially in genitalia and the relationship is confirmed by barcode. It appears that in this genus, as in the related *Semioscopis*, wing venation shows substantial infrageneric variation. There is even variation in distance separating base of CuA_2 and CuA_1 in *L. lobella*.
   Enicostoma Stephens, 1829 has type species *geoffrella* Linnaeus, 1767, subsequently designated by Duponchel (1838). Fletcher's (1929) designation of *lobella* as

type species was therefore not acceptable, necessitating the description of a new genus *Luquetia* by Leraut (1991).

We include the description of *L. osthelderi* (Rebel, 1936) comb. n. from south-east Turkey, previously placed in *Semioscopis*.

5   *Luquetia lobella* ([Denis & Schiffermüller], 1775)

Tinea lobella [Denis & Schiffermüller], 1775: 138.
  Pyralis thunbergiana Fabricius, 1781: 284.
  Haemylis lugubrella Duponchel, 1838: 612.

DESCRIPTION. Wingspan 13.5–20 mm. Head dark fuscous. Labial palp segment 3 two-fifths length of segment 2, segment 2 smooth to slightly rough-scaled below grey-brown, slightly paler on inside in some specimens, segment 3 buff, with blackish brown rings at base and between half and three-quarters. Antenna greyish brown. Thorax dark fuscous. Forewing broad, greyish brown, darker at base and in costal third, with flecks of slightly darker scales all over wing; costa with paler grey-brown strigulae; deep brown scale tufts, one at one-third forming a partial fascia not reaching costa or dorsum, often broken into three spots, a smaller tuft at end of cell, sometimes broken into two spots; a series of deep brown spots at termen; fringe grey-buff, with two pale grey-brown lines. Hindwing light brownish grey, slightly darker towards apex; fringe light greyish brown, with two indistinct lines.

VARIATION. The scale tufts may be broken into spots or not, sometimes they are associated with some whitish scales, which may also be present on veins to termen.

SIMILAR SPECIES. The broad forewings and the two series of dark brown scale tufts placed more or less transversely are characteristic.

MALE GENITALIA. Socii small, conjoined to give truncate end of genitalia; gnathos round; valva broad, slightly curved, ventral side slightly angled at three-quarters, sacculus ending in well sclerotised digitate cuiller, slightly enlarged towards apex, extending across valva to about two-thirds; anellus lobes narrow; anellus almost round; saccus wide; aedeagus tapering to broad apex, a dense mass of small cornuti present.

FEMALE GENITALIA. Papilla analis long; posterior apophysis as long as papilla analis; anterior apophysis two-fifths length of posterior apophysis; segment VIII wide and short, sternite width:length 4:1, with a semicircular anterior bulge accommodating ostium, a field of microtrichia from ostium to posterior margin; ductus bursae nearly twice as long as nearly circular corpus bursae, with a single turn at posterior end; signum a transverse elliptical sclerotised plate with two pairs of large thorn-like teeth and a few smaller teeth.

DISTRIBUTION. Middle latitudes of Europe north to southern Scandinavia and south to northern Italy and North Macedonia, extending to far eastern Russia. Absent from Ireland, Finland, Iberian Peninsula and Mediterranean Islands.

BIONOMICS. Larva on *Prunus spinosa* L. preferring small suckers, in a web beneath a leaf, bending the leaf edges downwards, June to October (Heinemann, 1870). Further host-plants mentioned in literature are *Sorbus aucuparia* L. and *Crataegus* L. (Schütze, 1931, Biesenbaum, 2014). Pupa hibernates. Adult flies May to July, occasionally attracted to light.

6   *Luquetia orientella* (Rebel, 1893)

Epigraphia orientella Rebel, 1893: 45.
  Luquetia abchasiella Lvovsky, 1995: 147.

DESCRIPTION. Wingspan 18–22 mm. Head mid-brown. Labial palp segment 3 about one-third as long as segment 2, smooth, segment 2 with numerous scattered light brown scales, segment 3 with a broad light brown median band. Antenna light brown, flagellum ciliate, not or scarcely serrate. Thorax mid-brown, grey-white posteriorly. Forewing narrower than in *L. lobella*, costa weakly concave, apex subacute, termen oblique; white to grey-white, overlaid with abundant scattering of grey-buff scales except in cell; a blackish patch at base, not reaching dorsum; scale tufts consisting of both black and white scales form a pair of obliquely placed black dots in cell at one-third, and also in middle and at end of cell, these predominantly white, edged black; a series of black dots or dashes between veins at apex and around termen; a fuscous blotch from middle of costa extending to white spots in middle and at end of cell; fringe light grey or whitish grey with one fringe line. Hindwing with costa and dorsum almost parallel, apex subacute; light grey; fringe light grey with one fringe line. Abdomen light grey-brown.

VARIATION. There is little variation.

SIMILAR SPECIES. Forewing of characteristic shape; whitish ground colour and the dark markings are unique in the Depressariidae.

MALE GENITALIA. Socii large, quadrate, widely separated; gnathos oval; valva slightly curved, ventral side slightly angled at three-quarters, sacculus ending in digitate cuiller, not enlarged to apex, extending across valva to three-quarters or more; anellus lobes narrow; anellus goblet-shaped, bifid with wide shallow sinus; saccus narrower than in *L. lobella*; aedeagus strongly bent in middle, tapering to broad apex, a large number of small cornuti present.

FEMALE GENITALIA. Papilla analis long and narrow; posterior apophysis three-quarters length of papilla analis; anterior apophysis about three-quarters length of posterior apophysis; segment VIII wide and short, sternite width:length 4:1, with a slight anterior bulge around ostium; a field of microtrichia from ostium to posterior margin of sternite; ostium with lateral angles shortly extended; ductus bursae forming a spiral with about four to six turns, corpus bursae long, elliptic; signum a transverse elongate lightly sclerotised plate, nearly three times as wide as long, with numerous teeth.

DISTRIBUTION. Albania, North Macedonia, Greece, South Russia. In BOLD also with data from Turkey (without further details) and Georgia; described as *L. abchasiella* from West Caucasus.

BIONOMICS. Host-plant unknown. The examined specimens were light-trapped between 11th April and 19th June.

REMARKS. Scale tufts may be lost or flattened during setting.

Originally described in *Epigraphia*, which is a synonym of *Semioscopis* but pecten is absent in *Semioscopis*.

L. abchasiella Lvovsky, 1995 was placed in synonymy with *orientella* by Lvovsky (2006).

7   *Luquetia osthelderi* (Rebel, 1936) comb. n.

Epigraphia osthelderi Rebel, 1936: 79.

DESCRIPTION. Wingspan 20–22 mm. Head brown, rough scaled on vertex. Labial palp segment 3 one-third as long as segment 2, smooth, segment 2 with numerous scattered dark brown scales, segment 3 buff with a broad fuscous median band. Antenna fuscous, flagellum ciliate, weakly serrate towards apex. Thorax and tegulae deep brown. Forewing shape similar to *L. orientella*, costa straight, apex subacute, termen oblique; dull grey-buff, overlaid with scattered fuscous scales particularly along veins to termen; costa with a series of fuscous strigulae from one-quarter to two-thirds; a series of blackish terminal dots between vein-ends; a deep brown basal patch outwardly edged whitish, not reaching dorsum; two black spots placed obliquely in cell at one-third, inwardly edged by a whitish line, followed by a dark brown patch extending to a black and white mixed scale tuft near end of cell, beyond this a deep brown patch extending to costa and outwardly edged by a whitish line; fringe concolorous with forewing, without obvious fringe line. Hindwing with apex subacute; light grey-brown; fringe light grey-brown with one fringe line.

VARIATION. Only one specimen has been examined.

SIMILAR SPECIES. *L. orientella* is whiter, with oblique dots on forewing not associated with a darker patch.

MALE GENITALIA. Unknown.

FEMALE GENITALIA. Similar to *L. orientella*. Papilla analis long and narrow; posterior apophysis three-quarters length of papilla analis; anterior apophysis less than half length of posterior apophysis; field of microtrichia from ostium to posterior margin of sternite larger; ductus bursae forming a spiral with six turns, corpus bursae long, elliptic; signum very small, a lightly sclerotised plate with few teeth.

DISTRIBUTION. Only known from two specimens, collected in south-east Turkey, Taurus Mountains, Maraş.

BIONOMICS. Hostplant unknown. The type specimens were collected in May.

REMARKS. Male unknown. Rebel (1936) described this species from two females. He compared it with *Epigraphia steinkellneriana* and very briefly with *E. orientella*. Since *Epigraphia* is a synonym of *Semioscopis*, it has subsequently been placed in *Semioscopis*, but external appearance, female genitalia and barcode place it close to *orientella* in *Luquetia*.

We have examined the female specimen in NHMW, which we designate as lectotype to avoid confusion in the future. "Epigraphia | osthelderi Rbl. | Type ♀" "Syria sept. | Taurus | Marasch | 18.v.28 | L. Osthelder leg." Maraş (Marasch) was in Syria for a few years after World War I but was reclaimed by Turkey in 1923. Osthelder was apparently unaware of this.

The whereabouts of the other specimen collected by Osthelder is unknown.

Exaeretia Stainton, 1849

Exaeretia Stainton, 1849: 152.
   Type species: *Exaeretia allisella* Stainton, 1849; monotypy.
Depressariodes Turati, 1924: 175.
   Type species: *Depressariodes marmaricellus* Turati, 1924; monotypy.
Martyrhilda Clarke, 1941: 125.
   Type species: *Depressaria canella* Busck, 1904; original designation.
Levipalpus Hannemann, 1953: 297, syn. n.
   Type species: *Depressaria hepatariella* Lienig & Zeller, 1846; original designation.

DESCRIPTION. Head with ocelli present; vertex with raised scales, face with appressed scales. Labial palp strongly recurved, with segment 2 curved, thickened, furrowed beneath or not, segment 3 two-fifths to two-thirds length of segment 2, slender; antenna more or less strongly ciliate in male, often finely ciliate near apex in female, scape with pecten. Thorax with or without posterior crest. Forewing long, costa almost straight, apex rounded and termen curved (except *E. allisella*), dorsum three-quarters to four-fifths length of costa; veins CuA1 and CuA2 stalked. Hindwing with lobe at anal angle weak or absent.

MALE GENITALIA. Gnathos wider than in most *Agonopterix*; sacculus with dorsal process (cuiller) and terminal process; saccus generally more developed than in *Agonopterix*; aedeagus without basal sheath; cornuti numerous and small to minute. Tergite VIII with median sclerotisation. The potential taxonomic significance of this sclerotisation merits further investigation.

FEMALE GENITALIA. Signum weakly or strongly sclerotised, often large and elongate, sometimes with a handle-like process at posterior end, teeth narrowly triangular, numerous and rather small, absent along the longitudinal axis or in the centre, increasing in size toward the outer edge.

DISTRIBUTION. A mainly Holarctic genus with 55 described species (Lvovsky, 2013a), although nine of these are known from the Neotropical region. In Europe most species occur in central, north-east and east Europe. The genus is poorly represented in south-west and north-west Europe.

BIONOMICS. The larva of *E. ciniflonella* feeds on *Betula*; *E. lutosella* and *E. ledereri* on Rutaceae. The remaining European species with known host-plants feed on Asteraceae belonging to the tribe Anthemideae usually inside the stem, with the exception of *E. hepatariella* which feeds on *Antennaria* which belongs to Asteraceae tribe Gnaphalieae. As far as known, only *E. ciniflonella* hibernates as adult.

REMARKS. Close to *Agonopterix*, but generally differing in less thickened segment 2 of labial palp and relatively shorter segment 3; antenna ciliate; absence of anal lobe of hindwing; male genitalia with sacculus having a terminal process in addition to the cuiller; aedeagus with basal sheath and process absent; female genitalia with weakly sclerotised signum. *Levipalpus* Hannemann, 1953 has more strongly ciliate antenna, less thickened labial palp segment 2, sacculus with processes broad and weakly sclerotised and signum reduced. All of these are characters that show substantial variation within *Exaeretia* and no character or combination of characters gives clear distinction from *Exaeretia*. DNA barcodes support the view that *L. hepatariella* belongs to genus *Exaeretia*. We therefore treat *Levipalpus* as a synonym of *Exaeretia*.

As with *Agonopterix* and *Depressaria* we have divided the genus into species groups and subgroups. The European species fall into six groups of which three have only a single species while one group is divided into two subgroups.

Group 1 *hepatariella* group (8) Terminal process of sacculus reaching end of valva, cuiller a broad triangle.

8 *Exaeretia hepatariella* (Lienig & Zeller, 1846)

Depressaria hepatariella Lienig & Zeller, 1846: 282.

DESCRIPTION. Wingspan 17–21 mm. Head reddish brown. Labial palp segment 3 two-fifths length of segment 2, segment 2 dark reddish brown on outside, whitish buff admixed with reddish brown scales on inner side, segment 3 reddish brown mixed with grey on outside, pale grey-buff on inside. Antenna mid-brown, narrowly ringed dark brown. Forewing dull reddish brown to fuscous with scattered darker scales often organised into indistinct transverse bands, base deep brown or basal area sometimes buff to grey-buff, costa darker brown or paler in basal half, interrupted along costa by weak pale strigulae; a white dot at end of cell, sometimes obsolete; fringe slightly reddish grey-brown, with two rather indistinct fringe lines.

Hindwing pale brownish grey; fringe pale grey-buff, without fringe line. Abdomen grey, weakly tinged reddish brown.

VARIATION. Ground colour varies from red-brown through darker brown to a dull dark fuscous; the basal area and inner half of costa may be clearly paler than rest of wing, or not; the white cell dot can be conspicuous or barely visible. Alpine specimens tend to be brighter coloured than those from the north of Europe.

SIMILAR SPECIES. Some forms of *E. lutosella* appear similar, but have forewing with black dots in cell at one-third.

MALE GENITALIA. Uncus small, narrow, distinct, in wide deep sinus between socii; gnathos small, circular; valva slightly curved, nearly parallel-sided, apex rounded or obliquely truncate, sacculus very well-developed, half as wide as valva at base, narrow in middle, with apical expansion extending to end of valva or even beyond, with broad triangular sclerotisation representing cuiller; transtilla lobes unusually small; anellus broadly cordate with wide distal sinus; aedeagus long, curved in middle, with a compact group of minute cornuti.

FEMALE GENITALIA. Sternite VIII about twice as wide as long, posterior margin notched in middle, anterior margin extended as a broad truncated triangle; ostium opening in anterior one-third of sternite; ductus bursae short, not clearly distinct from elliptic or clavate corpus bursae, signum small, stellate with marginal teeth.

GENITALIA DIAGNOSIS. In male shape of socii and uncus in combination with very broad terminal process of sacculus distinct. In female U-shaped start of ductus bursae in combination with very small signum separates it from rest of *Exaeretia* species.

DISTRIBUTION. Northern Europe from Scotland, the Netherlands and Poland to Fennoscandia and the Baltic countries and northern Russia. Also in the mountains of central Europe from France to Romania. Recently found in Russian Altai.

BIONOMICS. Larva on *Antennaria dioica* (L.) Gaertner, overwintering in the bud that will produce the flowering stem the following year. In Denmark Per Falck in Buhl *et al.* (1989) observed the larva feeding from May to July, initially in the flower-head, then eating through the stem before going down to the leaf rosette. Heckford (2004) noted that in Scotland it feeds on the leaves from a silken tube that goes down into the ground. Flies end of July to end of August.

REMARKS. This species was placed in the monotypic genus *Levipalpus* by Hannemann (1953), but the given differences have little significance as they are in characters that show wide variation within *Exaeretia*, including the proportions of the labial palp segments and their thickening, antennal ciliation and the form of the saccular processes. Barcode results support its inclusion in *Exaeretia*.

Harper *et al.* (2002) give several differences between males and females suggesting significant dimorphism. The only difference between males and females that we have detected is a tendency for the forewing to be narrower in outer half in females, but we hardly consider this to constitute dimorphism.

Group 2 *allisella* group (9)

Forewing with termen sinuous.

9   *Exaeretia allisella* Stainton, 1849

Exaeretia allisella Stainton, 1849: 152.

Depressaria lechriosema Meyrick, 1928: 467.

DESCRIPTION. Wingspan 14–24 mm. Head deep brown. Labial palp segment 3 two-fifths length of segment 2; segment 2 creamy buff on inner side, grey-brown to deep brown on outer side with scales buff-tipped; segment 3 grey-brown to deep brown on outer side with scales buff-tipped, extreme tip whitish buff. Antenna deep brown, scarcely ciliate. Thorax deep brown, slightly paler posteriorly. Forewing with termen sinuous, weakly so in male; pale ash-grey, irregularly flecked with small groups of light grey-brown scales, these absent in subterminal area and in dorsal side of cell; base whitish buff on dorsum, elsewhere deep grey-brown, basal two-fifths of costa blackish or dark greyish brown, an oblique brownish ochreous to deep brown stripe extending from two-fifths on costa towards three-quarters on dorsum, but not reaching dorsum, sharply separated from grey ground colour anteriorly and gradually fading to a weak suffusion posteriorly, the boundary touching end of cell where it forms a blackish brown dash; an obliquely placed pair of dark brown dots in cell at one-quarter, often obscure; fringe pale grey with darker grey-brown basal and terminal lines. Hindwing whitish, grey towards apex; fringe whitish grey with three light grey-brown lines. Abdomen grey-brown.

VARIATION. There is little variation apart from lighter or darker brown suffusion in the area beyond the oblique marking.

SIMILAR SPECIES. The sinuous termen and the oblique demarcation on the forewing are characteristic.

MALE GENITALIA. Uncus well-developed, socii oval, longer than uncus; gnathos medium-sized, round; valva slightly curved, tapering to rounded apex, sacculus with terminal process stout, curved, not projecting beyond valva, cuiller a short spine; anellus with broad base, expanding distally, bifid to about two-fifths with deep narrow sinus; aedeagus stout, hardly tapering, with abundant small cornuti.

FEMALE GENITALIA. Sternite VIII more than twice as wide as long, posterior margin with wide triangular excavation, anterior margin with large expansion accommodating ostium, ostium densely covered with microtrichia; ductus bursae short with expansion at entry to corpus bursae containing a circular sclerite with some spicules; signum with many rows of small triangles either side of smooth area, large, occupying from half to nine-tenths length of elliptical corpus bursae.

GENITALIA DIAGNOSIS. In male sacculus with terminal process stout, outcurved, not projecting beyond valva (may be less distinct due to intraspecific variability). In female anterior margin of sternite with large expansion accommodating ostium is unique.

DISTRIBUTION. Northern and parts of Central Europe from Ireland, Belgium, Germany, Czechia, Slovakia and Poland northwards to Fennoscandia and the Baltic countries, also in Switzerland and the Carpathians, but not recorded from France, Italy or Austria. Also reported from Russian Altai and Far East Russia (Amur region) (Lvovsky, 2013a).

BIONOMICS. Larva in rootstock of *Artemisia vulgaris* L. from late summer, hibernating at the base of the host-plant in a young shoot or in roots. In spring in a young shoot ascending to some 20 cm (Sonderegger, *in litt.*), sometimes with frass visible at the centre of the plant. Also recorded from *Artemisia campestris* L. (Lvovsky, 2013a). Pupa from May or June to July. Imago from July to September, rarely seen.

REMARKS. *Exaeretia liupanshana* Liu & Wang, 2010 was described from Mt. Liupan, Ningxia Huizu Autonomous Region, China. This species is indistinguishable in habitus and male genitalia from *E. allisella*; the main difference presented was primarily in the size of the signum in female (Liu & Wang, 2010). Four females collected at the same place in Switzerland (Ardez, Graubünden) show high variability in this character which makes it likely that the difference in the size of signum falls within the intraspecific variability. With only the information given in the original description of *E. liupanshana* and without type material, a decision on possible synonymy cannot be made here.

Group 3 *ciniflonella* group (10)

Forewing veins R$_4$ and R$_5$ stalked, stalk shorter than free part. Gnathos small; cuiller reaching costal margin of valva; aedeagus short. Larva on trees, mainly *Betula*. Adult hibernates.

10   *Exaeretia ciniflonella* (Lienig & Zeller, 1846)

Depressaria ciniflonella Lienig & Zeller, 1846: 280.
  Depressaria klamathiana Walsingham, 1881: 314.
  Depressaria smolandiae Palm, 1943: 27.
  Martyrhilda isa Clarke, 1947: 5.

DESCRIPTION. Wingspan 15–25 mm. Head with vertex pale grey, more or less mixed with light brown scales, frons whitish. Labial palp with segment 3 two-thirds length of segment 2, outer side of segment 2 brown, whitish at apex, inner side whitish grey with some brown scales, segment 3 whitish grey, apical half blackish except

extreme apex, sometimes with a basal dark ring. Antenna pale grey with narrow
dark brown rings. Thorax grey, more or less overlaid with brown. Forewing whitish
grey to grey-buff, with scattered groups of grey-brown scales, terminal dots darker
grey, dorsal and central area to two-thirds wing length overlaid with orange-brown;
basal field and first one-third of costa light grey-buff or not differentiated from
rest of wing; blackish marks in cell consisting of a pair of obliquely placed dots at
two-fifths, the outer sometimes slightly elongate, a dot or very short streak in fold,
two short streaks in mid-wing and a ring surrounding a whitish dot at end of cell;
some blackish streaks between costa and cell; fringe grey-brown with two indistinct
fringe lines. Hindwing grey, grey-brown towards termen; fringe light grey-brown
with two indistinct fringe lines.

VARIATION. Some specimens have sharply demarcated whitish forewing base
extending along base of costa, but in other specimens this is indistinct or absent;
the abundance and intensity of the scattered groups of grey-brown scales varies as
does the depth of coloration of the orange-brown overlay, which may be very well-
marked or inconspicuous. Some specimens have a weakly marked V-shaped fascia
at three-quarters wing-length, others have the outer costal area heavily marked
dark grey.

SIMILAR SPECIES. The black streaks in cell and between cell and costa are not
found in other European *Exaeretia*.

MALE GENITALIA. Uncus forming an apiculus in wide sinus between semicircular,
divergent socii; gnathos small, circular to ovate; valva slightly curved, parallel-
sided, tapering from end of sacculus to obtuse apex, sacculus with small pointed
terminal process not crossing margin, cuiller arising from broad triangular base,
slender, digitate, reaching costal margin of valva; anellus bifid to about one-fifth,
with wide triangular sinus; aedeagus short, with compact group of very numerous
small cornuti.

FEMALE GENITALIA. Sternite VIII width:length 5:2, anterior margin slightly
extended, ostium adjacent to anterior margin, a furrow from ostium to posterior
margin, very fine microtrichia on either side of furrow; ductus bursae short, as long
as ovate corpus bursae, signum large, elliptic, half as long as corpus bursae, with
three or four rows of teeth.

GENITALIA DIAGNOSIS. In male cuiller slender, rather straight and ± reaching costa of valva (not always distinct), distinct in combination with very narrow
uncus. Female similar to *Exaeretia lepidella* group, furrow from ostium to posterior
margin of plate appears helpful, but may be not always clear. Ductus bursae covered
with tiny dots in central part (a feature not found in *E. lepidella* group) and with
transverse folds before it meets corpus bursae.

DISTRIBUTION. Widespread in the boreal region of Eurasia and North America.
From Scotland to Russia, Germany, Czechia, Poland and the Alps. Not recorded from
France, Denmark or Lithuania.

BIONOMICS. Larva in spun leaves of *Betula* L. in early summer (Jacobs, 1954). In Russia and North America also on *Alnus* L., *Salix* L. and *Populus* L. (Lvovsky, 1981, 2013a). Imago recorded from July to May. It can be found resting on trunks in autumn and spring.

REMARKS. *E. ciniflonella* is unusual within the genus in its bionomics and also in the forewing venation with stalk of R_4 and R_5 shorter than free part of the veins; in other *Exaeretia* the stalked part is longer than the free part.

Group 4 *lutosella* group (11–14)

Terminal process of sacculus projecting away from sacculus margin. Ductus bursae at least twice as long as corpus bursae.

preisseckeri subgroup (11)

Signum more than half as long as corpus bursae.

11   *Exaeretia preisseckeri* (Rebel, 1937)

Depressaria preisseckeri Rebel, 1937: 14.
   Martyrhilda gozmanyi Balogh, 1951: 25

DESCRIPTION. Wingspan 21–23 mm. Head with vertex buff, frons light brown. Labial palp segment 3 two-fifths length of segment 2, segment 2 grey-brown on outer side, buff on inner side, segment 3 buff, outer side with a few brown scales. Antenna dark brown, weakly ciliate towards apex. Thorax brown, becoming brown-buff posteriorly. Forewing broad, pale buff, with small groups of grey-brown scales scattered over whole wing except centre line of cell, more concentrated towards costa and termen, and more orange-brown in dorsal area; costa darker grey-brown and with 3–4 grey-brown spots in outer half; a dark brown dot in cell at one-third and another at end of cell, a less well marked dark brown dot in fold at one-third; fringe buff with two indistinct fringe lines. Hindwing pale buff; fringe pale brown with two indistinct fringe lines.

VARIATION. Forewing ground colour with slight variation, some specimens mid-brown. There is some variation in shape of socii.

SIMILAR SPECIES. *E. nebulosella* is superficially similar, but has narrower wings and more distinct spots around costa and termen.

MALE GENITALIA. Apically truncated or shallowly notched between socii which are not or hardly extended posteriorly and sometimes contiguous; gnathos large, rounded at apex, truncate at base; valva short, broad, straight, costa more or less straight, sacculus with terminal process digitate, slightly tapering, directed outwards from valva, cuiller shorter, digitate, variable in direction; anellus short and wide, with wide shallow distal sinus; aedeagus longer than tegumen, straight, cornuti minute, numerous in an elongate group.

FEMALE GENITALIA. Sternite VIII width:length 3:2, posterior margin straight, anterior margin with anterior extension, further expanded around ostium, two fields of microtrichia from ostium to posterior margin of segment; ductus bursae long, gradually widened into ovate corpus bursae, signum narrow, about half as long as corpus bursae with about four rows of small teeth on each side.

GENITALIA DIAGNOSIS. In male gnathos large, rounded with truncate base. Female rather similar to that of *E. lutosella* subgroup, but signum much longer.

DISTRIBUTION. Italy, Austria, Czechia, Hungary, Croatia, Bulgaria, Romania.

BIONOMICS. Host-plant unknown. Flies in June.

REMARKS. A single examined male from Croatia differs markedly in barcode and some details of male genitalia, but due to lack of further material its taxonomic status is unclear.

lutosella subgroup (12–14)

Signum small, less than half as long as corpus bursae.

12   *Exaeretia lutosella* (Herrich-Schäffer, 1854)

Depressaria lutosella Herrich-Schäffer, 1854: 122.

DESCRIPTION. Wingspan 18–25 mm. Head with vertex light brown mixed buff, sometimes darker or with admixture of grey. Labial palp segment 3 half as long as segment 2; segment 2 with outer side buff, mixed with brown, deep brown and dark grey-brown scales, inner side similar but with more buff scales, segment 3 buff to light cinnamon-brown, ringed dark grey-brown in middle. Antenna brown, scape dark brown, ciliate in male. Thorax grey-brown to cinnamon-brown, sometimes paler posteriorly. Forewing elongate, dull orange-brown, overlaid with grey-brown scales in small groups, in outer third of wing organised into narrow transverse bands; base of costa dark grey-brown often extending to middle of base; costa with indistinct spots; terminal spots blackish; black dots at one-third, two obliquely placed in cell, one in fold, a white spot at end of cell, weakly bordered rust-brown, area between costa and end of cell spot with indistinct grey-brown cloud; fringe grey-brown, fringe lines indistinct. Hindwing grey-brown; fringe grey-brown, with two lines. Abdomen grey-buff.

VARIATION. Ground colour of forewing varies considerably. Grey-brown scales on forewing very variable in extent, when in smaller quantity and more even distribution then wing has a smooth look, but when dark and strongly aggregated wing is mottled. Basal field is very variable, from almost all blackish brown, to dark only in costal half or pale grey-brown or obsolete. Dark cloud between end of cell spot and costa variable in strength. Terminal spots vary from indistinct to well-marked. Western European populations are rather different in appearance from those from the Balkan countries, which have shorter wings, ground colour often pale ochreous, only one cell dot at one-third, more extensively dark basal area and better developed dark area between white cell spot and costa.

SIMILAR SPECIES. *E. lutosella* is very variable, but always has a whitish dot at end of cell. *E. ledereri* has a pair of whitish dots obliquely placed in cell at one-third. *E. thurneri* has strongly developed black area between costa and white cell dot which is more or less surrounded by orange scales.

MALE GENITALIA. Socii widest at base; gnathos as wide as or wider than long; valva tapering from middle; terminal process of sacculus straight, not tapering, obtuse or acute at apex, cuiller not or slightly expanded to apex, extending no more than two-thirds across valva; anellus with rounded lateral protrusions beyond middle, often weakly developed, distal margin often ornamented with fine parallel folds, with wide, shallow sinus; aedeagus with abundant minute hair-like cornuti.

FEMALE GENITALIA. Sternite VIII twice as wide as long with semicircular expansion on anterior margin, closely enclosing ostium, two fields of microtrichia extending from posterior margin of ostium to posterior margin of sternite; ductus bursae long, gradually widening to small corpus bursae, signum a small irregular plate of variable shape, which can be longer than wide or wider than long,

GENITALIA DIAGNOSIS. In male the very shallow emargination of the anellus distinguishes *E. lutosella* from *E. thurneri*, but not always clearly from *E. ledereri*. Distal edges of anellus obtuse, with parallel folds, but without warts, in *E. thurneri* and *E. ledereri* these edges triangular, with warts, but nearly or completely without parallel folds. Female of *E. lutosella* has fields of microtrichia in mid-line of sternite VIII, which are not present in *E. thurneri* and *E. ledereri*.

DISTRIBUTION. Mediterranean areas from Portugal to Greece, including Balearic Islands, Corsica and Sardinia; Crimea (Lvovsky, 2013a). Also in North Africa and Israel.

BIONOMICS. Larva on *Ruta angustifolia* Pers. living in a silk tube on the soil attached to the basal leaves, in March and April. Pupa found under dead basal leaves of *R. angustifolia* (Corley, personal observation). Other *Ruta* species are undoubtedly used as the species can be common where *Ruta graveolens* L., *Ruta chalepensis* L. or *R. montana* (L.) L. are present, but not *R. angustifolia*. Imago can be found from May to October.

REMARKS. The external differences between western and eastern European populations of *E. lutosella* are not backed by clear differences in male or female genitalia and barcode differences are small. We do not consider the differences to merit recognition of separate taxa.

13   *Exaeretia thurneri* (Rebel, 1941)

Depressaria thurneri Rebel, 1941: 7.
  Exaeretia nigromaculata Hannemann, 1989: 398, syn. n.

DESCRIPTION. Wingspan 20–27 mm. Head grey-buff on vertex, with raised scales. Antenna dark fuscous. Labial palp segment 3 two-fifths length of segment 2, segment 2 with a few grey-brown scales, segment 3 whitish buff with a broad grey-brown band in the middle. Forewing pale ochreous to brownish buff, overlaid orange-brown in places, mainly close to base and around end of cell, with scattered blackish scales, particularly along veins in distal quarter; a blackish spot at base of costa often extending to fill basal area except for a grey-white dorsal edge; some blackish spots on costa, particularly around middle, and a larger blackish patch from costa extending to end of cell; a black dot in cell and another in fold at one-third, whitish dot of variable size at end of cell, ringed dark reddish brown, sometimes a very small white dot in cell at one-third; sometimes with blackish terminal spots; fringe grey, with one fringe line. Hindwing grey; fringe grey with one or more fringe lines.

VARIATION. Ground colour varies considerably, from palest buff through various shades of ochreous to brownish buff, usually overlaid orange-brown to some extent, but rarely all traces of ochreous or brown are absent, so the whole moth is whitish buff with grey to blackish markings. The size of blackish patch between costa and end of cell also varies; a white dot at one-third is occasionally present.

SIMILAR SPECIES. Some eastern forms of *E. lutosella* are similar, but never have such a distinct or dark cloud between cell and costa. In cases of doubt genitalia should be examined.

MALE GENITALIA. Uncus small, triangular or truncate; socii extended posteriorly, narrowest at base; gnathos rather large, broadly ovate; valva slightly tapered, costal margin weakly concave, obliquely truncate, forming acute apex, sacculus ending in long digitate process following margin of valva, cuiller shortly digitate, directed towards apex of valva; anellus longer than wide, widest above middle, distally forming two triangles with minute spinules at their tips and with rounded to triangular sinus between them; aedeagus slightly curved, cornuti numerous, hairlike, forming a fairly small group.

FEMALE GENITALIA. Sternite VIII half as long as wide, or a little longer, with wide expansion on anterior margin, enclosing oval ostium, without fields of microtrichia between ostium and posterior margin of segment; ductus bursae long, abruptly expanding to globose corpus bursae, signum small, round, densely covered with papillae.

GENITALIA DIAGNOSIS. For differences in male see under *E. lutosella*. Separation from *E. ledereri* is more difficult: in *E. thurneri* gnathos globose, angle between cuiller and distal process of sacculus usually less than 45°, in *E. ledereri* gnathos more or less elliptic, often pointed (best seen in lateral view), angle between cuiller and distal process of sacculus usually more than 45 °; distal edge of anellus triangular in both species, but in *E. thurneri* usually longer and more densely covered with warts than in *E. ledereri*. While intraspecific variability is low in *E. thurneri*, it is large in *E. ledereri* and some features may appear as in *E. thurneri* in some specimens. Female *E. lutosella* with fields of microtrichia in mid-line of segment VIII, *E. ledereri* shows no reliable differences.

DISTRIBUTION. Bosnia, North Macedonia, Greece, Bulgaria, Romania. Also in Turkey (Asian part).

BIONOMICS. Host-plant unknown. Flies in August and September.

REMARKS. Hannemann (1953) placed this species in *Agonopterix* after dissecting a male in NHMW. The source of this misplacement was an interchange of labels used for slides with numbers directly in series: the slide of the type of *Exaeretia* (at that time in genus *Depressaria*) *thurneri* got the label which had been prepared for the slide of the type of *Agonopterix* (at that time *Depressaria*) *conciliatella* (Rebel, 1892) and vice versa. This was clarified by comparison of the affected slides and external features of types and non-type specimens of both species, all stored in NHMW (Buchner, 2015b).

Hannemann described *Exaeretia nigromaculata* based on three males from Greece in ZMUC and mentions an additional male from Turkey in the Staudinger collection. Both his description of the species and his figure of the male genitalia quite clearly belong to *E. thurneri*. In view of the muddle over slide labels described above, it is not surprising that he considered *nigromaculata* to be a new species.

14   *Exaeretia ledereri* (Zeller, 1854)

Depressaria ledereri Zeller, 1854: 248.
  Depressaria homochroella Erschoff, 1874: 100
  Depressaria xyleuta Meyrick, 1913: 115
  Depressaria leviella Amsel, 1935: 294

DESCRIPTION. Wingspan 20–29 mm. Head with vertex buff mixed brown, frons whitish buff. Labial palp with segment 3 about half length of segment 2, outer side of segment 2 with mixed buff, cinnamon-brown, dark brown and grey scales, inner side whitish buff with a few brown and grey scales, segment 3 outer side cinnamon-brown to middle, with median ring of darker brown, inner side and all of apical part whitish buff. Antenna brown, ringed darker brown, scarcely ciliate. Thorax greyish buff; tegulae deep brown anteriorly on inner and outer sides, reddish brown in middle, becoming paler posteriorly. Forewing elongate; ground colour brownish buff with scattered groups of grey-brown scales particularly in dorsal and terminal areas, other groups of pale reddish brown scales mainly in basal and dorsal areas; extreme base whitish buff except near costa, sharply demarcated outwardly by fine black line; two whitish dots arranged obliquely at one-third in mid-line and another at end of cell, each surrounded by small patch of rust-brown; a blackish dot in fold at one-third; extreme base of costa dark brown, subcostal area to one-third ochreous, blackish brown spots along whole length of costa, smaller dark dots between veins to termen, usually weak and often absent; small dark grey cloud at one-third between fold and rusty area, and another larger patch from one-half to three-quarters extending from costa to rusty area around end of cell; a weakly marked V-shaped fascia at three-quarters; fringe brown at base, light grey-brown with two or more fringe lines. Hindwings brownish grey; fringe greyish buff beyond grey-brown basal line. Abdomen buff.

VARIATION. The description above relates to typical specimens, but there is a range of forms in which the ground colour is paler buff, more or less overlaid with pale ochreous or pale grey-brown, but with grey and rusty coloration more or less lacking; costal spots often absent or pale brown; basal area not distinguishable or not demarcated; white dots sometimes not surrounded with rusty or orange-brown; dot in fold may be absent.

SIMILAR SPECIES. Characterised by the three white dots in cell, two placed obliquely at one-third and one at end of cell. *E. praeustella* has white scales in these positions, but always associated with black scales.

MALE GENITALIA. Uncus triangular, short; socii more or less quadrate, often narrowest at base, widely separated; gnathos narrowly ovate; valva slightly concave on costal side, ventral side more or less abruptly curved just beyond end of terminal process, sacculus with terminal process digitate, extending beyond margin of valva, cuiller smaller and narrower, curved towards costa of valva, but only reaching just beyond half-way across; anellus bifid, forming two triangles separated by triangular sinus; aedeagus small, cornuti very small, numerous.

FEMALE GENITALIA. Sternite VIII twice as wide as long, anterior margin with wide triangular expansion enclosing rounded ostium, ostium extended narrowly through to posterior margin of plate, microtrichia absent; ductus bursae long, slender; corpus bursae narrowly obovate, signum elliptical, less than one-quarter length of corpus bursae, with about four rows of teeth.

GENITALIA DIAGNOSIS. See under *E. lutosella* and *E. thurneri*.

DISTRIBUTION. Spain, Croatia, North Macedonia, Greece, Bulgaria, Romania, Crimea. Also in North Africa, widespread in Asia.

BIONOMICS. Larva on *Haplophyllum suaveolens* (DC.) G. Don (Rutaceae) (Savchuk & Kajgorodova, 2013). Flies from end of April to beginning of October (Lvovsky, 2013a).

Group 5 *lepidella* group (15–20)
Terminal process of sacculus following margin of sacculus or slightly incurved over valva (except *E. lvovskyi*). Ductus bursae short, much less than twice as long as corpus bursae.

15   *Exaeretia praeustella* (Rebel, 1917)

Depressaria praeustella Rebel, 1917: 19.

DESCRIPTION. Wingspan 14–20 mm. Head grey-brown on vertex, frons light grey-brown with some dark brown scales. Labial palp with segment 3 three-fifths length of segment 2, outer side of palp dull cinnamon-brown, inner side of segment 2 light grey-buff admixed with cinnamon scales, segment 3 whitish buff. Antenna brown, ciliate in male, very weakly ciliate towards apex in female. Forewing ochreous brown, extreme base dark brown, costa and a cloud along costal margin of cell expanding and enveloping end of cell fuscous, veins to termen broadly fuscous, costa with weakly greyish ochreous strigulae, dorsal and terminal parts of wing with a few scattered groups of darker scales, an obliquely placed pair of deep brown dots followed by a few whitish scales in cell at one-third, the one on costal side much smaller, and another, preceded by whitish scales at end of cell; fringe with brown basal line and grey-brown outer line. Hindwing light grey-brown; fringe light grey-brown with two lines.

VARIATION. There is some variation in the extent to which the dark brown coloration penetrates the apical and subterminal areas.

SIMILAR SPECIES. The elongate fuscous patch between costa and cell and the three black dots all associated with white scales are characteristic.

MALE GENITALIA. Uncus scarcely developed; sinus between socii wide and shallow; gnathos large, circular; valva with costa slightly concave, sacculus with terminal process digitate, directed towards apex of valva, cuiller digitate, perpendicular to sacculus or slightly curved towards valva apex; anellus narrow in proximal half, beyond middle with rounded lateral protrusions, distally with broad shallow sinus, central area scarcely sclerotised; aedeagus as long as tegumen, cornuti of moderate size, numerous in two lines.

FEMALE GENITALIA. Sternite VIII width:length 3:1, anterior margin with medial expansion much wider than ostium, microtrichia present in low number; ostial opening extending through to posterior margin of sternite; ductus bursae gradually expanding to orbicular corpus bursae, signum elliptical, variable in size, one-third to one-sixth length of corpus bursae, with up to four rows of teeth each side of a smooth centre, sometimes continuing towards ductus bursae as a series of fragments.

GENITALIA DIAGNOSIS. In male an area of weak sclerotisation in centre of anellus which may look like a hole. In female ductus bursae with a section which shows transverse folds. In *E. lepidella* group this feature is also found in *E. mongolicella*, differences are in extent of microtrichia around ostium (much more in *E. mongolicella*) and shape of signum.

DISTRIBUTION. South Sweden, Finland, Baltic States, Poland, Romania, Ukraine and Russia, extending to Mongolia.

BIONOMICS. Larva in spun leaves of *Artemisia campestris* L. and *A. laciniata* Willd. (Palm, 1989) in June and July. Flies from late July to September.

16   *Exaeretia nebulosella* (Caradja, 1920)

Depressaria nebulosella Caradja, 1920: 130.

DESCRIPTION. Wingspan 16–19 mm. Head light grey-buff. Labial palp segment 3 two-thirds length of segment 2, pale buff, segment 2 with some light brown scales, segment 3 with or without grey median ring. Antenna grey-buff. Thorax grey-buff. Forewing pale buff, overlaid pale dull ochreous yellow at base, in mid-line through cell and along veins to costa and termen; small basal area cream, remainder of wing except between blackish cell dots at two-fifths and three-fifths with scattered single scales and small patches of light grey-brown, giving a mottled effect to the wing; a series of grey spots along costa and termen; fringe pale buff with grey fringe line. Hindwing whitish grey with fine grey terminal line; fringe whitish grey with one or two grey fringe lines. Abdomen grey-buff.

VARIATION. Most specimens seen have buff coloration, but there is also a form with near white ground colour and pale grey markings particularly around costal and terminal margins and along veins to termen.

SIMILAR SPECIES. Wing pattern similar to *E. preisseckeri*, which has broader forewing, and *E. niviferella*, which has markings more distinct.

MALE GENITALIA. Uncus not developed or very small; socii wide, separated by narrow sinus; gnathos circular; valva parallel-sided with broad rounded apex, sacculus with short triangular terminal process directed towards apex, cuiller short, slightly expanded and curved outwards to rounded apex, extending about two-thirds across valva; anellus widest beyond middle, lunate protrusions towards side of anellus plate, distal margin with shallow median depression; aedeagus slightly curved, slender, cornuti small, in two rows, very numerous.

GENITALIA DIAGNOSIS. Genitalia quite different from *E. preisseckeri* and *E. niviferella*. In male *E. lepidella* very similar, based on two slides of each species. We recognised these differences: *E. lepidella* with longer terminal process of sacculus which follows margin of sacculus, and cuiller more slender, while *E. nebulosella* has terminal process shorter and directed slightly inwards, towards middle of apex of valva, cuiller wider. An *E. nebulosella* genitalia slide from a type specimen, made by Hannemann, shows intermediate features. The female of *E. nebulosella* is unknown.

DISTRIBUTION. South Russia (Saratov, Astrakhan and Cheliabinsk provinces), North-western Kazakhstan, near Uralsk (Lvovsky, 2013a).

BIONOMICS. Host-plant unknown. Adults fly in May.

REMARKS. *E. nebulosella* and *E. lepidella* share barcode. As host-plants are unknown for both species and the female of *nebulosella* is unknown, it remains unclear whether *lepidella* and *nebulosella* should be treated as two species or only one. This might be clearer if females were available. Externally they are reasonably distinct and there are small differences in male genitalia, so we prefer to treat them as separate species.

17   *Exaeretia lepidella* (Christoph, 1872)

Depressaria lepidella Christoph, 1872: 19.

DESCRIPTION. Wingspan 15–20 mm. Head with vertex pale cream, frons whitish buff. Labial palp segment 3 nearly half length of segment 2; outer side of segment 2 pale buff, inner side and whole of segment 3 pale cream. Antenna buff with brown rings near base, browner distally, ciliate in male. Thorax pale cream. Forewing pale cream, tinged pale ochreous at base and in costal area; dark brown spot at one-third and another at end of cell; light grey-brown terminal dots and faint dots on costa; fringe white. Hindwing white, pale greyish buff towards termen; fringe white dorsally, pale buff towards apex with indistinct fringe line. Abdomen pale buff.

VARIATION. There is very little variation.

SIMILAR SPECIES. *E. buvati* cannot be separated externally.

MALE GENITALIA. Uncus obsolete; socii large, with proximal margin straight, at a right angle to outer margin, apically rounded with narrow sinus between them; gnathos large, circular; valva slightly tapering, apex rounded, sacculus with stout terminal process directed along ventral margin of valva, cuiller more slender, of even thickness, curved outwards, directed towards apex of valva, extending 60–67% of valva width; anellus lobes well developed; anellus with a pair of outwards directed protrusions beyond middle, distal margin with shallow emargination; aedeagus long, curved, abundant small cornuti in two rows.

FEMALE GENITALIA. Sternite VIII width:length 3:1, posterior margin emarginate, anterior margin with wide bulge; ostium drop-shaped, ostial opening extending towards posterior margin, microtrichia in and around ostium; ductus bursae slightly expanding towards nearly orbicular corpus bursae, signum elliptic, large, at least one-half length of corpus bursae, with four or five rows of teeth each side of central line.

GENITALIA DIAGNOSIS. See *E. praeustella*, *E. mongolicella* and *E. lvovskyi* for distinguishing features; *E. nebulosella* is best distinguished by external appearance, *E. buvati* by distribution or barcode. In female ductus bursae without transverse folds, but with longitudinal folds especially in proximal part, a feature only shared by *E. lvovskyi* in *lepidella* group but this species without microtrichia. *E. buvati* not distinguishable by female genitalia. Females of *E. nebulosella* unknown.

DISTRIBUTION. Russia: Orenburg district. Also in Asian part of Russia (Cheliabinsk and Saratovsk districts, Altai Republic) and Kazakhstan.

BIONOMICS. Host-plant unknown.

REMARKS. See remarks under *E. nebulosella* and *E. buvati*.

18   *Exaeretia buvati* Nel & Grange, 2014

Exaeretia buvati Nel & Grange, 2014: 52.

DESCRIPTION. Wingspan 18–20 mm. Head pale cream. Labial palp segment 3 two-fifths length of segment 2, pale cream, segment 2 with a few fuscous scales on outer side. Antenna light brown with narrow fuscous rings, ciliate in male. Thorax pale cream. Forewing pale cream, tinged pale ochreous at base and in costal area; conspicuous blackish dot at one-third and another at end of cell, scattered fuscous scales mainly between veins to costa and termen often present; light grey-brown terminal dots and faint spots on costa; fringe pale cream. Hindwing uniformly pale yellowish grey; fringe concolorous.

VARIATION. Females slightly smaller than males, with narrower more acute fore-wing. The extent of scattered fuscous scales between the veins is variable and they can be almost absent.

SIMILAR SPECIES. *E. lepidella* cannot be separated externally or by female genitalia.

MALE GENITALIA. Very similar to *E. lepidella*, based on only two slides of *E. lepidella* and three of *E. buvati* we recognised these differences: *E. buvati* with socii less expanded laterally; cuiller narrowed beyond middle then slightly expanded, less curved and therefore directed more steeply towards costa of valva and ending closer to costa, 75–80% of valva width. See also Remarks.

FEMALE GENITALIA. No reliable differences from *E. lepidella*.

GENITALIA DIAGNOSIS. For comparison see under male genitalia above.

DISTRIBUTION. France. Known only from two localities in the Pyrenees.

BIONOMICS. Host-plant unknown. Flies in June.

REMARKS. Our preparations of *E. buvati* and *E. lepidella* show slight variation in each species, suggesting that the differences between these two species may be less reliable than indicated above. In the original description Nel & Grange (2014) show the apical part of the cuiller with a recurved barb, but this is not evident in our preparations.

The difference between barcodes of *E. buvati* and *E. lepidella* is 2.63%, which gives support to the hypothesis that these are two separate species.

19   *Exaeretia mongolicella* (Christoph, 1882)

Depressaria mongolicella Christoph, 1882: 15.
   Depressaria leucostictella Rebel, 1917: 21
   Depressaria exquisitella Caradja, 1920: 132, syn. n.
   Exaeretia amurella Lvovsky, 1990: 644, syn. n.

DESCRIPTION. Wingspan 13–22 mm. Head brown. Labial palp with segment 3 half length of segment 2, buff with strong mixture of deep reddish brown scales throughout, but fewer on segment 3. Antenna dark brown, weakly ciliate in male. Thorax dark brown to grey-buff. Forewing mid to dark brown, sometimes pale buff or pale grey-buff, with scattered small groups of dark grey-brown scales; elongate patch in middle of cell, fold and V-shaped fascia at three-quarters ochreous brown or buff-brown, sometimes pale buff or grey-buff; a short deep brown streak in middle of wing base; an obliquely placed pair of deep brown dots in mid-line at two-fifths; whitish spot at end of cell, incompletely ringed deep brown to blackish, the deep brown area extending into angle of fascia; deep grey-brown on narrow costal margin, on veins to costa and termen, terminal spots and between fold and dorsum; fringe rather dark grey-brown, with indistinct fringe line. Hindwing pale grey-brown, almost white towards base; fringe light grey-brown, with two fringe lines. Abdomen buff or light brown.

VARIATION. The first oblique dot, dark costa and veins are inconspicuous in specimens with darkest ground colour and the pale fascia is incomplete in these specimens and may be obscure in some other specimens also. There are also paler forms with ochreous buff, grey-buff or pale grey ground colour of head, palp, thorax and forewing, sometimes with weak markings, or with conspicuously dark patches around oblique dots, beyond whitish dot and basal streak. The whitish dot at end of cell can be reduced in size.

SIMILAR SPECIES. *E. praeustella* has similar coloration to typical forms of *E. mongolicella*, but the cell dots are all associated with a few white scales. Forms with ochreous ground colour can be similar to such forms of *E. indubitatella*.

MALE GENITALIA. Uncus not developed or a broad shallow triangle; socii wider than long to circular, divergent, widely separated; gnathos large, circular; valva short, slightly curved, slightly tapering to broad rounded apex, sacculus ending in digitate process, directed towards apex of valva, cuiller narrowed in middle, wider apical part parallel-sided, slightly curved outwards, extending three-quarters across valva; anellus with a pair of lateral triangular or rounded protrusions beyond middle, distal margin bifid forming two triangles separated by sinus of varying depth; aedeagus slightly curved, numerous small cornuti in a compact group occupying half length of aedeagus.

FEMALE GENITALIA. Sternite VIII width:length about 5:2, posterior margin slightly angled inwards toward centre, anterior margin with a rectangular or triangular expansion around ostium and furrow from ostium to posterior margin, ostium, furrow and adjacent part of sternite with abundant microtrichia; ductus bursae shorter than ovate corpus bursae, signum large, elongate, half to entire length of corpus bursae, with numerous small spines all over, and a short extension with large spines at posterior end.

GENITALIA DIAGNOSIS. In male anellus typically with apical sinus deeper and lateral protrusion longer, more distinct than in the other species of the *E. lepidella* group, subcentral lobes usually absent, only in some specimens present but rather inconspicuous. For female see *E. praeustella*.

DISTRIBUTION. Poland, Lithuania, Russia, extending to Mongolia.

BIONOMICS. In stems of *Artemisia vulgaris* L. (Palm, 1989). Flies from June to August.

REMARKS. Lvovsky & Stanescu (2019) recently referred *Depressaria exquisitella* Caradja, 1920 to the genus *Exaeretia*. They also placed *Exaeretia amurella* Lvovsky, 1990 in synonymy with *E. exquisitella*. We fully agree with this synonymy, but we can find no discernible difference between male genitalia of *E. exquisitella* and those of *E. mongolicella*. We therefore treat *E. exquisitella* and *E. amurella* as junior synonyms of *E. mongolicella*.

20   *Exaeretia lvovskyi* Buchner, Junnilainen & Nupponen, 2019

Exaeretia lvovskyi Buchner, Junnilainen & Nupponen, 2019: 11.

DESCRIPTION. Wingspan 15–18 mm. Head light grey, interspersed with darker grey and brownish scales. Labial palp segment 3 half as long as segment 2, segment 2 slender, whitish, with a variable proportion of dark grey scales, sometimes forming a subapical blackish ring, segment 3 often with grey or blackish ring before apex. Antenna grey-buff, thin, diameter in the female about 2/3 of that in the male. Thorax whitish, irregularly mixed with some medium to dark grey scales, usually darker at front. Forewing extremely elongate, termen oblique; ground colour whitish with weakly demarcated medium to dark grey pattern, at base in middle and on dorsal side, a fascia from dorsum at one-third darker and angled inwards in costal half more or less concealing dark grey cell spot, a cloud in costal half around a darker spot at three-fifths and most of terminal area leaving two whitish lines parallel with termen, the inner sometimes with a branch to costa, making a narrow V-shaped fascia; a dark grey spot at base of costa and a series of more or less distinct spots along costa, strongly marked around apex and between veins to termen; fringe light grey, with one or two darker lines. Hindwing uniformly pale greyish, only the edge a little darker; fringe with an indistinct dark subbasal line. Abdomen pale to medium grey.

VARIATION. Some specimens show a distinct black subapical ring on segment 3 of labial palp, which is diffuse or even absent in others. The proportion of white in the forewing varies, being wider and clearer in some individuals than in others. The dark central dots are nearly invisible in some individuals and distinct in others.

SIMILAR SPECIES. The unusually narrow wings distinguish this species from all other European *Exaeretia* species.

MALE GENITALIA. Uncus obsolete; socii with anterior margin straight, ending in a near right-angle, posterior margin almost straight then apically rounded, separated by wide deep sinus; gnathos slightly longer than wide; valva short and broad, ventral margin angled inwards at middle of stout terminal process of sacculus which extends well beyond margin of valva, cuiller short, triangular or digitate, reaching two-thirds to three-quarters across valva; anellus widest at two-thirds length, distal margin deeply notched; aedeagus two-thirds length of valva, cornuti small, numerous becoming larger towards apex, as long as width of aedeagus.

FEMALE GENITALIA. Sternite VIII width:length 3:1, anterior margin with wide anterior step, ostium round, anterior to centre of sternite, microtrichia absent; ductus bursae short, gradually expanding to broadly elliptic corpus bursae; signum round with posterior projection and several rows of small teeth.

GENITALIA DIAGNOSIS. Differs from other species of *E. lepidella*-group in male with gnathos longer than wide (all others globose); in female the only species without microtrichia.

DISTRIBUTION. Only known from Russia: South Ural, Altai Mountains, Buryatia.

BIONOMICS. Host-plant unknown. Fresh moths have been collected at the end of June, at altitudes between 700 and 2200 metres.

Group 6 *niviferella* group (21–23)
Gnathos bilobed. Signum paddle-shaped with posterior 'handle'.

21 *Exaeretia stramentella* (Eversmann, 1844)

Yponomeuta stramentella Eversmann, 1844: 566.
  Depressaria culcitella Herrich-Schäffer, 1854: 127.

DESCRIPTION. Wingspan 15–18mm. Head with vertex blackish, frons golden brown. Labial palp with segment 3 nearly half length of segment 2, blackish except pale yellow terminal two-thirds of segment 3. Antenna blackish, not ciliate. Thorax blackish. Forewing pale straw-yellow, a patch across base, a dot at one-third in mid-line and another at end of cell blackish; some weak grey marks on costa and terminal dots blackish; fringe pale straw-yellow. Hindwing whitish at base, pale grey-brown towards apex, a series of grey-brown dashes around termen; fringe whitish buff with indistinct fringe line. Abdomen pale straw-yellow.

VARIATION. There is variation in the strength of the dark markings on fore-wing costa and around termen. The blackish markings may sometimes be dark chocolate-brown.

SIMILAR SPECIES. The blackish thorax and wing base contrasting with pale straw-coloured ground colour is a unique combination in European Depressariidae.

MALE GENITALIA. Uncus not developed; socii semicircular with narrow sinus between them; gnathos broadly heart-shaped with notched posterior margin; valva tapering, slightly curved towards broadly rounded apex, sacculus with stout, slightly curved, narrowly triangular terminal process, directed towards apex of valva, cuiller short, with stalked expanded apex, slightly curved outwards, extending three-fifths across valva; anellus widening posteriorly with pair of lateral horns, distal margin emarginate; aedeagus tapering throughout, with two ranks of small cornuti.

FEMALE GENITALIA. Sternite VIII width:length about 3:1, anterior and posterior margins with similar curve, ostium not quite adjacent to anterior margin, semicircular with anterior lip, ostial area with abundant microtrichia; ductus bursae short, corpus bursae ovate, signum large, two-thirds length of corpus bursae, paddle-shaped with long, slender posterior extension, with several rows of teeth, largest on margin.

GENITALIA DIAGNOSIS. Males of *E. niviferella* group with gnathos more or less divided. Note: an undivided gnathos can appear divided if it is damaged by too much pressure during preparation. Females with paddle-shaped signum. Distinguishing the species of the *E. niviferella* group is easy by external features, but genitalia differences between the three species are unreliable. Male *E. niviferella* has more slender cuiller than the other two species.

DISTRIBUTION. Central Europe from north Italy and Germany to Slovakia and Hungary, Montenegro, North Macedonia, Crimea, South Russia.

BIONOMICS. The larva lives in a tube of leaves spun to the stem of *Tanacetum corymbosum* (L.) Schultz in May. Flies in July.

REMARKS. The synonymy of *E. culcitella* with *E. stramentella* was recently established by Sinev *et al.* (2017).

In recent decades this species has disappeared from many places, although the host-plant is still present.

22   *Exaeretia niviferella* (Christoph, 1872)

Depressaria niviferella Christoph, 1872: 20.

DESCRIPTION. Wingspan 18–23 mm. Head with vertex light cinnamon-brown, scales tipped white. Labial palp segment 3 just over half length of segment 2; segment 2 outer side whitish mixed with cinnamon-brown scales, inner side similar but paler, segment 3 whitish buff on outer side, whiter on inner side. Antenna whitish near base, light brown distally, barred dark brown on anterior side, strongly ciliate in male, weakly ciliate distally in female. Thorax and tegulae mid-brown anteriorly, becoming whitish posteriorly. Forewing whitish with grey-brown transverse bands of irregular size, width and length; dark grey-brown spot in cell at one-third and another at end of cell; base of wing, series of spots on whole length of costa and dots around termen dark grey-brown; fringe whitish with two pale orange-brown fringe lines. Hindwing whitish buff, greyish buff posteriorly, especially on veins, outer half of costa and terminal dots grey-brown; fringe whitish, with two indistinct fringe lines. Abdomen buff, with a few brown scales.

VARIATION. The size and depth of colour of the cell spots is variable.

SIMILAR SPECIES. Characterised by grey-brown wing pattern on white ground colour and regular spots all around costa and termen. The markings are more distinct than in *E. nebulosella*.

MALE GENITALIA. Uncus a minute triangle; socii semicircular; gnathos nearly as wide as socii, consisting of two circular contiguous lobes; valva with costa straight, ventral margin gently and evenly curved, sacculus ending in stout triangular process directed towards apex of valva, cuiller digitate, small, extending just over half way across valva; anellus broad, distal margin with wide rounded sinus; aedeagus slender, basal part tapering before bend, numerous small cornuti in a compact group.

FEMALE GENITALIA. Sternite VIII width:length about 2:1, with broad rounded anterior bulge; ostium not quite adjacent to anterior margin, with anterior lip, ostial area with abundant microtrichia; ductus bursae short, half as long as elliptical corpus bursae, signum medium-sized, about one-third length of corpus bursae, obovate with posterior projection and several rows of small teeth.

GENITALIA DIAGNOSIS. Male and female see *E. stramentella*.

DISTRIBUTION. Ukraine, Russia (Lvovsky & Anikin, 2009). Also in Turkey (Asian part), Kyrgyzstan.

BIONOMICS. Host-plant unknown. Flies in May and June (Lvovsky, 2013a).

23 *Exaeretia indubitatella* (Hannemann, 1971)

Martyrhilda indubitatella Hannemann, 1971: 263.

DESCRIPTION. Wingspan 22 mm. Head dark grey-brown. Labial palp segment 3 half as long as segment 2, segment 2 dark grey-brown, inner and outer side with scattered whitish scales, segment 3 pale yellow, almost white, with black ring at base and before the off-white apex. Antenna blackish grey, finely and densely ciliate. Thorax grey mixed with blackish scales, tegulae concolorous, edged pale. Forewing elongate; ground colour dull grey-brown to dull ochreous, with scattered black to dark brown scales, more or less organised into transverse rows; extreme base blackish brown; costa and termen with dark grey spots; a pair of obliquely placed black dots at one-third, often joined into a curved mark, a small black dot in fold, a white dot at end of cell outwardly edged black; fringe concolorous. Hindwing very pale grey to mid-grey; fringe concolorous, sometimes with a fringe line.

VARIATION. Terminal dots sometimes absent. Oblique dots at one-third often fused. One examined specimen with ochreous buff ground colour, first oblique dot very small and dot at end of cell black.

SIMILAR SPECIES. The form with oblique dots joined is highly characteristic. Other forms may be confused with other species and are likely to need genitalia examination.

MALE GENITALIA. Uncus obsolete; socii divergent, from broad base; gnathos consists of two contiguous circular lobes; valva curved, slightly tapering, costal margin angled towards end of ventral margin, sacculus with terminal process triangular, stout, directed towards apex of valva, cuiller falcate, tapering to point, directed towards apex of valva; anellus wide, distally with pair of spreading lateral horns, distal margin with pair of triangles separated by V-shaped sinus; aedeagus slender, hardly tapered, with two rows of medium-sized cornuti.

FEMALE GENITALIA. Sternite VIII width:length about 2:1, with broad rounded anterior bulge; ostium not quite adjacent to anterior margin, with anterior lip, ostial area with abundant microtrichia; ductus bursae short, one-quarter as long as pyriform corpus bursae, signum large, about one-half length of corpus bursae, elliptic with long posterior projection extending into neck of corpus bursae, with several rows of small teeth.

GENITALIA DIAGNOSIS. Male and female see *E. stramentella*.

DISTRIBUTION. Russia (Ural Mountains); more widespread in Asia.

BIONOMICS. Host-plant unknown. Specimens have been collected from late June to August.

REMARKS. *Martyrhilda indubitatella* was described from two males collected in Mongolia in August.

Agonopterix Hübner, [1825]

Agonopterix Hübner, [1825]: 410.
   Type species: *Tinea signella* Hübner, 1796; subsequent designation.
Epeleustia Hübner, [1825]: 410.
   Type species: *Tinea liturella* Hübner, 1796; subsequent designation.
Pinaris Hübner, [1825]: 411.
   Type species: *Tinea gilvella* Hübner, 1796; subsequent designation.
Tichonia Hübner, [1825]: 412.
   Type species: *Tinea atomella* [Denis & Schiffermüller], 1775; subsequent designation.
Haemylis Treitschke, 1832: 235.
   Type species: *Tinea hypericella* Hübner, [1817]; subsequent designation.

Agonopteryx Stephens, 1834: 201, misspelling.
Syllochitis Meyrick, 1910: 462.
    Type species: *Syllochitis petraea* Meyrick, 1910; monotypy.
Ctenioxena Meyrick, 1923a: 611.
    Type species: *Ctenioxena cryptipsila* Meyrick, 1923a; monotypy.

Lvovsky (2013b) has separated 11 Asian species into a separate subgenus *Subagonop-*
terix, characterised by absence of antennal pecten, labial palp with appressed
scales, together with some differences in venation and male and female genitalia.
All European species remain in subgenus *Agonopterix* and the description that follows refers only to the typical subgenus.

DESCRIPTION. Head with ocelli present; crown with raised scales, face with
appressed scales. Labial palp strongly recurved, with segment 2 curved, thickened,
furrowed beneath, segment 3 three-fifths to four-fifths length of segment 2, slender.
Antenna not ciliate, scape with pecten. Thorax with or without posterior crests.
Forewing long, costa almost straight, apex rounded and termen curved or rarely
apex right-angled to weakly falcate, dorsum three-quarters to four-fifths length
of costa; veins CuA1 and CuA2 stalked. Hindwing with lobe at anal angle present,
sometimes weakly so; veins CuA_1 and M_3 connate or approximated. Abdomen dorsoventrally flattened. Forewing pale to dark ochreous or any shade of brown, rarely
whitish, characteristically with pale base demarcated from rest of wing by a partial
fascia and cell with a pair of obliquely placed dots followed by further markings
to end of cell, and with a larger dark spot on anterior margin of cell at one-half
wing length, a series of dark spots or dots along costal margin and around termen
between vein ends. There is much variation in these characteristic markings and
any or all may be absent in some species.
    MALE GENITALIA. Socii well developed; uncus small or absent; spined gnathos
usually narrow; valva entire, with weak indentation near end of sacculus; sacculus
quite short, never reaching middle of valva, with process (*cuiller*) perpendicular to
ventral margin of valva; transtilla present; aedeagus short to medium length, never
as long as valva, lightly curved, usually with numerous minute cornuti, with a basal
process on inner side.
    FEMALE GENITALIA. Anterior and posterior apophyses not very elongate;
ostium on eighth sternite; ductus seminalis arises close to ostium; ductus bursae
membranous; corpus bursae with more or less quadrate signum with thorn-like
teeth, signum rarely absent.
    DISTRIBUTION. Holarctic, with a very small number of species elsewhere.
    BIONOMICS. The majority, but by no means all species overwinter as adults, and
all have larvae feeding in spring and early summer. Larvae feed in spun leaves, leaf
lobes or margins, among flowers or in shoot tips, often making neat tubes on leaves.
Only a few families of flowering plants are utilised as host-plants, with Apiaceae

being the most used, followed by Asteraceae and Fabaceae. In addition there are a few European species on Rutaceae, Salicaceae and Clusiaceae. Outside Europe a few additional plant families are also used. One species, *Agonopterix squamosa* has host-plants in both Apiaceae and Asteraceae.

REMARKS. *Agonopterix* species differ from *Depressaria* in the stalked forewing veins CuA1 and CuA2 and in some genitalia characters, but in almost all cases can be readily recognised by the wing markings. *Exaeretia* differs mainly in genitalia characters, including the sacculus with apical process in addition to cuiller and with different signum. They are more difficult to separate on wing markings. The labial palp characters of *Exaeretia* are too inconsistent to be used to separate the genera.

While the large genus *Depressaria* can readily be divided into distinct groups based on genitalia characters, the same is not true for the even larger genus *Agonopterix*, where genitalia differences are small throughout the genus. As a result there has been little attempt to divide the genus up, and relationships between species have been hard to find, except in one or two sub-groups such as the straw-coloured species related to *A. kaekeritziana*, which also lack most of the characteristic forewing markings. The order of species in many works shows little sign of relationship between successive species. Only Palm (1989) classified the species into groups using host-plant family preference as the character defining each group.

This does bring some sense of order to the genus and remains the best means of dividing the genus, but has the disadvantage that if the host-plant is unknown, it may not be possible to place a species in a group. Recently DNA barcoding has begun to show that the Palm groups do have some validity and the principle is followed here with some modifications.

We recognise four main groups based on host-plant families. One of these is divided into two parts, with species feeding on Apiaceae and Rutaceae respectively. Within the two largest groups, we have been able to recognise a number of subgroups with species sharing morphological characters backed up by evidence from barcodes. This is not entirely satisfactory since there are a few subgroups with some slightly anomalous species. There are also some species which appear to be quite isolated as they do not fit readily into any subgroup. To maintain consistency of treatment each of these is placed in its own subgroup.

Forewing Markings of *Agonopterix* Species

There are a number of markings on the forewing of *Agonopterix* which although not all present in every species, are so frequent within the genus that it will simplify the descriptions that follow if they are described and illustrated here. To avoid lengthy and repetitive description of some of these features in the species diagnoses, the terms used for these are given here in italics.

Extreme base of wing with a small dark spot on costa; basal area of wing often paler in colour than remainder of forewing (*basal field*, 1), this coloration some-times extending as a band along proximal part of costa, the basal area demarcated by at least a dark spot usually close to base and close to dorsum (*subdorsal spot*) and usually this is extended to form a dark partial fascia (*basal fascia*, 2) from dorsum to dorsal margin of cell or sometimes to anterior margin of cell, on its outer side the basal fascia gradually fades into the ground colour of forewing; a series of dark spots of varying size, distribution and with irregular shape along costa (*costal spots*, 3); a series of small dark spots between veins close to terminal margin (*terminal spots*, 4); a pair of obliquely placed dark dots at one-third to two-fifths wing length (*oblique dots*, 5), the proximal (*first*, 5a) on costal margin of cell, the distal (*second*, 5b) beyond it and in cell, the two sometimes joined into a curved mark, (in species where the first oblique dot is absent throughout and therefore nothing is present which appears oblique, for the "*second oblique dot*" the alternative term "*inner cell dot*" is used), a dot in the cell at one-half of wing length (*median cell dot*, 6) and another at end of cell (*end of cell dot*, 7), term for both "*distal pair of cell dots*", term for all four dots but excluding plical dot "*central wing dots*", one or more of the cell dots frequently associated with or replaced by small whitish marks; sometimes a dot in fold beneath oblique dots (*plical dot*, 8); a spot of variable shape but most often round (*blotch*, 9) overlapping costal margin of cell at one-half, not usually extending to costa. The ground colour of the forewing is rarely uniform, normally with some darker areas and some paler areas, often related to the extent of addi-tional colouring with mixed in scales of some colour more or less different from ground colour. Very commonly there is a scattering of darker scales over the wing, sometimes very few or very localised, sometimes along veins, but in other speci-mens these scattered scales may be much more abundant and found over nearly the whole wing surface, often banded together into irregularly shaped transverse lines of varying length.

Many *Agonopterix* species when newly emerged exhibit some pink colora-tion, not only on the forewings, but also on other parts of the body, particularly

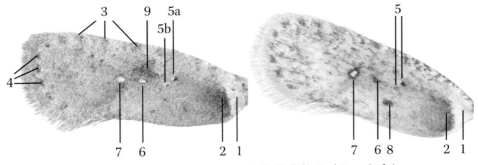

FIGURE 15   Forewing markings in *Agonopterix*: *A. multiplicella* (left), *A. alpigena* (right)

the legs. In the majority of cases this fades quite quickly, but because *Agonopterix* are frequently reared from larvae, the pink coloration is commonly mentioned in descriptions.

Typical male genitalia. Socii large, attached to posterior margins of tegumen, each with greater area than gnathos which is elliptic or lanceolate with obtuse apex; uncus projects as a small triangle between the socii but may be absent; tegumen widest at point of origin of gnathos arms, with a deep rounded anterior sinus; valva widest at base, with slight curve, costal margin straight or weakly concave, ventral margin convex, usually with a slight indentation at end of sacculus, with more curvature at about two-thirds length resulting in more tapering distal one-third; cuiller arising from end of sacculus; transtilla of various thickness, transtilla lobes overlapping posterior end of anellus or not; anellus goblet-shaped with short proximal stalk and wider distal plate, usually with median notch in posterior margin, anellus lobes overlapping edges of anellus; saccus short, only just extending beyond base of valvae. Aedeagus usually about half length of valva from base on ventral side to apex, curved beyond middle and tapering to fine point from about three-quarters; base with a short sclerotised sheath with a process, commonly of similar length to sheath, on concave margin of aedeagus; vesica with very large number of extremely small cornuti.

Problems in *Agonopterix*

Agonopterix is the sixth largest genus of European Microlepidoptera with 99 species. Any large genus inevitably presents some problems of identification and species delimitation. It is appropriate to consider what these problems are.

There has been no systematic review of the genus on a Europe-wide scale. The nearest available is the series of papers by Hannemann (1953, 1954, 1957, 1958a, 1958b, 1959, 1976a, 1976b, 1982, 1983, 1989, 1990) in which he looked at male genitalia mainly of western Palaearctic species of *Agonopterix*, *Exaeretia* and *Depressaria*. After the first papers he also included female genitalia for a number of species. Hannemann's work is immensely valuable, the drawings are for the most part of good quality and he recognised a large number of new species as well as reducing a number of names to synonymy. However the descriptive text accompanying these papers is often inadequate. In the earliest papers there is no description or figure of the habitus of the moth, so as an identification guide this is nearly impossible to use. Moreover it contains mistakes, which have only become evident during the preparation of the present work. Often he did not examine type specimens and some of his genitalia figures are from specimens collected far from the type locality, which has sometimes caused problems. Apart from Hannemann's papers, there are good accounts of the Depressariid fauna of North Europe (Palm, 1989) and the British Isles (Harper *et al.*, 2002). Central Europe was covered by Hannemann (1995) and there are various works on the species of Russia by Lvovsky, although Central Europe and Russia have substantially more species than these authors recognised. There has been no publication covering the rich fauna of southern Europe.

Compared with *Depressaria*, in which many species look very similar, *Agonopterix* species show a greater range of colour, wing shape, and markings. However the amount of variation that occurs within species, in some much more than others, blurs the boundaries between species. As an example, typical forms of *A. nervosa* are very distinctive, but there are also reasonably common forms with longitudinal streaking that are extremely similar to *A. umbellana*. J.R. Langmaid (pers.comm.) remarked on the difficulty of constructing a key to species for Harper *et al.* (2002), even for the rather limited British fauna. Almost every species shows some degree of variation and this problem extends beyond the habitus of the moths to the genitalia and the larvae.

Some *Agonopterix* larvae are quite distinctive and occasionally larval characters have been used as a means of separating species (e.g., Rymarczyk *et al.*, 2013). Certainly larval features can be useful, as when more than one species is present on a host-plant at the same time making it possible to know which of two species one is collecting, but they also vary within species regionally and sometimes even in a single locality. *A. furvella* larvae of very different coloration were present in two consecutive years at a single site in Austria (PB). *A. nanatella* larvae in England and Portugal look significantly different (MC). Our view is that larval characters may add to the features separating two species, but can rarely be used as the primary feature.

Host-plant is another feature that sometimes separates species, but there are more species that are oligophagous than those that are monophagous, so unless the host-plants are in different families, this is another character for distinguishing species that must be used with caution. The choice of host-plants can also vary regionally.

Group 1 *impurella* group (Species 24–28)
Species feeding on Clusiaceae or Salicaceae.

impurella subgroup (24–27)

Barcodes suggest that this subgroup of species with larvae feeding on Clusiaceae or Salicaceae is more closely related to *Exaeretia* than to *Agonopterix*. It resembles *Exaeretia* in the labial palpi with segment 2 slightly thickened but not tufted and the absent or very small anal lobe on the hindwing, but these characters do occur in a few other *Agonopterix* species. All except *A. impurella* have broadly elliptic gnathos, also unusual in *Agonopterix*, but otherwise male and female genitalia characters place these species firmly within *Agonopterix*. Cuiller with papillae or small ridges on outer side distally; aedeagus with sheath reaching nearly to middle.

24   *Agonopterix impurella* Treitschke, 1835

Haemylis impurella Treitschke, 1835: 178.
  Tinea hypericella auct. nec Hübner, 1817: pl. 66, fig. 441.

DESCRIPTION. Wingspan 14–18 mm. Head greyish white, orange-brown above eyes. Labial palp greyish white, with some brown scales, particularly on outside of segment 2, segment 2 weakly furrowed, nearly smooth-scaled, segment 3 with ring at one-quarter and apical half blackish. Antenna with scape dark fuscous, flagellum greyish white with dark fuscous rings. Thorax and tegulae greyish white with brown scales anteriorly and posteriorly on thorax. Forewing mid to dark brown mixed with some orange-brown scales, basal area, except on dorsum, pale greyish white extending along costa to apex as a band of varying width, with two narrow partially interrupted angled lines crossing dark brown area to reach dorsum at one-half and three-quarters, some flecks of light brown or orange-brown scales particularly along costa; oblique dots black, the first smaller, immediately before first whitish crossline, followed by a black dot on outer edge of crossline and another small white dot at end of cell, more or less surrounded by a blackish area; a series of white dots alternating with blackish terminal spots; fringes dark grey at base, pale grey with pink tinge in outer half. Hindwing whitish grey, darker grey towards apex; fringes light grey. Abdomen grey-buff to grey.

VARIATION. Variation is mainly in the depth of the brown forewing coloration, partly due to the extent of orange-brown scales. The inner narrow whitish crossline may be almost obsolete except adjacent to the oblique dots. The oblique dots and other blackish dots in cell are hardly visible in the darkest specimens. The white dot at end of cell may be much reduced.

SIMILAR SPECIES. The broad white area from base to apex and two partial crosslines contrasting with dark brown remainder of forewing, together with series of white terminal dots is unique in the family.

MALE GENITALIA. Socii small, slightly divergent; gnathos lanceolate, about three times as long as wide; valva with costa straight, ventral margin narrowed from two-thirds to rounded end; cuiller straight, about 67% of valva width, slightly narrowed in middle, apex rounded, outer margin papillose in distal part; anellus lobes narrow; anellus longer than wide, notched at apex; aedeagus with sheath nearly reaching middle, with a group of numerous small cornuti.

FEMALE GENITALIA. Anterior apophysis two-thirds or a little more of length of posterior apophysis; sternite VIII width:length 3:1, anterior margin more or less convex, posterior margin concave; ostium large, close to anterior margin extending across most of sternite length, with microtrichia posteriorly; ductus bursae straight, slightly wider in posterior two-fifths; corpus bursae ovate or elliptical, signum small, wider than long, constricted in middle or separated into two parts, with a few teeth on each.

GENITALIA DIAGNOSIS. In male best single feature is outline of gnathos in lateral view, tapering in distal 2/3 with ventral and dorsal edges nearly straight. Female differs from other species of *impurella* group and most other *Agonopterix* species by presence of microtrichia posterior to ostium.

DISTRIBUTION. From northern Italy and Switzerland northwards and eastwards to Sweden, Russia and Romania. Asia: Lvovsky (2019): widespread in Russia, reaching Far East; Japan (Japan Moths, 2002–2023). Not listed for China in Li (2010).

BIONOMICS. Larvae on *Hypericum perforatum* L., *H. pulchrum* L., *H. tetrapterum* Fries (Sonderegger, *in litt.*) and probably other *Hypericum* spp., usually on underside of a leaf, which is constricted or longitudinally rolled, several leaves are used during development. Avoids dry and fully sun-exposed habitats. From April to July or rarely August, emerging predominantly in July and August (rarely in September), flying until April (rarely May) of next year.

REMARKS. This species was known as *impurella* Treitschke, 1835 for many years (e.g. Hannemann, 1953) until Bradley (1966) declared that it was conspecific with *hypericella* Hübner, 1817 which therefore had priority. This was apparently accepted without question and the name *hypericella* was used by Hannemann (1976) in his Catalogue of Depressariidae. Why Bradley believed that Hübner's illustration of *hypericella* was the same as *impurella* is unclear, as it unmistakably shows *A. liturosa* with conspicuous black curved mark in the cell (joined oblique dots) and a distinct pale line bordering the basal field, both characters absent from *impurella*. We therefore reinstate *impurella* as the correct name for this species.

25   *Agonopterix liturosa* (Haworth, 1811)

Depressaria liturosa Haworth, 1811: 508.
  Tinea liturella Hübner, 1796: 410, nec [Denis & Schiffermüller], 1775: 137.
  Tinea hypericella Hübner, 1817: pl. 66, fig. 441.
  Agonopterix huebneri Bradley, 1966: 224.

DESCRIPTION. Wingspan 16–21 mm. Head creamy ochreous to grey-buff or orange-brown. Labial palpus whitish buff to creamy ochreous, segment 2 deep brown to dark grey or blackish on outer side, segment 3 with black apex. Antenna dark fuscous, narrowly ringed blackish. Thorax pale ochreous-buff, with cinnamon-pink broad median stripe, to grey-buff or orange-brown; tegulae dark reddish brown to deep fuscous. Forewing reddish-brown, more or less extensively overlaid dark grey, particularly at base and in costal third extending to three-fifths wing length; subdorsal spot black; basal field concolorous with rest of wing, narrowly edged cream to orange-yellow; costal spots blackish fuscous, separated by blocks of grey to buff, with one larger pale spot at two-fifths, extending to costal margin of cell followed by a blackish fuscous patch extending to two-thirds; some additional grey to buff dots along costal margin of cell and towards costa at three-quarters; terminal spots black, often united to form a line; black oblique marks at one-third fused to form a comma-shaped mark, sometimes interrupted posteriorly, usually narrowly edged orange-brown; a pair of whitish buff dots at end of cell, sometimes with associated black dot; black plical dash sometimes present; fringe reddish brown, more or less tinged fuscous, often with two fringe lines. Hindwing light grey, darker towards termen; fringe light grey with one or more fringe lines. Abdomen fuscous.

VARIATION. There is much variation in the extent of dark grey overlying ground colour, sometimes rendering almost the entire wing deep grey-brown; most of the markings also vary, particularly the oblique curved mark in cell and the dot at end of cell.

SIMILAR SPECIES. Characterised by pale thorax contrasting with dark tegulae, rich reddish brown forewing more or less overlaid dark fuscous, oblique dots forming comma-shaped mark and pale spot below costa at two-fifths. Worn specimens can be confused with *A. conterminella* (*q.v.*).

MALE GENITALIA. Socii small, divergent; gnathos elliptic, about 1.7 times as long as wide; valva with costa weakly concave, ventral margin curved abruptly at two-thirds, apex rounded; cuiller straight, about 67% of valva width, stout, apex rounded, outer margin papillose in distal half; anellus lobes apparently absent; anellus with distal half nearly circular with wide V-shaped notch; aedeagus with sheath nearly reaching middle, with a group of numerous small cornuti.

FEMALE GENITALIA. Anterior apophysis three-quarters of length of posterior apophysis; sternite VIII width:length 5:2 to 2:1, anterior margin slightly convex in middle part, posterior margin with small median notch; ostium close to anterior margin, with a furrow distally extending to posterior margin of sternite; ductus bursae straight, slightly widening towards long ovate or elliptical corpus bursae, signum large, rhombic, slightly wider than long, with numerous teeth.

GENITALIA DIAGNOSIS. Male with cuiller straight, cuiller in *A. conterminella* expanded distally; no reliable genitalia differences from *A. arctica*. Female without reliable differences from *A. conterminella*. See also *A. impurella* and *A. arctica*.

DISTRIBUTION. Nearly all Europe extending south to north Portugal and Greece, but rare in the south and absent from Mediterranean islands and parts of the Balkans. Asia: we have seen only records from Turkey and Kyrgyzstan; western part of Russia to Altai Republic (Lvovsky, 2019); China (Zhu *et al.*, 2023).

BIONOMICS. Larva in spun shoot tips of *Hypericum perforatum* L. (Hannemann, 1995); *H. undulatum* Willd. (Corley *et al.*, 2007); *H. maculatum* Crantz (Sonderegger, *in litt.*); *H. hyssopifolium* Chaix (Rymarczyk *et al.*, *in litt.*), spring to May or June, pupation usually in the spinning, emerging May to July, flying until autumn, not hibernating. Altitudinal range from lowlands to above 2000 m.

REMARKS. Larvae of *Lathronympha strigana* (Fabricius, 1775) (Tortricidae) can often be found in similarly spun shoot tips of *Hypericum*.

26   *Agonopterix conterminella* (Zeller, 1839)

Depressaria conterminella Zeller, 1839: 196.

DESCRIPTION. Wingspan 17–21 mm. Head buff to grey-brown, sometimes almost white, face buff. Labial palp buff, segment 2 heavily mixed with dark fuscous scales on outer side and ventrally, segment 3 mid-brown with broad blackish brown ring beyond middle or rarely occupying almost whole of segment, tip buff. Antenna fuscous with narrow deep brown rings. Thorax light grey-buff with scattered deep brown scales, more rarely whitish with only anterior half of tegula brown. Forewing fuscous, dark fuscous or deep brown, often with some reddish brown areas, particularly in dorsal third and near costa at two-fifths and three-quarters, usually with some darker and paler areas and with scattered dark scales, particularly along veins, elsewhere often organised into transverse lines; basal field buff to mid or deep brown, basal fascia a dentate whitish line, not quite reaching costa, outwardly deep brown to one-fifth; costal spots dark fuscous separated by blocks of grey-buff often with reddish tinge; terminal spots blackish, usually fused to form a line; first oblique black dot at one-third, second extended into short streak, separate, contiguous or most often fused with first to form a comma-shaped mark; a buff or whitish transversely elongate dot at end of cell; the black streak in cell usually outlined red-brown, followed by an area mixed with pale grey scales that may extend towards costa; plical dot blackish, elongate; blotch absent or faint; fringe grey-brown with weak fringe line. Hindwing light grey-brown, slightly darker towards termen, with one or two faint fringe lines. Abdomen light grey-brown.

VARIATION. There is much variation in the ground colour and degree of mottling and extent of reddish-brown coloration. Some specimens are almost uniformly dark grey-brown. The black dot and streak at one-third can be separate to completely fused. Occasionally there is an additional blackish mark and/or a whitish dot in cell at one-half.

SIMILAR SPECIES. The dark coloration and transversely elongate pale dot at end of cell are characteristic. The more reddish brown forms of *conterminella* resemble *A. liturosa*, but that has tegula of different colour from thorax.

MALE GENITALIA. Socii small, slightly divergent; gnathos elliptic, nearly twice as long as wide; valva slightly curved, apex rounded; cuiller slightly incurved, 67–75% of valva width, more or less swollen on outer side in distal half, apex rounded, outer margin papillose in distal part; anellus lobes narrow; anellus longer than wide, notched at apex, posterior margin thickened; aedeagus with sheath nearly reaching middle, cornuti inconspicuous and very small. Variation: the cuiller varies considerably in the shape of the distal bulge, sometimes it is no more than a gradual thickening towards a truncate apex.

FEMALE GENITALIA. Anterior apophysis slightly over half of length of posterior apophysis; sternite VIII width:length 2:1, anterior margin slightly convex, posterior margin with median notch; ostium close to anterior margin, with a furrow distally extending to posterior margin of sternite; ductus bursae straight, with a slightly wider section posterior to middle, corpus bursae ovate, signum large, wider than long, rounded laterally, with numerous teeth.

GENITALIA DIAGNOSIS. Male see *A. liturosa*. Female without reliable differences from *A. liturosa*. See also *A. impurella* and *A. arctica*.

Diagnostic differences between fresh *A. liturosa* and *A. conterminella*, in order of reliability

	A. liturosa	*A. conterminella*
Tegulae	darker than thorax	concolorous with thorax
Basal field	concolorous with ground colour nearly throughout	paler than ground colour, at least in dorsal part
Forewing coloration	reddish brown to copper-brown proportions usually distinct in fresh specimens	reddish brown to copper-brown proportions indistinct to absent even in fresh specimens
Forewing shape	rather narrow, length : width ratio (excluding fringes) usually 1 : 0.32–0.35, apex rather sharply rounded	rather broad, length : width ratio (excluding fringes) usually 1 : 0.35–0.39, apex more gently rounded

DISTRIBUTION. Northern and Central Europe south to Italy and Romania; Holarctic.

BIONOMICS. Larva in spun shoot tips of *Salix caprea* L. (Hannemann, 1953); *S. aurita* L., *S. cinerea* L., *S. viminalis* L., *S. daphnoides* Vill. and *S. repens* L. (Palm, 1989); *Salix alba* L., *S. fragilis* L. and *S. triandra* L. (Harper *et al.*, 2002); *S. purpurea* L. (Sonderegger, *in litt.*), in spring until May or early June, pupation on the soil, emerging May to July, flying until August or September, not hibernating.

REMARKS. According to Hannemann (1995) adults fly until June of the next year, but this is contradicted by the authorities mentioned above. We have not seen moths taken in early spring.

27   *Agonopterix arctica* (Strand, 1902)

Depressaria arctica Strand, 1902: 7.
   Depressaria nordlandica Strand, 1920: 67.

DESCRIPTION. Wingspan 15–19 mm. Head grey-buff. Labial palp segment 2 weakly furrowed, nearly smooth-scaled, pale grey-buff with some light brown scales, segment 3 with few brown scales. Antenna grey-buff, ringed fuscous. Thorax fuscous to dark fuscous, posteriorly with more grey-brown scales. Forewing greyish fuscous, with some scattered dark fuscous scales, mainly in costal half and towards termen, forming small spots on costa, some grey scales mainly in middle of cell extending to costa; basal field grey-buff to costa but not extending along it, outwardly broadly bordered dark fuscous; oblique dots at one-third blackish, well marked, small plical dot present, a blackish cloud adjacent to end of cell, sometimes following a white dot at end of cell, but this is commonly absent; a series of dark grey-fuscous terminal dots.

VARIATION. Ground colour varies a little with some specimens more brown and some more grey. The basal field and basal fascia are usually weakly developed. The white dot at end of cell is possibly absent more often than present.

SIMILAR SPECIES. The relatively small size, almost smooth labial palps, drab coloration, inconspicuous basal field and lack of a blotch are characteristic.

MALE GENITALIA. Socii small, slightly divergent; gnathos shortly elliptic, about 1.25 times as long as wide; valva slightly curved, costa weakly concave, ventral margin with increased curve at two-thirds, apex rounded; cuiller straight, reaching 66–75% of valva width, stout, apex rounded, distal half papillose; anellus lobes narrow; anellus widest near base, with shallow apical notch; aedeagus with sheath nearly reaching middle, cornuti scarcely visible.

FEMALE GENITALIA. Anterior apophysis three-fifths length of posterior apophysis; sternite VIII width:length 4:1; ostium almost as long as sternite; ductus bursae straight, corpus bursae ovate, signum small, with large teeth.

GENITALIA DIAGNOSIS. Male without reliable genitalia differences from *A. liturosa*. Female differs from *A. liturosa* and *A. conterminella* by ostium almost as long as sternite and by smaller signum.

DISTRIBUTION. Norway, Sweden, Finland and north Russia. Listed for China in Zhu *et al.* (2023), but the depicted moth is clearly not this species.

BIONOMICS. *Salix myrsinites* L. and *S. myrtilloides* L. (Palm, 1989). Has been reared on *Salix* sp., larva found end of June, emerging middle of July. Detailed data on phenology not available, but likely similar to *A. conterminella*.

REMARKS. Hannemann (1995) lists *Vaccinium myrtilloides* Michx. and *V. uliginosum* L. as host-plants. *V. myrtilloides* Michx. is a North American species, where *A. arctica* is not found. Benander (1964) pointed out that this was an error for *Salix myrtilloides* L. The record from *Vaccinium uliginosum* by Benander (1929) was later corrected by him to be from *Salix myrtilloides* (Benander, 1965).

ocellana subgroup (28)

A. ocellana is treated as part of Group 1 because it feeds on Salicaceae, but it does not appear to be closely related to other species in the group. Labial palp segment 2 is rough scaled ventrally, as in most other *Agonopterix*. In male genitalia, socii are large, gnathos narrow, cuiller without microsculpture, aedeagus sheath short.

28   *Agonopterix ocellana* (Fabricius, 1775)

Pyralis ocellana Fabricius, 1775: 652.
   Tinea characterella [Denis & Schiffermüller], 1775: 137.
   Tinea signella Hübner, 1797: pl. 12, fig. 80.
   Depressaria signosa Haworth, 1811: 508.

DESCRIPTION. Wingspan (19–)21–25 mm. Head buff to ochreous-buff. Labial palp pale buff, segment 2 with rough scales and a furrow on ventral side, with a few scattered brown scales on outer side, segment 3 with dark grey-brown scales at base and at two-thirds, rarely forming complete rings, extreme apex dark grey-brown. Antenna buff, narrowly ringed ochreous-brown in proximal half. Thorax buff to ochreous-buff with a few scattered brown scales. Forewing buff to ochreous-buff, pale basal field extending beneath costa, basal fascia ochreous-brown to dark brown reaching dorsal margin of cell; a sprinkling of ochreous-brown and deep brown scales, particularly on veins, costal spots dark grey; terminal spots grey; first oblique dot at one-third, second extended into a short streak and frequently joined to first, followed by a short ferruginous streak extending to end of cell where it expands and forms a circle surrounding a whitish buff dot; plical dot slightly elongate, ferruginous; blotch grey to blackish, more or less triangular, adjacent to ferruginous streak; fringe buff to orange-buff, without fringe lines. Hindwing pale buff, grey towards termen, fringe pale buff, without or with up to three faint fringe lines. Abdomen buff to orange-buff.

VARIATION. There is some variation in ground colour from pale buff to pale ochreous-buff and a few specimens show a distinct grey tinge; the first oblique dot often fused with streak representing second, creating a single short curved mark. The ferruginous markings in the cell vary in colour and may be bright red.

SIMILAR SPECIES. The series of ferruginous markings in the cell ending in a circle resemble those of *A. oinochroa*, but that is a smaller species usually with darker wings and almost without black markings. See also *A. subtakamukui* and *A. pseudoferulae*.

MALE GENITALIA. Socii well-developed, extending posteriorly; gnathos lanceolate with blunt apex, 2.5 to 3 times as long as wide, extending beyond socii; valva tapering in distal third to rounded apex, costa almost straight; cuiller straight, reaching 75% of valva width, from broad triangular base, slightly narrowed in middle, apex rounded; anellus lobes short; distal half of anellus parallel-sided, posterior margin truncate, emarginate; saccus extending well beyond base of valvae; aedeagus with short basal sheath and short process.

FEMALE GENITALIA. Anterior apophysis nearly three-quarters of length of posterior apophysis; sternite VIII width:length 4:1, anterior margin with wide, shallow median extension, posterior margin with median notch; ostium near middle of sternite, posteriorly extended to notch in sternite; ductus bursae straight, gradually expanding to ovate corpus bursae, signum large, wider than long, rounded laterally, with short anterior projection and longer posterior projection, teeth numerous.

GENITALIA DIAGNOSIS. Male well characterised by combination of short, straight, blunt cuiller, anellus with wide V-shaped incision, anellus lobes not exceeding basal half of anellus, gnathos hardly overtopping socii and stout aedeagus. *A. subtakamukui* has anellus lobes reaching top of anellus, gnathos overtopping socii by far. In female in contrast to distinct external appearance, clear differences are lacking from many other *Agonopterix* species. The externally similar *A. subtakamukui* has large triangular extension on anterior margin of sternite VIII.

DISTRIBUTION. Most of Europe, but absent from Mediterranean islands and in south-east recorded only from Greece and Romania. We have seen only a few records from Asia; Turkey (Koçak & Kemal. 2009); widespread throughout Russia (Lvovsky, 2019); Japan (Japan Moths, 2002–2023); North Africa (Hannemann. 1995).

BIONOMICS. Larva on any of several species of *Salix*, including *S. cinerea* L., *S. caprea* L., *S. aurita* L., *S. repens* L., *S. purpurea* L., *S. viminalis* L. and *S. alba* L. (Harper *et al.*, 2002); *S. myrsinites* L. (Palm, 1989); *S. eleagnos* Scop. and *S. fragilis* L. (Rymarczyk *et al.*, *in litt.*). It has been reared from *Populus tremula* L. (Lepiforum e.V. 2006–2023). Larvae from spring until June or July in spun shoots or folded leaves, emerging June to July, rarely August, flying until May of following year.

REMARKS. We follow Palm (1989) in placing *ocellana* in the *impurella* group because its host-plant is *Salix*, but it is a misfit in this group, differing in many ways. Barcode suggests it might be better placed with the Fabaceae-feeding species of Group 4.

Group 2A (29–76)
    Species feeding on Apiaceae.

fruticosella subgroup (29–30)
    Pupa with characteristic pair of bulges on head; larvae on *Bupleurum* spp.

29   *Agonopterix fruticosella* (Walsingham, 1903)

Depressaria fruticosella Walsingham, 1903: 267.

DESCRIPTION. Wingspan 20–23 mm. Head light grey-brown, scales tipped whitish buff. Labial palp segment 2 whitish buff, heavily marked dark grey-brown on outer side, segment 3 pale pink, tipped blackish, sometimes with a few mid-brown scales on outer side. Antenna dark grey-brown, narrowly ringed blackish. Thorax light grey-brown. Forewing rather broad due to strong curve at base of costa, light grey-brown sometimes tinged olive, scales tipped pale buff; small groups of blackish scales all over wing, particularly on veins running to termen; base sometimes with subdorsal spot, basal field buff, not or hardly distinct; blackish costal and terminal spots and oblique dots variably developed, often difficult to distinguish from overall blackish speckling; yellow-buff dots in cell just before and just after one-half, the inner smaller and often obsolete, the outer sometimes partially ringed grey-brown, or sometimes obsolete; fringe light grey-brown tinged pink, with fringe line. Hindwing grey, slightly darker towards termen; fringe grey with faint lines. Abdomen grey.

VARIATION. Ground colour varies slightly with some specimens darker and also some with more evident pink tinge; the abundance of dark speckles is sometimes greatest along the veins running to the termen; the median white cell spot may be absent or clearly visible.

SIMILAR SPECIES. Characterised by the almost obsolete oblique dots and the abundant black speckles over the whole forewing which is distinctly broad due to the steeper curve at base of costa. *A. rigidella* is generally smaller, with forewing narrower, more reddish brown, often without black speckling (although sometimes this is abundant), with two or three whitish cell dots and often with blackish rings on labial palp.

MALE GENITALIA. Apparent differences from *A. rigidella*: socii with an obtuse angle between nearly straight basal and outer sides; uncus very small, not nearly reaching end of socii; spinulose gnathos 1.5 times as long as wide, more rounded at apex; valva slightly narrower, costal side straighter, apex broadly pointed; cuiller similar in shape, but only about four-fifths as long as width of valva, occasionally *rigidella* has similar length of cuiller; transtilla lobes slightly longer and narrower; anellus with apical notch one-third length of anellus.

FEMALE GENITALIA. Anterior apophysis two-thirds of length of posterior apophysis; sternite VIII width:length 5:3, anterior margin with broad triangular bulge, broadly truncated at apex of triangle, with truncated margin undulate, posterior margin with median notch; ostium slightly distal to middle of sternite; ductus bursae at most 1.5 times as long as corpus bursae; corpus bursae ovate, signum Christmas tree-shaped, with small anterior projection, main part wide, with several rows of teeth, anterior margin straight, posterior margin gradually contracting into triangular posterior extension, acute, with a few detached teeth beyond apex.

GENITALIA DIAGNOSIS. Differences from male *A. rigidella*, see description of male genitalia above. In female best differences from *A. rigidella* in shape of signum and anterior margin of sternite VIII.

DISTRIBUTION. Portugal, Spain, Sardinia.

BIONOMICS. Larvae on *Bupleurum fruticosum* L., in shoot tips (Corley *et al.*, 2020). April to May, emerging May to June. In Portugal found on north-facing slopes at 550 m. It is not known in which stage this and the next species spend the winter.

REMARKS. See under *A. rigidella* (below).

30   *Agonopterix rigidella* (Chrétien, 1907)

Depressaria rigidella Chrétien, 1907a: 90.
   Depressaria rebeli M. Hering, 1936: 342.

DESCRIPTION. Wingspan 14.5–22.5 mm. Head light grey-brown with buff-tipped scales, through to brick red or darker brown, darker forms without buff-tipped scales, face pale buff. Labial palp segment 2 pale buff to pinkish buff, heavily speckled dark grey-brown and cinnamon-pink on outer side and ventrally, segment 3 pink with dark grey-brown rings at base, beyond middle and at tip. Antenna deep grey-brown. Thorax light grey-brown with buff-tipped scales to brick-red or darker brown without buff tips. Forewing narrower than in *A. fruticosella*, due to weaker curve at base of costa, light grey-brown with buff-tipped scales, to brick red or darker brown without pale tips, with a scattering of grey-brown scales, mainly in terminal one-fifth; basal field buff in paler specimens, demarcated by a dark grey-brown partial fascia reaching dorsal margin of cell, or concolorous with rest of wing in darker specimens and only indicated by dark grey-brown subdorsal spot; costa with a dark brown spot at extreme base; costal spots brown, rather weak, costal band paler than rest of wing; terminal spots blackish; oblique dots blackish, the second often with associated white scales on its outer margin; white or whitish median and end of cell dots, partly edged or surrounded dark grey-brown, the inner sometimes absent; fringe light grey-brown, often tinged pink, with two fringe lines. Hindwing grey, fringe pale grey with two or more faint lines. Abdomen light grey.

VARIATION. Portuguese specimens are smaller than those from France. The black rings on the labial palp are less strongly marked in French specimens and sometimes almost absent. Coloration of head, thorax and forewing is concolorous in any specimen, but varies from light grey-brown with all scales tipped buff to a uniform dark brick red or darker brown with scales not pale tipped. Scattered dark scales vary considerably in quantity, and appear to be most abundant in darker brown specimens from central and southern Spain and central Portugal. Specimens from France and north-east Spain are generally lighter coloured. The basal field is often more evident in paler specimens with well developed basal fascia. There may be from one to three whitish dots in cell. The terminal dots may be very faint, well marked or sometimes coalescing to form a dark line. The pink coloration present in fresh specimens quickly fades, as in other *Agonopterix* species. In the female genitalia, the signum is quite different from that of *A. fruticosella*, except in one French specimen with intermediate characters.

SIMILAR SPECIES. *A. fruticosella* (q.v.).

MALE GENITALIA. Socii with margin rounded; uncus almost reaching as far as end of socii; gnathos about twice as long as wide, broadly pointed; valva broad, costal margin slightly concave, ventral margin curved at about three-fifths, tapering to obtuse apex; cuiller slightly S-shaped, gradually tapering to pointed apex, usually almost reaching costal margin or sometimes just exceeding it; transtilla thickened in central half, transtilla lobes broad, rounded; anellus lobes shortly incurved over upper half of anellus; anellus 1.5 times as long as wide, with a waist at one-third from base, apex with a V-shaped notch about one-fifth of anellus length; aedeagus 3.5 times as long as greatest width, basal process short.

FEMALE GENITALIA. Similar to *A. fruticosella* but anterior apophysis about four-fifths length of posterior apophysis; sternite VIII width:length 5:2, anterior margin with wide triangular bulge, with a narrow notch at apex of triangle, posterior margin with wide shallow excavation; ostium slightly distal to middle of sternite; signum with main part much wider than long, nearly parallel-sided with three rows of teeth, but with a median bulge on posterior margin extending into a narrow spine-like portion of considerable length, sometimes also with small anterior projection.

GENITALIA DIAGNOSIS. See under *A. fruticosella*.

DISTRIBUTION. Portugal, Spain, France.

BIONOMICS. Larva on *Bupleurum rigidum* L. in a tube formed from three or four narrow leaves spun together; a single record of larvae on *B. falcatum* L. (Corley *et al.*, 2020). Pupation takes place in the spinning. Both fresh and worn specimens have been taken in early October.

REMARKS. *A. rigidella* was reduced to synonymy with *A. fruticosella* by Hannemann (1958b). Corley *et al.* (2020) restored *A. rigidella* to species rank, based on differences in biology and barcode and small differences in morphology. The genitalia are very similar, but there are good morphological characters separating the two species (see under *A. fruticosella*) and there are biological differences with the larvae feeding on different species of *Bupleurum*. *B. fruticosum* is a shrub with broad leaves, but *B. rigidum* is a herb with grass-like leaves. In Algarve *A. fruticosella* has been found in montane woodland with a northerly aspect, while *A. rigidella* lives in xerothermic limestone grassland, yet the significantly larger larvae of *A. fruticosella* complete their development a few weeks earlier than those of *A. rigidella*.

A. rebeli appears to be a form without rings on labial palp (Corley *et al.*, 2020).

adspersella subgroup (31–38)

Key to female genitalia of *adspersella* subgroup (31–38)

1a	Anterior margin of sternite VIII strongly bulging over its whole length; pair of folds absent, only shallow median excavation may be present	*A. leucadensis*
1b	Anterior margin of sternite VIII convex but not strongly bulging; pair of folds present	2
2a (1b)	Inner ends of folds-pair forming finger-like protrusions with an anvil-shaped excavation in between, very close (distance about 1/12–1/15 sternite width). Ductus bursae in distal part not expanding, may show longitudinal rims; central part with oblique ridges (usually indistinct); proximal part gradually expanding to diameter of corpus bursa, with longitudinal rims	*A. cadurcuella*

2b	Inner ends of folds-pair not forming finger-like protrusions, more distant	3
3a (2b)	Distance between inner ends of folds-pair about one-quarter sternite width, larger than in rest of this group, structure of ductus bursae similar to *A. cadurciella*	*A. olusatri*
3b	Distance between inner ends of folds-pair about 1/6–1/10 sternite width, structure of ductus bursae different from *A. olusatri* and *A. cadurciella*	4
4a (3b)	Distance between inner ends of folds-pair about 1/6–1/8 sternite width, ductus bursae in a short distal or subdistal part with oblique ridges (more or less distinct); the proximal remainder with longitudinal undulate rims, without transverse folds	5
4b	Distance between inner ends of folds-pair about 1/8–1/10 sternite width (sometimes even less), ductus bursae in distal 2/3 not widened, with longitudinal ridges (may include a few transverse folds), proximal 1/3 much wider, with numerous transverse folds	6
5a (4a)	Ductus bursae in distal 1/4 with distinct oblique ridges, the proximal remainder with longitudinal undulate rims, without transverse folds, not widened until it meets corpus bursae	*A. cervariella*
5b	Ductus bursae in a short distal section without distinct structures, followed by a short section with oblique ridges (not always distinct); the proximal remainder with longitudinal undulate rims, without transverse folds, gradually widened in its course	*A. paracervariella*
6a (4b)	Ductus bursae in distal 2/3 not widened, with longitudinal ridges but without transverse folds, proximal 1/3 much wider, with numerous distinct transverse folds. Distance between folds-pair in anterior margin of sternite VIII about 1/10 sternite width or slightly less, these folds moderately sclerotised, often with shallow excavation in between (all features show overlap with *A. thapsiella* / *A. chironiella*!)	*A. adspersella*

6b   Ductus bursae in distal 2/3 not widened, with   *A. thapsiella,*
       longitudinal ridges and sometimes also with a few   *A. chironiella*
       transverse folds, wider proximal section tends to be
       shorter, less wide and with less distinct transverse
       folds than in *A. adspersella*. The folds-pair in anterior
       margin of sternite VIII with a tendency to be more
       strongly sclerotised and closer together than in *A.
       adspersella*, between these folds often a proximal
       projection (all features show overlap with *A.
       adspersella*!)

31   *Agonopterix olusatri* Buchner & Corley, 2019

Agonopterix olusatri Buchner & Corley, 2019: 2
    Depressaria chironiella auct. nec Constant, 1893

DESCRIPTION. Wingspan 22.5–26.5 mm. Head fuscous, scales tipped buff, face buff. Labial palp pinkish buff, segment 2 speckled fuscous on outer side and ventrally, segment 3 with a partial dark fuscous ring at base, a complete ring at two-thirds and tip. Antenna fuscous. Thorax light grey-brown with slight pinkish tinge. Forewing light brownish buff with faint pink tinge, with a scattering of dark fuscous scales; subdorsal spot dark fuscous, often faint or absent; basal field buff, tinged pink; basal fascia broad, dark fuscous reaching dorsal margin of cell; costa with deep brown dot at base; costal spots blackish, extending along whole length of costa; terminal spots blackish; oblique dots blackish, another at end of cell, occasionally associated with one or two whitish scales; blotch dark grey, indistinct; plical dot blackish; fringe light brownish buff, with two fringe lines. Hindwing light grey-buff, greyer towards termen, with a few dark grey spots between veins; fringe pale grey-buff with two or more fringe lines. Abdomen grey.

VARIATION. Ground colour sometimes buff with pink tinge, to mid-brownish buff with weak pink tinge; there is considerable variation in the amount of fuscous overlay; the blotch is weak at best, but can be very faint. There is a form in Andalusia and Gibraltar, where the host-plant is *Ferula tingitana*, with cream ground colour and black scales scattered all over wings.

SIMILAR SPECIES. *A. adspersella* and *A. thapsiella* have slightly broader wings, a whitish dot usually present at end of cell, blotch usually absent or very weak and ground colour less heavily overlaid with fuscous; but some specimens of *A. thapsiella* and especially *A. chironiella* have a distinct blotch and ground colour overlaid with fuscous, furthermore the whitish dot can be present in *A. olusatri* and absent in *A. chironiella*, so for safe determination dissection may be necessary.

MALE GENITALIA. Socii large, ovate, narrower posterior one-third free from uncus, not reaching end of narrowly ovate (twice as long as wide) gnathos; end of valva variable in width; cuiller variable, 80–100% width of valva, straight or curved outwards or slightly curved inwards, apex rounded; anellus lobes covering lateral margins of anellus and projecting beyond it; anellus with distal part nearly orbicular with narrow V-shaped sinus to about one-third; aedeagus with a wide thin-walled area extending in a U-shape almost down to basal sheath; minute cornuti in a dense mass.

FEMALE GENITALIA. Anterior apophysis just over half length of posterior apophysis; sternite VIII width:length 5:2, anterior margin bulging, rounded, a section in middle with lateral folds, posterior margin with broad median notch; ostium near middle of sternite, a broad channel posterior to ostium extending to marginal notch of sternite; ductus bursae gradually expanding to ovate corpus bursae; signum with narrow lateral wings with two rows of teeth, short median anterior and posterior projections.

GENITALIA DIAGNOSIS. In male extent of thin-walled area of aedeagus separates this species from all other *adspersella* subgroup species. In female best differentiating feature from all *A. adspersella* subgroup species is distance between lateral folds in anterior margin of sternite VIII: space between inner edges of folds about one-quarter width of sternite VIII (which is about equalling or even exceeding width of ostium), while in the compared species, this distance is about one-tenth width of sternite VIII (which is much less than width of ostium).

DISTRIBUTION. Canary Islands (Lanzarote), Portugal, Spain, France, Sardinia, Sicily, Malta, Greece (Crete and Samos), Cyprus. Also reported from Morocco and Israel.

BIONOMICS. Larvae on *Ferula communis* L., *F. tingitana* L., *Foeniculum vulgare* Mill. and *Smyrnium olusatrum* L. (Buchner & Corley, 2019). Larvae found in March in Portugal emerged end of April to middle of May; in Morocco one larva was found end of January. Collected adults have been found second half of April to June and September to January. Most likely adults aestivate and eggs are laid in autumn or winter.

REMARKS. Constant (1893) described *Depressaria chironiella* from France, with a series of moths reared from larvae on *Opopanax chironium* (L.) W.D.J. Koch. Subsequently the name became attached to a species found in south-west Europe. However, Constant's *chironiella* has a distinct white dot at end of cell and is close to *A. thapsiella*. Likewise his description of the larva also agrees with *thapsiella* and not with *olusatri*.The species that had been recognised as *A. chironiella* in south-west Europe is different from Constant's *A. chironiella* and has now been described as *A. olusatri*. The form of *olusatri* in southern Spain and Gibraltar with cream ground colour and heavy blackish speckling looks different, but has no difference in barcode from typical *olusatri*.

32   *Agonopterix leucadensis* (Rebel, 1932)

Depressaria leucadensis Rebel, 1932: 55.

DESCRIPTION. Wingspan 18–23 mm. Head light brownish buff. Labial palp grey-buff, segment 2 with some light brown to fuscous scales on outer and ventral sides, segment 3 with a dark grey, usually incomplete ring at one-half, extreme apex black. Antenna grey-buff. Thorax and tegula light brownish buff. Forewing, buff with faint brown tinge to grey-buff, with scattered blackish scales, more numerous in posterior part of wing, there often weakly organised into irregular transverse rows; basal field creamy buff, weakly to strongly demarcated; costa with blackish spots, sometimes best developed in basal half of wing, blackish terminal dots; a pair of black oblique dots of variable size, sometimes one absent, an inconspicuous plical dot, and a larger dot at end of cell; fringe buff. Hindwing very light grey, slightly darker posteriorly; fringe whitish grey with weak line. Abdomen buff to light brown.

VARIATION. The costal and terminal spots may be very well developed, large and intensely black, forming a continuous series from base of costa round the termen. In such specimens the scattered blackish scales on the forewing are also more conspicuous. The development of the oblique dots is variable, from faint to conspicuous, sometimes one is absent.

SIMILAR SPECIES. Similar to *A. adspersella*, but spot at end of cell without white centre. Externally cannot be safely distinguished from *A. adspersella* subsp. *pavida*, but with very different genitalia in both sexes. *A. olusatri* has longer wings, oblique dots and dot at end of cell approximately similar in size and normally has blotch.

MALE GENITALIA. Similar to *A. adspersella*, but cuiller slender, distal two-fifths sickle-shaped with narrowly pointed apex, just exceeding costa of valva; anellus divided nearly to base; aedeagus abruptly tapered in apical quarter, apex less slender than in *adspersella*.

FEMALE GENITALIA. Anterior apophysis half length of posterior apophysis; sternite VIII width:length 2:1, anterior margin strongly bulging, almost semicircular, with a shallow median excavation, posterior margin with small notches either side of median projection from margin of posteriorly placed ostium; ductus bursae gradually expanding to middle, then of even width to ovate corpus bursae; signum wider than long, lateral wings tapering, with few teeth, short median anterior and posterior projections, a few smaller teeth in central area.

GENITALIA DIAGNOSIS. In male details of cuiller separate this species from other *adspersella*-group species. Female differs from other *adspersella* subgroup species by absence of lateral folds in anterior margin of sternite VIII.

DISTRIBUTION. Greece including Crete.

BIONOMICS. Reared in May from *Scandix pecten-veneris* L. (Crete, Sonderegger, *in litt.*, one specimen); two reared specimens without host-plant information (G. Baisch). Collected adults have been found end of April to beginning of July and October to November. Obviously adults aestivate and eggs are laid in autumn or winter.

33   *Agonopterix adspersella* (Kollar, 1832)

Haemylis adspersella Kollar, 1832: 92.
  Depressaria feruliphila Millière, 1866: 209.
  Depressaria amanthicella Heinemann, 1870: 157.
  Depressaria reichlini Heinemann, 1870: 173.
  Depressaria crassiventrella Rebel, 1891: 627.
  Depressaria sabulatella Turati, 1921: 334.
  Depressaria rubripunctella Amsel, 1935a: 295.
  Depressaria karmeliella Amsel, 1935b: 127.

DESCRIPTION. Wingspan 18–26.5 mm. Head light grey-brown, scales buff-tipped, face buff. Labial palp buff, segment 2 outer side with abundant fuscous scales, ventral side sometimes pinkish, segment 3 slightly pink-tinged, with a few fuscous scales at base, a dark fuscous ring beyond middle and blackish at tip. Antenna dark fuscous. Thorax light grey-brown, scales buff-tipped. Forewing uniformly light grey-brown to light brown with scattered fuscous scales; basal field and base of costa, sometimes a broad costal band as far as two-fifths grey-buff to light brownish buff; basal fascia dark fuscous extending to dorsal margin of cell, slightly tinged fuscous beyond fascia; costa with deep brown spot at base, costal spots blackish, most developed in proximal half of costa; terminal spots blackish; oblique dots at one-third, the first blackish, the second smaller, black, deep brown or rusty brown, a white dot at end of cell, outwardly or completely surrounded by blackish, deep brown or rusty brown scales; plical dot black, deep brown or rusty brown; blotch between cell and costa faint grey, or more often obsolete; fringe light fuscous, sometimes tinged pink, without line. Hindwing light grey, darker towards termen; fringe light greyish with faint line. Abdomen light grey.

VARIATION. Forewing colour quite variable, typically light grey-brown to light brown; scattered fuscous scales vary from rather few to quite numerous; the white dot at end of cell is outwardly dark edged or completely ringed, occasionally there are no white scales in that position; the dark oblique dots, plical dot and that at end of cell are sometimes partly or entirely rusty brown; sometimes the plical dot and terminal dots are absent.

SIMILAR SPECIES. *A. adspersella* is characterised by uniform forewing ground colour with relatively few scattered fuscous scales, giving the forewing a clean appearance; median cell dot and blotch between cell and costa are absent. See *A. thapsiella* and *A, chironiella* for further discussion under those species. Other similar species are *A. olusatri*, *A. cadurciella*, *A. paracervariella* and *A. leucadensis*. Subspecies *pavida* (described below) is more likely to be confused with *A. leucadensis*.

MALE GENITALIA. Socii large, ovate with broad, rounded apex, attached to uncus to about three-quarters of their length; gnathos overtopping socii, 2 to 2.5 times as long as wide; valva with costa concave, ventral margin strongly curved at two-thirds, distal third tapered to narrow rounded apex; cuiller stout, straight or with slight curve in basal half, reaching costal margin of valva or shortly crossing it, apex rounded, sometimes slightly expanded; anellus lobes large, reaching close to posterior end of anellus or even exceeding it; anellus with short stalk then as wide as long, apex broadly rounded with shallow emargination; aedeagus stout, thin-walled area not reaching basal sheath, very numerous small cornuti in conspicuous group.

FEMALE GENITALIA. Anterior apophysis three-fifths length of posterior apophysis; sternite VIII width:length 7:3, anterior margin bulging, with or without a shallow median excavation, a pair of small outwardly curved folds near middle, posterior margin with small median projection from margin of posteriorly placed ostium; anterior margin of tergite concave; ductus bursae slightly expanding, near corpus bursae abruptly expanded into short section with transverse folding; corpus bursae elliptical; signum wider than long, rhombic or with lateral wings tapering, short median anterior and posterior projections, marginal teeth not larger than teeth in central area.

GENITALIA DIAGNOSIS. Male similar to *A. thapsiella* and *A. chironiella*, not always safely distinguishable. Female very similar to and not safely distinguishable from *A. thapsiella* and *A. chironiella* (these two inseparable), but with tendencies: *A. adspersella*: the two lateral folds in anterior margin of sternite VIII moderately sclerotised, area in between not distinctly projecting, ductus bursae without transverse folds in caudad half, transverse folds in craniad half distinct. *A. thapsiella, A. chironiella*: the two lateral folds in anterior margin of sternite VIII distinctly sclerotised, area in between distinctly projecting, ductus bursae sometimes with transverse folds in caudad half, transverse folds in craniad half less distinct.

DISTRIBUTION. Southern Europe from Spain to Greece and Bulgaria, most of the larger Mediterranean islands and southern parts of central Europe north to Czechia.

BIONOMICS. Larvae on *Bupleurum falcatum* L., *Athamanta cretensis* L. and *Meum athamanticum* Jacq. (Hannemann, 1953); *Bupleurum petraeum* L. (Hannemann, 1995); *Pastinaca sativa* L., *Pimpinella saxifraga* L., *Seseli libanotis* (L.) Koch., *Bupleurum stellatum* L., *Scandix pecten-veneris* L. and *Trinia glauca* (L.) Dumort. (Sonderegger, *in litt.*); *Seseli gummiferum* Sm. (Savchuk & Kajgorodova, 2017); *Daucus carota* L., *Orlaya daucoides* (L.) Greut., *Foeniculum vulgare* Mill., *Prangos trifida* (Mill.) Herrnst. & Heyn, *Scandix australis* L. and *Seseli tortuosum* L. (Rymarczyk *et al., in litt.*); *Bupleurum frutescens* L. (van Nieukerken, reared specimen); *Laserpitium siler* L. (Buchner, personal observation). Collected adults have been found from May to November. Obviously adults do not hibernate and eggs are laid in autumn. Specimens collected in Mediterranean area and collected March, April and December may refer to *A. thapsiella*.

LARVAL DESCRIPTION. Head black, prothoracic plate and usually whole of first thoracic segment black, body green with three darker green lines, pinacula more or less green, anal plate green, thoracic legs black, or second and third pair partly black, prolegs green.

REMARKS. *Agonopterix adspersella*, mainly found in the Alps and *A. thapsiella* found in Mediterranean areas, have long been treated as separate species. The first occurs at altitudes up to 2400 m and has different host-plants from the second species which is found in southern Europe, mainly at low altitude. Furthermore there appear to be some differences in the larvae. However, there are no reliable external differences between the adult moths, no differences in male or female genitalia and no significant difference in barcode. Separation of two species based on geographical location, host-plant and small differences in larvae is hardly acceptable when not backed by morphological differences. The situation is further complicated by the existence of *A. chironiella* and some other populations within the complex which show very weak morphological differences, barcode clusters with very small p-distance and at least sometimes differences in larval coloration and specialised host-plants.

Our initial proposal was to treat the *adspersella*-complex as a single species, which had the advantage of avoiding taxonomic complications. We found this to be an unsatisfactory treatment. The specialisation of some populations on single species or genera of host-plant, together with larval differences has brought us round to the conclusion that this is a very actively evolving group. So with a gradually changing situation, it is impossible for us to construct a complete picture of relationships within the complex or to define clearly recognisable species, but we can try to do so. Here we recognise three species, *adspersella*, *thapsiella* and *chironiella*. We could add further species to this, but at present we have insufficient knowledge of these, so we prefer not to do so.

Without knowledge of larva and host-plant it may be impossible to identify some specimens with certainty. In such cases we recommend referring them to *A. adspersella/thapsiella*.

Rymarczyk *et al.* (2013) use the name *adspersella* for the species we treat here as *A. cadurciella*. For a detailed argument explaining our decision see Buchner & Šumpich (2020). Rymarczyk *et al.* (2013) use larval characters to separate *thapsiella* and *feruliphila*, which is the name they use for our *adspersella*.

Specimens reared from *Athamanta cretensis* tend to be particularly small.

Depressaria crassiventrella was described from a single female from Croatia. Genitalia show no difference from typical *adspersella*.

Hannemann (1958a) synonymised *rubripunctella* Amsel, 1935a, *karmeliella* Amsel, 1935b and *amanthicella* Heinemann, 1870 with *adspersella*.

Agonopterix adspersella subspecies *pavida* (Meyrick, 1913) stat. rev.

Depressaria pavida Meyrick, 1913: 114.

Type locality: Turkey, Taurus Mts.

DESCRIPTION. Similar to the typical subspecies, but series of about six blackish spots along whole length of costa of equal size; end of cell dot without white centre or with faint pale dot of ground colour.

MALE GENITALIA. Differs from subspecies *adspersella* by mass of cornuti which occupies half to nearly two thirds length of aedeagus compared to only one-third in typical subspecies.

FEMALE GENITALIA. Indistinguishable from subspecies *adspersella*.

MATERIAL EXAMINED. Holotype: ♀, Turkey, Taurus Mts, 1906, NHMUK010293014, gen. prep. 9463 C.F.G. Clarke, ex coll. Meyrick, coll. NHMUK

Paratypes (arranged according to collection date):

1 ♂, Turkey, Diyarbakır Province, without date, gen. prep. DEEUR 5742 P. Buchner, ex coll. Staudinger, coll. ZMHB.

1 ♂, Turkey, Gaziantep Province, Kadirli, 700 m, 10.vii.1987, gen. prep. DEEUR 2716 P. Buchner, DNA barcode id. TLMF Lep 21993, M. Fibiger leg., coll. RCKL

1 ♂, 1 ♀, Turkey, Mersin Province, 10 km southeast Arslanköy, 1300 m, 11.vii.1987, DEEUR 2717 (♂) & gen. prep. DEEUR 3120 P. Buchner, DNA barcode id. TLMF Lep 21950 (♀), M. Fibiger leg., coll. RCKL.

1 ♂, Greece, Evro, Alexandropolis, Kirki, 400 m, 25.x.1987, gen. prep. DEEUR 5474 P. Buchner, DNA barcode id. TLMF Lep 26110, Moberg & Hillman leg., coll. ZMUC.

1 ♂, Turkey, Antalya Province, Imrasan geçidi north of Akseki, 1525 m, 11.–14.vii.1996, gen. prep. DEEUR 5389, M. Schepler leg., coll. ZMUC.

REMARKS. The external similarity with *A. leucadensis*, is supported by barcode, however male and female (one specimen only) genitalia quite different from *A. leucadensis* and with only minor differences from *A. adspersella*.

34   *Agonopterix thapsiella* (Zeller, 1847)

Depressaria thapsiella Zeller, 1847: 838.
   Depressaria linolotella Chrétien, 1929: 193.

DESCRIPTION. Externally the appearance of the moths falls within the variation of *A. adspersella*, except that some specimens have a weak dark blotch between middle of cell and costa, the plical dot is often absent and specimens from some populations have patches of light grey suffusion overlying the ground colour.

VARIATION. Similar to that of *A. adspersella*, sometimes with the small differences given above under description. See also the Portuguese population described under Remarks.

SIMILAR SPECIES. See under *A. adspersella*.

MALE GENITALIA. Similar to *A. adspersella*, but valva sometimes wider in distal half and cuiller more variable, sometimes not reaching costa of valva and often curved inwards in proximal half; aedeagus more slender.

FEMALE GENITALIA. Similar to *A. adspersella* but with some small differences. Anterior apophysis two-fifths length of posterior apophysis; sternite VIII width:length 3:1, anterior margin bulging, with a shallow median excavation enclosing a pair of small outwardly curved folds, which are closer together than in *adspersella*; ductus bursae slightly expanding from middle, without abrupt expansion before corpus bursae; signum wider than long with lateral wings tapering, short median anterior and posterior projections, with marginal teeth and a few teeth in central area including one considerably enlarged pair.

GENITALIA DIAGNOSIS. Male not reliably separable from *A. adspersella*. Female inseparable from *A. chironiella* and not reliably separable from *A. adspersella* (*q.v.*).

DISTRIBUTION. Portugal, Western Mediterranean countries, Sardinia, Sicily, Slovenia, Greece including Crete, Bulgaria.

BIONOMICS. Larva on *Thapsia garganica* L. (Hannemann, 1953); *T. villosa* L. (Carvalho & Corley, 1995). Adults usually emerge in April to June and can sometimes be found during the winter months, one moth collected in March.

LARVAL DESCRIPTION. Head black, prothoracic plate black, yellow on anterior edge and sometimes with a yellowish median longitudinal line, remainder of first thoracic segment yellow-green, body green, lines faint or absent, pinacula dark, anal plate green, thoracic legs green or the first pair black or partly black, prolegs green.

REMARKS. In north Portugal there is a population which we tentatively refer to *thapsiella*, with larvae on *Pimpinella villosa* Schousb. Externally the moths resemble *thapsiella*, but forewing colour is a darker brown. Larvae differ in head dark brown, prothoracic plate without yellowish anterior margin, body with three darker green lines, pinacula concolorous with body.

Separation of *A. thapsiella* and *A. chironiella*: As barcode and genitalia are not useful here, separation relies on small differences in external appearance (of species known to be variable), associated with different host-plant. Larval morphology also shows differences, but range of larval variability insufficiently known. Additionally the value of the last character is open to question since differences in larval appearance within a single species are known to occur in some other Depressariidae. Nevertheless we prefer to keep these as separate species provisionally. We suspect that these are species in the making, but reproductive isolation is not yet complete. More evidence is needed before a decision is made on the status of these taxa.

35   *Agonopterix chironiella* (Constant, 1893)

Depressaria chironiella Constant, 1893: 392

DESCRIPTION. Wingspan 20–27 mm. Externally the appearance of the moths is similar to *A. thapsiella*, but differing slightly in ground colour light brown, with slightly darker brown patches between cell and costa from mid-wing and around end of cell and in terminal area; basal field hardly differing in colour from rest of wing; oblique and cell dots never rusty brown; plical dot usually absent; end of cell dot nearly always dark-ringed; a distinct round dark grey blotch between middle of cell and costa.

VARIATION. The amount of brown suffusion around end of cell and in terminal area is variable and may be nearly absent; the round blotch at mid-wing beneath costa is variable in strength.

SIMILAR SPECIES. *A. thapsiella* and *A. adspersella* but these rarely have darker brown patches on forewing or significant development of grey blotch.

MALE GENITALIA. Similar to *A. adspersella*, but gnathos shorter, twice as long as wide, only just exceeding socii; valva slightly shorter and broader, cuiller slightly sinuate, reaching costal margin of valva, rounded at apex; anellus divided to middle with narrow sinus.

FEMALE GENITALIA. Similar to *A. adspersella* but with some small differences. Anterior apophysis two-fifths length of posterior apophysis; sternite VIII width:length 3:1, anterior margin bulging, with a shallow median excavation enclosing a pair of small outwardly curved folds, which are closer together than in *adspersella*; ductus bursae slightly expanding from middle, without abrupt expansion before corpus bursae; signum wider than long with lateral wings tapering, short median anterior and posterior projections, with marginal teeth and a few teeth in central area including one considerably enlarged pair.

GENITALIA DIAGNOSIS. Male compared with *A. adspersella* in description above. Female inseparable from *A. thapsiella* and not reliably separable from *A. adspersella* (*q.v.*).

DISTRIBUTION. France, Italy, Croatia, Greece.

BIONOMICS. Larvae have only been recorded from *Opopanax chironium* (L.) W.D.J. Koch (Constant, 1893). Reared specimens from Gargano region (Central Italy) emerged early May. Phenology likely similar to *A. thapsiella*.

LARVAL DESCRIPTION. Head black, prothoracic plate black, pale on anterior edge and with a pale median longitudinal line, remainder of first thoracic segment yellow-green, body green, without lines, pinacula black, anal plate green, thoracic legs yellowish, prolegs green.

REMARKS. Constant's (1893) description of adult and larva of *Depressaria chironiella* shows considerable similarity with *A. thapsiella*. Recently reared moths from

Opopanax chironium, the host-plant of Constant's *chironiella* have the same bar-code as *thapsiella*.

The status of *A. chironiella* and *A. thapsiella* is discussed in more detail under Remarks for *A. thapsiella* (above).

The name *chironiella* was for many years erroneously used to refer to an unde-scribed species mainly occurring in south-west Europe. That species has now been described as *A. olusatri*. Literature references to *chironiella* may well refer to *olusatri*.

36   *Agonopterix cervariella* (Constant, 1884)

Depressaria cervariella Constant, 1884: 251.

DESCRIPTION. Wingspan 21–26 mm. Head grey-brown with cinnamon tinge. Labial palp creamy buff, segment 3 buff with pink tinge, extreme tip blackish. Antenna blackish in basal part, brown beyond middle. Forelegs black above. Thorax and tegula grey-brown with cinnamon tinge. Forewing light grey-brown, overlaid with cinnamon beyond basal field in dorsal half and on costal and dorsal sides of end of cell, most of the wing surface with thinly scattered blackish scales; basal field buff, extending onto base of costa; a black dot at base of costa, costal margin pinkish; a series of blackish costal spots, mainly in proximal half, terminal dots weak, grey; oblique dots black, end of cell with a white dot ringed light brown, usually with a few black scales on outer margin; plical dot absent; fringe light brown with pinkish tinge. Hindwing light grey, darker posteriorly; fringe light grey. Abdomen grey buff.

VARIATION. *A. cervariella* shows little variation, but the pinkish cinnamon color-ation which can pervade the whole upper surface, soon fades.

SIMILAR SPECIES. The pinkish cinnamon coloration of head, thorax, forewings and segment 3 of labial palp (with extreme tip blackish) is characteristic.

MALE GENITALIA. Similar to *A. adspersella*, but socii with posterior one-third to one-quarter free from uncus and slightly convergent; gnathos twice as long as wide, reaching end of socii or projecting a little beyond; cuiller tapering slightly beyond middle, sometimes bent in middle, apex rounded, reaching 80–95% width of valva; anellus plate divided to one-quarter or one-half; aedeagus slender, angled in mid-dle, process of basal sheath long, reaching middle of aedeagus.

FEMALE GENITALIA. Anterior apophysis half length of posterior apophysis; sternite VIII width:length about 5:2, anterior margin bulging, rounded, with a shal-low median excavation between a pair of well-separated small outwardly curved folds, posterior margin with median excavation; anterior margin of tergite concave; ostium slightly distal to middle of sternite; ductus bursae with some oblique ridges in distal one-quarter; corpus bursae elliptical, signum wider than long, approximately rhombic or triangular, but with all angles more or less rounded, teeth numerous.

GENITALIA DIAGNOSIS. In male details of aedeagus (length of basal process, angled in middle) are distinct, but most similar to *A. paracervariella* (*q.v.*). For female see key p. 000).

DISTRIBUTION. France, Austria, also recorded from Italy, Switzerland, Germany, Croatia, Bosnia-Herzegovina, but probably extinct in several of these countries. The only post-1950 record we have seen is from Austria. Turkey, Konya province (Koçak & Kemal, 2009).

BIONOMICS. Larva on *Peucedanum cervaria* (L.) Lapeyr. (Hannemann, 1953). Specimens reared in Austria emerged in June and July. We have not seen specimens collected earlier or later.

REMARKS. *A. cervariella* shares barcode with the next species, but they are very different in habitus and female genitalia. There are also differences in male genitalia but these are less obvious. In addition the two species have different host-plants.

37   *Agonopterix paracervariella* sp. nov.

Type locality: Switzerland, Ticino.

DESCRIPTION. Wingspan 20–27 mm. Head white to whitish grey. Labial palp whitish, segment 3 creamy white, a few fuscous scales ventrally at middle, extreme tip blackish. Antenna fuscous. Thorax and tegula whitish to whitish grey. Forewing whitish to whitish grey, largely suffused pale grey-buff, scattered fuscous scales mainly in terminal part of wing, sometimes more extensive and organised into lines between veins; basal field white to whitish grey, without dark posterior edge, extending onto costa; a black dot at base of costa, costal spots grey, small, often faint grey dots round apex sometimes becoming short dashes along termen; oblique dots and plical dot black, white spot at end of cell small, edged blackish outwardly, plical dot sometimes present; fringe creamy white to pale grey. Hindwing whitish grey, darker posteriorly; fringe whitish grey. Abdomen whitish grey to whitish buff.

VARIATION. Ground colour can be white or pale grey-white; the overall coloration depends on the depth of colour of the grey-buff suffusion and the extent of the scattered fuscous scales. The outer oblique dot may be obsolete.

SIMILAR SPECIES. The near white coloration is unusual in *Agonopterix*. *A. cadurciella* has black ring on labial palpus, forewing usually has basal field with dark posterior edge.

MALE GENITALIA. Similar to *A. cervariella*, but socii widely separated; gnathos just over half as wide as long, slightly exceeding socii; cuiller straight or incurved to weakly sinuate, narrower in distal half, slightly expanded at apex, reaching 95% width of valva; anellus plate round, with V-shaped division to beyond middle; aedeagus strongly curved at three-fifths, process of basal sheath long, reaching beyond middle of aedeagus.

FEMALE GENITALIA. Anterior apophysis three-fifths length of posterior apophysis; sternite VIII width:length about 5:2, anterior margin bulging, somewhat triangular, at anterior angle a short truncate extension between a pair of oblique folds, posterior margin with wide, shallow median excavation; anterior margin of tergite concave; ostium distal to middle of sternite; ductus bursae with some oblique ridges in distal one-quarter; corpus bursae elliptical, signum wider than long, triangular, but with all angles more or less rounded, a weak anterior projection, teeth mainly marginal.

GENITALIA DIAGNOSIS. Best differences in male from *A. cervariella* visible in outline of uncompressed aedeagus in lateral view: broader, with distinct expansion just distal to angle. For differences in female from other *adspersella*-group species see key.

DISTRIBUTION. Switzerland, Italy and Slovenia.

BIONOMICS. Larva on *Athamanta cretensis* L. (Sonderegger, *in litt.*) and *Laserpitium peucedanoides* L. (Klimesch). We have only seen specimens collected in July and September.

MATERIAL EXAMINED. Holotype: ♂, Switzerland, Ticino, Sonvico, Denti della Vecchia, 1400 m, reared from *Athamanta cretensis*, e.l. 15.vi.2011, gen. prep. DEEUR 0601 P. Buchner, DNA barcode id. TLMF Lep 06566, P. Sonderegger leg., will be deposited in coll. NMBE.

Paratypes (arranged according to collection date):

1 ♂, 1 ♀, Italy, Alpi Giulle, Altiplano Montasio, reared from *Laserpitium peucedanoides*, 18.vii.1950, gen. prep. DEEUR 0894 (♂), 5152 (♀) P. Buchner, J. Klimesch leg., coll. ZSM

2 ♀, Switzerland, Ticino, Sonvico, Denti della Vecchia, 1400 m, reared from *Athamanta cretensis*, e.l. 26.vii.2011, gen. prep. DEEUR 7702, 7709 P. Buchner, DNA barcode id. TLMF Lep 26219, 26225, P. Sonderegger leg., museum id. GBIFCH00723204, GBIFCH00723211, will be deposited in coll. NMBE

1 ♂, Slovenia, Kamniske Alpe, Jezersko, 1000 m, reared from *Athamanta cretensis*, 18.ix.2012, gen. prep. DEEUR 6358 P. Buchner, J. Skyva, coll. NMPC

ETYMOLOGY. The species name refers to the close relationship with *A. cervariella*.

REMARKS. See under *A. cervariella* (above).

38   *Agonopterix cadurciella* (Chrétien, 1914)

Depressaria cadurciella Chrétien, 1914: 159.

Agonopterix adspersella sensu Rymarczyk *et al.*, 2013, nec Kollar, 1832

DESCRIPTION. Wingspan 20–27mm. Head creamy white to buff. Labial palp creamy white to pale buff, segment 3 with a black ring in middle. Antenna dark fuscous. Thorax creamy white to buff or light brown. Forewing whitish grey, overlaid with patches of light brown, especially outside basal field, around spot at end of cell and around three-quarters wing length, scattered blackish scales over much of wing, occasionally following veins in terminal area; basal field concolorous with thorax, extending onto base of costa, sometimes edged dark fuscous, often by two dots; a black dot at base of costa, a series of black or dark grey spots on costa, smaller beyond middle, a series of subterminal dots; oblique and plical dots black, dot at end of cell white, edged blackish except on proximal margin; fringe grey, often tinged light brown. Hindwing light grey, darker towards apex; fringe whitish grey. Abdomen grey-buff.

VARIATION. The extent to which the ground colour of forewing is overlaid with pale brown varies as does the amount of blackish speckling. The basal field may be sharply demarcated by a dark fuscous line or by two dots, or simply by a change to the pale brown colour.

SIMILAR SPECIES. *A. paracervariella* is usually whiter, lacks areas of pale brown on the forewing, has basal field less well demarcated and weaker costal spots, labial palpus without black ring. Forms with light brown patches confluent and covering most of whitish grey ground colour (e.g. fig. 38a) resemble *A. adspersella* and are indistinguishable, if whitish ground colour is covered completely (fig. 38d).

MALE GENITALIA. Similar to *adspersella*, gnathos about twice as long as wide, extending well beyond socii; valva with distal third of ventral margin slightly concave, end of valva quite narrow, rounded, cuiller straight or sinuate, gradually tapering to blunt apex, reaching 110–115% of valva width; anellus plate with incision to at least one-half.

FEMALE GENITALIA. Anterior apophysis half to two-thirds length of posterior apophysis; sternite VIII width:length about 5:2, anterior margin bulging, somewhat triangular, anteriorly truncate with an excavation as wide as ductus bursae between a pair of small folds, ostium slightly distal to middle of sternite, open distally to margin of sternite; ductus bursae narrow posteriorly, gradually expanding to ovate corpus bursae; signum wider than long, posterior and anterior projections absent or variably developed, teeth mainly marginal, but present in varying number in central part.

GENITALIA DIAGNOSIS. Best difference in male from other *adspersella* subgroup species is shape of cuiller: tapering to very slender apex, exceeding costa of valva, usually slightly sinuate. For differences in female from other *adspersella* subgroup species see key.

DISTRIBUTION. Spain, France, Italy, Hungary, Croatia, Montenegro, Albania and North Macedonia; Sicily. Also found in Turkey (Asian part).

BIONOMICS. Larva has been reared from *Seseli pallasii* Besser, *Seseli leucospermum* Waldst. & Kit., *Seseli montanum* L., *Seseli tortuosum* L. (Rymarczyk *et al.*, *in litt.*, listed as *A. adspersella*), *Peucedanum officinale* L. (R. Heckford and S. Beavan, pers. comm.). The adult specimens we have seen were collected from early June to September.

REMARKS. According to Chrétien (1914) larvae were found on a *Hyoseris* species (Asteraceae) in June. All other species of the *adspersella* group with known host-plants feed on Apiaceae. It seems probable that Chrétien made a mistake. See also remarks under *A. adspersella*. Hannemann's (1958a) drawing of the male genitalia of the holotype show a different cuiller from other specimens. This may represent natural variation.

rotundella subgroup (39–40)

39   *Agonopterix nodiflorella* (Millière, 1866)

Depressaria nodiflorella Millière, 1866: 214.
   Agonopterix inoxiella Hannemann, 1959: 38, syn. n.

DESCRIPTION. Wingspan 13–20 mm. Head ochreous-buff, face creamy buff. Labial palp pale ochreous-buff with a few light brown scales on outer side of segment 2. Antenna dark fuscous. Thorax ochreous-buff, without posterior crests. Forewing buff to ochreous-buff with a slight scattering of light brown to fuscous scales, sometimes forming lines on veins particularly in terminal area; a blackish spot at base of costa sometimes present; a blackish subdorsal spot; terminal spots blackish; oblique dots and dot at end of cell blackish; fringe ochreous-buff, without line. Hindwing light grey, darker towards termen; fringe grey-buff with fringe line. Abdomen grey-buff.

VARIATION. Some specimens have a weak streak of darker scales from subdorsal spot through cell. Specimens from Canary Islands often more heavily streaked with brown scales and sometimes with additional dot in middle of cell and plical dot.

SIMILAR SPECIES. Hindwings are usually narrower and more pointed than is generally the case in *Agonopterix*. *A. rotundella* is similar in coloration and markings, but lacks second oblique dot; specimens with distinct streaks of brown scales are not found in *A. rotundella*.

MALE GENITALIA. Socii ovate, obtuse, directed laterally, together wider than tegumen; gnathos about twice as long as wide, extending beyond socii; valva slightly expanded in middle due to convex costa, cuiller stout, straight or nearly so, apex rounded, sometimes slightly expanded, reaching 75–80% width of valva; anellus lobes rather narrow, not covering much of anellus; anellus plate inverse bell-shaped, the distal angles curving out and slightly sclerotised, posterior marhin with slight depression; aedeagus gradually tapering, slightly curved, cornuti inconspicuous; process of basal sheath as long as sheath or shorter.

FEMALE GENITALIA. Anterior apophysis half to two-thirds length of posterior apophysis; sternite VIII width:length about 3:1, anterior margin with low triangular bulge, posterior margin concave; ostium in or slightly distal to middle of sternite;

ductus bursae long; corpus bursae small; signum a small irregularly shaped plate with a few small teeth.

GENITALIA DIAGNOSIS. In male laterally directed socii hardly overtopped by gnathos and stout, more or less straight cuiller characterise *A. nodiflorella* and *A. rotundella*. Female can be characterised by combination of small corpus bursae with small signum, anterior margin of sternite VIII with low triangular bulge and a diffuse, often faint curved fold below ostium, central section running along margin. Genitalia differences between *A. nodiflorella* and *A. rotundella* are only tendencies and unreliable for safe determination. *A. purpurea* female also very similar.

DISTRIBUTION. Canary Islands; Southern Europe from Portugal to Greece, Romania and Ukraine, including most of the larger Mediterranean islands; not recorded from much of the Balkan Peninsula; Asia: Turkey. Lvovsky (2019): Crimea, Lebanon.

BIONOMICS. Larva on *Ferula communis* L. (Hannemann, 1953); *Ferula glauca* L. (Rymarczyk *et al., in litt.*); in Canary Islands on *Foeniculum vulgare* Mill. (Agassiz, pers. comm.). Moths have been collected from end of April to February, eggs are probably laid in autumn and winter.

REMARKS. Misidentification of *A. rotundella* as *A. nodiflorella* (e.g. Derra, 1989) is due to incorrect figures in Hannemann (1953, 1995). See under *A. rotundella*, below. Although worn specimens of these two species may not be possible to identify, when fresh they should not cause problems. Additionally they can be separated by host-plant of larvae and by barcode.

40   *Agonopterix rotundella* (Douglas, 1846)

Depressaria rotundella Douglas, 1846: 1270.
   Depressaria peloritanella Zeller, 1847: 837.

DESCRIPTION. Wingspan 14–18.5mm. Head light ochreous, face buff. Labial palp pale buff, segment 2 ochreous-buff ventrally. Antenna fuscous. Thorax ochreous-buff, more ochreous anteriorly. Forewing light buff to ochreous-buff, with scattered brown or fuscous scales over much of the wing surface, sometimes organised into lines; terminal spots fuscous; subdorsal spot dark fuscous; blackish first oblique dot and end of cell dot; fringe pale buff to ochreous-buff, without fringe line. Hindwing greyish buff, grey posteriorly; fringe whitish buff to light grey, with a sometimes faint fringe line. Abdomen greyish buff, sometimes tinged mid-brown.

VARIATION. Ground colour of forewing from pale buff to ochreous-buff, with admixture of darker scales varying from extensive to very few.

SIMILAR SPECIES. Characterised by small size, buff forewing coloration and cell markings restricted to two dark dots. *A. nodiflorella* (*q.v.*). See Remarks also.

MALE GENITALIA. Similar to *nodiflorella*, socii less widely divergent, gnathos 2 to 2⅓ times as long as wide; cuiller reaching 75–85% width of valva; anellus plate

V-shaped with wide apical incision reaching nearly to middle of plate, distal angles rounded; aedeagus with process of basal sheath short.

FEMALE GENITALIA. Anterior apophysis half to two-thirds length of posterior apophysis; sternite VIII width:length about 3:1, anterior margin with low triangular bulge, posterior margin concave; ostium in or slightly distal to middle of sternite; ductus bursae long; corpus bursae small, signum a small irregularly shaped plate with a few small teeth.

DISTRIBUTION. South and west Europe extending north to Belgium, Scotland and Ireland and east to Greece; not recorded from Mediterranean islands except Crete. Asia: Turkey, Armenia and Georgia.

BIONOMICS. Larva on *Daucus carota* L. (Hannemann, 1953), with single records from *Distichoselinum tenuifolium* (Lag.) García-Martín & Silvestre (Corley, 2005) and *Orlaya grandiflora* (L.) Hoffm. (R. Heckford and S. Beavan, pers. comm.). Moths hibernate and fly until March or April. In southern areas moths start to hatch at end of April, but usually in July and August in the northern part of its range.

REMARKS. Hannemann (1953, 1995) figured the male genitalia of *A. medelichensis* Buchner, 2015 under the name *rotundella*. As a result of this *A. rotundella* was recorded from Russia (Lvovsky & Knyazev, 2013). A further consequence is that true *rotundella* has been misidentified as *nodiflorella* (e.g. Derra, 1989), because the male genitalia of *rotundella* and *nodiflorella* are actually very similar, therefore using Hannemann's figures for identification of *rotundella* leads to *nodiflorella*. Further details can be found in Buchner (2015).

purpurea subgroup (41)

Shows some affinities with *rotundella* subgroup, but also some differences.

41   *Agonopterix purpurea* (Haworth, 1811)

Depressaria purpurea Haworth, 1811: 511.
   Tinea vaccinella Hübner, 1813: pl. 62, fig. 416.
   Agonopterix banatica Georgesco, 1965: 113, syn. n.

DESCRIPTION. Wingspan 11–15.5 mm. Head light fuscous, scales tipped buff, face whitish buff to buff. Labial palp whitish buff, segment 2 more or less heavily mixed red-brown and fuscous on outer side and ventrally, segment 3 with blackish rings at base, two-thirds and extreme tip. Antenna light fuscous, narrowly ringed deep brown. Thorax fuscous, most scales pale-tipped, often pinkish on anterior margin of tegulae and posterior crest of thorax. Forewing reddish-brown in dorsal two-thirds of its width; costal one-third deep brown mixed with pale grey-tipped fuscous scales, which at two-fifths extend to middle of cell; costal spots deep brown to blackish, separated by pale grey, sometimes edged pinkish; basal field light fuscous with scales tipped white, enclosing one to three blackish dots, demarcated by a white line from dorsum, angled along costal margin of cell to about one-sixth; oblique dots blackish, sometimes joined, with some white scales adjoining on posterior side of dots; a small whitish dot in cell at one-half, sometimes absent and another at end of cell, often dark brown edged; blotch dark fuscous, usually merging with dark brown area towards costa; fringe reddish brown, without fringe line. Hindwing grey, darker towards termen, fringe light grey, darker towards apex, without fringe line. Abdomen fuscous, light grey laterally.

VARIATION. Occasionally the reddish brown coloration is absent leaving the ground colour light brownish. There is variation in the proportion of dark brown in the costal third of forewing. The white dot in middle of cell may be absent.

SIMILAR SPECIES. Characterised by small size, reddish-brown coloration of most of forewing and clear whitish line demarcating basal field.

MALE GENITALIA. Socii not laterally directed, not wider than tegumen; gnathos about 2⅓ times as long as wide; valva slender, gradually tapering from base to apex, cuiller short, somewhat incurved, stout, tapering near apex to a short outwards directed hook, reaching about 75% valva width; anellus plate wider than long, apical margin truncate with median notch; aedeagus about 40% length of valva, curved at middle then tapering to blunt apex, cornuti in a small group; process of basal sheath divided almost to its insertion.

FEMALE GENITALIA. Anterior apophysis half length of posterior apophysis; sternite VIII width:length about 3:1, anterior margin slightly bulging in form of a low triangle or with median one-third shortly projecting, posterior margin slightly concave; ostium in middle of sternite; ductus bursae long; corpus bursae small, signum small, wider than long, with small posterior projection, teeth quite numerous.

GENITALIA DIAGNOSIS. In male combination of shape of valva (slender, gradually tapering to apex) and cuiller (short, stout, somewhat incurved, ending in a short outwards directed hook) is distinct. Female very similar to A. nodiflorella and A. rotundella, but ostium slightly closer to proximal edge of sternite VIII and signum slightly larger.

DISTRIBUTION. Almost all Europe north to southern Finland; absent from a few countries and some islands. Asia: Turkey, Armenia. Lvovsky (2019): Crimea and European Russia.

BIONOMICS. Larva on *Torilis japonica* (Houtt.) DC. (Hannemann, 1953); *Chaerophyllum temulum* L., *Silaum silaus* (L.) Schinz & Thell., *Anthriscus sylvestris* (L.) Hoffm. and *Daucus carota* L. (Palm, 1989); *D. muricatus* (L.) L. (Corley, 2015); *Tordylium maximum* L. and *Torilis arvensis* (Huds.) Link (Rymarczyk *et al., in litt.*). Emerging May to July, flying until April of following year.

REMARKS. We have not seen any specimen referred to *Agonopterix banatica* Georgesco, 1965. Georgesco compared the genitalia of this new species with *A. purpurea*, but we consider the differences to fall within the natural variation found in most *Agonopterix* species. Forewing length is given as 10 mm, but without remark on any size difference from *A. purpurea*, so this may have been an error. From the description we consider *A. banatica* to be synonymous with *A. purpurea*.

curvipunctosa subgroup (42–43)
Wide sacculus allowing short stout cuiller to reach costal margin of valva; anellus with pair of basal sclerites; aedeagus long, almost straight.

42   *Agonopterix curvipunctosa* (Haworth, 1811)

Depressaria curvipunctosa Haworth, 1811: 511.
  Tinea zephyrella Hübner, [1813]: pl. 62, figs. 414, 415.
  Depressaria granulosella Stainton, 1854: 94.
  Depressaria amasina Mann, 1861: 192.
  Depressaria turbulentella Glitz, 1863: 40.
  Agonopterix kotalella Amsel, 1972: 135. syn. nov.

DESCRIPTION. Wingspan 15–19 mm. Head light brownish buff, face buff. Labial palp buff, segment 2 with scattered dark fuscous scales, mainly ventrally and towards apex, segment three with one or two dark fuscous rings, one close to base, often fragmentary or absent, and one from half to three-quarters always present. Antenna light fuscous with narrow deep brown rings. Thorax light brownish buff. Forewing light brownish buff to brown, with scattered dark fuscous scales all over, often giving appearance of fuscous coloration; costal spots dark fuscous separated by light brownish buff spots; terminal spots fuscous; basal field brownish buff; basal fascia deep brown, reaching two-thirds across wing; oblique dots black, the second more or less elongate, separate, contiguous or fused to make a curved mark, plical dot slightly elongated; a larger black dot at end of cell, rarely with whitish dot at its centre; blotch grey, often rather faint; fringe light fuscous with two more or less faint fringe lines. Hindwing light greyish buff, darker towards termen, fringe greyish buff with one or more faint fringe lines. Abdomen light greyish brown.
    VARIATION. Varies in the darkness of the ground colour and amount of scattered fuscous scales, some specimens from Croatia weakly marked and with pinkish buff ground colour; oblique dots from well separated to joined to make a single curved mark. A whitish centre of end of cell dot is sometimes present.
    SIMILAR SPECIES. Characterised by small size and fuscous coloration with limited markings. Smaller than *A. vendettella* but overlapping, and normally lacking the whitish dot at end of cell. *A. heracliana* and *A. capreolella* may have similar fuscous ground colour, but have three white dots in cell.

MALE GENITALIA. Socii divergent, but not wider than tegumen; gnathos lanceolate, acute 2 to 2.5 times as long as wide, extending beyond socii; valva long with narrow apex, stout round-ended cuiller short but reaching costal margin of valva due to wide sacculus; transtilla not thickened in middle; anellus lobes narrow, not overlapping anellus; anellus obpyriform, base with pair of sclerites, apex broad, shallowly notched, a Y-shaped area of plate more strongly sclerotised; aedeagus long, nearly two-thirds length of valva, straight or slightly curved, with a few external teeth near apex, on opposite side from long process of basal sheath, cornuti forming a conspicuous group.

FEMALE GENITALIA. Anterior apophysis about half length of posterior apophysis; sternite VIII width:length about 3:1, anterior margin slightly convex in middle; ostium slightly distal to middle of sternite, antrum strongly sclerotised ventrally; ductus bursae with elongate sclerite laterally in posterior end, long, slender, not expanding to ovate corpus bursae; signum with long narrow parallel-sided wings with marginal teeth, posterior and anterior projections short.

GENITALIA DIAGNOSIS. See *A. vendettella.*

DISTRIBUTION. Nearly all Europe but rare in north and absent from Ireland, Finland and the Baltic countries and a few countries and islands in the south. Asia: widespread, eastwards at least to India (specimen in NHMUK).

BIONOMICS. Larva on *Anthriscus caucalis* Bieb., *A. sylvestris* (L.) Hoffm. (Hannemann, 1953); *Chaerophyllum temulum* L., *Angelica archangelica* L. and *Seseli libanotis* (L.) W.D.J. Koch (Palm, 1989); *Selinum silaifolia* (Jacq.) Beck and *Opopanax chironium* (L.) W.D.J. Koch (Rymarczyk *et al., in litt.*). Emerging starts in May in the south and in July in the northern part of its range, flying until May or June of the following year.

REMARKS. Larval characters of populations in England on *Anthriscus caucalis* and in Croatia on *Opopanax chironium* are quite different (R. Heckford and S. Beavan, pers. comm.).

Hannemann (1976a) placed *kotalella* Amsel, 1972 in synonymy with *A. vendettella,* but it belongs to *curvipunctosa.*

43 *Agonopterix vendettella* (Chrétien, 1908)

Depressaria vendettella Chrétien, 1908b: 259.
   Depressaria iliensis Rebel, 1936b: 96.

DESCRIPTION. Wingspan 16.5–21 mm. Head dark fuscous, scales tipped light brownish buff, face creamy buff. Labial palp pale buff, segment 2 with scattered dark fuscous scales on outer side, segment 3 ringed blackish at base, beyond middle and at extreme apex. Antenna ochreous-buff with dark fuscous rings or all dark fuscous. Thorax dark brown anteriorly, light brownish buff elsewhere. Forewing mid to dark brown, more or less mixed buff, with a scattering of dark fuscous scales particularly in outer one-third; costal and terminal spots dark grey; basal field buff, extended in a broad band along costa to near mid-wing; basal fascia deep brown reaching anterior margin of cell, fading outwardly; a small blackish dot in basal field on anterior margin of cell; oblique dots blackish, the outer often elongated, sometimes joined to form a curved mark; a whitish dot at end of cell, usually edged blackish posteriorly; blotch deep brown to dark fuscous, almost lost in darker specimens and sometimes obsolete in paler ones; fringe greyish fuscous, sometimes with scales tipped dull pink. Hindwing light grey, darker at terminal margin; fringe light grey-brown with fringe line. Abdomen greyish buff.

VARIATION. The forewings vary considerably in depth of ground colour and the extent of the admixture of both paler and dark scales resulting in some more uniform dark specimens and others that are more mottled. The blotch varies considerably in depth of coloration, and the oblique dots can be small or large, the second elongate or round, joined to the first or not.

SIMILAR SPECIES. *A. curvipunctosa* (q.v.). The more mottled specimens resemble some forms of *A. heracliana*, but these have additional white dots in cell.

MALE GENITALIA. Similar to *curvipunctosa*, but socii not spreading laterally; cuiller short but reaching costal margin of valva or just beyond, due to wide sacculus; transtilla thickened in middle; anellus with unsclerotised apical part hardly notched, base with pair of sclerites, larger than in *curvipunctosa*; aedeagus with weak curve, external teeth on distal third on same side as basal process (smaller than in *A. curvipunctosa* and often absent).

FEMALE GENITALIA. Similar to *A. curvipunctosa* but anterior apophysis about three-fifths to two-thirds length of posterior apophysis; antrum strongly sclerotised ventrally, posterior margin appearing more deeply excised; posterior end of ductus bursae slightly expanded and with patchy sclerotisation, remainder long, slender, not expanding to ovate or elliptical corpus bursae.

GENITALIA DIAGNOSIS. Characters of genitalia of both sexes in *curvipunctosa* group unique in European *Agonopterix*. In male *A. curvipunctosa* has lower side of basal sclerites evenly rounded, external teeth on aedeagus opposite to basal process, basal process more than one-third aedeagus length. *A. vendettella* has lower side of basal sclerites acute, external teeth on aedeagus on same side as basal process, basal process up to one-third aedeagus length. Differences in female: *A. curvipunctosa*: colliculum about as long as broad, caudal margin with shallow excavation, ductus below colliculum dilated but not forming a globose structure. *A. vendettella*: colliculum longer than broad, caudal margin deeply excavated, ductus between colliculum and the second sclerotisation swollen and appearing globose in standard preparations.

DISTRIBUTION. Canary Islands, Madeira, Iberian Peninsula, France, Italy, Corsica and Sardinia. North Africa: Morocco.

BIONOMICS. Larvae on *Smyrnium olusatrum* L. (Hannemann, 1953) but the species has also been reared once from *Conium maculatum* L. (Corley, 2015) and from *Oenanthe pteridifolia* Lowe on Madeira (Buchner & Karsholt, 2019). Imago from April onwards, flying until April of the following year.

REMARKS. The drawing of male genitalia in Klimesch (1985: 144, fig. 4) clearly does not show this species.

Depressaria iliensis Rebel, 1936 was described from Sardinia based on a single male in NHMW. This was dissected by P. Huemer, but was found to be female and clearly belongs to *A. vendettella*. The synonymy was published in Buchner & Karsholt (2019). According to Hannemann (1976a) *A. kotalella* Amsel is a synonym of *A. vendettella*, but this is incorrect, it is a synonym of *A. curvipunctosa*.

alpigena subgroup (44–55)

Anellus with characteristic triangular apical processes.

Key to male genitalia of *alpigena* and *angelicella* subgroups

1a	Anellus extremely elongated, cuiller clearly exceeding costa	*A. socerbi*
1b	These features not both present	2
2a (1b)	Gnathos appears broadly elliptic to nearly circular (length:width ratio < 1–1.5:1, rarely nearly 2:1) in standard preparation (for this type of gnathos that means gnathos turned aside and seen therefore from lateral view)	3
2b	Gnathos elongate (length:width ratio 2:1 or longer)	5
3a (2a)	Cuiller with bulge on inner side near middle, then bent inward and tapering to a rather sharp tip (fig. 53f), but sometimes this feature present only in a reduced form (fig. 53i)	*A. selini*
3b	Cuiller blunt, S-shaped or bent outwards, not clearly bulged near middle.	4
4a (3b)	Gnathos orbicular with tendency to be broader than long, cuiller S-shaped, transtilla lobes large, overlapping transtilla and anellus	*A. lessini*
4b	Gnathos broadly elongate-elliptic (length:width ratio 1.3–1.5:1, rarely nearly 2:1), transtilla lobes small and therefore not overlapping transtilla and anellus, cuiller slightly bent outwards (not S-shaped)	*A. angelicella, A. paraselini* (differences see description)

5a (2b)	Cuiller swollen at or slightly distal to middle (sometimes whole basal half very broad), S-shaped with pointed tip and therefore similar to that of *A. selini*	*A. ordubadensis*
5b	Cuiller different	6
6a (5b)	Cuiller distinctly S-curved, anellus processes triangular, rather large (distance fold to tip 0.05–0.08 mm), with triangular incision in between. Transtilla lobes large, overlapping transtilla.	7
6b	Cuiller bent outwards, not S-shaped (rarely slightly so). Anellus processes small to large (distance fold to tip 0.02–0.09 mm). Transtilla lobes small (not overlapping transtilla) or large (overlapping transtilla).	8
7a (6a)	Gnathos exceeding socii by up to half of its length, transtilla lobes nearly touching in standard preparation (gap 0–0.02mm), transtilla distinctly broadened in middle	*A. kayseriensis*
7b	Gnathos exceeding socii by nearly full length, transtilla lobes with rather wide gap in between (gap 0.05–0.06mm), transtilla not or only slightly broadened in middle	*A. cachritis*
8a (6b)	Transtilla lobes small, not or only slightly overlapping transtilla, ratio vertical:horizontal diameter about 1 or only slightly more	9
8b	Transtilla lobes large, clearly overlapping transtilla, ratio vertical:horizontal diameter about 2 or only slightly less. Group of species with similar genitalia, determination must be based on combination of several features. Barcoding is the best method to get safe determination results.	10
9a (8a)	Triangular anellus processes rather large (distance fold to tip 0.05–0.07 mm), gnathos not or scarcely exceeding socii, sacculus ending at about 1/3 valva length	*A. galicicensis*
9b	Triangular anellus processes rather small (distance fold to tip 0.02–0.03 mm), gnathos clearly exceeding socii, sacculus ending at about 1/4 valva length	*A. parilella*

10a (8b)	Cuiller stout in proximal half, only slightly curved outwards at middle and more slender in distal half; anellus processes rather large (distance fold to tip 0.05–0.07 mm), more or less triangular with triangular incision in between; transtilla distinctly broadened in middle; gnathos not or scarcely exceeding socii	*A. ferulae*
10b	Cuiller usually strongly bent outwards, rather long and thin, may be sharply pointed; triangular anellus processes small to medium sized (distance fold to tip 0.02–0.05 mm) with shallow incision in between; transtilla not or only slightly broadened in middle; gnathos considerably exceeding socii	*A. coenosella*
10c	Cuiller usually bent outwards, usually blunt, tends to be shorter than in *A coenosella* but may overlap; anellus processes large (distance fold to tip 0.05–0.09 mm), more horn-like than triangular with a semicircular incision in between; transtilla moderately broadened in middle (overlap with *A. coenosella*).	*A. alpigena,* *A. richteri* (differences see description)
10d	Cuiller usually slightly bent outward over its whole length and slightly broader than in *A. alpigena* and *A. richteri*, with a tendency to be sharply bent outward at the tip, anellus processes tend to be smaller than in *A. alpigena*, but overlapping in size, gnathos tends to exceed socii by up to half of its length	*A. langmaidi*

Female genitalia of species in *alpigena* and *angelicella* subgroups are very similar, and in combination with intraspecific variability, safe determination based only on female genitalia is rarely possible. Further information should be included, preferably external appearance and also host-plant (if known); distribution can be helpful to exclude some species. Barcoding allows safe determination in most cases.

44 *Agonopterix alpigena* (Frey, 1870)

Depressaria alpigena Frey, 1870: 248.
  Depressaria sileris Pfaffenzeller, 1870: 320.
  Depressaria salevensis Frey, 1880: 353.
  Depressaria ragonoti Rebel, 1889: 308.
  Depressaria seraphimella Chrétien in L'Homme, 1929: 194.

DESCRIPTION. Wingspan 18–22 mm. Head light brown, orange-brown to fuscous, scales tipped buff. Labial palp pale buff, segment 2 outer and ventral sides with some dark grey scales, outer side of segment 3 with some light brown scales. Antenna light fuscous, weakly ringed light brown near base. Thorax and tegula ochreous to orange-brown often with darker brown or grey-brown scales anteriorly. Forewing creamy yellow, orange-brown or dull light to grey-brown always with scattered dark fuscous scales; basal field creamy yellow sometimes more or less suffused with ground colour, not usually reaching costa and outwardly demarcated by a blackish fascia, shading outwards into ground colour; fuscous costal spots small and numerous; fuscous terminal spots; black oblique dots occasionally followed by a few pale scales, a plical dot, a white median dot, often black-edged, sometimes white scales obsolete, a white dot edged black at end of cell, sometimes lying in or adjacent to a fuscous to blackish patch which sometimes extends to costa and on the dorsal side frequently has a tail extending towards tornus; fringe concolorous with ground colour. Hindwing grey, darker posteriorly, fringe light grey, with fringe line. Abdomen greyish buff.

VARIATION. The forewing ground colour is extremely variable from creamy yellow through orange-brown to quite dark fuscous; the amount of fuscous scaling is also very variable, occasionally indicating veins in terminal third of wing. The basal field is usually, but not always, cut off from the costa by a darker band on costa. The dark area around the white spot varies in intensity, sometimes forming a large black patch between cell and costa.

SIMILAR SPECIES. According to the general coloration, *A. alpigena* resembles many different species, particularly others in the *alpigena* group and *A. paraselini*; yellow specimens with large black patch resemble *Exaeretia thurneri*, but basal field never dark and labial palp clearly different. (All forms of *alpigena* retain the pale basal field.)

MALE GENITALIA. Socii rounded, almost as wide as long; gnathos ellipsoid, but in preparations twisted to one side, appearing to have one side flat or nearly so, from slightly shorter than socii to exceeding them; valva wide, ventral margin more strongly curved at three-fifths, apex fairly wide, cuiller curved outwards, often strongly so from middle, apex rounded or sometimes strongly tapered, 65–80% of valva width; anellus lobes long with median bulge overlapping anellus; anellus broadly shield-shaped with pair of postero-lateral claw-like sclerotisations and pair of round-ended diverging apical processes; aedeagus slightly more than half length of valva, slender, slightly curved, abundant cornuti in a narrow group, process from basal sheath longer than sheath, patent.

FEMALE GENITALIA. Anterior apophysis half to three-fifths length of posterior apophysis; sternite VIII width:length 2:1, anterior margin convex, nearly straight in middle, posterior margin straight or slightly excavated in middle; ostium towards posterior margin of sternite; ductus bursae long, slender, slightly expanding to ovate corpus bursae; signum elliptical to irregular, slightly wider than long with numerous teeth of variable size.

GENITALIA DIAGNOSIS. Male, see description and key (p. 120); most similar to *A. richteri* (*q.v.*). Female, see general remark on *alpigena* and *angelicella* subgroups.

DISTRIBUTION. Mountain areas from France to Slovenia, also in Croatia, North Macedonia and Bulgaria.

BIONOMICS. Larvae on *Laserpitium siler* L. (Hannemann, 1995); *L. nestleri* Soy.-Will., *L. gallicum* L. and *L. latifolium* L. (Rymarczyk *et al.*, 2015a). Emerging May and June, the latest specimens we have seen were collected 13 September.

REMARKS. Chrétien's description of *Depressaria seraphimella* was published by L'Homme (1929), based on a single reared specimen. Rymarczyk *et al.* (2015a) were able to show that it was synonymous with *Agonopterix alpigena*.

The gnathos is somewhat flattened; in preparations the gnathos arms usually twist so that the gnathos appears to varying degree in lateral view. Depending on the amount of twisting the length to width ratio is variable, so is not given here. The same problem occurs in other species of the *alpigena*-group.

45   *Agonopterix richteri* sp. nov.

Type locality: Bulgaria, Sandanski.

DESCRIPTION. Wingspan 18–20 mm. Head bright brick-red on crown. Labial palp whitish buff. Antenna fuscous, weakly ringed grey. Thorax and tegula bright brick-red, orange-ochreous posteriorly. Forewing bright brick-red with admixture of dull ochreous scales and a very few scattered dark fuscous scales, most of wing except costa dulled by weak grey tinge; basal field orange, edged ochreous-buff, outwards demarcated dull brown; costa clear bright brick-red, a blackish spot at extreme base and a series of small blackish spots mainly between one-quarter and one-half; oblique dots black followed by a few cream-coloured scales, a faint plical dot, median dot cream partly edged black, dot at end of cell cream, edged black, set in a slightly darker area; fringe mixed grey and brick-red. Hindwing grey, darker posteriorly; fringe light grey with more or less distinct line. Abdomen light grey.

VARIATION. Very limited material seen, but the coloration may be dulled by a deeper grey tinge to parts of the forewing; the cream or white scales associated with the oblique and median dots may be reduced or absent; a series of dull grey terminal dots is sometimes present.

SIMILAR SPECIES. *A. socerbi* and *A. kayseriensis* have similar bright brick-red coloration, in *A. socerbi* scattered dark scales more numerous, diffuse patch around median and end of cell dots usually darker and pale basal field not so distinctly extending along costa as a broad pale stripe; in *A. kayseriensis* white parts of oblique and cell dots usually larger, but all these features overlapping; *A. alpigena* is never so brightly coloured and also has more scattered dark scales.

MALE GENITALIA. Similar to *alpigena*, but gnathos slightly narrower, extending beyond socii, valva slightly shorter, cuiller ridged at apex, 85–95% of valva width; anellus with apical processes sometimes more acute; aedeagus half length of valva.

FEMALE GENITALIA. Similar to *A. alpigena* but anterior apophysis two-thirds length of posterior apophysis; sternite VIII with anterior margin convex, but middle one-third concave, posterior margin excavated in middle; ostium slightly distal to middle of sternite; ductus bursae long, slender, gradually expanding to ovate corpus bursae; signum in shape of a quarter sector of a circle, with anterior projection, teeth numerous.

GENITALIA DIAGNOSIS. Male see description above for differences from *A. alpigena*. Female, see general remark on *alpigena* and *angelicella* subgroups.

DISTRIBUTION. Bulgaria, Greece.

BIONOMICS. Host-plant unknown. Fresh specimens collected July, worn specimens at end of September.

MATERIAL EXAMINED. Holotype: ♂, Bulgaria, Sandanski, 24.ix.2011, gen. prep. Ignác Richter under DEEUR 1218, DNA barcode id. TLMF Lep 07137, Ignác Richter leg., will be deposited in coll. TLMF

Paratypes (arranged according to collection date):

1 ♂, Greece, Pangeo, Kokkinochori, 1500m, 7.ix.1994, gen. prep. DEEUR 2998 P. Buchner, K. Larsen leg., coll. RCKL

1 ♀, Bulgaria, Sandanski, 24.ix.2011, gen. prep. DEEUR 1031 P. Buchner, DNA barcode id. TLMF Lep 07090, L. Srnka leg., coll. RCLS

1 ♂, 3 ♀, Bulgaria, Sandanski, 24.ix.2011, gen. prep. DEEUR 1143 (♂) P. Buchner, DEEUR 1156, 1157 & 1158 (♀), Ignác Richter leg., coll. RCIR

1 ♀, Bulgaria, Tuzlata, 50m, 29.ix.2011, DEEUR 9671, L. Srnka leg., coll. RCLS

2 ♀, Bulgaria, Ilindenci, 27.ix.2011, DEEUR 1156, 1155 & 1159, Ignác Richter leg., coll. RCIR

1 ♀, Bulgaria, Ilindenci, 17.vii.2014, DEEUR 4461, J. Junnilainen leg., coll. RCJJ

1 ♂, Bulgaria, Blagoevgrad, Rupite, 21.ix.2017, DEEUR 6484, J. Junnilainen leg., coll. RCJJ.

ETYMOLOGY. The species is named after Ignác Richter, who collected part of the type series.

46   *Agonopterix coenosella* (Zerny, 1940)

Depressaria coenosella Zerny, 1940: 43.

DESCRIPTION. Wingspan 16.5–21 mm. Head dull chestnut-brown to fuscous, Labial palp fuscous, paler on inner side, segment 3 with weak dark median ring. Antenna fuscous. Thorax concolorous with forewing. Forewing from dull chestnut-brown to light brown or fuscous; scattered blackish scales mainly in costal half and towards termen, small blackish spots on costa; basal field paler than rest of forewing, sometimes extended shortly along costa, variable in coloration, demarcated by dark brown basal fascia fading into ground colour distally; oblique dots deep brown, often edged posteriorly by a few whitish scales, inner cell spot mostly obscure or obsolete, occasionally whitish partly edged dark brown, end of cell spot conspicuous, whitish, narrowly edged dark brown; terminal dots weak or absent; fringe concolorous with wing. Hindwing grey, darker towards apex; fringe light grey-buff with weak fringe line. Abdomen grey-buff.

VARIATION AND SIMILAR SPECIES. Very variable externally. Ground colour may be rather uniformly dark greyish (predominant in Ural mountains), usually scales not pale tipped but sometimes distinctly so, grey ground colour often more or less replaced by fuscous to chestnut-brown, the last resembling other chestnut-brown to reddish brown species, e.g. *A. kayseriensis*; forms with strongly mottled forewings (e.g. specimens from Romania) have a strange appearance, not usually found in *Agonopterix alpigena* group. Fuscous forms resemble *A. cachritis* which always has three white dots; some browner forms resemble *A. heracliana*, but that has plical dot and terminal spots and usually a more mottled forewing. Depending on individual appearance it may resemble further species, so it is not recommended to determine *A. coenosella* based on external appearance only.

MALE GENITALIA. Similar to *alpigena*, but gnathos slightly narrower, extending beyond socii, ventral margin of valva slightly concave at end of sacculus, cuiller with acute apex, 75–85% of valva width; anellus with apical processes short, round-ended, strongly divergent; aedeagus about half length of valva.

FEMALE GENITALIA. Similar to *A. alpigena* but anterior apophysis three-fifths to two-thirds length of posterior apophysis; sternite VIII width:length 5:2, anterior margin convex, nearly straight in middle, posterior margin straight or with a median excavation; ostium distal to middle of sternite; signum variable, often elliptical, slightly wider than long, but sometimes kite-shaped, teeth numerous, of variable size.

GENITALIA DIAGNOSIS. Male, see description and key (p. 120). Female, see general remark on *alpigena* and *angelicella* subgroups.

DISTRIBUTION. South Russia, Romania (Kovács & Kovács, 2020). Widespread in Western Asia, we have seen specimens from Turkey, Russia, Tajikistan, Uzbekistan, Kazakhstan, Iran and Afghanistan.

BIONOMICS. Host-plant unknown. Moths have been collected from May to September.

47   *Agonopterix kayseriensis* Buchner, 2020

Agonopterix kayseriensis Buchner, 2020a: 2.

DESCRIPTION. Wingspan 13.5–17 mm. Head pale yellowish brown. Labial palp segment 2 pale yellowish on inner side, with some medium to dark brown scales on outer side, segment 3 pale yellowish, with some darker scales on outer side. Antenna dark greyish brown, distinctly more stout in males than in females. Thorax and tegulae with posterior crest, pale yellow to brick-red with a tendency to be darker in frontal and central parts. Forewing ground colour rich brick-red when fresh fading to pale yellowish buff, often with more yellow areas between cell and dorsum and beyond cell where the pale area posteriorly forms a wide angle separated from terminal area with plentiful scattered reddish-brown to fuscous scales; a blackish spot is usually present at base of costa, irregular groups of blackish scales present or not along costa; light fuscous terminal dots often present; basal field pale yellow, on dorsal side edged brown, on costal side extending into a broad stripe along costa fading into ground colour at or before one-third; a blackish plical dot, oblique dots white, edged blackish on inner edge, white median and larger white end of cell dot more or less ringed blackish and surrounded by distinct diffuse dark reddish brown to blackish field; fringe similar to ground colour of forewing. Hindwing rather dark greyish, paler at base, fringe similar to ground colour, with a diffuse darker fringe line near the base. Abdomen medium grey.

VARIATION. On forewings contrast of basal field against ground colour, number of irregularly interspersed dark scales and costal spots, size of cell dots and extent of dark area around these dots all vary. Ground colour of forewing soon fades.

SIMILAR SPECIES. The yellow basal field extending to the costa should distinguish this species from other reddish-brown species in the *alpigena* subgroup such as *A. socerbi*, further details under *A.richteri* and *A. coenosella*.

MALE GENITALIA. Most similar to *cachritis*, but gnathos only exceeding socii by half its length, valva with ventral margin concave in distal one-third, giving more slender distal part; transtilla thickened.

FEMALE GENITALIA. Similar to *A. alpigena* but anterior apophysis three-fifths length of posterior apophysis; sternite VIII width:length 5:2, anterior margin convex, middle one-third in a slight excavation, straight; ostium proximal to middle of sternite; signum variable, rhombic, sometimes with anterior and posterior angles produced, or wider than long, numerous teeth of moderate size.

GENITALIA DIAGNOSIS. Male, see description and key (p. 120). Female, see general remark on *alpigena* and *angelicella* subgroups.

DISTRIBUTION. Romania, Greece, European Turkey; also in Asiatic Turkey.

BIONOMICS. Moths have been collected from June to September. Has been reared from an undetermined Apiaceae, host-plant species unknown.

48   *Agonopterix cachritis* (Staudinger, 1859)

Depressaria cachritis Staudinger, 1859: 237.
  Depressaria epicachritis Ragonot, 1895: CVI.

DESCRIPTION. Wingspan 16–18.5 mm. Head light fuscous. Labial palp light fuscous, segment 3 with a blackish ring at one-quarter and another just before apex. Antenna light fuscous. Thorax light fuscous. Forewing light fuscous to light brown, sometimes overlaid with brown scales in parts, always with some scattered brown or fuscous scales in terminal third; weak dark fuscous spots on costa; terminal dots inconspicuous; basal field buff to light brown, hardly reaching costa, broadly edged blackish brown from dorsum to cell; three white spots in cell, all at least partly edged blackish, the first at one-third partly edged by the inconspicuous oblique dots, the second often small or obsolete, the third at end of cell larger and more clearly edged blackish; plical dot present; fringe concolorous. Hindwing light grey, darker towards termen; fringe light grey. Abdomen buff.

VARIATION. Ground colour varies with extent of brown scales over the light fuscous background. The middle white spot is weak and often lost.

SIMILAR SPECIES. The uniform dull light grey-brown forewing without blotch, with two or three white dots and poorly expressed oblique dots are characteristic. Other species of similar coloration with two or three white dots have wings more mottled with darker scales and blotch present.

MALE GENITALIA. Similar to *alpigena*, but socii almost orbicular; gnathos narrower, extending beyond socii by almost its whole length; valva broad, but tapering in distal one-third to quite acute apex, cuiller S-shaped, curving outwards at middle and inwards towards apex, about 85% of valva width; anellus with apical processes long, fairly acute; aedeagus slender, curved beyond middle, of almost uniform width until last quarter.

FEMALE GENITALIA. Similar to *A. alpigena* but anterior apophysis about three-fifths length of posterior apophysis; sternite VIII width:length 5:2, anterior margin straight, posterior margin excavated in middle; anterior margin of tergite nearly straight; ostium distal to middle of sternite; signum wider than long with narrow parallel-sided wings, slightly contracted in mid-line between wings, teeth few, mainly marginal.

GENITALIA DIAGNOSIS. Male, see description and key (p. 120). Female, see general remark on *alpigena* and *angelicella* subgroups.

DISTRIBUTION. South Spain, Portugal. There are no reliable records since 1886 from Spain, so the species may be extinct there. One specimen stored in NHMW is labelled "Hispania, Madrid, 10.vi.1914, Knitschke", but the locality appears unlikely, so the label data are questionable. Unexpectedly this species was recently found in South-west Portugal (Nunes *et al.*, 2024).

BIONOMICS. According to Staudinger (1859) larvae were found in February and March on *Cachrys laevigata* Lam. on coastal sand dunes. Moths emerged in May. Examined moths from Spain with information on month of collection were taken in August, the very fresh specimen from Portugal on 15 May.

REMARKS. *Cachrys laevigata* is now treated as a synonym of *Prangos trifida* (Mill.) Herrnst. & Heyn, which does not grow on sand dunes. The sand dune plant is *C. libanotis* L. and it is likely that this is the true host-plant.

Depressaria epicachritis (Ragonot, 1895: 106) was described from "Haute-Syrie" (now part of Turkey). According to Hannemann (1958b) it is synonymous with *A. cachritis*. We have not examined this, but question the synonymy (Nunes *et al.*, 2024). Hannemann (1976a, 1996) mentions *A. cachritis* from North Africa, Syria and Sardinia, but we have seen no specimens from outside the Iberian Peninsula.

49   *Agonopterix ferulae* (Zeller, 1847)

Depressaria ferulae Zeller, 1847: 840.

DESCRIPTION. Wingspan 14.5–21.5 mm. Head dark fuscous or deep cinnamon pink, face creamy buff. Labial palp pale buff, segment 2 with numerous fuscous scales on outer side, rough scales deep cinnamon pink; segment 3 pink-tinged with scattered or numerous dark fuscous scales in basal half, a long blackish ring beyond middle and a black tip. Antenna dark fuscous, markedly thicker in males than in females. Thorax dark fuscous anteriorly, remainder either creamy buff or ochreous-cinnamon. Forewing with basal field concolorous with thorax, rest of fore-wing fuscous to deep fuscous brown, sometimes with pale grey or cinnamon-tipped scales in outer third of wing, whole wing with scattered blackish scales, inconspic-uous in darkest specimens; base of costa with a deep brown spot, blackish costal spots evident against narrow pink costal edge, but inconspicuous when pink edge is absent; terminal spots blackish; oblique dots black, the first larger, the second often absent, a very small median dot and a larger dot at end of cell, all four dots may be associated with whitish dots, the outermost being largest; plical dot variously devel-oped, sometimes a short dash; fringe dark fuscous, sometimes cinnamon-tipped, with one fringe line. Hindwing grey, darker towards termen, fringe fuscous with indistinct line. Abdomen fuscous.

VARIATION. Among Portuguese specimens with deep brown forewings there are two colour forms: one has fuscous head with creamy buff thorax and basal field, the other has deep cinnamon pink head and ochreous-cinnamon thorax and basal field. Specimens from further east in Europe generally have less dark forewing coloration. Forewing ground colour varies from fuscous through to very dark fuscous brown, with other dark markings barely visible in the darkest specimens. The black and the whitish cell dots all vary in size and some, particularly the second oblique dot, may be absent, the white dots varying from one to four.

SIMILAR SPECIES. The plain dark forewing contrasting with the cream or orange thorax and wing base shows similarity to several species including *A. alpigena* and *A. furvella*, but these have more reddish brown forewing and *A. furvella* is much larger. *A. pseudoferulae* has an orange streak between white dots in cell. *A. galicicensis* and *A. langmaidi* (*q.v.*) are similar.

MALE GENITALIA. Similar to *alpigena*, gnathos shortly (up to half its length) extending beyond socii; valva with ventral margin strongly curved at about two-thirds, distal one-third slightly concave, cuiller stout in proximal half, tapering, curved outwards at middle and more slender in distal half, apex with inner side straight and outer side curved inwards, 75–85% of valva width; transtilla strongly thickened, transtilla lobes overlapping transtilla, proximal margin inclined; anellus lobes usually with posterior bulge; anellus widest at middle with apical processes broadly acute; aedeagus slender, curved, basal process much wider than aedeagus.

FEMALE GENITALIA. Similar to *A. alpigena* but anterior apophysis three-fifths to three-quarters length of posterior apophysis; sternite VIII width:length from 3:1 to 4:1, anterior margin weakly convex, posterior margin angled in towards slightly excavated middle; ostium round, slightly distal to middle of sternite, antrum as long as ostium; signum as wide as long, irregularly shaped, with few teeth or elliptical, wider than long, with mainly small teeth and some larger teeth in central part of plate.

GENITALIA DIAGNOSIS. Male, see description and key (p. 120). Female, see *A. galicicensis* and general remark on *alpigena* and *angelicella* subgroups.

DISTRIBUTION. South Europe from Portugal to North Macedonia and Greece, including most of the larger Mediterranean Islands, but not recorded from Crete. Morocco (Hannemann, 1953); Israel (Lvovsky *et al.*, 2016).

BIONOMICS. Larvae feed gregariously on *Ferula communis* L. in the leaf sheaths or in spun leaves (Hannemann, 1953) and on *F. glauca* L. (Rymarczyk *et al., in litt.*). Moths can be found from May onwards to autumn.

REMARKS. The two colour forms are found together and have been reared from a single spinning. Forms with ochreous-cinnamon thorax and basal field and dark fuscous forewings are distinct, forms with creamy buff thorax and basal field resemble other *Agonopterix* species and dissection is recommended for determination.

We have recently come to the conclusion that two populations of supposed *ferulae* with significantly different barcode are better treated as separate species. The descriptions of *A. galicicensis* sp. n. and *A. langmaidi* sp. n. follow.

50   *Agonopterix galicicensis* sp. n.

Type locality: North Macedonia, Prespa Pass.

DESCRIPTION. Wingspan 13.5–15.5 mm. Head fuscous tinged chestnut-brown, face creamy buff. Labial palp pale buff, segment 2 with numerous fuscous scales on outer side, rough scales chestnut brown; segment 3 buff with a blackish ring at base, long blackish ring beyond middle and a blackish tip. Antenna dark fuscous, not thicker in males than in females. Thorax ferruginous-brown anteriorly gradually fading posteriorly, remainder ochreous-ferruginous to ochreous. Forewing with basal field concolorous with rear of thorax, rest of forewing fuscous strongly tinged chestnut-brown, scattered blackish scales in posterior one-third of wing, inconspicuous against dark wing colour; base of costa with a deep brown spot, blackish costal spots present, often faint; terminal spots blackish; oblique dots black, a small median dot and a larger dot at end of cell, all four dots usually associated with white scales, frequently the amount of white is greater than the amount of black; plical dot present; fringe concolorous with forewing, with one fringe line. Hindwing grey, darker towards termen, fringe fuscous with indistinct line. Abdomen grey to fuscous.

VARIATION. Few specimens seen but there is some variation in the extent and depth of darker ferruginous coloration of the thorax and in the number and size of the white cell markings, normally four on each wing, but may be fewer and may differ between one wing and the other.

SIMILAR SPECIES. Closely resembles *A. ferulae*, but the coloration of the thorax and wing base is stronger with more ferruginous-ochreous tinge; the ground colour of the forewings has a reddish tinge, resulting in chestnut-brown rather than the dark fuscous of *A. ferulae*. Antenna of male *A. ferulae* is often thickened compared with female, but this is not the case in *A. galicicensis*.

MALE GENITALIA. Similar to *A. ferulae* but gnathos shortly extending beyond socii; cuiller less tapering, stouter and curving outwards in distal half, obliquely truncated at apex, 92–94% of valva width; transtilla slightly thickened, transtilla lobes not overlapping transtilla, proximal margin partly transverse; anellus lobes without posterior bulge; anellus widest at middle, apical processes well developed.

FEMALE GENITALIA. Very similar to *A. ferulae* but ostium distal to middle of sternite, more elongate, reaching posterior margin, antrum short, not reaching near anterior margin of sternite.

GENITALIA DIAGNOSIS. Male, see description and key (p. 120). Female, see general remark on *alpigena* and *angelicella* subgroups.

DISTRIBUTION. Only known from North Macedonia.

BIONOMICS. Host-plant unknown. Moths have been taken at around 1600 m in July and August.

MATERIAL EXAMINED. Holotype: ♂, North Macedonia, Prespa pass, 1600m, viii.1977, gen. prep. DEEUR 1641 P. Buchner, Fr. Zürnbauer leg., coll. TLMF

Paratypes (arranged according to collection date):

2 ♂, 1 ♀, North Macedonia, Prespa pass, 1600m, viii.1977, DEEUR 1638 (♂), DEEUR 1639 (♀), DEEUR 2122 (♂) with DNA barcode id. TLMF Lep 19192, Fr. Zürnbauer leg., coll. TLMF

1 ♀, North Macedonia, Galicica NP, 1580m, 23.vii.2013, gen. prep. DEEUR 2343 P. Buchner, DNA barcode id. TLMF Lep 17694, B. Skule & C. Hviid leg., coll. ZMUC

1 ♂, 7 ♀, North Macedonia, Galicica NP, 25.vii.2015, gen. prep. DEEUR 7260 P. Buchner (♂), gen. prep. DEEUR 7259 P. Buchner (♀), DEEUR 7261 (♀) with DNA barcode id. TLMF Lep 26027, DEEUR 7256, 7257, 7262, 7263 (♀), Ignac Richter leg., coll. RCIR

1 ♀, North Macedonia, Galicica NP, Asan Gjura, 27.viii.2021, DEEUR 9269, Ignac Richter leg., coll. RCIR

ETYMOLOGY. The species is named after the Galičica National Park in North Macedonia, one of the localities in which the species has been found.

REMARKS. Barcode differs from both *A. ferulae* and *A. langmaidi* by 3.9%.

51   *Agonopterix langmaidi* sp. n.

Type locality: Italy, Sicily, Mistretta.

DESCRIPTION. Wingspan 15–22.5 mm. Head buff to fuscous, face creamy buff. Labial palp pale buff, segment 2 with outer side mainly fuscous; segment 3 buff with a fuscous ring at base, a weak fuscous ring beyond middle and a black tip. Antenna fuscous, not thickened in males. Thorax buff to fuscous or darker brown, sometimes darker anteriorly. Forewing with basal field concolorous with head and posterior part of thorax, outwardly curved to extend shortly along dorsum, rest of forewing light to dark fuscous, scattered blackish scales in posterior one-third of wing, inconspicuous against dark wing colour; base of costa with a deep brown spot, blackish costal spots present, often faint; terminal spots blackish, small and faint; oblique dots black, small, sometimes followed by a few whitish buff scales, a small median dot with or without whitish buff scales adjacent and a larger whitish buff dot at end of cell, partially edged blackish; plical dash present; fringe concolorous with forewing. Hindwing grey, darker towards termen, fringe fuscous. Abdomen grey to fuscous.

VARIATION. Few specimens seen and these not in very good condition, but there is some variation in coloration of head, thorax and forewings. The proportions of black and whitish buff in the cell markings vary, often only the end of cell dot is really conspicuous.

SIMILAR SPECIES. Resembles *A. ferulae* and *A. galicicensis*, but the coloration of the thorax and basal field is duller. Antenna of male *A. ferulae* is usually thickened compared with female, but this is not the case in *A. langmaidi*.

MALE GENITALIA. Similar to *A. ferulae* but gnathos extending well beyond socii; cuiller almost parallel-sided, apex rounded or hooked outwards, with microsculpture on inner margin, 89–93% of valva width; transtilla not thickened, transtilla lobes just overlapping transtilla, proximal margin inclined; anellus lobes with posterior bulge; anellus narrowest at three-fifths length, apical processes relatively small; basal process of aedeagus not or slightly wider than aedeagus.

FEMALE GENITALIA. Not available.

GENITALIA DIAGNOSIS. Male, see description and key (p. 120).

DISTRIBUTION. Only known from Sicily and Crete.

BIONOMICS. Host-plant unknown. Moths have been taken at around 1100 m in Sicily in September, in Crete at 600 m in September and also in Crete, Omalos plateau, in October.

MATERIAL EXAMINED. Holotype: ♂, Italy, Sicily, Mistretta, 1100m, 10.ix.2015, gen. prep. DEEUR 4720 P. Buchner, DNA barcode id. TLMF Lep 23367, T. Nupponen leg., coll. RCTN.

Paratypes:

1 ♂, Italy, Sicily, Mistretta, 3 km S, 1060–1110m, 10.ix.2015, T. Nupponen leg., gen. prep. DEEUR 4721 P. Buchner, DNA barcode id. TLMF Lep 19363, coll. RCTN.

1 ♂, Greece, Crete, 5 km NE Hora Sfakia, 600 m, 26.ix.2006, gen. prep. DEEUR 4697 P. Buchner, DNA barcode id. TLMF Lep 19221, T. Nupponen leg., coll. RCTN

1 ♂, Greece, Crete, Omalos plateau, Chania, 3.x.2016, gen. prep. DEEUR 6714 P. Buchner, DNA barcode id. TLMF Lep 26312, K. Larsen leg., coll. RCKL

ETYMOLOGY. The species is named in memory of Dr John Langmaid (1934–2022), eminent British microlepidopterist, who had a special interest in Depressariidae and gave MC considerable help and encouragement when he began working on this book.

REMARKS. Barcode differs from *A. ferulae* by 3.6% and from *A. galicicensis* sp. n. by 3.9%.

A. ferulae was described by Zeller from specimens he reared from *Ferula* in Sicily. Although none of the type series of *ferulae* in NHMUK has been dissected the specimens we have examined from several countries more closely resemble Zeller's *ferulae* externally than *langmaidi*. We have been able to confirm from recently collected specimens that both *A. ferulae* and *A. langmaidi* occur on Sicily. We therefore see no reason to suppose that Zeller's *ferulae* might actually be *langmaidi*.

52   *Agonopterix lessini* Buchner, 2017

Agonopterix lessini Buchner, 2017a: 82.

DESCRIPTION. Wingspan 16–20 mm. Head with face creamy buff, vertex cinnamon with scales tipped creamy buff. Labial palp creamy buff, segment 2 outer and ventral sides cinnamon mixed with a few fuscous scales, segment 3 with fuscous scales, sometimes forming two distinct rings. Antenna thicker in male than female, dark fuscous above, creamy buff below. Thorax and tegula pale cream, red-brown anteriorly, extending backwards to varying extent, often mainly red-brown throughout. Forewing red-brown, with some yellowish buff scales, mainly in terminal one-third, and with some scattered dark grey scales, costa and terminal quarter often heavily mixed dark grey, some blackish costal and terminal spots usually present; basal field more or less concolorous with thorax, cream, often more or less suffused orange-brown, reaching costa or not, basal fascia dark red-brown outwardly fading into ground colour; plical and oblique dots black, the latter sometimes followed by a few whitish scales; median dot usually blackish, indistinct, sometimes absent, occasionally with whitish centre, end of cell dot large, white, edged black, set in a darker grey-shaded patch which can be indistinct or absent; fringe reddish brown, tipped grey, grey towards tornus. Hindwing light to dark grey; fringe light to dark grey. Abdomen grey.

VARIATION. The extent of yellowish-buff or darker grey scales can alter the overall forewing colour from dull brick-red to fuscous, but there is always some dull brick-red coloration; white markings associated with the various cell dots are very variable, sometimes there can be 3 or 4 white marks, in other specimens only the end of cell dot is white. One specimen from Crete has forewing beyond cream basal field uniform grey with a hint of purple, scales tipped grey-buff, spots along costa and termen absent.

SIMILAR SPECIES. *A. selini, A. paraselini* and *A. parilella* are all similar externally, but of these species only *A. selini* tends to have blackish elements in third segment of labial palp widely reduced, in *A. selini* and *A. lessini* diffuse patch around end of cell dot absent or, if present, roundish, while in *A. paraselini* and *A. parilella* (only exceptionally in *A. selini*) diffuse patch extending toward tornus as a narrow stripe, median dot usually present in *A. selini* and *A. lessini,* absent in *A. paraselini* and *A. parilella. A. socerbi, A. richteri* and *A. kayseriensis* are also quite similar, but basal field tends to extend along costa as broad pale stripe. Forms of *A. alpigena* may also be very similar, although usually larger, scattered blackish scales more numerous and diffuse patch around end of cell dot absent, but all these features overlapping, therefore safe determination may require dissection.

MALE GENITALIA. Socii large, extending well beyond width of tegumen, usually exceeding gnathos; gnathos cushion-shaped, in natural position narrowly elliptic, in preparations under pressure of cover slip and due to twisting of gnathos arms, approximately orbicular, but always slightly asymmetrical, slightly wider than long and equalling socius in size; valva curved throughout, distal one-third almost parallel-sided, apex broadly rounded, cuiller gradually tapering to rounded apex, curving outwards in middle and often with a hint of an inwards curve at apex, 75–80% of valva width; transtilla strongly thickened in middle; anellus of *alpigena* type, but almost round with low apical processes ending in a right-angle; aedeagus slender, curved, basal process long.

FEMALE GENITALIA. Anterior apophysis three-quarters length of posterior apophysis; sternite VIII width:length 5:2, anterior margin slightly convex, nearly straight in middle, posterior margin angled in towards middle; anterior margin of tergite weakly concave; ostium sharply outlined, forming an incomplete circle at one-third sternite length from anterior margin; ductus bursae long, slender, slightly expanding to ovate corpus bursae; signum wider than long, elliptical, teeth quite small.

GENITALIA DIAGNOSIS. Male, see description and key (p. 120). Female differs from other species in *alpigena* and *angelicella* subgroups by ostium sharply outlined, forming an incomplete circle, and projections of ductus bursae more narrow than in the compared species, often extremely narrow.

DISTRIBUTION. France, Italy, Slovenia, Croatia, Romania and Greece including Crete; Crimea (Savchuk & Kajgorodova, 2020). Also in Asiatic Turkey.

BIONOMICS. Larvae have been found on *Ferulago campestris* (Besser) Grecescu (Buchner, 2017a) and in Crimea on *Ferulago galbanifera* W.D.J. Koch (Savchuk & Kajgorodova, 2020). Moths have been collected from May to October, one record from Greece on 24. April.

REMARKS. See under *A. selini*. The publication of *A. lessini* dates from late 2017, in spite of the date of September 2016 on the relevant issue of *Gortania*.

Ferulago campestris and *F. galbanifera* are often treated as synonymous (e.g. Euro+Med, 2006).

53   *Agonopterix selini* (Heinemann, 1870)

Depressaria selini Heinemann, 1870: 167.

DESCRIPTION. Wingspan 16–20 mm. Head creamy buff, often more or less mixed rusty brown. Labial palp creamy buff, segment 2 with some grey-brown scales mainly on outer side, segment 3 unringed. Antenna dark grey-brown, narrowly ringed buff in proximal part. Thorax and tegula pale buff, rusty brown anteriorly, extending backwards to varying extent. Forewing red-brown, with some yellowish buff scales, mainly in terminal one-third, and with some scattered dark grey scales, costa and terminal quarter often heavily mixed dark grey, some blackish costal and terminal spots present; basal field creamy buff, sometimes tinged ochreous, reaching costa or not, basal fascia outwardly dark red-brown fading into ground colour; oblique and plical dots black, first oblique dot usually followed by a few whitish scales; median dot present, sometimes with white centre, end of cell dot large, white, edged black, in some specimens set in a darker grey-shaded patch; fringe reddish brown tipped grey near apex, grey towards tornus. Hindwing light grey; fringe light grey. Abdomen pale grey-buff.

VARIATION. The extent of yellowish buff or darker grey scales can alter the overall forewing colour from dull brick-red to fuscous, but there is always some dull brick-red coloration; the grey colour in the costa and terminal areas is lost in some specimens; white markings associated with the various cell dots are very variable, sometimes there can be 3 or 4 white marks, in other specimens only the end of cell spot is white; the grey shade around end of cell, if present, rarely continues as narrow stripe toward tornus.

SIMILAR SPECIES. *A. paraselini* has only one white dot at end of cell and this may be absent. *A. parilella* has a dark ring on segment 3 of labial palp and median dot usually absent. More details under *A. lessini*.

MALE GENITALIA. Similar to *lessini*, but socii extending much wider than tegumen, gnathos behaving in same way beneath cover-slip, but smaller than socius; basal excavation of tegumen rather short; valva tapering in distal one-third, apex quite broadly rounded, but narrower than in *lessini*, cuiller typically tapering from broad base to middle (inner margin curved, outer straight) then with a bulge on inner side, before contracting to quite narrow more or less pointed apex, outer edge in distal half somewhat concave, 75–80% of valva width, however there is variation in the amount of curvature and shape and amount of the bulge; transtilla very deep; anellus with apical processes more or less acute; aedeagus slightly tapering from base to apex, basal process long.

FEMALE GENITALIA. Anterior apophysis three-fifths length of posterior apophysis; sternite VIII width:length 5:2, anterior margin convex, with median bulge round ostium, posterior margin slightly concave; ostium near anterior margin of sternite, extending into bulge, distally extended; ductus bursae long, slender, slightly expanding to ovate corpus bursae; signum wider than long, variable in shape and in size and number of teeth.

GENITALIA DIAGNOSIS. Male, see description and key (p. 120). Female differs from other species in *alpigena* and *angelicella* subgroups (except *A. ordubadensis*) by ostium near anterior margin of sternite, extending into bulge, distally extended. In some specimens this feature is less clear, here number of turns of ductus spermatheca can be helpful: 7–9 (usually 8) turns, but in the externally similar *A. parilella* 4–5 turns.

DISTRIBUTION. Spain, mountains of Europe from France to Ukraine, extending north-eastwards to Scandinavia and the Baltic countries and in some Balkan countries south to Greece. Asia: Turkey. The record from Crimea (Lvovsky, 2019) may refer to *A. lessini* which is not mentioned in this paper.

BIONOMICS. *Peucedanum palustre* (L.) Moench, *P. oreoselinum* (L.) Moench *Ligusticum lucidum* Mill. and *Selinum carvifolia* (L.) L. (Buchner, 2017a); *Selinum silaifolia* (Jacq.) Beck (Rymarczyk *et al.*, *in litt.*). Moths have been collected from May to September, one record on 30 April from Greece.

REMARKS. Buchner (2017a) found that material under the name *A. selini* included three species: *A. selini* and two previously undescribed species, *A. lessini* and *A. paraselini* with differences in male and female genitalia, barcode, distribution and host-plants.

54   *Agonopterix ordubadensis* Hannemann, 1959

Agonopterix ordubadensis Hannemann, 1959: 34.

DESCRIPTION. Wingspan 18–24 mm. Face buff, vertex buff to mid-brown. Labial palp buff, segment 2 outer and ventral sides with predominantly dark and light brown scales. Antenna grey-buff to dark grey. Thorax and tegula yellowish buff. Forewing yellowish buff, more or less overlaid deep brown to dark grey, scattered deep brown to blackish scales mostly in costal area but sometimes predominant in terminal area of wing; costa with small blackish spots, termen with blackish spots; basal field yellowish buff, extending onto costa, basal fascia deep brown to blackish, outer edge gradually fading into ground colour; a pair of black oblique dots at one-third, a whitish dot at end of cell partially or fully edged blackish; a dark brown to dark grey patch between these dots and around end of cell dot, extending towards costa; fringe yellowish buff to mid-brown. Hindwing pale grey, darker towards apex; fringe pale grey-buff. Abdomen yellowish grey.

VARIATION. There is great variation in the extent of the dark brown to dark grey patches, at minimum an area at base of wing in dorsal half and a patch from middle of cell surrounding end of cell dot, but often much more extensive covering most of the costal half of the wing and sometimes following the veins in the remainder of the wing. The white dot at end of cell is always small, but may be absent, represented by a black dot. The black oblique dots can be obscured by the dark patch. A median dot is usually obscured by dark scales, but is occasionally white.

SIMILAR SPECIES. *A. putridella* has some similar forms, but the white dot at end of cell is always larger. *A. putridella* subsp. *scandinaviensis* has a dark patch which lies around end of cell dot but not extending to costa.

MALE GENITALIA. Socii large, longer than wide, little exceeding tegumen in width; gnathos narrowly elliptic, exceeding socii by half its length; valva strongly tapering in distal two-thirds, apex rounded but quite narrow, cuiller stout, straight in proximal half, curving outwards from middle then inwards towards pointed apex, 80–85% of valva width; transtilla strongly thickened; anellus similar to *selini*; aedeagus of nearly equal width to four-fifths, basal process long. Cuiller appears shorter and stouter in some preparations, which correlates with a wider sacculus.

FEMALE GENITALIA. Only one female examined, similar to *A. selini* (*q.v.*).

GENITALIA DIAGNOSIS. Male, see description and key (p. 120). Female, see general remark on *alpigena* and *angelicella* subgroups.

DISTRIBUTION. Russia (Orenburg). Asiatic Turkey, Armenia and Azerbaijan.

BIONOMICS. Host-plant unknown. Checked specimens were collected from June to September.

REMARKS. The species was described from one male from Azerbaijan.

55   *Agonopterix socerbi* Šumpich, 2012

Agonopterix socerbi Šumpich, 2012 in Šumpich & Skyva, 2012: 162.

DESCRIPTION. Wingspan 15–18 mm. Head reddish brown, scales tipped buff; face ochreous. Labial palp pale buff, segment 2 heavily mixed with grey-brown and dark brown scales on outer and ventral sides, segment 3 with some grey-brown scales in basal half, a blackish ring beyond middle, tip black. Antenna grey-brown. Thorax bright red-brown, with posterior crest. Forewing bright red-brown, often tinged pale orange beneath costa, between cell and dorsum and in middle of wing at four-fifths, with scattered blackish scales, sometimes more plentiful, then forming dotted lines along veins towards termen; basal field orange-brown, delimited by dark brown to red-brown fascia fading into ground colour; costal and terminal spots blackish; oblique dots black usually followed by a few whitish buff scales, plical dot brownish, median dot blackish, more or less concealed by diffuse dark brown to dark grey patch extending just beyond white end of cell dot; fringe grey, tinged orange-brown. Hindwing light grey; fringe light grey. Abdomen light grey-buff.

VARIATION. The red-brown coloration may be lacking leaving ground colour pale grey-buff with slight overlay of pale orange, especially in basal field and base of costa. There is some variation in the amount of orange suffusion on the forewing; the amount of scattered blackish scales is variable; oblique dots sometimes without following whitish scales, median dot sometimes may have a few associated white scales.

SIMILAR SPECIES. Characterised by bright red-brown coloration with some orange suffusion and orange-brown basal field. Most similar to *A. richteri* and *A. kayseriensis,* details under *A. richteri*.

MALE GENITALIA. Socii not very large, not exceeding tegumen in width; gnathos elliptic, extending beyond socii; valva gradually narrowed from base, more so in distal one-third after curve in ventral margin, apex rounded but quite narrow, cuiller slender, slightly curved outwards, incurved at apex, acute, unusually long, 110–130% of valva width; transtilla strongly thickened in middle; anellus long, almost reaching transtilla, parallel-sided or contracted in middle with a bulge beyond middle, apical processes long, horn-like; aedeagus slender, nearly parallel-sided to four-fifths, basal process long.

FEMALE GENITALIA. Similar to *A. lessini* but anterior apophysis two-thirds length of posterior apophysis; sternite VIII width:length 3:1, anterior margin weakly convex, middle one-third sometimes slightly indented, posterior margin straight with deep excavation in middle; ostium proximal to middle of sternite, sharply outlined with posterior extension; signum wider than long, more or less elliptical, with numerous teeth.

GENITALIA DIAGNOSIS. Male, see description and key (p. 120). Female, see general remark on *alpigena* and *angelicella* subgroups.

DISTRIBUTION. Slovenia, Italy.

BIONOMICS. Larvae on *Ligusticum lucidum* Mill. (Sonderegger, *in litt.*). Checked specimens were collected from June to September.

angelicella subgroup (56–58)
Sacculus short, ending closer to base of valva than in *alpigena* group.
Male genitalia of this subgroup are keyed together with *alpigena* group (p. 120).
Female genitalia are mostly not reliably separable, as in the *alpigena* group.

56   *Agonopterix angelicella* (Hübner, 1813)

Tinea angelicella Hübner, [1813]: pl. 49, fig. 337.

DESCRIPTION. Wingspan 15–19.5mm. Head ochreous, sometimes tinged cinnamon-pink, face buff. Labial palp pale ochreous-buff, more or less tinged cinnamon-pink, segment 2 with scattered dark brown scales on outer side. Antenna light fuscous, narrowly ringed deep brown. Thorax ochreous, more or less tinged cinnamon-pink, sometimes light brown anteriorly. Forewing usually ochreous more or less tinged cinnamon-pink, more strongly beyond basal fascia and along costa, with scattered fuscous scales over most of wing but mainly absent from an area surrounding oblique dots and extending towards mid cell dot and towards costa; subdorsal spot brown; costal spots grey; terminal dots fuscous, sometimes obscure; basal field pale ochreous, often suffused with darker colours of forewing, not reaching costa; basal fascia mid-brown; oblique dots blackish, the second smaller and sometimes absent, followed by another blackish dot at end of cell, more or less obscured in a dark grey patch which extends towards costa and as a narrow curved line towards tornus; plical dot blackish, often elongate; fringe light ochreous-buff, often tinged cinnamon-pink, with fringe line, sometimes very faint. Hindwing whitish-buff, light grey posteriorly; fringe whitish buff with one or two faint fringe lines, sometimes absent. Abdomen light ochreous-grey.

VARIATION. The extent and depth of the cinnamon-pink coloration is variable and may be entirely absent, leaving ground colour buff to ochreous-buff; the abundance of scattered fuscous scales is very variable and may extend into an overlay of fuscous, particularly right across base, including basal field, and another area on costa at one-half.

SIMILAR SPECIES. *A. astrantiae* (q.v.); *A. parilella* (q.v.). Typical forms recognisable by combination of small size (not reaching 20 mm wingspan), forewings medium ochreous with scattered fuscous scales all over wing well developed, median cell dot absent, end of cell dot black, surrounded by a diffuse dark area extending to costa or nearly so and as a narrow diffuse stripe to tornus, diffuse pale area between cell dots and oblique dots in costal half. If specimens are very pale or very dark these features may become obscure.

MALE GENITALIA. Socii large, exceeding tegumen in width; gnathos not or only just exceeding socii, elliptic in natural position, under pressure of cover slip and due to twisting of gnathos arms, more broadly elliptic and often asymmetrical; valva with ventral margin concave at end of sacculus, distal part weakly tapering, apex rounded, quite wide, cuiller more or less stout, basal half tapering, curved slightly outwards just beyond middle, apical third sometimes with bark-like fissures on inner side, apex rounded, 65–68% of valva width; transtilla moderately thickened in middle; transtilla lobes rather small, not overlapping transtilla or anellus; anellus plate orbicular, posterior margin entire or with wide shallow sinus; aedeagus slender, slightly tapering, curved beyond middle, with long group of minute cornuti, basal process long.

FEMALE GENITALIA. Anterior apophysis three-fifths to two-thirds length of posterior apophysis; sternite VIII width:length 4:1, anterior margin weakly convex, posterior margin weakly concave; ostium slightly distal to middle of sternite; ductus bursae long, gradually expanding to ovate corpus bursae; signum wider than long, more or less elliptical, teeth varying in size and number.

GENITALIA DIAGNOSIS. Male, see description and key (p. 120). Female, see general remark on *alpigena* and *angelicella* subgroups.

DISTRIBUTION. Middle latitudes of Europe from Ireland and France to Finland in the north, Italy, Croatia, Ukraine and Russia; absent from the Baltic countries and most of the Balkans. Asia: we have seen specimens from Russian Altai, records from further east may represent different subspecies or closely related but different species.

BIONOMICS. Larvae gregarious on leaves of *Angelica sylvestris* L. (Hannemann, 1953); also recorded on *Pastinaca sativa* L., *Pimpinella saxifraga* L., *Aegopodium podagraria* L. and *Heracleum* (Palm, 1989); *Peucedanum palustre* (L.) Moench

(Harper *et al.*, 2002); *P. ostruthium* (L.) W.D.J. Koch, *Laserpitium latifolium* L. and *L. krapfii* Crantz (Sonderegger, *in litt.*). We have seen specimens collected May to end of September. Most reared specimens emerged in June.

57   *Agonopterix paraselini* Buchner, 2017

Agonopterix paraselini Buchner, 2017a: 94.

DESCRIPTION. Wingspan 14–18 mm. Head creamy buff. Labial palp creamy buff, segment 2 outer side with a few dark grey scales, segment 3 with dark grey scales in middle, sometimes forming a distinct ring. Antenna thicker in male than female, dark fuscous above, paler below. Thorax and tegula creamy buff, sometimes red-brown or fuscous anteriorly, occasionally mainly red-brown throughout. Forewing red-brown, with some yellowish buff scales, mainly in terminal one-third, and with some scattered dark fuscous scales, some blackish costal and terminal spots usually present; basal field more or less concolorous with thorax, cream, often more or less suffused orange-brown, reaching costa or not, dark brown basal fascia outwardly dark red-brown fading into ground colour; oblique dots black, rarely followed by a few whitish scales, a plical dot usually present, median dot absent, end of cell dot large, white, edged black, set in a darker grey-shaded patch which usually extends narrowly to tornus; fringe reddish brown, tipped grey, grey towards tornus. Hindwing light grey; fringe light grey with one or more fringe lines. Abdomen grey-buff.

VARIATION. The extent of yellowish-buff or darker grey scales can alter the overall forewing colour from dull brick-red to fuscous, but there is always some dull brick-red coloration; plical dot, or short line, usually present; usually only the end of cell spot is white. Labial palp segment 3 with or without a blackish ring.

SIMILAR SPECIES. Similar to *A. selini* and *A. lessini* (*q.v.*) and several other species, but smaller than most, with usually only end of cell dot white, plical dot present: a dark area extends from around end of cell dot towards tornus, as in *A. angelicella*, which is very closely related; it shares barcode and has nearly identical genitalia, but is well separated by biology, external appearance also is different (rarely overlapping): forewings dark red-brown, scattered fuscous scales rather indistinct, end of cell dot with white centre, no distinct diffuse pale area between cell dots and oblique dots in costal half.

MALE GENITALIA. Very similar to *angelicella*, but cuiller with whole of apical one-third finely fissured like bark, 70–75% valva width, transtilla more thickened in middle, distal margin of anellus without sinus.

FEMALE GENITALIA. Very similar to *A. angelicella*, anterior apophysis two-thirds length of posterior apophysis; sternite VIII width:length 3:1.

GENITALIA DIAGNOSIS. Male, see description and key (p. 120). Female, see general remark on *alpigena* and *angelicella* subgroups.

DISTRIBUTION. Mountain areas from France to Poland and Slovenia. Asiatic Turkey.

BIONOMICS. Larvae have only been found on *Peucedanum cervaria* (Buchner, 2017a). Larvae collected in May produced adults in June and July. Adults have also been taken in summer up to 16 September, but not in autumn or spring, suggesting that the species does not overwinter as adult.

REMARKS. The publication of *A. paraselini* dates from late 2017, in spite of the date of September 2016 on the relevant issue of *Gortania*.

This species shares barcode with *A. angelicella*.

58   *Agonopterix parilella* (Treitschke, 1835)

Haemylis parilella Treitschke, 1835: 178.
   Haemylis humerella Duponchel, 1840: 619.

DESCRIPTION. Wingspan (14–)15.5–18 mm. Head ochreous-buff, face ochreous-buff. Labial palp buff, segment 2 more or less heavily speckled fuscous on outer side, segment 3 ochreous-buff with some light fuscous scales at three-quarters forming incomplete ring. Antenna dark fuscous. Thorax ochreous-buff, darker brown anteriorly. Forewing from orange-brown with cinnamon tinge to fuscous or deep brown, with scattered blackish scales, more numerous in terminal area; costal spots blackish, often separated by spots of cinnamon or light to dark brown with cinnamon tinge; terminal spots fuscous; basal field pale ochreous to ochreous-buff with deep brown spot at base of costa; basal fascia deep brown, gradually less dark outwardly; oblique dots blackish, surrounded by mid-brownish orange patch in some specimens, small black median dot, white dot at end of cell ringed black; plical dot black; fringe mid-brown to fuscous, sometimes with fringe line. Hindwing light grey, darker towards termen; fringe light grey with two fringe lines. Abdomen grey to brownish grey.

VARIATION. Some specimens much darker than others, with dark markings mostly obscured by dark ground colour. Plical dot and median dot often absent. White spot at end of cell occasionally absent, represented by a large black dot.

SIMILAR SPECIES. *A. selini* and some other species of *A. alpigena* group, details under *A. lessini*. Forms with end of cell dot surrounded by a diffuse dark area extending as a narrow diffuse stripe toward tornus may resemble *A. angelicella* and *A. astrantiae* at first sight, but these species never have such dark forewings as some forms of *A. parilella*. *A. angelicella* does not have a white dot. *A. astrantiae* is larger and median cell dot is present.

MALE GENITALIA. Socii rather variable in size, just exceeding tegumen in width or not; gnathos elliptic, 2.5–3 times as long as wide, far exceeding socii; valva gradually tapering from middle to quite slender apex, cuiller parallel-sided with slight outwards curve at middle or straight, apex with curve on outer side, apex rounded obtuse, bark-like fissures weakly developed, 80–88% of valva width; transtilla thickened in middle; anellus and transtilla lobes as in *angelicella*; aedeagus slightly tapering throughout, curved from before middle, basal process long, sometimes reaching one-half.

FEMALE GENITALIA. Similar to *A. angelicella* but anterior apophysis about three-fifths length of posterior apophysis; sternite VIII width:length 3:1, anterior margin weakly convex on either side of wide, slightly concave median part, posterior margin angled in towards middle; ostium in middle of sternite; signum wider than long, variable in shape, three-sided with anterior side curved, or narrow and constricted in middle, teeth varying in size and number.

GENITALIA DIAGNOSIS. Male, see description and key (p. 120). Female, see general remark on *alpigena* and *angelicella* subgroups.

DISTRIBUTION. Middle latitudes of Europe from France, Italy, Croatia and Romania north to Sweden and the Baltic countries; Russia. Asiatic Turkey.

BIONOMICS. *Peucedanum cervaria* (L.) Lapeyr., *Peucedanum oreoselinum* (L.) Moench., *Selinum carvifolia* (L.) L., *Seseli libanotis* (L.) W.D.J. Koch (Hannemann, 1953). According to Hannemann (1995) moths hibernate and eggs are laid in spring. We have only seen specimens collected May to September and found no evidence for hibernation.

ciliella subgroup (59–61)
Valva with costal margin curved towards apex at five-sixths, saccus well developed.

59   *Agonopterix ciliella* (Stainton, 1849)

Depressaria ciliella Stainton, 1849: 161.
  Depressaria annexella Zeller, 1868: 416.

DESCRIPTION. Wingspan (19–)22–26 mm. Head mid-brown, scales tipped buff; face light brownish-buff. Labial palp segment 2 buff with some reddish-brown scales on brush and at apex on inner side, segment 3 buff with deep brown rings at base and from half to three-quarters, sometimes tipped deep brown. Antenna light grey-brown, narrowly ringed deep brown at least in basal half. Thorax mid-brown, becoming buff posteriorly and in posterior half of tegula. Forewing mid-chestnut brown often with yellowish buff areas, particularly along costa, widened at two-fifths to touch oblique dots and often forming an indistinct angled fascia beyond end of cell; costal and terminal spots dark brown, often alternating with pale spots; basal field yellowish buff extending onto costa; basal fascia dark brown, fading into sub-basal area; an irregular darker brown cloud in costal half of wing from half to three-quarters, and another in subterminal area, some dark brown scales scattered throughout, but mainly along veins in terminal area; oblique dots black, with adjacent white scales, sometimes lacking next to first dot, a white median dot and another at end of cell with some black edging; fringe concolorous with forewing to buff, with two indistinct fringe lines. Hindwing light greyish-buff, often darker towards apex; fringe pale buff with five fringe lines, the second sometimes very faint. Abdomen grey-brown.

VARIATION. Forewing coloration varies in intensity and some specimens have a reddish tinge; the intensity of the darker brown markings is very variable and they may be absent; the first white dot and rarely the second may be entirely replaced by black.

SIMILAR SPECIES. Similar to *A. heracliana*, but larger and more richly coloured. The five fringe lines of the hindwing are characteristic, although not always very clear. The extra-limital *A. caucasica* Karsholt, Lvovsky & Nielsen, 2006 is also similar.

MALE GENITALIA. Socii of moderate size, outer sides nearly equal; gnathos lanceolate, 3.5–4 times as long as wide, far exceeding socii; valva nearly parallel-sided to two-thirds, then ventral margin curved, apical third gradually tapering, costal margin curved at five-sixths, resulting in strong taper giving narrow apex, cuiller quite variable, slightly incurved near base, erect from middle onwards, or straighter with a narrowed section near base, or with a slight bulge in basal one-half, apex rounded, 63–67% of valva width; transtilla not to moderately thickened in middle; anellus orbicular with faint V-shaped incision; saccus more developed than in most *Agonopterix*; aedeagus tapering near base, then hardly tapering till close to apex, minute cornuti in two long rows, not always distinguishable, basal process quite short.

FEMALE GENITALIA. Anterior apophysis half to two-thirds length of posterior apophysis; sternite VIII width:length 4:1, anterior margin convex, posterior margin angled in towards an excavation in middle; ostium near posterior margin of sternite; ductus bursae long, slightly expanding to ovate or elliptical corpus bursae; signum wider than long, rather variable in shape, often slightly constricted in middle, teeth rather uniform in size.

GENITALIA DIAGNOSIS. For genitalia differences from *A. orophilella* and *A. heracliana* see under those species.

DISTRIBUTION. Throughout northern Europe, extending south-west to northern Portugal, Spain and Italy and south-east to Romania, Russia (Ural Mts.). We have not seen specimens collected outside Europe. The record from Asiatic Turkey (Buchner, 2017c) refers to *A. orophilella*. According to Lvovsky (2019) widespread in Russia. According to Hannemann (1995) in Siberia, Far East Russia and North America. Hodges (1974) mentioned single records from Nova Scotia, Canada in 1915 and 1943. He wondered if the species was established or would spread further through eastern North America. There are no records in BOLD from these regions and it is not listed from Japan in Japan Moths (2002–2023).

BIONOMICS. Larvae on *Daucus carota* L., *Selinum carvifolia* (L.) L. and *Heracleum sphondylium* L. (Hannemann, 1953); *Peucedanum palustre* (L.) Moench and *Carum* L. (Benander, 1965); *Angelica sylvestris* L., *A. archangelica* L., *Sium latifolium* L., *Cicuta virosa* L. and *Pimpinella saxifraga* L. (Palm, 1989); *Silaum silaus* (L.) Schinz & Thell., *Pastinaca sativa* L., *Aegopodium podagraria* L. and *Meum athamanticum* Jacq. (Harper *et al.*, 2002); *Astrantia major* L., *Laserpitium halleri* Crantz, *Chaerophyllum aureum* L., *C. villarsii* Koch, *Seseli libanotis* (L.) Koch, *Peucedanum ostruthium* (L.) Koch., *P. venetum* (Sprengel) Koch and *Pimpinella nigra* L. (Sonderegger, *in litt.*); *Oenanthe silaifolia* M. Bieb. (Corley, 2015); *Pimpinella major* (L.) Huds., *Laserpitium latifolia* L., *L. siler* L., *Anthriscus sylvestris* (L.) Hoffm., *Ligusticum lucidum* Mill., *L. ferulaceum* All. and *Peucedanum alsaticum* L. (Rymarczyk *et al., in litt.*). The adult moths hibernate. In northern Portugal MC found larvae of this species in May 2013 which produced adults in June. The same site was visited in July 2013 when much smaller larvae were found, suggesting a second generation.

60   *Agonopterix orophilella* Rymarczyk, Dutheil & Nel, 2013

Agonopterix orophilella Rymarczyk, Dutheil & Nel, 2013: 16.

DESCRIPTION. Wingspan 16–20 mm. Head dark fuscous, scales sometimes tipped buff, face whitish with some brown scales. Labial palp segment 2 whitish buff, outer side mixed grey, segment 3 whitish with two blackish rings. Antenna with scape grey mottled white, flagellum weakly ringed grey and brown. Thorax and tegulae dark fuscous, paler posteriorly. Forewing narrow, buff, more or less heavily overlaid fuscous to dark fuscous; basal field grey-brown, basal fascia deep brown, inconspicuous against ground colour, costa grey or grey-brown with blackish brown costal spots; four white dots in cell, the first two following blackish oblique dots, frequently joined, the median dot slightly elongate, smaller than the end of cell dot, both more or less edged black; a weak V-shaped fascia of slightly paler scales beyond end of cell; blackish terminal dots; fringe reddish brown with a brown line. Hindwing light grey, darker towards apex; fringe light grey.

VARIATION. On the forewing the amount and depth of the fuscous overlay varies; the white dots following the oblique dots may be separate or fused.

SIMILAR SPECIES. Similar to darker forms of *A. heracliana*, but forewing distinctly narrower.

MALE GENITALIA. Similar to *ciliella*, but socii more strongly divergent, valva shorter, broader at apex, cuiller with slight bulge in middle; anellus with lateral bulge at middle, and small sclerotised lateral process at three-quarters; aedeagus shorter, with slightly longer basal process.

FEMALE GENITALIA. Similar to *A. heracliana*. Sternite VIII with anterior margin convex laterally, straight in middle, posterior margin angled inwards; ostium in middle of sternite; corpus bursae small, ovate, signum wider than long, elliptical, teeth numerous, small.

GENITALIA DIAGNOSIS. In male cuiller as in *A. ciliella*, socii (wider expanded laterally) and anellus (sclerotised lateral processes at three-quarters) and with clear differences in gnathos from both *A. ciliella* and *A. heracliana*. Only one female specimen available, which shows no differences compared to *A. heracliana*.

DISTRIBUTION. France (Savoie, Hautes Alpes). One record from Asiatic Turkey, 50 km NE Erzurum, 1600 m, in Buchner (2017c) reported as *Agonopterix ciliella* in error.

BIONOMICS. Larva on *Anthriscus sylvestris* (L.) Hoffm. (Rymarczyk *et al.*, 2013). Occurs at altitudes of 1600–2300 m. Larvae full grown in July. The adult has been taken at light at end of July.

REMARKS. The species is known from few specimens, mostly reared from larvae.

61   *Agonopterix heracliana* (Linnaeus, 1758)

Phalaena (Tortrix) heracliana Linnaeus, 1758: 532.
  Phalaena punctata Clerck, 1759: pl. 2, fig. 15.
  Pyralis applana Fabricius, 1777: 294.
  Phalaena cerefolii Retzius, 1783: 45, *nomen invalidum.*
  Tinea cicutella Hübner, 1796: 39, pl. 12, fig. 79.

DESCRIPTION. Wingspan (15–)18–23(–24) mm. Head light brownish buff, face whitish buff. Labial palp pale buff, segment 2 with scattered fuscous scales on outer side and ventrally, segment 3 greyish fuscous ringed at base, two-thirds and at apex. Antenna light fuscous with narrow dark brown rings. Thorax light brownish buff, sometimes darker, anteriorly brown. Forewing mid-brown to fuscous, often with weak reddish tinge, with scattered fuscous scales; costal spots fuscous separated by light fuscous buff; terminal spots fuscous, often weak; basal field buff, more or less mottled fuscous; basal fascia brown reaching costal margin of cell, outwardly gradually fading into ground colour; oblique dots black, with white scales on outer edge, particularly of second dot; white or whitish buff dots at one-half and end of cell, narrowly ringed dark fuscous; blotch fuscous often indistinct; fringe mid-brown to fuscous with faint reddish tinge, with two fringe lines. Hindwing whitish buff, greyer towards termen, fringe whitish buff, fringe with two fringe lines, sometimes very faint. Abdomen light fuscous buff.

VARIATION. Ground colour from mid-brown with weak reddish tinge to fairly dark fuscous. The palest specimens with mid-brown overlaid in large patches. Sometimes the ground colour is uniformly fuscous obscuring most of the markings except the white dots. Iberian specimens more variegated light and dark fuscous without any reddish tinge.

SIMILAR SPECIES. The white scales adjacent to the second oblique dot together with the two white dots beyond in cell are characteristic, but this feature is shared with several other species: *A. ciliella* (larger and more richly coloured), *A. capreolella* (smaller), *A. cotoneastri* (usually slightly smaller), *A. scopariella* (apex square); diffuse paler areas interspersed in mid-brown forewing are typical for *A. ciliella* subgroup (although not found in all specimens), but usually not found in the compared species. *A. perezi* (*q.v.*). Ratio antenna-length:forewing-length (80% in males, 75% in females) is larger than in any similar species, where it does not exceed 70% (usually clearly less).

MALE GENITALIA. Similar to *ciliella*, gnathos 3–3.5 times as long as wide; valva from broad base slightly tapering to two-thirds then more strongly tapering, cuiller usually slightly inclined inwards, weakly S-shaped, hooked outwards at acute apex, 66–70% of valva width; transtilla lobes rarely overlapping anellus; aedeagus usually shorter and more tapering, two rows of cornuti closely approximated and often appearing as one row.

FEMALE GENITALIA. Anterior apophysis half to three-fifths length of posterior apophysis; sternite VIII width:length 3:1, anterior margin bulging, middle part straight or curved, posterior margin angled in towards middle; ostium placed at one-third sternite length from anterior margin; ductus bursae long, gradually expanding, not clearly distinct from small elongate corpus bursae; signum rather small, wider than long, variable in shape, often slightly constricted in middle, teeth usually numerous, variable in size.

GENITALIA DIAGNOSIS. Male genitalia separated well from the former two species by shape of cuiller (hooked outwards at acute apex). In female best distinguishing features compared with *A. ciliella* are found in ostium: in *A. ciliella* it is in or slightly distal to centre of sternite VIII and tends to be circular, in *A. heracliana* it is proximal of centre and tends to be elongated distally. However intraspecific variability and differences in preparation can result in some specimens being impossible to name reliably. Another feature is worth mentioning: (sample size: 5 *ciliella* and 7 *heracliana*): in *A. ciliella* the section of ductus bursae lying between approximately one-fifth and two-fifths (measured from posterior end) is more thick-walled with oblique and transverse folds and ridges and tends to become stained more intensely with chlorazol black than the section from two-fifths onwards. In *A. heracliana* this section is not structurally differentiated.

DISTRIBUTION. Nearly all Europe but absent from several countries in the south and from all Mediterranean islands. We have not seen specimens collected outside of Europe, also no data in BOLD from there. Widespread in western Russia, reaching Altai region (Lvovsky, 2019); Japan (Japan Moths, 2002–2023); North Africa (Hannemann, 1995).

BIONOMICS. Larvae most frequently found on *Anthriscus sylvestris* (L.) Hoffm. (Harper *et al.*, 2002) but also found at times on many other species of Apiaceae including *Aegopodium podagraria* L., *Chaerophyllum temulum* L., *Conopodium majus*

(Gouan) Loret, *Daucus carota* L., *Heracleum sphondylium* L., *Meum athamanticum* Jacq., *Myrrhis odorata* (L.) Scop., *Pastinaca sativa* L., *Oenanthe* spp., *Torilis japonica* (Houtt.) DC., *Angelica sylvestris* L., *Silaum silaus* (L.) Schinz & Thell., *Smyrnium olusatrum* L., *Sison amomum* L. and *Ligusticum scoticum* L. (Harper *et al.*, 2002); *Seseli libanotis* (L.) W.D.J. Koch (Benander, 1965); *Carum carvi* L., *Chaerophyllum aureum* L., *C. villarsii* Koch, L. and *Laserpitium latifolium* L. (Sonderegger, *in litt.*); *Conium maculatum* L. (Huisman, 2012); *Sanicula europaea* L. (W. Stark, *in litt.*); Palm (1989) adds *Peucedanum*. In the Iberian Peninsula the main host-plant is *Oenanthe crocata* L. (Corley, personal observation). Flying from June (rarely May) until May of the next year. It comes to house lights, sometimes in numbers on mild nights in winter, mainly after New Year, but is not commonly attracted to mercury vapour light.

REMARKS. Iberian specimens show differences in ground colour, but genitalia and larval appearance are not different from those from more northern parts.

The complex nomenclatorial history of this species, which was long known as *applana* Fabricius, is examined in detail by Karsholt *et al.* (2006).

perezi subgroup (62)

62   *Agonopterix perezi* (Walsingham, 1908)

Depressaria perezi Walsingham, 1908: 957.

DESCRIPTION. Wingspan 18–21 mm. Head with face yellow-buff, vertex buff to reddish brown. Labial palp buff, segment 2 with mix of fuscous and sometimes cinnamon scales on outer and ventral sides, segment 3 with dark grey rings above base and at two-thirds, extreme apex black. Antenna fuscous. Thorax and tegula darker anteriorly, yellowish buff to orange brown, scales at least tipped yellowish buff. Forewing apex rounded, light grey-brown to bright chestnut-brown, with scattered blackish scales, more or less extensively mixed buff to yellowish buff; sometimes with larger patches of grey or darker brown; costa with a series of dark grey spots, terminal dots usually weak, dark grey; basal field sometimes extending onto costa for a short distance, buff to bright yellow-buff, sometimes with reddish-brown tinge, delimited by a blackish partial fascia outwardly fading into ground colour, inwardly edged with bright buff or yellow line extending parallel with costa to one-sixth, a small black dot detatched from basal fascia near its costal end; oblique dots black, separate or joined, each with an associated buff to yellow spot on posterior edge, often coalescing into one spot, median and end of cell dots buff to yellow, the median smaller or sometimes absent, both sometimes partly outlined red-brown; greyish blotch more or less distinct; fringe concolorous with outer part of forewing, with line. Hindwing pale grey, slightly darker towards apex; fringe pale grey with several faint lines. Abdomen buff.

VARIATION. Forewing coloration very variable. The median cell dot may be absent.

SIMILAR SPECIES. The oblique dots and associated pale spot or spots are usually more conspicuous than the median and outer cell dots. The detached black dot costad of the black bar delimiting basal field and the yellowish line that runs from the basal field along costal margin to one-sixth are unusual features. The other very variable *Agonopterix* of Macaronesia is *A. conciliatella*, but this has right-angled forewing apex.

MALE GENITALIA. Socii elliptical, erect; gnathos with basal unspined portion about one-quarter total length, not exceeding socii; valva narrow, elongate, costal margin curved at two-thirds, tapering to quite narrow rounded apex, cuiller well distanced from base, paddle-shaped, with more or less slender middle and gradually expanding distal half with broadly rounded apex, 88–93% of valva width; transtilla moderately thickened; transtilla lobes not overlapping anellus; anellus lobes as high as wide with crenate margin; anellus elliptic with narrow, deep incision; aedeagus curved in middle, numerous minute cornuti forming a short to long patch.

FEMALE GENITALIA. Anterior apophysis three-fifths length of posterior apophysis; sternite VIII width:length about 4:1, anterior margin slightly indented in middle part, with microtrichia, a protruding lip partly covering ostium, posterior margin with median excavation; ostium near anterior margin of sternite; ductus bursae slightly expanding to large elliptical corpus bursae; signum large, more or less rhombic, some edges concave, anterior and posterior angles produced, teeth numerous, large.

GENITALIA DIAGNOSIS. In male best single feature is paddle-shaped cuiller (compare with *A. rutana* and *impurella* and *rotundella* subgroups). In female anterior margin of sternite VIII with microtrichia and a protruding lip partly covering ostium are distinctive.

DISTRIBUTION. Canary Islands and Madeira.

BIONOMICS. Larva on *Ruta pinnata* L.f. Baez (1998) adds *Drusa oppositifolia* DC. Aguiar & Karsholt (2006) mention *Oenanthe pteridifolia* Lowe but this resulted from a misidentification of *Agonopterix vendettella* (Buchner & Karsholt, 2019). Moths from April to April (rarely May) of the following year.

REMARKS. There appears to be a degree of uncertainty regarding host-plants for this species. *Drusa oppositifolia* is not known from Canary Islands, but Baez (1998) concerns Canary Islands and not Madeira. This plant is not mentioned under *A. perezi* in Aguiar & Karsholt (2006), furthermore it belongs to Asteraceae, not Apiaceae or Rutaceae.

putridella subgroup (63–65)
Socii almost contiguous to slightly overlapping (not true for *hippomarathri*).

63   *Agonopterix putridella* ([Denis & Schiffermüller], 1775)

Tinea putridella [Denis & Schiffermüller], 1775: 138.

DESCRIPTION. Wingspan 14–18.5 mm. Head ochreous-buff, face pale ochreous-buff. Labial palp segment 2 buff with scattered dark brown scales on outer side, segment 3 pale ochreous-buff. Antenna grey-brown with narrow dark brown rings in proximal half. Thorax pale ochreous-buff sometimes with some grey-brown scales in mid-line. Forewing pale ochreous-buff; basal field ochreous-buff, basal fascia a short dark line reaching halfway across, sometimes interrupted or reduced to a spot, outwardly tinged greyish brown; in a common form, all veins except the subcostal grey to dark grey-brown, this coloration more extensive in cell; terminal spots dark grey-brown, extending around apex onto costa; oblique dots black, a whitish dot ringed black at end of cell, sometimes a black median dot; plical dot blackish; blotch absent; fringe light grey, with one fringe line. Hindwing light greyish buff, darker towards termen, veins darker; fringe pale buff with two weak fringe lines. Abdomen ochreous-buff.

VARIATION. There is variation in the extent and intensity of the darkened area in middle of wing and of the dark veins; the median cell dot can be well-developed to absent. Rarely the whole forewing is blackish fuscous with all markings obscure. Some specimens from the south of France have the whole forewing blackish fuscous except the basal field and sometimes basal half of costa; the cell dots are obscure except end of cell dot which is pale; labial palps, front of thorax and forelegs are dark fuscous. In another form the veins are scarcely marked and the ground colour more or less overlaid with light brown.

Specimens from Norway, Sweden and Finland are smaller and show other external differences. We treat these as a separate subspecies (see below).

SIMILAR SPECIES. The black-ringed end of cell dot distinguishes the typical subspecies from forms of *A. umbellana* and *A. nervosa* with conspicuously marked veins; *A. yeatiana* has conspicuous blotch. The dark form from south France resembles *A. ferulae*, but that has three white dots in cell. See also subspecies *scandinaviensis* and *A. quadripunctata*.

MALE GENITALIA. Socii wider than long, with narrow gap between them; gnathos lanceolate or elliptic, 2.5–3 times as long as wide, exceeding socii; valva broad, ventral margin curved at three-fifths, distal one-third straight or concave, apex with obtuse or acute point, cuiller more or less slender, S-shaped, curving inwards in basal half, outwards beyond middle and inwards at rounded to slightly pointed apex, 75–92% width of valva; transtilla strongly thickened; anellus more or less orbicular, with a V-shaped apical incision; aedeagus slightly curved at middle, minute cornuti forming a long row, basal process strongly developed.

FEMALE GENITALIA. Anterior apophysis half to three-fifths length of posterior apophysis; sternite VIII width:length about 3:1, anterior margin nearly straight, posterior margin with median excavation; ostium in middle of sternite; ductus bursae long, gradually expanding to small ovate corpus bursae; signum rather small, wider than long, more or less elliptical, teeth on plate larger than marginal teeth.

GENITALIA DIAGNOSIS. Determination of the species of *putridella* group using only male genitalia is problematic, because some of the features mentioned in descriptions are tendencies rather than hard facts. Best single feature to separate *A. quadripunctata* from the other two species is relative length of gnathos: not or only just overtopping socii. Differences between *putridella* and *hippomarathri* are mainly found in length of cuiller: more than 80 % of valva width usually only found in *putridella*, but length of 70–80 % can be found in both species, such genitalia remain doubtful. External appearance, host-plants and barcode are different. For female genitalia, see under *A. hippomarathri*.

DISTRIBUTION. Very scattered in Europe and absent from many countries. England, France, Germany, Austria, Czechia, Hungary, Montenegro, Greece, Romania, Ukraine and Russia. Asia: Turkey, Russian Altai Mts.

The population in Norway, Sweden and Finland is here treated as a separate subspecies *scandinaviensis*.

BIONOMICS. Larva on *Peucedanum officinale* L. (Hannemann, 1953); *P. alsaticum* L. (PB); *Trinia glauca* (L.) Dumort. (Sonderegger, *in litt.*). Larva of subsp. *scandinaviensis* on *Seseli libanotis* (L.) W.D.J. Koch (Palm, 1989). Moths fly from May to October, not hibernating.

REMARKS. The dark form from the south of France looks very different, but no clear differences in genitalia are evident.

There are two barcode clusters with p-distance of about 2.3 %. One represents only *A. putridella*, sequenced specimens are from France, Greece, Turkey and Russian Altai Mts. The second is shared with *A. agyrella* (Russia: Tannu-Ola Mts.), *A. archangelicella* (Far East Russia), *A. septicella* (Far East Russia) and *A. buryatica* (Russia, Buryatia, Mt. Khulugaisha), *A. putridella* specimens representing this sequence are from Austria and Russia (North Caucasus). *A. putridella* subsp. *scandinaviensis* also shares this sequence.

Agonopterix putridella subspecies *scandinaviensis* subsp. n.

Type locality: Sweden, Gotland, Näs

DESCRIPTION. Similar to the typical subspecies, but usually smaller, wingspan 15–17 mm; veins only marked towards termen, basal fascia often weak, median cell dot close to black end of cell dot, which is normally without white centre, usually surrounded by a dark grey cloud.

MALE GENITALIA. Indistinguishable from subspecies *putridella*.

FEMALE GENITALIA. Posterior apophysis unusually short, shorter than papilla analis.

MATERIAL EXAMINED. Holotype: ♂, Sweden, Gotland, Näs, e.l. 9.vii.1956, gen. prep. DEEUR 2498 P. Buchner, I. Svensson leg., coll. ZMUC

Paratypes (arranged according to collection date):

1 ♂, Sweden, Oeland, Hulterstad, e.l. 11–21.vii.1973, DEEUR 3811, I. Svensson leg., coll. ZMUC

2 ♂, Sweden, Gotland, leg. larva 7.vi.1994 on *Seseli libanotis*, e.l. 6.vii.1994 (DEEUR 2496) & 12.vii.1994 (DEEUR2495), H. Hendriksen leg., coll. ZMUC

1 ♀, Finland, Kirkkonummi, Porkkala, 31.vii.2002, DNA barcode id. Lepid Phyl 15553, DEEUR 3849, T. & M. Mutanen leg., coll. RCMM

1 ♀, Finland, Finström, Tjudö, larva iv.2006 on *Seseli libanotis*, DNA barcode id. Lepid Phyl 24742, gen. prep. DEEUR 3851 P. Buchner, P. Välimäki. & M. Mutanen leg., coll. RCMM

1 ♂, Finland, Finström, Tjudö, larva 19.vi.2006 on *Seseli libanotis*, DNA barcode id. Lepid Phyl 15554, DEEUR 3850, M. Mutanen leg., coll. RCMM

1 ♂, 1 ♀, Sweden, Gotland, Nar, leg. larva 25.vi.2006 on *Seseli libanotis*, e.l. 15.vii.2006, DEEUR 7387 (♂) & DEEUR 7388 (♀), H. Rowek, N. Savenkov & I. Savenkova leg., coll. ECKU

ETYMOLOGY. Named after Scandinavia, the part of Europe in which it occurs.

REMARKS. We treat this as a subspecies of *A. putridella* because there is no difference in male genitalia or barcode. Specimens determined as *A. quadripunctata* from Norway, Sweden and Finland reared from *Seseli libanotis* belong here. They are illustrated under *quadripunctata* in Palm (1989: Plate 2, figs. 10, 12).

64 *Agonopterix quadripunctata* (Wocke, 1857)

Depressaria quadripunctata Wocke, 1857: 117.

DESCRIPTION. Wingspan 16–18.5(–20) mm. Head with face creamy buff, vertex yellowish buff. Labial palp buff to yellowish buff, unmarked. Antenna light brown. Thorax and tegula yellowish-buff. Forewing yellowish buff, with scattered fuscous or mid-brown dots all over wing surface, sometimes organised to define veins in distal one-third of wing, sometimes a weak light brown wash extends over much of the wing surface except the costal strip; a black dot at extreme base of costa, costal spots weak or absent, a series of blackish dots from three-quarters round apex and termen; basal field weakly defined, concolorous with rest of wing but without scattered dots, delimited by a darker brown area fading outwardly into ground colour; oblique dots blackish, a median more or less diffuse dot and another similar but larger at end of cell, sometimes with minute white centre; a diffuse irregular brown streak sometimes present from middle of cell to beyond end of cell; fringe concolorous with forewing. Hindwing whitish buff to grey, usually darker towards apex; fringe whitish buff to pale grey.

VARIATION. Ground colour shows little variation, but is sometimes partly overlaid by light brown; brown streak in and beyond cell frequently not present; one or other oblique spot and median cell spot are sometimes absent. When the end of cell dot has a white centre it is generally a single scale.

SIMILAR SPECIES. The four black dots in cell give this species its name. Although characteristic they are not always equally developed and sometimes not strongly marked. *A. putridella* subsp. *scandinaviensis* usually has a grey to blackish cloud around end of cell dot.

MALE GENITALIA. Similar to *putridella*, socii contiguous or even overlapping; gnathos elliptic, 2–3 times as long as wide, not or just exceeding socii; valva less broad than *putridella*, cuiller with stout erect basal half, distal half more slender, abruptly curved outwards, apex slightly expanded, incurved, 75–80% of valva width.

FEMALE GENITALIA. Anterior apophysis three-fifths length of posterior apophysis; sternite VIII width:length nearly 4:1, anterior margin weakly convex, posterior margin concave; ostium slightly distal to middle of sternite; ductus bursae long, gradually expanding to small ovate corpus bursae; signum rather small, wider than long, more or less elliptical, teeth mainly small, but a few larger on plate.

GENITALIA DIAGNOSIS. For males see under *A. putridella*. Females see under *A. hippomarathri*.

DISTRIBUTION. Locally distributed in Eastern Europe. Sweden, Finland, Estonia, Poland and Romania. Flight period from May to end of summer; we have not seen specimens collected later than 20 August.

BIONOMICS. Larvae on *Selinum dubium* (Schkuhr) Leute (Palm, 1989).

REMARKS. Both *A. quadripuntata* and *A. putridella* subsp. *scandinaviensis* are present in Sweden and Finland, the latter having been previously identified as *quadripunctata*, e.g. by Palm (1989). As a result *Seseli libanotis* has been included erroneously as a host-plant of *quadripunctata*.

65   *Agonopterix hippomarathri* (Nickerl, 1864)

Depressaria hippomarathri Nickerl, 1864: 3.

DESCRIPTION. Wingspan (14–)15–18(–19) mm. Head buff, face creamy white. Labial palp whitish buff, segment 2 with outer side and rough scales mixed mid-brown, segment 3 unringed, occasionally with a few mid-brown scales. Antenna light brown with narrow dark fuscous rings. Thorax buff. Forewing pale buff with usually weak light brown tinge, with thinly scattered blackish scales including a few on costa; costal spots dark fuscous, rather weak, beginning at three-fifths and extending around termen; a blackish spot at base of costa; basal field pale buff, extending onto costa for variable distance; basal fascia dark brown; oblique dots blackish, the first larger; median cell dot absent, end of cell dot white, always edged with some blackish scales, but not always fully ringed; fringe buff. Hindwing grey, slightly darker towards termen; fringe light greyish buff, with weak line.

VARIATION. There is some variation in ground colour, largely dependent on amount of overlying light brown coloration; exceptionally the whole forewing except basal field has dark fuscous ground colour.

SIMILAR SPECIES. Characterised by small size, pale buff basal field demarcated by curved fascia hardly reaching dorsum, absence of blotch and median cell dot, and by relatively large white spot at end of cell, partially edged blackish. *A. nodiflorella* and *A. medelichensis* have basal field hardly evident and end of cell dot not white.

MALE GENITALIA. Similar to *putridella*, but socii not contiguous or overlapping; gnathos 2–3 times as long as wide, extending well beyond socii; valva usually with costal edge angled towards apex at about five-sixths, cuiller similar to *quadripunctata*, but less strongly curved outwards at middle and not so slender in distal half, about 75% of valva width.

FEMALE GENITALIA. Anterior apophysis a little over half to two-thirds length of posterior apophysis; sternite VIII width:length from 3:1 to 4:1, anterior margin straight or weakly convex, posterior margin concave with median excavation; ostium in middle of sternite; ductus bursae long, slightly expanding to small ovate corpus bursae; signum variable, often small, wider than long, approximately rhombic, but sometimes very narrow, teeth varying in number according to size of plate.

GENITALIA DIAGNOSIS. For males see under *A. putridella*. Female genitalia in *putridella* group very similar throughout, some differences found in number of turns of ductus spermatheca: about 3–4 in *A. putridella*, 2–3 in *A. quadripunctata* and 1–2 in *A. hippomarathri*.

DISTRIBUTION. A mainly southern European species extending from Spain to Greece, Romania and Ukraine, and north to Czechia and Hungary; Corsica and Sardinia. We have not seen specimens collected from outside Europe. Russia, Altai Republic (Lvovsky, 2019).

BIONOMICS. *Seseli hippomarathrum* Jacq. (Hannemann, 1953); *S. austriacum* (Beck) Wohlf., *Laserpitium peucedanoides* L. (Hannemann, 1995); *Trinia glauca* (L.) Dumort., *Athamanta turbith* (L.) Brot. (Sonderegger, *in litt.*); *Seseli longifolium* L. and *S. montanum* L. (Rymarczyk *et al., in litt.*). Emerging starts in May, peaking in June, flying until October.

astrantiae subgroup (66–68)
Valva long with distal one-fifth nearly parallel-sided before abrupt contraction to rounded or obtuse apex, cuiller incurved, shaped like half a boomerang; anellus small; anellus lobes touching one another or nearly so.

66   *Agonopterix astrantiae* (Heinemann, 1870)

Depressaria astrantiae Heinemann, 1870: 165.
  Depressaria isabellina Klemensiewicz, 1898: 173.

DESCRIPTION. Wingspan (18–)19–22 mm. Head and face ochreous. Labial palp ochreous-buff, segment 2 with variable amount of ferruginous brown on outer side, segment 3 sometimes with a weak ferruginous ring at three-quarters. Antenna ochreous-brown to light fuscous with narrow dark brown rings. Thorax ochreous-buff, tinged ferruginous to chestnut-brown anteriorly or more extensively. Forewing pale ochreous-buff to ochreous, tinged cinnamon-pink to reddish brown in some parts, sometimes heavily, particularly in costal third, around cell dots and in subterminal area, with a scattering of dark fuscous scales; costa deeper pink, spots dark grey; terminal spots grey; basal field pale buff with a short spike along costal margin of cell, almost separated from costa by a dark brown dot at base of costa and a narrow darker margin; basal fascia deep ferruginous, fading to ground colour outwardly; oblique dots dark brown, followed by creamy buff dots, the median small, the end of cell dot larger, more or less edged brown, from which a brown line runs obliquely outwards towards dorsum, gradually curving towards tornus; plical mark slightly elongate, dark brown; brownish blotch present, often faint; fringe buff, more or less heavily tinged cinnamon-pink, with faint fringe line. Hindwing pale buff, greyer towards termen; fringe whitish buff with faint fringe line. Abdomen greyish ochreous.

VARIATION. The ground colour may be lighter or darker.

SIMILAR SPECIES. *A. angelicella* is similar in coloration, but smaller and has a dark spot at end of cell and lacks pale dots in the cell and the curved brown line leading to tornus is very weakly developed. *A. parilella* (*q.v.*).

MALE GENITALIA. Socii large, broad, strongly divergent, separated by narrow gap, extending beyond greatest width of tegumen; gnathos elliptic, about 2.5 times as long as wide, hardly extending beyond socii; valva elongate, distal one-fifth weakly tapering or almost parallel-sided, apex subacute, cuiller erect at base, curving inwards above, inner margin gently curved, outer margin with bulge at two-fifths, beyond middle gradually tapering to narrow rounded apex, the whole structure somewhat reminiscent of half a boomerang, about 90% of valva width; transtilla thickened; anellus lobes narrowly separated; anellus unusually small, distal edge with thickened margin, emarginate; aedeagus about half length of valva, with two rows of minute cornuti, not always visible as distinct.

FEMALE GENITALIA. Anterior apophysis about half length of posterior apophysis; sternite VIII width:length from 3:1 to 4:1, anterior margin bulging, posterior margin concave; ostium in middle of sternite; ductus bursae long, expanding to ovate corpus bursae; signum wider than long, narrowly rhombic, anterior and posterior angles produced, teeth numerous.

GENITALIA DIAGNOSIS. For males and females see under *A. cnicella*.

DISTRIBUTION. Middle latitudes of Europe south to Italy, Croatia and Romania and north to Fennoscandia. Uncommon in western parts but present in Ireland, England and France. We have not seen specimens collected from outside Europe.

BIONOMICS. Larva on leaves of *Astrantia major* L. (Hannemann, 1953), probably also on other Central European *Astrantia* species; *Sanicula europaea* L. (Benander, 1965). Emerging in June and July, flying until autumn.

67   *Agonopterix cnicella* (Treitschke, 1832)

Haemylis cnicella Treitschke, 1832: 237.

DESCRIPTION. Wingspan (16–)17.5–20.5(–22) mm. Head reddish brown, scales brownish buff tipped, face creamy buff. Labial palp creamy buff, segment 2 with scattered red-brown scales, rough scales pinkish brown, segment 3 buff to pink, red-brown to dark fuscous at base, a fuscous ring beyond middle sometimes present. Antenna ochreous-fuscous, narrowly ringed dark fuscous, to dark fuscous throughout. Thorax red-brown, scales sometimes tipped greyish buff, posteriorly mainly greyish buff. Forewing dull reddish brown with some admixture of grey-buff scales and dark fuscous scales, the latter mainly in costal third and along veins to termen; costal spots dark fuscous; basal field whitish buff, boundary closer to base on dorsum than on costa with a short spike along anterior margin of cell, field enclosing some red-brown to dark fuscous dots, including one close to dorsum and two on costa; oblique dots white, another white dot in cell at one-half and a larger one at end of cell, each preceded by a few dark red-brown scales; a weakly marked fuscous line from end of cell towards dorsum curving to tornus; blotch fuscous; a weakly marked pale angled fascia is often visible beyond cell; fringe red-brown tinged fuscous, with faint fringe line. Hindwing grey; fringe grey with a basal line and sometimes some additional faint fringe lines. Abdomen pale ochreous-grey.

VARIATION. Forewing coloration is nearly always reddish brown, occasionally reddish brown is absent, leaving forewing paler or darker. The extent of grey-brown and dark fuscous admixture varies considerably. The size of the white dots also varies, and the inner three dots may be much reduced.

SIMILAR SPECIES. The whitish buff basal field with spike along cell margin contrasting with red-brown remainder of forewing and the creamy white cell dots combine to make a unique combination of characters. *A. melancholica* (*q.v.*) lacks reddish coloration, but is otherwise very similar.

MALE GENITALIA. Similar to *astrantiae*, but socii not strongly divergent, well separated, not extending beyond greatest width of tegumen; gnathos extending beyond socii, 2.5–3.5 times as long as wide; valva variable but apical part similar to *astrantiae*, cuiller stouter at base without bulge on external margin but tapering beyond middle to narrow rounded apex, 88–95% of valva width; anellus lobes contiguous or overlapping; anellus small, posterior margin shallowly concave; aedeagus tends to be slightly longer than in *astrantiae*.

FEMALE GENITALIA. Anterior apophysis half to two-thirds length of posterior apophysis; sternite VIII width:length from 3:1 to 5:1, anterior margin bulging but straight in middle one-third, posterior margin inclined inwards to median excavation; ostium distal to middle of sternite, posterior margin close to excavation; ductus bursae long, expanding to ovate corpus bursae; signum wider than long, with narrow parallel-sided wings and short anterior posterior projections, teeth mainly marginal, but a single row in middle of each wing.

GENITALIA DIAGNOSIS. Best features to distinguish male genitalia of *A. astrantiae* and *A. cnicella* are size of anellus lobes and outline of cuiller. Relative length of gnathos is usually also different, but specimens of *A. cnicella* with gnathos not overtopping socii are found. Lateral expansion of socii relative to tegumen width is also different, but a problematic feature for determination, because it is seriously influenced by preparation details. All other features overlap. Genitalia description of *A. melancholica* is based on 2 specimens, it is doubtful whether there are any reliable differences from *A. cnicella*. In females, most characters overlap to some extent and are therefore not useful. The difference in signum shape (wings tapering to apex in *astrantiae*, parallel-sided in *cnicella*) needs to be tested with a larger sample.

DISTRIBUTION. One of the commonest species in Mediterranean countries and islands, extending east to the Black Sea countries and north along the Atlantic coast to England and Norway; in Central Europe north to Poland. North Africa: Morocco. Asia: Turkey, Armenia, Syria, Iran, Georgia, Russia.

BIONOMICS. Larvae on *Eryngium campestre* L. (Hannemann, 1953); *E. maritimum* L. (Harper *et al.*, 2002). Rymarczyk *et al.* (*in litt.*) mention a single occurrence on *Seseli tortuosum* L. Emerging in May and June, flying until November.

REMARKS. See under *A. melancholica*.

68   *Agonopterix melancholica* (Rebel, 1917)

Depressaria melancholica Rebel, 1917a: 21.
  Depressaria funebrella Caradja, 1920: 131.

DESCRIPTION. Wingspan (17–)21–23 mm. Head dark grey-brown, scales tipped buff. Labial palp buff, segment 2 with outer and ventral sides mixed with light brown and grey-brown scales, segment 3 with grey ring at base. Antenna grey-brown. Thorax grey-brown, darker anteriorly. Forewing dark grey-brown or brown, mixed with black scales; basal field grey-buff, reaching costa, with short spike along costal edge of cell, delimited by blackish partial fascia reaching cell; costa grey-buff, interrupted by blackish spots, terminal spots small; oblique dots black, small, especially the second hardly distinct from background, plical dot present as an indistinct short dash, median cell dot sometimes with white centre, sometimes absent, a larger white dot at end of cell weakly blackish-edged; blackish brown blotch obscure against dark ground colour; an indistinct slightly paler angled fascia at three-quarters; fringe grey-buff. Hindwing light grey; fringe light grey. Abdomen grey-buff.

VARIATION. There is variation in forewing ground colour from dark brown to dark grey. Pale scales associated with the median dot and less often the oblique dots can be present.

SIMILAR SPECIES. Characterised by dark grey or brown coloration with one or sometimes two white dots, and labial palp segment 3 with grey basal ring. Dark forms of *A. heracliana* have 3 to 4 white dots. Dark forms of *A. putridella* have pale thorax.

MALE GENITALIA. Similar to *cnicella*, socii separated by narrow gap; gnathos lanceolate, 3–4 times as long as wide; cuiller 82–95% of valva width; aedeagus stouter than in *cnicella*.

FEMALE GENITALIA. Similar to *cnicella* but anterior apophysis less than half as long as posterior apophysis; sternite VIII width:length 7:2, anterior margin bulging, without flattened part in middle.

GENITALIA DIAGNOSIS. Differences in male from *A. cnicella*, see description, but these characters may not be reliable. Female genitalia see description, but this is based on one specimen only.

DISTRIBUTION. Romania, Russia: Orenburg and Uralsk, also Omsk in Asian Russia.

BIONOMICS. Host-plant unknown. Checked specimens have been collected in July and August. Phenology probably similar to *A. cnicella*.

REMARKS. Specific differences between *A. melancholica* and *A. cnicella* are small, as they have nearly identical male genitalia and incomplete separation by barcode. It is usually larger than *A. cnicella* (but wingspans overlap), also differing in the dark coloration of the forewings, without red tinge, absence of reddish scales around median and end of cell dots (only visible in fresh *A. cnicella*) and the smaller or absent median cell dot. Female genitalia apparently different, but this is based on one specimen only; host-plant unknown, but unlikely to clarify the relationship with *A. cnicella*.

For now, we retain *melancholica* as a separate species from *cnicella*, but we have considerable doubts regarding this decision.

A. funebrella (Caradja, 1920) which has been recorded from Saratov and Orenburg districts of south Russia, has identical male genitalia to *A. melancholica*. We have not seen authentic material nor do we have a barcode, but we are treating it as a synonym of *A. melancholica*, following Lvovsky (2006).

alstromeriana subgroup (69)

69   *Agonopterix alstromeriana* (Clerck, 1759)

Phalaena alstromeriana Clerck, 1759: pl. 10, fig. 1
   Tinea monilella [Denis & Schiffermüller], 1775: 138.
   Tinea puella Hübner, 1796: pl. 12, fig. 82.
   Haemylis albidella Eversmann, 1844: 570.

DESCRIPTION. Wingspan 16–20 mm. Head white. Labial palp white, segment 2 with a few mid-brown scales on outer side, segment 3 sometimes with a mid-brown partial ring at base, with another at two-thirds, tip dark brown to black. Antenna pale buff with narrow dark brown rings. Thorax white, brownish at posterior margin. Forewing white, extensively overlaid light grey-buff, with a few scattered dark fuscous scales; some dark fuscous or black costal spots, particularly at one-half; terminal spots black, often reduced to two; basal field white, extending along costa to two-fifths; basal fascia deeper grey-buff; oblique dots black, followed by a small white patch; an elongate bright ferruginous spot in cell at one-half surrounding a small white dot, followed by a small white area; blotch grey, merging into costal spots; fringe buff with three or four fringe lines. Hindwing light grey, slightly darker towards termen; fringe whitish or whitish buff, with three fringe lines. Abdomen light greyish buff.

VARIATION. This species shows very little variation.

SIMILAR SPECIES. Very distinctive, characterised by pure white head, thorax and basal field, and elongate ferruginous spot in cell.

MALE GENITALIA. Socii triangular (in lateral view), but outline in standard preparation depends on expansion of the connecting membrane; gnathos lanceolate or elliptical, 2.5–3 times as long as wide, exceeding socii; valva elongate, nearly parallel-sided in basal two-thirds, tapering distally to rounded or obtuse angled apex, sacculus short, cuiller of moderate thickness to stout in basal two-thirds, distal one-third more slender, apex rounded, almost straight or with slight outward curvature below and inward near apex, about 70% of valva width; transtilla not thickened; anellus with stalk as long as plate, plate with V-shaped incision to middle; aedeagus less than half as long as valva, basal process strongly developed.

FEMALE GENITALIA. Anterior apophysis about three-fifths length of posterior apophysis; sternite VIII width:length from 3:1 to 4:1, anterior margin convex or with bulge restricted to middle one-third, posterior margin inclined inwards to median excavation; anterior margin of tergite concave; ostium slightly proximal to middle of sternite; ductus bursae long, expanding to small ovate corpus bursae; signum wider than long, more or less elliptical, teeth quite small.

GENITALIA DIAGNOSIS. Male genitalia characterised by unusual shape of anellus in combination with outline of cuiller (anellus may be similar in *A. petasitis*, *q.v.*). Female genitalia without specific distinct features, although one feature may be mentioned: the paler area surrounding the ostium is not circular, but with a little triangular projection directed laterally on either side in distal half. Such projections are also found in quite unrelated species, especially in *A. assimilella*, but their presence can exclude many *Agonopterix* species.

DISTRIBUTION. Nearly all European countries; Sardinia and Sicily, but not recorded from other Mediterranean islands. Asia: We have checked specimens from Turkey, Kazakhstan, Kyrgyzstan and Tajikizstan. Caucasus, Siberia, Morocco (Hannemann, 1995). North America (Hodges *et al.*, 1983). New Zealand (BOLD).

BIONOMICS. Larva on *Conium maculatum* L. (Hannemann, 1953). Moth emerging from end of June to August, hibernating, can be found in spring still very fresh, flying until May or June.

REMARKS. Presumably introduced in North America and New Zealand.

yeatiana subgroup (70–71)
Valva long and slender, cuiller with expanded apical part bearing microsculpture.

70   *Agonopterix capreolella* (Zeller, 1839)

Depressaria capreolella Zeller, 1839: 196.
  Depressaria caprella Stainton, 1849: 157.

DESCRIPTION. Wingspan 14–16(–17) mm. Head light fuscous, face buff. Labial palp buff, segment 2 with scattered dark brown scales on outer side and ventrally, segment 3 light fuscous, dark fuscous at base, forming a ring beyond middle and sometimes at tip. Antenna light fuscous with narrow deep brown rings. Thorax light brownish buff to light fuscous. Forewing light fuscous to mid-grey brown, with weak reddish tinge; with scattered dark fuscous scales; costal spots dark fuscous separated by buff; terminal dots blackish; basal field buff; basal fascia deep brown, reaching margin of cell, outwardly shading into ground colour; oblique dots blackish, usually accompanied by a few white scales, white dots in cell at one-half and at end of cell, sometimes surrounded by a small dark fuscous cloud; fringe greyish buff, more fuscous towards apex, with three fringe lines. Hindwing light grey, becoming darker posteriorly; fringe whitish buff, grey towards apex, with one fringe line. Abdomen light greyish brown.

VARIATION. The ground colour varies in darkness of coloration and depth of reddish tinge; the basal field is sometimes very weakly marked.

SIMILAR SPECIES. Most similar to *A. cotoneastri*, which tends to have area around white dots not darkened, to be slightly more reddish brown and larger, but all these features overlap. Best distinguished by genitalia, if not reared. Smaller than *A. heracliana* and less variegated.

MALE GENITALIA. Socii rounded, rather small; gnathos narrowly elliptical, about 3 times as long as wide, its whole length beyond socii; valva elongate, costal margin with slight hump in first one-third, concave at two-thirds, ventral margin with slight curve at one-half, straight beyond, apical part of valva parallel-sided, narrow apex rounded, cuiller with more or less oval, papillose or punctate plate placed obliquely on short stout stalk, apex directed inwards, 70–75% of valva width; transtilla slightly thickened; anellus plate rounded, slightly wider than long, distally with pair of slightly divergent sclerotised triangles; aedeagus rather short, less than half as long as valva.

FEMALE GENITALIA. Anterior apophysis two-fifths to just over half length of posterior apophysis; sternite VIII narrow, width:length about 5:1, anterior margin convex, middle part slightly raised, posterior margin inclined inwards; ostium slightly distal to middle of sternite; ductus bursae long, expanding to ovate corpus bursae; signum wider than long, usually with narrow parallel-sided wings, sometimes wings less narrow, with or without anterior and longer posterior projections, teeth mainly marginal but numerous on plate when wings are wider.

GENITALIA DIAGNOSIS. Male with rather short cuiller with terminal swelling, long and thin gnathos, valva very narrow in distal third, this combination of features distinct in European *Agonopterix*. Female with ostium rather large, in middle of sternite, proximally surrounded by a curved fold which touches proximal edge of sternite VIII and fades away before it reaches half-way to distal edge of sternite. Compare *A. yeatiana* and *A. cotoneastri*.

DISTRIBUTION. Nearly all Europe but absent from a few countries, mainly in the Balkans and from Mediterranean islands except Sardinia. Asia: Turkey, Armenia, Iran, Russia (Caucasus, Siberia: Buryatia). Widespread in western Russia, eastward to Irkutsk region (Lvovsky, 2019). China (Zhu *et al.*, 2023).

BIONOMICS. Larva on *Pimpinella saxifraga* L., *Sium latifolium* L., (Hannemann, 1953); *Peucedanum cervaria* (L.) Lapeyr. (Sonderegger, *in litt.*), *Daucus carota* L., *Falcaria vulgaris* Bernh., *Silaum silaus* (L.) Schinz & Thell. (Palm, 1989). Moth emerging end of May to July, hibernating, flying until April or May of the following year.

71   *Agonopterix yeatiana* (Fabricius, 1781)

Pyralis yeatiana Fabricius, 1781: 286.
  Phalaena albidana Donovan, 1806: 50.
  Depressaria putrida Haworth, 1811: 509, nec [Denis & Schiffermüller], 1775.
  Depressaria ventosella Stephens, 1834: 198.
  Depressaria atricornella Mann, 1855: 564.
  Depressaria oglatella Chrétien, 1915: 341.
  Depressaria fuscovenella Rebel, 1917: 18.

DESCRIPTION. Wingspan 17–22. Head ochreous-buff to light ochreous-brown, face buff. Labial palp buff, segment 2 with scattered dark fuscous scales on outer side and ventrally, segment 3 with a few dark fuscous scales near base, a ring beyond one-half and extreme apex dark. Antenna light to dark fuscous, with narrow deep brown rings. Thorax buff to light ochreous-brown. Forewing buff to light ochreous-buff, scattered fuscous to deep brown scales, mainly along veins; basal field buff, basal fascia dark brown, from dorsal spot extended across to margin of cell; terminal spots blackish; oblique dots black, a white to pale buff dot at end of cell, narrowly ringed or partially ringed dark fuscous; blotch dark fuscous, round or more or less triangular, sometimes with a small weak mid-brown cloud extending to pale dot at end of cell; fringe ochreous-buff to light brownish buff, with faint fringe line. Hindwing whitish buff, grey posteriorly; fringe pale buff, sometimes tinged grey, with faint fringe line. Abdomen light grey to grey-brown.

VARIATION. Forewing ground colour from buff to pale ochreous-brownish; scattered fuscous dots sometimes very few, but in some specimens marking all veins. Whitish parts of end of cell dot sometimes inconspicuous.

SIMILAR SPECIES. Usually recognisable by combination of pale coloration with conspicuous blotch, black oblique dots, median cell dot absent, end of cell dot whitish with dark ring, but darker specimens also occur.

MALE GENITALIA. Similar to *capreolella*, gnathos 2.5–3.5 times as long as wide, exceeding socii by about half its length; valva without hump on costal margin, cuiller shoe-shaped on short stout stalk, face towards costal margin of valva finely ridged and punctate or papillose, 73–83% of valva width; anellus slightly wider than long, posterior margin nearly straight, but wide V-shaped incision on ventral face; aedeagus moderately stout, tapering in distal third, weakly curved, with cornuti arranged in three adjacent rows, basal process long.

FEMALE GENITALIA. Similar to *A. capreolella*, anterior apophysis about half length of posterior apophysis; sternite VIII narrow, width:length about 4:1, anterior margin convex; ostium in middle of sternite, with a conspicuous curved fold on each side; signum wider than long, nearly rectangular, teeth usually in three rows.

GENITALIA DIAGNOSIS. In male cuiller with "shoe-shaped" terminal part is distinctive. See also *A. cotoneastri* and *A. rutana*. In female ostium rather large, in middle of sternite, surrounded by a curved fold which touches proximal edge of sternite VIII, more distinct than in *A. capreolella* and nearly reaching distal edge of sternite.

DISTRIBUTION. Canary Islands; south and middle latitudes of Europe, including most larger Mediterranean islands. Absent from a few countries. Asia: Turkey. North Africa (Hannemann, 1995).

BIONOMICS. Larva on *Daucus carota* L. (Hannemann, 1995); also recorded on *Sium latifolium* L. (Benander, 1965); *Cicuta virosa* L., *Angelica archangelica* L., and *Carum* L. (Palm, 1989); *Chaerophyllum temulum* L., *Peucedanum palustre* (L.) Moench., *Silaum silaus* (L.) Schinz & Thell., *Apium graveolens* L. and *Oenanthe crocata* L. (Harper *et al.*, 2002); *Oenanthe pimpinelloides* L., *Crithmum maritimum* L. and *Pimpinella peregrina* L. (Rymarczyk *et al., in litt.*). Moth emerging end of May to July, hibernating, flying until May of the next year.

REMARKS. Mann's (1855) *Depressaria atricornella* has frequently been misspelt as *altricornella*.

silerella subgroup (72)

72   *Agonopterix silerella* (Stainton, 1865)

Depressaria silerella Stainton, 1865: 221.

DESCRIPTION. Wingspan 19–25 mm. Head pale buff to light brown. Labial palp pale buff with some light grey scales on outer side of segment 2. Antenna grey-brown. Thorax light brown anteriorly, buff posteriorly. Forewing buff with faint pink tinge when fresh, to light brown in some specimens, with scattered blackish scales some-times partly organised along veins; basal field buff to pale brown delimited from dorsum to middle by dark brown partial fascia; costal spots blackish, not strongly marked, terminal spots blackish; first oblique dot conspicuous, black, second very small or more often absent, median black dot in cell larger than end of cell dot, plical dot present; poorly defined grey-brown patches in middle of terminal area and towards end of dorsum, absent in palest specimens; fringe buff to light brown. Hindwing light grey, fringe pale grey with weak fringe line. Abdomen buff.

VARIATION. Ground colour is pale buff in Austrian specimens, but light brown in those from France. The extent of the scattered blackish scales is very variable. Some specimens have a grey-brown shade in parts, particularly beyond end of cell extending towards termen and near end of dorsum. The second oblique dot is com-monly absent, but may be present and very small.

SIMILAR SPECIES. *A. silerella* is characterised by central forewing dots without any white parts, reduced or absent second oblique dot, median cell dot conspicu-ously larger than end of cell dot, presence of plical dot. Species in the *adspersella* complex have well developed costal spots, two distinct oblique dots and end of cell dot larger than median cell dot.

MALE GENITALIA. Socii with long attachment to tegumen, longer than wide, about equalling width of tegumen; gnathos lanceolate to elliptic, 2–2.5 times as long as wide, exceeding socii; valva broad, ventral margin curved at three-fifths, taper-ing to broad, rounded apex, cuiller with erect basal half, curved outwards at middle, distal half usually more slender, curved in again at narrow, obtuse apex, 96–110% of valva width; transtilla slightly thickened; anellus lobes very narrow, indistinct; anel-lus plate wider than long, distal margin truncate with shallow V-shaped incision; aedeagus quite stout, three-fifths to two-thirds length of valva, bent at middle, taper-ing in distal half, cornuti forming a short patch, basal process strongly developed.

FEMALE GENITALIA. Anterior apophysis half length of posterior apophysis; sternite VIII width:length about 2:1, anterior margin convex, middle part with sem-icircular bulge; ostium in semicircular bulge, the paler area surrounding ostium is not closed distally, but is connected with the pale V-shaped area formed by inwardly inclined posterior margin; ductus bursae long, expanding to quite large elliptical corpus bursae; signum wider than long, narrowly elliptical, teeth quite numerous.

GENITALIA DIAGNOSIS. In male the long, slightly S-curved cuiller tapering in distal half resembles *A. cadurciella*. Best difference is shape of anellus lobes (much narrower), further differences in transtilla (thickened), socii and cuiller (both longer). In female semicircular bulge of sternite VIII enclosing ostium is similar in *A. ligusticella* (*q.v.*), see also *A. medelichensis, A. subtakamukui, A. guanchella*.

DISTRIBUTION. Mountain areas of Central Europe from France to Slovenia and Germany south to Montenegro. Asia: Russia, Altai Republic (Lvovsky, 2019).

BIONOMICS. Larva on *Laser trilobum* Borkh. (Hannemann, 1953); *Laserpitium siler* L. (Sonderegger, *in litt.*). Moth emerging June to August, hibernating, flying until May of the following year.

ligusticella subgroup (73)

73   *Agonopterix ligusticella* (Chrétien, 1908)

Depressaria ligusticella Chrétien, 1908a: 126.
    Agonopterix tripunctaria Buchner, 2015a: 102.

DESCRIPTION. Wingspan 17–18.5 mm. Head light yellowish to reddish ochreous, face yellowish. Labial palp yellowish, a few scattered reddish brown scales on outer side of segment 2. Antenna dark brown, a few yellowish scales on scape. Thorax and tegula yellowish buff, tinged flesh-coloured, or more reddish brown, posterior crest absent. Forewing pale yellowish buff to pale reddish brown, mixed all over with light brown scales, with flesh-coloured tinge, without evident pattern but with a tendency to be grouped, sometimes forming irregular transverse lines; basal field not marked; cell with three black dots, oblique dots with first often larger than second, third dot at end of cell usually associated with some paler scales; terminal spots light brown, inconspicuous; fringe concolorous with end of wing. Hindwings light grey, slightly darker terminally. Abdomen light to dark grey-brown.

VARIATION. French specimens are a richer brown colour than those from further east, which have some flesh colouring when fresh.

SIMILAR SPECIES. Buchner (2015a) mentions specimens misidentified as *A. nodiflorella* and *A. ? rotundella*. Neither of these species exhibits any flesh colour. *A. nodiflorella* has a subdorsal dot near base of forewing, streaky markings associated with some of the veins and acute hindwings. *A. rotundella* has only one dot at one-third in cell.

MALE GENITALIA. Socii rather small, erect; gnathos elliptical, 2.5–3 times as long as wide, exceeding socii; valva broad, costal margin with slight bulge near base, then concave, ventral margin curved at three-fifths, distal one-third tapering to broad rounded apex, cuiller slender, gradually curving inwards or straightening at middle, rarely straight throughout, apical one-sixth with swelling on one side terminating in acute hook on other side, directed inwards or outwards depending on twisting under pressure of cover glass, 95–110% of valva width, crossing costal margin or not depending on curvature of distal half of cuiller; transtilla slightly thickened in middle; anellus tulip-shaped, distal margin with wide shallow notch; aedeagus half as long as valva, slightly bent at middle.

FEMALE GENITALIA. Anterior apophysis half to three-fifths length of posterior apophysis; sternite VIII width:length about 2:1, anterior margin convex, middle part with semicircular bulge, posterior margin inclined inwards; ostium proximal to middle of sternite, partly within semicircular bulge; ductus bursae long, gradually expanding to elliptical corpus bursae; signum wider than long, variable, nearly parallel-sided and narrow, or with distal margin curved, teeth variable in size and number.

GENITALIA DIAGNOSIS. In male the long cuiller with apical swelling, terminating in acute hook is nearly unique in European *Agonopterix*, only similar in *A. socerbi* which has completely different anellus and *A. lidiae* (q.v.). Female with semicircular bulge of sternite VIII enclosing ostium is similar to *A. silerella* (q.v.), also see *A. medelichensis, A. subtakamukui, A. guanchella*.

DISTRIBUTION. France, Italy, Slovenia, Croatia, Greece and Romania; also in Asiatic Turkey.

BIONOMICS. In France larva on flowers of *Ligusticum lucidum* Mill. in July (Chrétien, 1908). Moths emerged in August. Burmann reared this species from *Ferulago campestris* Besser in northern Italy (Buchner, 2015a). Adults have been found from June to August and again in March and April, indicating that it hibernates in the adult stage.

REMARKS. *A. ligusticella* was reared by Chrétien from flowers of *Ligusticum lucidum* in Pyrénées Orientales, where it was refound in 2012 by F. Rymarczyk and M. Dutheil. In MNHN there is no specimen among Chrétien's type specimens. Hannemann (1953, 1958) did not illustrate the genitalia of this species, so presumably he was also unable to find any specimens. The lack of genitalia figures led PB to describe *A. tripunctaria* as a new species. Only later when he was able to examine the F. Rymarczyk and M. Dutheil material, it became clear that *ligusticella* was an earlier name for *A. tripunctaria* Buchner, 2015. The synonymy was published in Buchner (2020b). Leraut evidently was able to locate a type specimen since he illustrates the female type in Leraut (2023).

Burmann identified his specimens reared from *Ferulago* as *A. nodiflorella*, probably misled in part by the shared host-plant.

medelichensis subgroup (74)

74 *Agonopterix medelichensis* Buchner, 2015

Agonopterix medelichensis Buchner, 2015: 107.

DESCRIPTION. Wingspan 14.5–18 mm. Head ochreous to rusty brown, face silvery grey. Labial palp yellowish, segment 2 with a few brown scales in furrow. Antenna dark brown. Thorax and tegula grey-brown, posterior crest absent. Forewing buff, sometimes tinged light brown or pale ochreous-yellow, with scattered brown or blackish scales over the wing, more numerous towards termen; basal fascia not or weakly indicated, sometimes a brown subdorsal spot present; oblique dots black, the first often larger than the second, end of cell dot black, dots not associated with paler scales; light brown to blackish terminal spots present, often extending round apex to outer part of costa; fringe concolorous with forewing ground colour. Hindwing grey or light grey-brown, darker towards apex; fringe grey, with two lines. Abdomen grey-buff to ochreous-buff.

VARIATION. Ground colour of head, thorax and forewing is variable, from light grey-buff, to more ochreous or light brown.

SIMILAR SPECIES. *A. rotundella* has only one dot at one-third; *A. nodiflorella* has strong subdorsal spot and usually has the scattered brown or blackish scales organised into streaks between the veins in the terminal one-third of forewing.

MALE GENITALIA. Socii longer than wide, erect, well separated; gnathos narrowly elliptic, about 3 times as long as wide, exceeding socii; valva with costa weakly concave, ventral margin curved at three-fifths, distally strongly tapered to rounded or obtuse apex, cuiller erect to slightly incurved, stout to three-fifths, then tapering to narrow rounded or obtuse apex which is weakly hooked outwards, 80–90% of valva width; transtilla slightly thickened; anellus plate about as wide as long, with deep, wide V-shaped excavation; aedeagus rather stout, gradually curved, weakly tapered, less than half length of valva, cornuti inconspicuous, basal process short.

FEMALE GENITALIA. Anterior apophysis half to three-fifths length of posterior apophysis; sternite VIII width:length 3:1 to 4:1, anterior margin straight except middle part with semicircular bulge, posterior margin slightly inclined inwards; ostium proximal to middle of sternite, partly or entirely within semicircular bulge; ductus bursae long, gradually expanding to small ovate corpus bursae; signum wider than long, usually nearly parallel-sided and narrow, sometimes constricted in middle, teeth rather small, variable in number.

GENITALIA DIAGNOSIS. In male stout and rather long cuiller tapering in terminal quarter to obtuse apex weakly hooked outwards is distinct. In *A. purpurea* and *A. alstromeriana* outline similar but shorter, best difference from *A. purpurea* is outline of valva, and from *A. alstromeriana* shape of anellus. In female semicircular bulge enclosing ostium in middle of proximal margin of sternite VIII rises from a flat anterior margin of sternite. *A. silerella* and *A. ligusticella* have similar bulge, but more prominent as it rises from a convex anterior margin; sternite VIII width:length ratio is also different.

DISTRIBUTION. Italy, Austria (not found since 1927), Czechia, Hungary, Slovenia, Croatia, Albania, Greece. Asia: Turkey, Russia (Omsk oblast).

BIONOMICS. Larvae have been found on *Trinia glauca* (L.) Dumort and *Seseli hippomarathrum* Jacq. Reared moths emerged in July; after hibernation flying in spring until May.

REMARKS. *A. medelichensis* was long overlooked due to consistent misidentification. It was stored under *A. rotundella* in NHMW, with the result that Hannemann (1953, 1995) illustrated the male genitalia of this species as *A. rotundella*, which led to further misidentifications. Independently it has also been misidentified as *A. nodiflorella* and even as the very different *A. hippomarathri*, which has end of cell dot white, probably because of the shared host-plant.

irrorata subgroup (75–76)
Socii with long hair-like scales.

75   *Agonopterix irrorata* (Staudinger, 1871)

Depressaria irrorata Staudinger, 1871b: 241. Pl. 3, fig. 7.
  Depressaria anthriscella R. Brown, 1886: lii.

DESCRIPTION. Wingspan 16–20 mm. Head buff to ochreous-buff, frons pale buff. Labial palp pale buff, segment 2 with numerous light brown scales on outer side. Antenna dark fuscous, underside of scape ochreous. Thorax ochreous-buff to orange-ochreous, with posterior crests. Forewing ochreous-buff to light orange-ochreous with abundant mixture of scattered light ochreous-brown to cinnamon-pink scales, irregularly organised into transverse lines, a few darker brown or blackish scales towards termen; basal field not distinct; basal fascia obsolete, sometimes represented by a blackish subdorsal dot; a small brown spot at extreme base of costa; only the second oblique dot developed, black, median dot small, black or black with white centre, close to larger white end of cell dot with black margin; a rather faint light brown to cinnamon-pink line sometimes runs from end of cell dot towards dorsum, curving towards termen; terminal dots often faint, brown; fringe bright cinnamon-pink, more rarely concolorous with wing. Hindwing grey, slightly darker towards termen; fringe light greyish buff with basal fringe line.

VARIATION. Cinnamon-pink coloration may be extensive or completely absent. The scattered blackish scales are sometimes absent, or may be all over the wing. The median dot may be all black or have white centre. The cell dots are occasionally light brown rather than black and therefore inconspicuous.

SIMILAR SPECIES. The orange-ochreous forewings with bright cinnamon-pink fringe are almost unique within the genus. *A. nervosa* has subfalcate apex; *A. atomella* has such fringe, but has obvious basal field. The median cell dot much closer to the white end of cell dot than to the black second oblique dot at one-third is also a characteristic feature of *irrorata*.

MALE GENITALIA. Socii erect, longer than wide; gnathos lanceolate or narrowly elliptic, 2.5–3.5 times as long as wide, exceeding socii; valva long, narrow, costal margin nearly straight, ventral margin curved at one-half to give long tapering distal half and obtuse apex, cuiller straight, moderately thick with rounded apex, 70–95% of valva width; transtilla hardly thickened; anellus plate slightly wider than long, posterior margin with two widely separated sclerotised humps; aedeagus less than half as long as valva, rather stout, curved at middle, cornuti in a small group.

FEMALE GENITALIA. Anterior apophysis about two-thirds length of posterior apophysis; sternite VIII width:length 4:1, anterior margin slightly convex with more bulging middle part, posterior margin slightly inclined inwards with large median excavation; ostium in middle of sternite, not lying within bulge; ductus bursae long, of rather uniform width; corpus bursae small, ovate, signum somewhat wider than long, rather irregularly shaped, with numerous small teeth.

GENITALIA DIAGNOSIS. Male best distinguished by exceptionally long hair-like scales on socii (shared only with *A. graecella*, q.v.). If these are lost or cleaned away, genitalia are difficult to characterise. Best remaining feature differing from most other *Agonopterix* species: socii broadly elliptic, longer than wide. Final decision should be confirmed by combination of shape of valva (long, narrow), cuiller (nearly straight, stout, blunt), anellus (posterior margin with two sclerotised humps) and anellus lobes (narrow). Descriptions of female genitalia of *A. irrorata* and *A. graecella* are based on only two specimens each, so described differences may appear larger than they are. Beside this, there are no reliable features for safe separation from many other *Agonopterix* species.

DISTRIBUTION. Southern Europe from France to Greece, Romania and Ukraine including Crimea; most larger Mediterranean islands except Sardinia. Asia: Lebanon; Turkey, Syria, Israel (Lvovsky *et al.*, 2016); Caucasus (Lvovsky, 2019).

BIONOMICS. Larvae on *Anthriscus sylvestris* (L.) Hoffm. (Lvovsky *et al.*, 2016); *Daucus carota* L., *Oenanthe pimpinelloides* L., *Pastinaca sativa* L., *Pimpinella pere-grina* L. and *Smyrnium olusatrum* L. (Rymarczyk *et al.*, *in litt.*). Checked specimens were collected from April to November. We found no evidence for hibernation in the adult stage.

76   *Agonopterix graecella* Hannemann, 1976

Agonopterix graecella Hannemann, 1976: 233. Pl. 13.

DESCRIPTION. Wingspan 23–25 mm. Head whitish buff. Labial palp whitish buff, segment 2 with a few buff scales at base or more extensively brown-scaled, segment 3 with or without blackish tip. Antenna grey-brown. Thorax buff, sometimes tinged ochreous. Forewing narrow, termen oblique, pale buff, terminal area usually greyer, scattered blackish scales present, sometimes plentiful and often forming dotted lines along veins; basal field slightly paler, extending onto costa to about two-fifths; basal fascia brown, very weak or stronger and sometimes gradually fading into ground colour over much of dorsal part of wing; extreme base of costa with a blackish dot, outer half of costa and termen with series of black spots or short dashes; first oblique dot blackish, second very small or absent, white median and end of cell dots, both finely edged blackish and often set in a small patch of dark brown scales, this extending towards costa in some specimens as a blotch; sometimes with a diffuse brown line from end of cell dot towards dorsum, slightly curved outwards; fringe grey-buff. Hindwing whitish grey; fringe pale grey-buff. Abdomen whitish buff.

VARIATION. The amount of scattered blackish scales varies greatly, as also does the development of the blotch and the brown area demarcating the basal patch.

SIMILAR SPECIES. Characterised by narrow forewing, pale buff ground colour and two white dots usually in a small blackish patch.

MALE GENITALIA. Socii longer than wide; gnathos elliptic, twice as long as wide, shortly exceeding socii; valva long, costal side weakly concave, at five-sixths slightly curved to meet ventral margin in obtuse apex, cuiller broad-based, erect, stout, slightly tapered from three-fifths, apex rounded to obtuse, 65–70% of valva width; transtilla slightly thickened; anellus plate wider than long, posterior margin with slightly sclerotised lunules at each end; aedeagus half as long as valva, angled just beyond middle, well-developed band of minute cornuti, basal process moderately long.

FEMALE GENITALIA. Anterior apophysis about two-thirds to three-quarters length of posterior apophysis; sternite VIII width:length about 4:1, anterior margin inclined towards slight bulge in middle, posterior margin inclined inwards with large median excavation; ostium distal to middle of sternite, not or only slightly lying within bulge; ductus bursae long, slightly expanded to ovate corpus bursae, signum wider than long, variable, constricted in middle or with anterior and posterior projections, each lateral half with a slightly enlarged part, ending in a more or less parallel-sided narrow part, teeth small, rather few.

GENITALIA DIAGNOSIS. Male similar to its closest relative *A. irrorata*. Differences in gnathos (broader, not or nearly not overtopping socii), valva (broader), cuiller (usually shorter). Female very similar to *A. irrorata* (*q.v.*).

DISTRIBUTION. Italy, Montenegro, North Macedonia, Albania and Greece.

BIONOMICS. Host-plant unknown, but likely an Apiaceae. Checked specimens have been collected in July and August.

pseudoferulae subgroup (77–78)
These two species show some similarities, but are not very closely related.

77   *Agonopterix pseudoferulae* Buchner & Junnilainen, 2017

Agonopterix pseudoferulae Buchner & Junnilainen in Buchner, Corley & Junnilainen, 2017: 135.

DESCRIPTION. Wingspan 19–23. Head mid-brown, scales tipped buff, face creamy buff. Labial palp segment 2 creamy buff on innerside, outer and ventral sides rust-brown, mixed with a few blackish scales, segment 3 pale creamy buff, with dark grey rings at base and beyond middle, apex blackish. Antenna dark brown. Thorax and tegula mid-brown, darkest anteriorly, scales tipped creamy buff, posterior crest present. Forewing long with rounded apex, mid-brown with a varying proportion of creamy-buff scales and scattered blackish scales; basal field creamy buff, extending shortly onto costa, where it is interrupted by blackish costal spots; blackish terminal spots; basal fascia black to deep brown, fading posteriorly; oblique dots black, close or touching, proximally edged brick-red, distally edged white, these white parts sometimes confluent, an orange to brick-red streak, usually interrupted, extending just beyond the white end of cell dot and enclosing median dot, which can be white or reduced to a deeper red section of the brick-red streak, a dark grey blotch present between median dot and costa; fringe concolorous with forewing. Hindwing light grey, darker towards apex: fringe light grey with one line. Abdomen grey to grey-buff.

VARIATION. The middle white spot may be absent and the dark blotch can be very indistinct. Occasionally the whole insect has a more grey appearance.

SIMILAR SPECIES. Named for its resemblance to paler forms of *A. ferulae*, but differing in the presence of brick-red scales in the cell, a feature shared by *A. subtakamukui*, which has right-angled forewing apex and terminal dashes rather than dots.

MALE GENITALIA. Socii longer than wide, with long attachment to tegumen; gnathos lanceolate to elliptical, anout three times as long as wide, exceeding socii; valva rather wide, variable in length, costal margin weakly concave, ventral margin curved at about three-fifths, strongly tapering to rounded or obtuse apex, sacculus long, cuiller straight or slightly sinuous, apex rounded, 86–94% of valva width; transtilla narrow; anellus plate nearly orbicular with deep V-shaped incision; aedeagus rather stout, slightly curved beyond middle, 60–70% of valva length, cornuti forming a short patch, basal process short.

FEMALE GENITALIA. Anterior apophysis about half length of posterior apophysis; sternite VIII width:length about 5:2, anterior margin broadly triangular, middle part slightly indented, triangular, edges strongly marked, posterior margin nearly straight, median excavation small; ostium slightly proximal to middle of sternite; ductus bursae long, posterior one-quarter narrow, then gradually expanded to ovate corpus bursae, signum wider than long, narrow, parallel-sided, teeth few, large, almost all marginal.

GENITALIA DIAGNOSIS. Male best characterised by anellus features: deep central incision with convex margins and without any terminal sclerotisations, anellus lobes semicordate. Other features are cuiller (nearly reaching costa, straight or slightly S-shaped, blunt) and aedeagus (very stout, weakly bent, basal process very short). In female anterior margin of sternite VIII triangular, bulging, with middle part slightly indented, triangular with edges strongly marked, is distinctive.

DISTRIBUTION. Italy including Sardinia and Sicily, Albania, Greece (Peloponnese).

BIONOMICS. Reared from larvae found in Italy in early April on *Elaeoselinum asclepium* (L.) Bertol. (Sonderegger, *in litt.*), emerging end of April. Captured moths from June to first half of December. Overwintering stage not known.

78   *Agonopterix subtakamukui* Lvovsky, 1998

Agonopterix subtakamukui Lvovsky, 1998a: 471.
  Agonopterix cluniana Huemer & Lvovsky, 2000: 135.

DESCRIPTION. Wingspan 20–22 mm. Head grey-brown mottled dark brown. Labial palp segment 2 buff with some dark brown scales on outer side, segment 3 two-thirds length of segment 2, buff with two blackish rings. Antenna grey-buff. Forewing long, apex right-angled; light greyish brown, overlaid reddish brown in parts, particularly near base, a few scattered blackish scales mainly in costal and distal parts of wing; basal field pale buff, finely edged blackish from dorsum to middle; some small blackish spots on costa, blackish spots on vein ends on termen and around apex; oblique dots in cell at one-third black, the second elongate, usually touching or completely joined to form a comma shape, often edged pale to whitish distally, median cell dot usually absent or sometimes present as small white dot, a white end of cell dot ringed orange-brown or black; an orange-brown streak from end of black comma to white dot, between this and costa a blackish blotch; fringes concolorous with ground colour, with a brown fringe line. Hindwing light grey, darker towards apex; fringes grey with one or more lines.

VARIATION. The reddish brown colour fades, as does the orange streak in the cell, and this feature may be very obscure in some specimens. A small white median cell dot is sometimes present, occasionally on one wing only.

SIMILAR SPECIES. The combination of right-angled forewing apex and orange-brown streak in cell is unique in European species. *A. ocellana* is usually slightly larger, ground colour not reddish-brown, streak in cell and around white spot red not orange.

MALE GENITALIA. Socii erect, narrow; gnathos lanceolate, 2.5–3 times as long as wide, exceeding socii by almost its whole length; valva with costal margin straight, ventral margin curved at three-fifths, then tapering to obtuse or sub-acute apex, cuiller straight or slightly curved outwards, apex rounded, 70–75% of valva width; transtilla thin; anellus lobes narrow; anellus plate slightly wider than long, distal outer edge strongly sclerotised, deeply bifid, sinus with concave edges; aedeagus rather stout, strongly bent just beyond middle, about two-thirds length of valva, basal process stout.

FEMALE GENITALIA. Anterior apophysis about two-fifths to three-fifths length of posterior apophysis; sternite VIII width:length about 5:3, large triangular extension on anterior side, with apex of triangle truncated, a pair of sinuous folds near mid-line of triangle, posterior margin concave with median emargination; ostium close to posterior margin of sternite; ductus bursae long, posterior one-quarter narrow, then expanded to ovate or elliptical corpus bursae, signum wider than long, wings parallel-sided or slightly tapering, short anterior and longer, slender posterior projections, teeth from few to many, medium-sized.

GENITALIA DIAGNOSIS. Male well separated from other European *Agonopterix* by combination of very narrow socii far overtopped by gnathos; anellus sclerotised at distal lateral edge, deeply incised, anellus lobes narrow; aedeagus straight and very stout in basal half, strongly bent just beyond middle. In female, most similar European species is *A. guanchella* (q.v.).

DISTRIBUTION. Western Austria, south-west Germany, Italy and Hungary. Far East Russia. Japan (H. Arashima, pers. comm.). Earliest specimen collected in Europe is from Hungary, 12 August 1968.

BIONOMICS. Hazumu Arashima (pers. comm.) reared this species from a species of Apiaceae in Japan in 2023 and 2024. Moths have been taken from early June to middle of September.

REMARKS. *A. cluniana* was described from Austria. The synonymy with *A. sub-takamukui* from the Russian Far East has only recently been recognised (Buchner, 2020b). *A. takamukui* (Matsumura, 1931), so far only known from Far East Asia, has very similar male genitalia (gnathos broader and overtopping socii only by about half of its length); reliable differences in female genitalia not yet found.

The first European record was as recent as 1968, suggesting the possibility that this is an introduction, but it is also possible that this species is an ice age relict in Europe (P. Huemer, pers. comm).

Due to the external similarity in some characters to *A. ocellana*, it was suggested that *Salix* might be the host-plant (Huemer & Lvovsky, 2000), but searching for larvae in areas where the moth occurs has been unsuccessful. Barcode results suggest that this species is more closely related to some of the Apiaceae feeding *Agonopterix* than to *A. ocellana* and this has now been confirmed with records from Japan of larvae on an Apiaceae species.

guanchella subgroup (79)

79 Agonopterix guanchella Buchner, 2022

Agonopterix guanchella Buchner, 2022: 396.

DESCRIPTION. Wingspan 25 mm. Head grey. Labial palp, segment 2 pale buff, segment 3 pale buff, blackish from one-half nearly to tip. Antenna fuscous. Thorax and tegula grey. Forewing elongate, dull grey, extensively speckled with scattered blackish scales, some more concentrated to form small dark patches mainly near dorsum and termen; basal field grey without blackish scales, outwardly demarcated by a blackish basal fascia fading into ground colour; a blackish dot at extreme base of costa, several blackish spots on costa, strongest towards apex and a series of terminal dots; a pair of blackish oblique dots and an elongated plical mark, white dots in middle and at end of cell, both with an incomplete ring of blackish scales, a diffuse dark grey blotch between median white dot and costa; fringe light grey. Hindwing light grey, slightly darker towards apex; fringe light grey with one line. Abdomen light grey-buff.

VARIATION. Description based on four specimens, no remarkable external variability found.

SIMILAR SPECIES. *A. conciliatella* which is common in the Canary Islands and extremely variable in wing markings and colour has wing apex forming a right angle. The termen of *A. guanchella* is oblique, with apex somewhat rounded.

MALE GENITALIA. Socii elliptic, widely separated, outer sides nearly parallel, uncus distinct, triangular, slightly overtopped by socii; gnathos elliptic, twice as long as wide, equalling socii; valva long, costal margin nearly straight, ventral margin in basal half nearly straight, curved at about three-fifths, then running straight to subacute apex; cuiller stout, tapering in basal half, apex rounded, very slightly curved inward over its whole length, about 82% of valva width; transtilla thin; transtilla lobes narrow; anellus as wide as long, posterior margin with indistinct V-shaped incision; aedeagus stout, gently curved near middle, tapering after two-thirds length, three-fifths valva length, basal process rather short, cornuti very small, numerous, in two not clearly separated groups.

FEMALE GENITALIA. Anterior apophysis about half length of posterior apophysis; sternite VIII width:length about 3:1, anterior margin strongly convex, middle part with additional semicircular bulge, slightly folded on each side, posterior margin strongly inclined inwards; ostium in middle of sternite, about half in bulge; ductus bursae long, gradually expanding to elliptical corpus bursae; signum wider than long, wings narrow, slightly tapering, with slender posterior projection, teeth few, marginal, quite large.

GENITALIA DIAGNOSIS. In male socii with outer margins parallel, uncus unusually well-developed. In female proximal margin of sternite VIII with large

semi-elliptic expansion and lateral rims accommodating ostium is distinctive. A
large proximal expansion is also present in *A. subtakamukui,* but with rims inside
the expansion and ostium near distal edge of sternite VIII. See also *A. silerella* and
A. ligusticella.

DISTRIBUTION. Canary Islands (Gran Canaria).

BIONOMICS. Host-plant unknown. The only male specimen was collected in
December, three females in June.

REMARKS. This species is not closely related to any other. As its host-plant is
unknown, it may not belong to this group, but from external appearance this seems
the most appropriate placement.

Group 2B *furvella*-group (80–82)
Species feeding on Rutaceae. See remarks under *A. furvella.*

80   *Agonopterix furvella* (Treitschke, 1832)

Haemylis furvella Treitschke, 1832: 239.

DESCRIPTION. Wingspan 23–25 mm. Head ochreous-buff, face creamy buff. Labial
palp creamy buff, segment 2 not much thickened, tufted furrow weakly developed,
with a few light brown scales near base, segment 3 with weak brown ring close
to apex. Antenna fuscous, narrowly ringed dark fuscous. Thorax ochreous-buff,
posterior crest mid-brown. Forewing broad, uniformly mid to dark brown with
orange-red tinge; dot at base of costa and terminal spots faint, fuscous; basal field
pale ochreous-buff; oblique dots blackish, the second less distal to the first than is
usual in the genus, often followed by small creamy ochreous dots, a conspicuous
creamy ochreous irregular-shaped spot at end of cell, finely edged blackish; fringe
light brown, darker near apex, with fringe line. Hindwing light grey; fringe greyish
ochreous, with basal fringe line. Abdomen light fuscous.

VARIATION. There is little variation. The creamy dots adjacent to the oblique
dots may be absent; small dark brown median cell dot sometimes present. We have
seen a small series of reared specimens from one place (Switzerland: Leuk) with all
specimens lacking the dark brown coloration, leaving the whole wing pale ochreous-buff with pink-tinged costa, reminiscent of *A. pupillana* but with abundant
scattered light brown scales, particularly along veins in posterior half and towards
dorsum; the cell markings are almost obsolete. *A. pupillana* has well developed cell
markings and is slightly smaller than *A. furvella.*

SIMILAR SPECIES. The large size, broad forewing and uniform forewing coloration almost without dark markings are characteristic of *A. furvella.*

MALE GENITALIA. Socii erect, longer than wide, just exceeding relatively well developed uncus; gnathos ovate, about twice as long as wide, usually not exceeding socii; valva with costal margin weakly concave, ventral margin gently curved from end of sacculus to rounded apex, cuiller variable, quite stout, distal one-third to two-fifths usually bent inwards, but sometimes straight, tapered to apex or not, about 70% of valva width; transtilla thin; anellus lobes wide and long; anellus plate as wide as long with deep incision; aedeagus rather stout, half as long as valva, basal process very short.

FEMALE GENITALIA. Anterior apophysis about three-fifths length of posterior apophysis; sternite VIII width:length between 3:1 and 4:1, anterior and posterior margins almost straight; ostium adjacent to posterior margin of sternite; ductus bursae not longer than large elliptical corpus bursae, signum wider than long, narrowly diamond-shaped, with slender anterior and posterior projections, teeth numerous, large.

GENITALIA DIAGNOSIS. Genitalia of *furvella* and *pupillana* not distinguishable. In males of the species pair combination of these three features is distinctive: unusually well developed uncus and socii longer than wide, exceeding gnathos; cuiller blunt with terminal one-third bent inwards; aedeagus with basal part and basal process very short. Exceptionally gnathos reaches end of socii and cuiller is straight and/or subacute. Well developed uncus also found in *A. perezi* (*q.v.*). In females there are no unique features for determination, but large corpus bursae not clearly separated from ductus bursae in combination with absence or near absence of triangular sclerites projecting from ductus bursae into ostium can exclude most *Agonopterix* species.

DISTRIBUTION. Southern Europe from France to Russia, extending northwards to Germany, Czechia and Slovakia; Corsica.

BIONOMICS. Larva on *Dictamnus albus* L. Moths have been taken from end of April to September. Two generations are likely: young larvae were collected early August, at end of August they were half grown and hibernated in this stage, next spring they quickly developed and moths emerged in May. These produced offspring which quickly developed and moths emerged in July (Buchner, 2013, observation in eastern Austria).

REMARKS. The three species in group 2A are placed together for convenience, but *rutana* is probably not closely related to *furvella* and *pupillana*, which differ from most *Agonopterix* in having rather slender segment 2 of labial palp with weak tuft and furrow, broad forewing and hindwing without lobe at anal angle. These characters are shared with *A. petasitis*, which is placed in group 3 as it has hostplants belonging to Asteraceae.

See also Remarks under *A. pupillana* (below).

81   *Agonopterix pupillana* (Wocke, 1887)

Depressaria pupillana Wocke, 1887: 62.
   Agonopterix dictamnephaga Rymarczyk, Dutheil & Nel, 2012: 14, syn. n.

DESCRIPTION. Wingspan 21–23 mm. Head creamy buff to ochreous-buff. Labial palp creamy buff, segment 2 not much thickened, tufted furrow weakly developed, with light brown scales on outer side in proximal half, segment 3 with weak dark brown ring close to apex. Antenna light fuscous. Thorax ochreous-buff. Forewing broad, uniformly creamy ochreous with a few scattered ferruginous scales, these sometimes plentiful towards dorsum and in terminal area along veins; margin pink from near base of costa round apex and termen to tornus; dark brown dot at base of costa, terminal spots grey-brown; oblique dots dark brown, that on dorsal side larger and only slightly more distal, a creamy ochreous roundish or irregular-shaped dot at end of cell, finely edged dark brown, set in a faint ferruginous cloud with extension towards tornus; fringe light grey, tinged pink with fringe line. Hindwing light grey; fringe light grey, with fringe line. Abdomen creamy buff.

   VARIATION. Variation is mainly in the amount of ferruginous scales scattered over the forewing, from almost absent to abundant in dorsal and terminal areas, often mainly following veins and then mixed pink. End of cell dot sometimes broadly dark brown ringed, leaving only a small creamy ochreous centre. The pink margin can be conspicuous or almost absent as in the figure of *A. dictamnephaga* in Rymarczyk *et al.* (2012).

   SIMILAR SPECIES. The yellowish coloration and absence of basal patch is reminiscent of species in the *pallorella* group, but these lack posterior crest on the thorax and never have a ringed whitish spot at end of cell.

   MALE GENITALIA. Preparations show socii more divergent than in *A. furvella* but this may be coincidence caused by selected specimens or preparation artefacts. All other features vary in the range of *A. furvella*.

   FEMALE GENITALIA. Similar to *A. furvella*.

   GENITALIA DIAGNOSIS. No clear genitalia differences from *A. furvella* have been found in either sex.

   DISTRIBUTION. France, Italy and Albania.

   BIONOMICS. Larva on *Dictamnus albus* L. Moths have been taken from May until September.

   REMARKS. Hannemann (1953) pointed out that the male genitalia of *A. pupillana* showed only small differences from *A. furvella* and that the two species had the same host-plant. He suggested the possibility that *A. pupillana* might just be a colour form of *A. furvella*. The barcode p-distance of 0.76 of these species indicate that they are closely related, but are sufficiently distinct to justify retention of *pupillana* at species level. *A. pupillana*, besides lacking the dark brown forewing coloration of *A. furvella* has pink edge to wing except on dorsum and the relative position of the oblique dots is different.

Rymarczyk *et al.* (2012) described *A. dictamnephaga*, noting that according to Hannemann (1953), *A. furvella* and *A. pupillana* had the same habitus. From this they concluded that their species with the whole forewing coloured yellow was distinct from either, further justified by small differences in the genitalia. Unfortunately they had misunderstood Hannemann who wrote under *A. pupillana* "*Diese Art zeigt den gleichen Bau wie A. furvella ...*" but he was referring to the appearance of the male genitalia, not the habitus of the moth. *A. pupillana*, like *A. dictamnephaga* has yellow forewings, very different from the brown-winged *A. furvella*. The distinctive features of the genitalia of *A. dictamnephaga* are very slight and lie in the shape of the cuiller and the valva, both more or less variable features in *Agonopterix* and besides they were comparing with Hannemann's drawing, not with an actual preparation. We hereby place *A. dictamnephaga* in synonymy with *A. pupillana*.

82   *Agonopterix rutana* (Fabricius, 1794)

Pyralis rutana Fabricius, 1794: 286.
　　Depressaria retiferella Zeller, 1850: 150.

DESCRIPTION. Wingspan 21–23.5 mm. Head fuscous with scales tipped mid-brown; face creamy buff. Labial palp segment 2 whitish with dark fuscous scales on outer side, rough scales dark fuscous mixed pale cinnamon, segment 3 pale cinnamon with a few dark fuscous scales at base, not forming complete ring. Antenna light fuscous with dark fuscous rings to entirely dark fuscous. Thorax dark fuscous, scales mainly tipped greyish buff. Forewing fuscous with many scales tipped dull cinnamon, others tipped pale grey to produce numerous narrow transversely orientated pale grey lines, some areas of wing darker fuscous, scattered deep fuscous scales mainly in costal and terminal areas of wing; a dark fuscous spot at base of costa; basal field of ground colour, demarcated by a narrow pale grey to greyish buff line from dorsum to costal margin of cell, extending as narrow buff line along costal margin of cell to about one quarter, similar lines in fold, through middle of cell to termen and sometimes along other veins to costa and termen, the latter often tinged cinnamon; oblique dots blackish; a whitish dot at end of cell; fringe grey fuscous with scales tipped dull pink, with one to two fringe lines. Hindwing grey, slightly darker towards termen; fringe very pale grey with five fringe lines. Abdomen light fuscous with scales tipped grey.

VARIATION. There is little variation beyond slight differences in ground colour and in how many veins to costa and termen are highlighted. Occasionally there is a white median cell dot.

SIMILAR SPECIES. The combination of deep brown coloration with pale transverse lines and longitudinal lines gives a reticulate appearance to the forewing which is unique in European *Agonopterix*, externally similar (although not closely related) to the Far East Asian *A. bipunctifera* Matsumura, 1931.

MALE GENITALIA. Socii well separated, attachment to tegumen quite short; gnathos ovate to lanceolate, 2–2.5 times as long as wide, exceeding socii; valva with costal margin concave, basal part tapering to end of sacculus, parallel-sided in middle part, tapering from two-thirds to obtuse or sub-acute apex, cuiller stout, slightly curving inwards, of even width or slightly narrower in middle part, apex rounded, 75–83% of valva width; transtilla not or slightly thickened; anellus plate orbicular, with short linear incision and microstructure reminiscent of warts (shared with *A. perezi*), narrow triangular lateral processes at base; aedeagus curved from two-fifths, about 70% length of valva, cornuti in a dense group, sheath weakly sclerotised and placed beyond base, process variably developed.

FEMALE GENITALIA. Anterior apophysis about three-fifths length of posterior apophysis; sternite VIII width:length about 3:1, anterior margin slightly convex, posterior margin almost straight; ostium in middle of sternite, half surrounded by semi-elliptical fold line reaching middle of sternite; ductus bursae twice as long as corpus bursae, slender, abruptly expanding before elliptical corpus bursae, signum wider than long, narrowly diamond-shaped, lateral ends rounded, with slender anterior and posterior projections, teeth numerous, mostly large.

GENITALIA DIAGNOSIS. In male cuiller differs from most other European *Agonopterix* species but may be similar in *A. impurella* group (gnathos different), *A. carduncelli* (anellus different) and *A. perezi* (different in many details). In female ostium partly surrounded by fold line, feature also found in *A. yeatiana* and *A. cotoneastri*, but folds are more diverging and do not extend as far in *A. rutana*. In *A. subumbellana* this fold line is very similar, but ostium is closer to posterior margin of sternite VIII than in *A. rutana*.

DISTRIBUTION. Southern Europe from Portugal to Greece, Bulgaria and Crimea, extending north to Switzerland; most larger Mediterranean islands. Asia: we have seen specimens from Armenia, Israel, Turkmenistan and Kyrgyzstan; Turkey (Koçak & Kemal. 2009); Caucasus, North Africa (Lvovsky *et al.*, 2016).

BIONOMICS. Larvae on leaves of *Ruta chalepensis* L. (Hannemann, 1953); *R. graveolens* L. (Sonderegger, *in litt.*); *R. angustifolia* Pers. and *R. montana* (L.) L. (Rymarczyk *et al.*, *in litt.*). Larvae have been collected in May and June, adults fly from June until May of the next year.

REMARKS. Apart from the shared host-plant family there is nothing to indicate any close relationship between *A. rutana* and the two previous species.

Group 3 (83–113)
Species feeding on Asteraceae

There are two main subgroups and four additional species in three subgroups

pallorella subgroup (83–98)
Feeding on tribe Cynareae. More or less ochreous-yellow species without demarcated basal field on forewing, with dot at two-fifths single or absent, not a pair of oblique dots; thorax usually with darker median line, without posterior crest; labial palp segment 2 with only the distal part rough-scaled and furrowed. Species 92–95 are included here but without complete certainty.

Male genitalia key for *pallorella* and *arenella* subgroups, including *A. cotoneastri*, *A. doronicella* and *A. petasitis*

1a	Cuiller distinctly expanded terminally or subterminally	2
1b	Cuiller without or with very indistinct terminal expansion (if aedeagus is nearly straight, cf. *A. multiplicella*, where rarely forms with scarcely swollen cuiller are found)	8
2a (1a)	Socii very widely separated, aedeagus nearly straight	*A. multiplicella*
2b	Socii not or moderately separated, aedeagus distinctly bent	3
3a (2b)	Cuiller club-shaped, expansion not exceeding twice diameter of basal part	*A. squamosa* (rare forms)
3b	Cuiller usually expanded to more than twice basal diameter, not simply club-shaped	4
4a (3b)	Basal process of aedeagus with pair of long digitate lobes expanding laterally (only clearly visible in ventral view)	*A. subumbellana*
4b	Basal process of aedeagus without such lobes	5
5a (4b)	Cuiller ± reaching costa, paddle-shaped, inner margin straight, outer margin with subterminal bulge.	*A. invenustella*
5b	Cuiller shorter, shape different	6

6a (5b)	Cuiller expanding from base to about three-quarters length, inner margin concave, outer margin convex, apex incurved, acute	*A. kuznetzovi*
6b	Distinct expansion of cuiller restricted to terminal half	7
7a (6b)	Cuiller swollen terminally with swelling more or less triangular, aedeagus rather symmetrical in ventral view	*A. pallorella*
7b	Cuiller swollen terminally with swelling more or less orbicular, aedeagus asymmetrical in ventral view with subterminal hook on right side	*A. arenella*
8a (1b)	Socii large, slightly overlapping and together forming an obtuse to subacute point, extending well beyond end of gnathos	*A. ferocella*
8b	Socii orbicular to elliptic, together not forming an obtuse to subacute point	9
9a (8b)	Cuiller as in figs. 111a, d	*A. cotoneastri*
9b	Cuiller different	10
10a (9b)	Cuiller rather short and more or less straight, ending with distinct outwardly directed hook	11
10b	Cuiller different	12
11a (10a)	Even base of gnathos exceeding socii	*A. abditella*
11b	Cuiller equalling or exceeding socii by up to half its length	*A. nanatella*
12a (10b)	Socii very small, overtopped by rounded uncus	*A. lidiae*
12b	Socii not remarkably small, uncus not overtopping socii, triangular or indistinct	13
13a (12b)	Valva parallel-sided over most of its length	14
13b	Valva not parallel-sided or only in a short section	15

14a (13a)	Socii with inner edges rather straight and contiguous, much expanded laterally; gnathos at least 2.5 times as long as wide	*A. ivinskisi*
14b	Socii medium-sized, not wider than long, well separated; gnathos broadly elliptical, about 1.5 times as long as wide	*A. doronicella*
15a (13b)	Anellus exceptionally long, reaching or exceeding transtilla, anellus lobes reduced	*A. carduella*
15b	Anellus of average *Agonopterix*-type, anellus lobes distinct	16
16a (15b)	Gnathos short, triangular with rounded apex	*A. xeranthemella*
16b	Gnathos different	17
17a (16b)	Gnathos lanceolate, broadest near base, outline slightly concave in distal half and tapering to a remarkably sharp tip	*A. laterella*
17b	Gnathos nowhere with concave outline	18
18a (17b)	Cuiller long (nearly reaching or exceeding costa), thin and clearly S-curved	19
18b	Cuiller different	21
19a (18a)	Gnathos equalling socii or exceeding by up to half of gnathos length	*A. bipunctosa,* *A. uralensis*
19b	Gnathos exceeding socii by most or full gnathos length	20
20a (19b)	Socii round, overtopped by most of gnathos length, cuiller nearly reaching or exceeding costa	*A. volgensis*
20b	Socii somewhat elongated, overtopped by full gnathos length, cuiller not quite reaching costa	*A. latipennella*
21a (18b)	Anellus obpyriform with distinct V-shaped incision and wide gap to transtilla; cuiller short (not exceeding two-thirds valva width), moderately stout and moderately bent inwards throughout	*A. petasitis*
21b	Features not present in this way	22

| 22a (21b) | Aedeagus abruptly bent near middle, asymmetrical in ventral view | 23 |
| 22a (21b) | Aedeagus rather evenly bent, not asymmetrical in ventral view | 25 |

| 23a (22a) | Canary Islands only; features as in figs. 108a–c | *A. cinerariae* |
| 23b | Widespread, features different | 24 |

| 24a (23b) | Anellus with U-shaped sclerotisation; basal process ending straight to slightly convex, not wider than aedeagus | *A. subpropinquella* |
| 24b | Anellus without U-shaped sclerotisation; basal process ending straight or slightly concave, lateral edges clearly exceeding width of aedeagus | *A. propinquella* |

| 25a (22a) | Cornuti (at least in part) unusually stout | *A. broennoeensis* |
| 25b | Cornuti tiny throughout (group of species with very similar male genitalia!) | 26 |

| 26a (25b) | Aedeagus usually strongly bent (about 90 °); upper edge of anellus usually with two distinct semicircular bulges with thickened margins, clearly exceeding anellus lobes; cuiller nearly straight, usually stout throughout and blunt (exceptionally aedeagus less bent and/or cuiller slightly expanded distally) | *A. squamosa* |
| 26b | Aedeagus not as strongly bent, anellus usually without distinct semicircular bulges or, if present, not clearly exceeding anellus lobes | 27 |

| 27a (26b) | Gnathos overtopping socii at least by more than half of its length; cuiller rather straight but ending slightly incurved | *A. carduncelli* |
| 27b | Gnathos overtopping socii by up to half of its length | 28 |

| 28a (27b) | Basal process of aedeagus long, ending convex, lateral edges exceeding width of aedeagus | *A. straminella* |
| 28b | Basal process of aedeagus shorter, ending straight or concave, lateral edges exceeding width of aedeagus or not | *A. kyzyltashensis*, *A. kaekeritziana* |

Female genitalia key for *pallorella* subgroup, including *A. latipennella, A. abditella* and *A. lidiae*

1a	Ductus bursae with a swelling and a loop in posterior 1/6	*A. broennoeensis*
1b	Ductus bursae with or without swollen sections, but without loop	2
2a (1b)	Inside the ostium, lateral to sclerites, a pair of diverging folds, forming a more or less incomplete "V" (incomplete, because the lowermost parts often invisible), these folds in upper half often more or less bent outwards (fig. 81); apart from the pair of diverging folds no further folds in sternite VIII	*A. pallorella*
2b	In ostium no such folds, but folds of different form may be present outside the ostium (e.g. figs. 83, 85)	3
3a (2b)	Signum absent	*A. carduncelli* *A. tschorbadjiewi*
3b	Signum present, but sometimes very small	4
4a (3b)	Signum very small (fig. 92c) *A. kaekeritziana* usually has a normal sized signum, but there are populations in Spain with exceptionally small signum.	*A. kaekeritziana* in part
4b	Signum medium-sized to large	5
5a (4b)	Sternite VIII without any kind of folds or bands described under 6a and 6b	*A. lidiae*
5b	Sternite VIII with folds or bands described under 6a and 6b	6
6a (5b)	Sternite VIII with a fold or band between ostium and anterior margin; this structure is usually running through, although sometimes weaker near middle or even interrupted, but not running into anterior margin on either side and ending there (e.g. fig. 83c)	7

6b	Sternite VIII with a fold on either side of ostium or in area anterior to ostium, running obliquely into anterior margin. Where meeting anterior margin, the fold is often somewhat swollen, bent outwards and overtopping margin (e.g. figs. 91, 92), but in other cases it simply ends there without swelling or overtopping margin (e.g. fig. 89c) Note: removing intersegmental skin connection with sternite VII is a critical step for these structures, because it can happen that parts of the folds are also removed	9
7a	Bend of fold normally with width less than 2.5 × depth (figs. 83a, c). In Europe only in most eastern parts	*A. subumbellana*
7b	Fold anterior to ostium forming a bend with width more than 2.5 × depth	8
8a (7a)	Ductus bursae with 3 sections of different structure (fig. 89c): distal third narrow and covered with tiny dots, central third expanded and with structure reminiscent of intestine, proximal third with irregular fine folds, gradually expanding to large corpus bursae	*A. squamosa* in part
8b	Ductus bursae not divided into 3 sections as described before	*A. invenustella*
9a (7b)	Anterior margin of sternite VIII steeply inclined outwards (angle of about 90 ° in between), rounded in middle part; folds on either side anterior to ostium running obliquely into anterior margin and ending there without overtopping it (fig. 95a)	*A. latipennella*
9b	Anterior margin of sternite VIII straight or inclined outwards, but with an obtuse angle in between (e.g. figs. 85, 92)	10
10a (9b)	Fold on either side anterior to ostium running obliquely into anterior margin and ending there without overtopping margin (fig. 89c)	*A. squamosa* in part

10b	Where folds on either side anterior to ostium meet anterior margin, they overtop it, and often they are somewhat swollen and bent outwards there (figs. 85, 87, 88, 92)	11
11a (10b)	Folds steep to perpendicular to anterior margin, part of fold which overtops margin usually longer than part inside sternite VIII, this structure forming a square (fig. 91a) to trapezoid (fig. 91b) protrusion	*A. volgensis*
11b	Folds not as steep to anterior margin, part of fold which overtops margin shorter than part inside sternite VIII	12
12a (11b)	Folds nearly parallel to anterior margin (figs. 87a, b)	*A. bipunctosa*
12b	Outer side angle between fold and margin about 25–35 °	*A. straminella, A. kaekeritziana, A. kyzyltashensis*

83   *Agonopterix subumbellana* Hannemann, 1959

Agonopterix subumbellana Hannemann, 1959: 42.

DESCRIPTION. Wingspan 20–22 mm. Head light brown, scales tipped buff. Labial palp buff. Antenna pale buff, darker distally. Thorax and tegula light brown mixed buff, with darker median line, paler posteriorly. Forewing elongate, termen oblique, buff with scattered fuscous scales, particularly towards costa; terminal dots blackish; all veins marked grey-brown; a blackish subdorsal dot and black dots in cell at one-third and at end of cell; a broad grey-brown streak from dorsum at one-sixth, ending just beyond and below end of cell dot; fringe pale buff. Hindwing rather narrow, light grey, darker towards apex; fringe light grey with distinct line. Abdomen buff.

VARIATION. Few specimens seen, mostly in poor condition.

SIMILAR SPECIES. *A. umbellana* has termen more or less at right angles to costa and lacks the dorsal streak of *subumbellana*.

MALE GENITALIA. Similar to *pallorella*, socii widely overlapping; gnathos only slightly tapering to broad rounded apex, shortly exceeding socii; valva more slender, costal margin almost straight, ventral margin slightly curved at one-half, distal half tapering to subacute apex, cuiller somewhat similar to *pallorella* but distal one-third curved outwards on inner margin, expanded apical part shorter, 68–75% of valva width; aedeagus bent just beyond middle, half as long as valva, basal process strong with pair of digitate lateral lobes.

FEMALE GENITALIA. Anterior apophysis about half length of posterior apophysis; sternite VIII width:length about 7:2, anterior margin convex, posterior margin slightly inclined inwards; a narrow curved fold line between anterior margin and ostium which is distal to middle of sternite; ductus bursae twice as long as rather small obpyriform corpus bursae; signum wider than long, narrowly elliptical, one pair of teeth larger than others.

GENITALIA DIAGNOSIS. In male cuiller terminally expanded, quadrate to triangular, very similar to *A. pallorella*, differing in longer anellus and presence of long digitate lateral lobes on basal process of aedeagus. In female ostium surrounded by a very similar fold line to that in *A. rutana*, but ostium is closer to distal end of sternite VIII and signum smaller; see also *A. cotoneastri* and *A. yeatiana*.

DISTRIBUTION. Russia; Asiatic Turkey, Armenia, Iran.

BIONOMICS. Host-plant unknown.

REMARKS. The species was described from one male from Armenia. Hannemann (1959) placed it next to *A. umbellana* based on external similarity only. We have placed it in this group because male genitalia suggest that it is more closely related to *A. pallorella*. Female genitalia and barcode are not helpful, therefore our placement must be seen as provisional. Knowledge of host-plant might shed some light on its relationships.

84   *Agonopterix pallorella* (Zeller, 1839)

Depressaria pallorella Zeller, 1839: 195.
  Depressaria subpallorella Staudinger, 1871a: 298.
  Depressaria divergella Caradja, 1920: 128.
  Agonopterix mikomoensis Fujisawa, 1985: 37.

DESCRIPTION. Wingspan 19–24 mm. Head ochreous to orange-ochreous, face pale buff. Labial palp ochreous-buff, segment 2 tinged light brown on outer side. Antenna deep brown. Thorax ochreous. Forewing ochreous to creamy yellow, veins indicated by grey-brown streaks or dots or not indicated, a broad ochreous-brown, dark brown or grey streak between dorsum and cell from one-sixth to two-thirds, sometimes extended upwards into terminal area; subdorsal spot deep brown to blackish; a black dot in cell at one-third and another at end of cell; terminal spots dark grey; fringe ochreous-buff. Hindwing pale grey-buff, slightly darker towards termen, veins marked grey; fringe whitish buff with one indistinct fringe line. Abdomen buff.

VARIATION. The subdorsal brown streak varies from ochreous-brown to quite deep brown and can be interrupted or absent; the dark dot at base of dorsum is sometimes absent; terminal dots sometimes faint or absent; the two dots in mid-wing can be absent or small and lost among the scattering of scales that is mainly but not exclusively along veins.

SIMILAR SPECIES. *A. bipunctosa* is similar when with subdorsal streak, but has a blackish dot at base of costa, a much weaker subdorsal streak and dark scales scattered over the forewing. *A. kyzyltashensis* and *A. kaekeritziana* also have subdorsal streaks, but of rather different shape. Less typical forms of *pallorella* could be mistaken for other species in the *pallorella* group and will need genitalia examination.

MALE GENITALIA. Socii large with long attachment to tegumen, usually close to each other or even just overlapping; gnathos long ovate, about twice as long as wide, often not exceeding socii, but occasionally so; valva with costal margin slightly concave, parallel-sided or tapering to two-thirds, where ventral margin is curved, then straight to obtuse apex, cuiller with inner margin straight or with slight bulge in middle, sometimes curved outwards at two-thirds, outer margin abruptly curved outwards just beyond one-half, doubling width of structure, then curved inwards at 70–80 °, margin in distal part slightly convex, apex rounded, the whole structure of characteristic hoof-like shape, 80–90% of valva width; transtilla thin; anellus abruptly widened above stout stalk, then tapering distally to two triangles separated by V or U-shaped excavation; aedeagus strongly curved near base with additional curve at three-fifths, about three-fifths length of valva, cornuti in two or three closely placed lines, basal process short, ending in divergent, nearly triangular lobes.

FEMALE GENITALIA. Anterior apophysis half to two-thirds length of posterior apophysis; sternite VIII width:length 5:2 to 3:1, anterior margin convex, middle part often slightly sinuate, posterior margin slightly inclined inwards, often excavated in middle; a pair of folds anterior to ostium forming a V, ostium slightly distal to middle of sternite; ductus bursae gradually expanding to large elliptical corpus bursae; signum wider than long, narrowly elliptical, with small anterior and posterior projections, teeth on plate mostly larger than marginal teeth, often one pair larger than others.

GENITALIA DIAGNOSIS. Males show similarities to *A. subumbellana* and *A. arenella*. Females have a combination of features not found in any other species: ostium large, broader than long, inside the ostium, lateral to sclerites, a pair of diverging folds, forming a more or less incomplete "V", these folds in upper half often more or less bent outwards, no further folds in sternite VIII (Note: the diverging folds sometimes are indistinct, especially if genitalia are not compressed dorsoventrally).

DISTRIBUTION. Almost all Europe, but absent from Mediterranean islands except Sicily. Asia: widespread, reaching Far East. North Africa: Tunisia.

BIONOMICS. Larvae on *Centaurea scabiosa* L., *C. nigra* L., *C. jacea* L. and *C. stoebe* L. (Hannemann, 1953); *Serratula tinctoria* L. (Harper *et al.*, 2002); *Centaurea vallesiaca* (DC.) Jordan (Sonderegger, *in litt.*); *C. aspera* L. and *C. paniculata* L. (Rymarczyk *et al., in litt.*). Emerging predominantly in May to July, flying until May, rarely June, of the following year.

REMARKS. Lvovsky & Stanescu (2019) synonymised *Depressaria divergella* Caradja, 1920 with *A. pallorella*.

85   *Agonopterix straminella* (Staudinger, 1859)

Depressaria straminella Staudinger, 1859: 238.
   Depressaria echinopella Chrétien, 1907: 276, syn. n.
   Depressaria liodryas Meyrick, 1921a: 392.
   Ctenioxena cryptipsila Meyrick, 1923a: 611.
   Agonopterix mendesi Corley, 2002: 26, syn. n.

DESCRIPTION. Wingspan (18–)20–25(–27) mm. Head pale ochreous, face creamy ochreous. Labial palp pale ochreous with a few light brown scales scattered on outer side of segment 2. Antenna with scape ochreous mixed dark grey-brown, flagellum dark grey-brown, sometimes ochreous at base. Thorax ochreous to deep ochreous. Forewing ochreous with scattered blackish scales and sometimes a few light brown scales, most numerous on veins in outer third of wing, costal area free of such scales to beyond mid-wing; subdorsal spot black or brown, blackish or brown dots in cell at one-third and at end of cell, often absent; fringe ochreous, tipped whitish. Hindwing whitish, weakly tinged ochreous-grey; fringe pale ochreous-grey. Abdomen pale ochreous-grey.

VARIATION. The extent of scattered dark fuscous and light brown scales on the forewing varies, with veins well indicated in outer third in some specimens and with very few such scales in others. The two dots in cell are frequently obscure or absent; additional dark markings rarely present.

SIMILAR SPECIES. Specimens with no dark dots in the cell area are rarer in other similar ochreous species, but identification will frequently require examination of genitalia.

MALE GENITALIA. Socii large with long attachment to tegumen, usually clearly separate; gnathos long ovate, about twice as long as wide, exceeding socii; valva with costal margin slightly concave, ventral margin curved at two-thirds, distal one-third straight or slightly concave, apex obtuse; cuiller stout, not or slightly tapering to rounded or subacute apex, usually slightly curving outwards, 75–80% of valva width; transtilla thin; anellus lobes larger than transtilla lobes; anellus orbicular, posterior margin with V-shaped notch; aedeagus curved in middle or up to three-fifths, about two-thirds length of valva, cornuti in a dense group, basal process long, usually ending in an obtuse to pointed triangle with lateral edges exceeding width of aedeagus

FEMALE GENITALIA. Anterior apophysis about half length of posterior apophysis; sternite VIII width:length about 5:2, anterior margin inclined outwards, middle part sometimes slightly bulging, with a shallow excavation, posterior margin straight or slightly inclined inwards, with small excavation or emargination in middle; a pair of short folds from either side of middle of anterior margin, diverging, ostium slightly distal to middle of sternite; ductus bursae expanding to large elliptical corpus bursae; signum wider than long, narrowly elliptical, with anterior and posterior projections, teeth on plate mostly larger than marginal teeth, often one pair larger than others.

GENITALIA DIAGNOSIS. Male not determinable by any single feature, *A. kyzyltashensis*, *A. squamosa* and *A. kaekeritziana* very similar. Distinguished by triangular end of basal process in combination with rather short, very slightly outcurved and very slightly tapering cuiller. In *A. kyzyltashensis* cuiller similar but slightly longer and basal process of aedeagus short, usually with concave ending. In *A. squamosa* cuiller usually not tapering, often expanding distally and aedeagus strongly curved over all its length in lateral view. *A. kaekeritziana* with remarkable intraspecific variability, but cuiller tends to be longer and slightly S-curved. Determination of these species should be cross-checked with external appearance. *A. broennoeensis* with generally similar male genitalia differs in clearly stouter cornuti. Female genitalia not safely distinguishable from *A. straminella*, *A. kaekeritziana* and *A. kyzyltashensis*. For differences from other species of *pallorella* subgroup see key (pp. 183–185).

DISTRIBUTION. South Portugal, Spain, France, Greece including Crete, Cyprus; Asia: Turkey, Syria, Israel, Jordan; North Africa: Algeria, Morocco, Tunisia. Lvovsky *et al.* (2016): Iran.

BIONOMICS. Larvae feed on *Centaurea sphaerocephala* L. (Corley, 2002), on the coast in March and early April, in terminal shoots or flower buds. Chrétien (1907) reared his *Depressaria echinopella* from larvae on *Echinops spinosissimus* Turra. Imago emerges in April, May or June, flying until early spring of the following year.

REMARKS. We can find no difference between *A. straminella* and *A. echinopella*, which is therefore synonymised here with *straminella* along with two species described by Meyrick which have been treated as synonyms of *echinopella*.

Agonopterix mendesi was described based on four specimens reared from *Centaurea sphaerocephala* on the western coast of Algarve, Portugal. Following comparison with a named specimen of *straminella* at NHMUK, it was concluded that it was not *straminella* but a new species. Unfortunately, the specimen in NHMUK was a misidentified specimen of *A. squamosa*. Corley (2002) mentions a Portuguese specimen in MNHN which he supposed belonged to *A. mendesi*, but which lacked a signum. According to Agenjo (1954) the corpus bursae of *straminella* lacks a signum, but the species in this group without signum is *A. carduncelli*, so these references appear to refer to the then undescribed *carduncelli*.

86   *Agonopterix carduncelli* Corley, 2017

Agonopterix carduncelli Corley, 2017 *in* Buchner, Corley & Junnilainen, 2017: 127.

DESCRIPTION. Wingspan 19.5–23 mm. Head dull ochreous-buff, face creamy buff. Labial palp segment 2 with only the distal part rough-scaled and furrowed, pale buff with scattered light brown scales, segment 3 pale buff or ochreous-buff. Antenna with scape dull ochreous-brown, proximal part of flagellum ochreous-buff, ringed grey-brown, distally grey-brown. Thorax dull ochreous-buff. Forewing pale ochreous-buff, with a variable amount of scattered light brown and blackish scales, particularly along veins towards termen and sometimes also in cell and between dorsum and fold; terminal spots dark grey-brown; subdorsal spot black or brown, a small dot in cell at two-fifths and usually another at end of cell; a faint grey-brown stripe stretching through subdorsal area ending in a wider patch below end of cell; fringe pale ochreous-buff with weak fringe line. Hindwing light grey, darker in outer half; fringe light greyish ochreous with indistinct fringe lines. Abdomen light greyish buff.

VARIATION. Some specimens have many more scattered dark scales than others. Sometimes the ground colour appears washed with indistinct grey streaks. The subdorsal spot can be distinct or dull pale brown; the cell dots can be very indistinct due to the abundance of scattered scales; the development of the subdorsal streak is variable.

SIMILAR SPECIES. *A. straminella* has black sub-basal dot and terminal dots together with paler hindwing, but often lacks cell dots. Other related species have better developed cell dots and different genitalia. Forms of *carduncelli* with subdorsal dark streak show some resemblance to *A. pallorella* (*q.v.*).

MALE GENITALIA. Socii smaller and less divergent than in related species; gnathos long ovate, about 2.5 times as long as wide, exceeding socii by most of its length; valva with costal margin slightly concave, ventral margin strongly curved at about three-quarters, apical one-quarter appearing truncated, apex obtuse; cuiller stout, not or slightly tapering to rounded or obtuse apex, curving outwards at middle, extreme apex often curving slightly inwards, 82–90% of valva width; transtilla slightly thickened in middle; anellus lobes not larger than transtilla lobes; anellus slightly longer than wide, posterior margin with narrow V-shaped notch; aedeagus slender, little tapering, curved near middle, about two-thirds length of valva, cornuti in a dense group, basal process with pointed lateral lobes.

FEMALE GENITALIA. Anterior apophysis slightly over half length of posterior apophysis; sternite VIII width:length 5:2 to 3:1, anterior margin slightly convex, middle part sometimes slightly sinuate, posterior margin with wide shallow excavation in middle; ostium distal to middle of sternite; ductus bursae twice as long as rather small elliptical corpus bursae; signum absent.

GENITALIA DIAGNOSIS. In male combination of cuiller ending slightly incurved but not tapering and gnathos overtopping socii by more than half to whole length is usually sufficient to separate from similar species. Female well characterised by absence of signum, but see *A. tschorbadjiewi*. There are populations of *A. kaekeritziana* in Spain with exceptionally small signum, therefore the possibility cannot be excluded that signum might be absent in some specimens.

DISTRIBUTION. Portugal, Spain, Sardinia, Sicily, Greece. Recently found in France (Nel *et al.*, 2022). North Africa: Morocco.

BIONOMICS. The larvae feed in the tips of young shoots of *Carthamus caeruleus* L. (Buchner *et al.*, 2017) in March. Adults emerged in captivity in May. Specimens taken at end of November were still in good shape, hibernation in adult stage is likely.

87   *Agonopterix bipunctosa* (Curtis, 1850)

Depressaria bipunctosa Curtis, 1850: 116.
  Depressaria perpallorella Morris, 1870: 52.

DESCRIPTION. Wingspan 18–22 mm. Head ochreous-buff, face pale ochreous-buff. Labial palp light ochreous-buff, outer side of segment 2 tinged dark grey-brown. Antenna dark grey-brown. Thorax ochreous-buff. Forewing ochreous-buff with a few scattered deep brown scales particularly on veins; black or blackish dots at extreme base of costa, close to base of dorsum, in cell at one-third and at end of cell and between vein ends around apex and termen; often with an ill-defined broad light brown streak between cell and dorsum from one-quarter to three-quarters; fringe pale buff. Hindwing whitish buff, slightly darker towards apex, with some grey spots around termen; fringe whitish buff, sometimes with an indistinct fringe line. Abdomen pale buff to ochreous-buff.

VARIATION. The frequency of dark scales on the veins is variable, the blackish subdorsal spot may be absent, that at base of costa also, but more rarely and the light brown subdorsal streak may also be absent.

SIMILAR SPECIES. The presence of a blackish spot at base of costa separates this species from all the similar species, except *A. squamosa* which occasionally has this spot.

MALE GENITALIA. Socii large; gnathos elliptic, about twice as long as wide, not or only shortly exceeding socii; valva with costal margin straight or slightly concave, ventral margin curved at about three-fifths, apical part tapering to obtuse or subacute apex; cuiller thin, strongly sinuous, slightly incurved at subacute apex, 82–96% of valva width; transtilla slender; anellus nearly orbicular, posterior margin without distinct notch; aedeagus tapering from broad base, weakly curved near middle, half as long as valva, cornuti numerous, hair-like, basal process short.

Female genitalia. Anterior apophysis about three-fifths length of posterior apophysis; sternite VIII width:length about 3:1, anterior margin strongly inclined outwards, a shallow excavation in middle, bordered by short folds running along margin, posterior margin strongly inclined inwards; ostium distal to middle of sternite; ductus bursae gradually expanding, two or three times as long as ovate to shortly elliptical corpus bursae; signum wider than long, narrowly elliptical or diamond-shaped, with anterior and posterior projections, plate with a few large teeth, marginal teeth small.

Genitalia diagnosis. Male with cuiller long (nearly reaching or exceeding costa), very thin and clearly S-curved is also found in *A. volgensis*, *A. uralensis*, *A. latipennella* and *A. lidiae*. *A. lidiae* differs by small socii, overtopped by rounded uncus. *A. latipennella* and *A. volgensis* differ by gnathos overtopping socii by more than half to whole length. *A. uralensis* without reliable differences in male genitalia, but external appearance similar to *A. kaekeritziana*. In female the folds on either side anterior to ostium are nearly parallel to anterior margin, see also key (pp. 183–185) to *pallorella* subgroup.

Distribution. Middle latitudes of Europe from north Portugal, Italy, Slovenia and Romania north to England, Sweden and south Finland, but not recorded from a number of countries within this range, notably absent from central European mountain areas. Asia: Russia (Altai Republic); Ural Mountains (Lvovsky, 2019).

Bionomics. Larvae on *Serratula tinctoria* L. (Harper *et al.*, 2002), rarely on *Centaurea scabiosa* L. (Palm, 1989). Checked specimens were taken from June to August.

88   *Agonopterix kyzyltashensis* Buchner & Šumpich, 2020

Agonopterix kyzyltashensis Buchner & Šumpich, 2020: 213.

Description. Wingspan 20–24 mm. Head pale yellowish buff. Labial palp pale yellowish buff, segment 2 with some grey-brown scales on outer and ventral sides. Antenna fuscous. Thorax and tegula ochreous anteriorly, pale yellowish buff posteriorly. Forewing elongate, pale yellowish buff, tinged with pale brown over most of wing surface especially between veins, except basal field and a wide costal stripe to about one-half; scattered blackish scales over most of wing; costal spots weakly developed, not distinguishable from scattered blackish scales; a series of fuscous terminal spots; a blackish subdorsal spot fading into a diffuse dark fuscous patch extending to base of cell, another diffuse dark fuscous patch from close to dorsal margin at one-fifth gradually expanding to fill more than half of wing width next to termen, more developed on veins in terminal area, often with incomplete interruption at 2/3, caused by the otherwise nearly invisible angled pale fascia; a black dot at

1/3, this often inconspicuous or even absent; a black dot at end of cell, just separate from dark fuscous area; fringe grey-brown. Hindwing light grey-buff, darker grey towards apex; fringe pale grey-buff.

VARIATION. The dark fuscous patch in dorsal part of wing varies somewhat in development and extent. The black dot at end of cell can be very inconspicuous.

SIMILAR SPECIES. *A. pallorella* also has a dark patch in the dorsal area, but usually not or only indistinctly extending beyond 2/3, fuscous patch from subdorsal spot to base of cell only rarely present, usually completely absent, wings broader.

MALE GENITALIA. Socii large; gnathos elliptic, about twice as long as wide, not or only shortly exceeding socii; valva with costal margin straight or slightly concave, ventral margin curved at about three-fifths, apical part tapering to obtuse or subacute apex; cuiller thin, strongly sinuous, slightly incurved at subacute apex, 82–96% of valva width; transtilla slender; anellus nearly orbicular, posterior margin without distinct notch; aedeagus tapering from broad base, weakly curved near middle, half as long as valva, cornuti numerous, hair-like, basal process short.

FEMALE GENITALIA. Anterior apophysis about two-thirds length of posterior apophysis; sternite VIII width:length about 3:1, anterior margin more or less strongly inclined outwards, a shallow excavation in middle, bordered by short folds directed inwards obliquely from margin, posterior margin slightly inclined inwards; ostium distal to middle of sternite; ductus bursae with a corrugated section in middle, 1.5 to 3 times as long as elliptical corpus bursae; signum wider than long, narrowly elliptical or diamond-shaped, with or without anterior and posterior projections, plate with or without a few large teeth, marginal teeth small.

SIMILAR SPECIES. Female genitalia of *A. straminella*, *A. kaekeritziana* and *A. kyzyltashensis* not safely distinguishable. For differences from other species of *pallorella* subgroup see key (pp. 183–185).

GENITALIA DIAGNOSIS. For male see under *A. bipunctosa*. For female see under *A. straminella* and key (pp. 183–185) to *pallorella* subgroup.

DISTRIBUTION. Russia (Chelyabinsk and Orenburg provinces, Altai Mts.), Kyrgyzstan (Tien Shan Mts.), Kazakhstan.

BIONOMICS. Host-plant unknown. Moths have been collected from July until September.

89   *Agonopterix squamosa* (Mann, 1864)

Depressaria squamosa Mann, 1864: 185. Pl. 4, fig. 13.
   Agonopterix budashkini Lvovsky, 1998a: 466, syn. n.

DESCRIPTION. Wingspan 19–24 mm. Head ochreous-buff. Labial palp pale buff. Antenna dark grey-brown. Thorax ochreous-buff. Forewing pale ochreous-buff with a scattering of ochreous-brown scales, mainly in outer one-third, or sometimes all over wing and organised into irregular transverse lines; a black dot in cell at one-third and another at end of cell; fringe pale ochreous-buff. Hindwing whitish, grey towards apex; fringe pale grey.

VARIATION. Extreme base of costa black, this feature sometimes indistinct or completely absent; the amount of scattered ochreous-brown scales varies from quite few to forming transverse lines over the whole forewing, more rarely there are larger areas of ochreous-brown near dorsum and between end of cell dot and tornus. *A. squamosa* rarely has a row of terminal spots, if present they are weakly developed; subdorsal spot usually completely absent, very rarely weakly developed.

SIMILAR SPECIES. Usually best recognised by the scattering of ochreous-brown scales, these rarely indistinct or absent.

MALE GENITALIA. Socii large; gnathos elliptic, about twice as long as wide, not or shortly exceeding socii; valva similar to that of *kyzyltashensis* but slightly longer; cuiller stout, not tapering, apex rounded slightly expanded or not, sometimes curved outwards beyond middle, 80–85% valva width; transtilla thin; anellus slightly longer than wide, posteriorly with two rounded lobes separated by mostly narrow V-shaped incision; aedeagus slender, strongly curved in middle part, three-fifths length of valva, cornuti numerous, in a dense group, basal process short.

FEMALE GENITALIA. Anterior apophysis half to two-thirds length of posterior apophysis; sternite VIII width:length 5:2 to 3:1, anterior margin slightly convex, the middle part around ductus bursae undulate, posterior margin nearly straight with median excavation; ostium distal to middle of sternite; ductus bursae with transverse ridges in middle, about equal in length to elliptical corpus bursae; signum wider than long, narrowly elliptical or diamond-shaped, with or without anterior and posterior projections, teeth numerous, one pair much enlarged.

GENITALIA DIAGNOSIS. For male see *A. straminella*. Female distinguished from similar species by combination of structure of ductus bursae (divided into three sections, central section with transverse ridges) and number of turns of ductus spermathecae (8–10, which is more than in other species of *pallorella* subgroup).

DISTRIBUTION. Southern Europe from Spain and France to Croatia, Greece, Bulgaria and Crimea; Sardinia and Sicily. Asia: Turkey, Iran. Lvovsky *et al.* (2016): Lebanon, Israel.

BIONOMICS. Larva in shoot tips of *Centaurea paniculata* L. (Rymarczyk *et al., in litt.*); *C. sterilis* Srev. (Lvovsky, 1998); also in basal rosettes of *Peucedanum officinale* L. (R. Heckford & S. Beavan, *in litt.*). Checked specimens have been taken from May until end of October.

REMARKS. As far as known, *A. squamosa* is unique among European Depressariidae in having host-plants belonging to Asteraceae and Apiaceae.

Agonopterix budashkini was described from several reared males and females from Crimea. It was differentiated from *A. squamosa* by grey coloration on parts of fore and mid-legs and slight differences in male genitalia. We find that these characters fall within the variation of *A. squamosa*, therefore *A. budashkini* is herewith synonymised with *A. squamosa*.

We have seen a single specimen from Spain that resembles *A. squamosa* externally, but with differences in male genitalia, including more clubbed cuiller, aedeagus straight to beyond middle and with long basal process. Furthermore it differs in barcode from both *A. squamosa* and *A. kaekeritziana* by over 1.7%. Additional material is needed to assess the status of this specimen.

90   *Agonopterix tschorbadjiewi* (Rebel, 1916)

Depressaria tschorbadjiewi Rebel, 1916a: (45).

DESCRIPTION. Wingspan 25 mm. Head pale buff. Labial palp pale buff, segment 2 with very few medium grey scales on outer side. [Rebel says it has brown and blackish scales on outer side of segment 2]. Antenna pale buff, ringed light brown. Thorax yellowish buff, overlaid with orange-brown scales in anterior half. Forewing yellowish buff, almost whole wing with numerous orange-brown scales more or less organised into irregular transverse rows and along veins to costa and termen, scattered blackish scales over most of wing, more numerous in fold and in terminal one-third, hardly distinguished from weak terminal spots; basal field almost free of orange-brown scales, extending very shortly along dorsum and costa, defined only by a blackish subdorsal dot; first oblique dot black, second obscure, brownish, a blackish dot at end of cell; fringe buff, more orange towards apex, more whitish towards tornus. Hindwing light grey; fringe whitish grey. Abdomen buff.

VARIATION. Only known from one specimen.

SIMILAR SPECIES. Differs from *A. squamosa* in the unusually numerous and distinct orange-brown scales scattered over almost whole wing, including a few black scales. *Agonopterix volgensis* smaller, with paler yellow coloration, paler cross markings and veins more conspicuously marked orange-ochreous.

MALE GENITALIA. Unknown.

FEMALE GENITALIA. Anterior apophysis half length of posterior apophysis; sternite VIII width:length 3:1, anterior margin strongly convex, the middle part slightly truncated, posterior margin slightly inclined inwards; ostium distal to middle of sternite; ductus bursae long, narrow, with prominent swelling at one-third length from sternite; corpus bursae narrowly elliptical, rather small, signum absent.

GENITALIA DIAGNOSIS. Female without signum and ductus bursae with swelling at one-third length from sternite should be distinctive, but this species is only known from holotype, therefore intraspecific variability is unknown.

DISTRIBUTION. Bulgaria.

BIONOMICS. Host-plant unknown.

REMARKS. Only known from the holotype female. Closely related to *A. squamosa*, a relationship supported by barcode.

91   *Agonopterix volgensis* Lvovsky, 2018

Agonopterix volgensis Lvovsky, 2018: 318.

DESCRIPTION. Wingspan 18–21 mm. Head pale ochreous-yellow, frons whitish. Labial palp pale ochreous-yellow. Antenna brownish buff. Thorax pale ochreous-yellow mixed orange-ochreous. Forewing pale ochreous-yellow, reticulate with ochreous-orange along veins to costa and termen and forming short transverse lines between veins; very fine dark grey dotted transverse lines also present between veins; basal field without orange-ochreous scales, subdorsal dot absent; blackish dot in cell at one-third and another at end of cell; fringe concolorous with wing; terminal dots absent. Hindwings almost white, glossy, hardly darker towards apex, fringe whitish. Abdomen whitish buff.

VARIATION. From the limited number of specimens seen there appears to be very little variation.

SIMILAR SPECIES. *A. tschorbadjiewi* (*q.v.*), *A. abditella*, *A. lidiae*. *A. volgensis* forewings with fine, distinct reticulate pattern, weak short greyish streak from end of cell dot toward termen and without blotch, *A. abditella* and *A. lidiae* with diffuse reticulate pattern, distinct greyish streak from end of cell dot towards termen and with blotch present.

MALE GENITALIA. Similar to *A. bipunctosa*. Socii large; gnathos narrow, lanceolate, 2.5 to 3.5 times as long as wide, exceeding socii by most of its length; valva with costal margin concave, ventral margin curved at about three-fifths, apical part concave, apex subacute; cuiller slender, strongly sinuous, slightly incurved at apex, sometimes crossing costal margin of valva, 95–110% of valva width; transtilla thin; anellus ovate, distal margin with linear incision; aedeagus tapering from base, gradually curved over much of its length, half as long as valva, cornuti less numerous than in most related species, hair-like, basal process of medium length.

FEMALE GENITALIA. Anterior apophysis three-quarters length of posterior apophysis; sternite VIII width:length 5:2, anterior margin convex, with a short truncated projection in middle, posterior margin straight with a median excavation; ostium close to middle of sternite; ductus bursae narrow, nearly twice as long as narrowly elliptical corpus bursae; signum wider than long, narrowly elliptical or diamond-shaped, with anterior and posterior projections, teeth large, numerous.

GENITALIA DIAGNOSIS. Male similar to *A. latipennella* (*q.v.*). For other species with similar male genitalia see remarks under *A. bipunctosa*. In female the short square to truncated projection in middle of anterior margin of sternite VIII is not present in other species of *pallorella* subgroup.

DISTRIBUTION. Russia (type locality Ulyanovsk Province), Ural Mts., Kazakhstan (Mugozhary Mts.)

BIONOMICS. Larva on *Centaurea*. Moths have been taken from June until September.

92   *Agonopterix kaekeritziana* (Linnaeus, 1767)

Phalaena (*Tortrix*) *kaekeritziana* Linnaeus, 1767: 876.
   Tinea liturella [Denis & Schiffermüller], 1775: 137.
   Phalaena sparmanniana Swederus, 1787: 101.
   Tinea flavella Hübner, 1796: pl. 14, fig. 97.
   Depressaria corichroella Turati, 1924: 173.

DESCRIPTION. Wingspan 19–23 mm. Head pale ochreous. Labial palp pale ochreous, segment 2 with proximal half of outer edge deep ochreous to fuscous. Antenna fuscous to deep fuscous with indistinct darker rings. Thorax pale ochreous, usually with deep ochreous median stripe from anterior margin to middle. Forewing pale creamy ochreous to ochreous, with deep ochreous streaks mainly between veins, in cell and fold, occasionally a few scattered fuscous scales; a deep brown subdorsal dot, sometimes absent; a blackish dot in cell at one-third and another at end of cell, a large fuscous spot on dorsal side of end of cell dot; a fuscous or deep ochreous streak between fold and dorsum from one-fifth to two-fifths; a weak series of fuscous dots between veins around termen; fringe concolorous with forewing, without fringe line. Hindwing whitish buff, grey towards termen, often with series of darker grey spots between veins around termen; fringe pale buff, with faint line. Abdomen pale ochreous.

VARIATION. Fringe near apex sometimes deeper ochreous and therefore darker than ground colour of forewing. Dark spot at base of costa usually absent, but sometimes present as pale brownish spot, one specimen has been checked with distinct black spot extending as fine blackish line along costa about 3× eye diameter. There is some variation in depth of ground colour and extent of deep ochreous streaks; subdorsal spot may be absent, as also the series of terminal spots. In the palest specimens even the fuscous spot may be obscure or absent.

SIMILAR SPECIES. Characterised by pale coloration, and the fuscous spot between dorsum and end of cell. *A. broennoeensis* and *A. uralensis* have similar fuscous spots but are narrower-winged. The fuscous spot continues as a paler subdorsal streak toward base but not reaching it, sometimes not continuous but separated into irregular patches, which may also be present in variable numbers distal and ventral to end of cell dot. Intensity of these fuscous elements varies widely, sometimes reduced to faint shadows. Rarely the fuscous scales form very small groups scattered over forewing, resembling the scattered ochreous-brown scales of *A. squamosa*, but usually presence of spot between dorsum and end of cell can distinguish such forms from that species. Some specimens from Spain (Granada) have all markings completely absent. We have also seen such specimens from *A. carduncelli* and *A. straminella* (all these determined by dissection and barcode).

MALE GENITALIA. Socii large, not always clearly separated; gnathos ovate, elliptic or lanceolate, about 2 to 2.5 times as long as wide, exceeding socii by about half its length; valva with costal margin slightly concave, ventral margin curved at two-thirds, distal one-third straight or slightly concave, apex obtuse to subacute; cuiller stout at base, sinuous to nearly straight, usually more or less bowed inwards in middle part, slightly tapering to subacute apex, 80–86% of valva width; transtilla thin; anellus longer than wide, widest at middle, distal margin bifid with narrow V-shaped incision; aedeagus strongly narrowed above base, then nearly parallel-sided to two-thirds, curved in middle or beyond middle, half as long as valva or a little more, cornuti in a more or less dense group, basal process short.

FEMALE GENITALIA. Similar to *A. bipunctosa* but anterior apophysis half to two-thirds length of posterior apophysis; sternite VIII width:length 2:1 to 5:2, posterior margin straight, with median emargination; ostium distal to middle of sternite; ductus bursae slightly expanded in anterior half, over twice as long as ovate corpus bursae; signum wider than long, variable in shape, teeth variable, usually one pair larger than the others.

GENITALIA DIAGNOSIS. For male and female see *A. straminella*.

DISTRIBUTION. Almost all Europe, but absent from Mediterranean islands except Sicily. Widespread in Asia, reaching Far East.

BIONOMICS. Larvae on *Centaurea jacea* L. (Hannemann, 1953); *C. scabiosa* L. (Benander, 1965); *C. nigra* L. and rarely *Knautia* (Palm, 1989); *C. nigrescens* Willd. (Sonderegger, *in litt.*); *C. paniculata* L., *C. pectinata* L., *C. uniflora* Turra and *Cyanus montanus* (L.) Hill (Rymarczyk *et al.*, *in litt.*). Lvovsky (1981) also mentions *Cirsium* and *Scabiosa columbaria* L. Has been reared a few times from *Anchusa officinalis* L. in Denmark, Ellinge Lyng, 1973 (E. Palm, specimen in ZMUC, confirmed by dissection and barcoding). Flying from summer until June of following year.

REMARKS. Records of larvae feeding on *Knautia* and *Scabiosa* require confirmation.

93   *Agonopterix broennoeensis* (Strand, 1920)

Depressaria broennoeensis Strand, 1920: 68.
   Agonopterix roseoflavella Benander, 1955: 54.

DESCRIPTION. Wingspan 18–22(–24) mm. Head buff. Labial palp buff, unmarked. Antenna fuscous. Thorax light ochreous-buff, without posterior crest. Forewing narrow, apex rounded, orange-ochreous, with a few scattered blackish scales, base and first half of costa light orange-buff, basal field paler than rest of wing; a brownish subdorsal spot; a blackish dot at extreme base of costa, cell with blackish dots at one-third and at end, a fuscous spot of more or less irregular shape, between end of cell and dorsum, a light fuscous streak between fold and dorsum from one-fifth to about one-half, often faint; a series of terminal dots, light brown to darker grey; a pink tinge of varying strength and extent, at least around apex and in outer margin of terminal area. Fringe ochreous with strong pink tinge from apex towards tornus, but light grey at tornus. Hindwing grey; fringe light grey.

VARIATION. The pink tinge is occasionally confined to the fringe close to the apex, or at the other extreme it can be present for the whole length of the costa and termen. Specimens seen from Russian Altai Mountains lack red colours and the fuscous spot and appeared to belong to an undescribed species, but barcode and genitalia are identical with European specimens. From Far East Russia (Primorije district) specimens have been seen looking like richly marked *A. kaekeritziana*. Such forms have not been seen in Europe, but they might possibly occur.

SIMILAR SPECIES. *A. broennoeensis* from Europe is characterised by narrow forewings, with orange-ochreous ground colour, fuscous spot between end of cell and dorsum and the pink tinge around the apex and termen. None of the other species in the group shows this combination of characters.

MALE GENITALIA. Socii large; gnathos ovate, about twice as long as wide, sometimes slightly wider, exceeding socii or not; valva with costal margin slightly concave, ventral margin curved at two-thirds, distal part straight or slightly concave, apex obtuse; cuiller stout, at base, slightly tapering to rounded or subacute apex, straight or more often bowed inwards in middle, about 85% of valva width; transtilla thin; anellus slightly wider than long, distal margin rounded to truncate with short V-shaped notch; aedeagus slightly tapering, weakly curved in middle, about three-fifths length of valva, cornuti numerous, larger than in related species, in an elongate group, basal process of medium length.

FEMALE GENITALIA. Anterior apophysis about five-eighths length of posterior apophysis; sternite VIII width:length 3:1, anterior margin strongly inclined outwards, a shallow excavation in middle, posterior margin nearly straight; ostium in middle of sternite; ductus bursae with a loop at one-sixth of its length from sternite, 2.5 times as long as ovate corpus bursae; signum wider than long, lateral lobes obtuse, anterior and posterior projections broad, with numerous teeth.

GENITALIA DIAGNOSIS. Differs from species with otherwise similar male genitalia (*A. straminella*, *A. kyzyltashensis*, *A. squamosa* and *A. kaekeritziana*) in clearly stouter cornuti. Female (description based on only one specimen) differs from other species of *pallorella* subgroup by presence of a loop in distal one-sixth of ductus bursae, although this may disappear if ductus is expanded during preparation.

DISTRIBUTION. Norway, Sweden, Finland and north Russia. Asia: Russian Altai Mts. and Far East Russia; Ural Mountains (Lvovsky, 2019).

BIONOMICS. Larva on *Saussurea alpina* (L.) DC. (Palm, 1989). Moths have been taken from June to September.

REMARKS. In northern Finland Marko Mutanen (pers. comm.) has found moths that are similar to *A. broennoeensis* but with darker ground colour and with different barcode, in an area where *Saussurea* is absent. This requires further investigation.

94   *Agonopterix uralensis* sp. nov.

Type locality: Russia, Moskovo

DESCRIPTION. Wingspan 20–21 mm. Head pale ochreous. Labial palp pale ochreous. Antenna dark fuscous. Thorax pale ochreous with deeper ochreous median stripe from anterior margin to middle, tegula pale ochreous. Forewing narrow, apex nearly right-angled, creamy yellow to pale ochreous, sometimes overlaid with deeper ochreous stripes, with scattered deep ochreous scales and a few scattered dark fuscous scales; faint brown terminal spots; a brown subdorsal spot, a blackish spot at base of costa, a black dot in cell at one-third, another dot at end of cell; a diffuse dark fuscous spot between end of cell spot and dorsum at three-quarters; a dull ochreous-grey streak between fold and dorsum from one-third to one-half; fringe pale ochreous. Hindwing creamy-grey, grey towards apex; fringe light grey.

VARIATION. The deeper ochreous median stripe on thorax can be indistinct. There is some variation in ground colour and abundance of blackish scattered scales. Cell dot at one-third can be difficult to distinguish from scattered dark scales.

SIMILAR SPECIES. *A. kaekeritziana* has similar dark marking between end of cell and dorsum, but has broader wing and usually no black dot at base of costa. *A. broenneensis* normally has some pink coloration near more rounded apex.

MALE GENITALIA. Socii medium to large, nearly contiguous; gnathos narrowly elliptical to lanceolate, about 2.5 times as long as wide, exceeding socii by about half its length; valva with costal margin concave, ventral margin gradually curved after middle, distal one-quarter sometimes slightly concave, apex obtuse or subacute; cuiller from more or less slender base, tapering, sinuous, slender, middle part strongly bowed inwards, distal two-fifths directed towards middle or end of costa, apex hooked inwards, acute, 80–85% of valva width; transtilla thin; anellus ovate, slightly longer than wide, distally narrowed, margin with narrow V-shaped or linear incision; aedeagus tapering beyond base, then nearly parallel-sided and gradually curved over much of its length, half as long as valva, cornuti minute, in one or two dense elongate groups, basal process rather short.

FEMALE GENITALIA. Not available.

GENITALIA DIAGNOSIS. Without reliable differences in male genitalia from *A. bipunctosa*, but external appearance of these species different. For further species with similar male genitalia see under *A. bipunctosa*.

DISTRIBUTION. Russia (Moscow; Chalk Hills).

BIONOMICS. Host-plant unknown.

MATERIAL EXAMINED. Holotype: ♂, Russia, Moskovo, 7.vii.2013, gen. prep. DEEUR 2200 P. Buchner, with DNA barcode id. TLMF Lep 17677, L. Srnka leg., will be deposited in coll. NMPC

Paratypes (arranged according to collection date):

7 ♂, Russia, South Ural near Kidriasovo, Chalk hills, 5.ix.2013, gen. prep. DEEUR 2187 P. Buchner with DNA barcode id. TLMF Lep 26077, DEEUR 2193 with DNA barcode id. TLMF Lep 17685, DEEUR 2197 with DNA barcode id. TLMF Lep 17681, gen. prep. DEEUR 2198 P. Buchner, DEEUR 2194, 2195 & 2196, L. Srnka leg., coll. RCLS

3 ♂, Russia, Orenburg district, Schibendy Valley 20 km S. Prokrova, 29.v.2003, gen. prep. DEEUR 4495 P. Buchner with DNA barcode id. TLMF Lep 19233, DEEUR 4600 with DNA barcode id. TLMF Lep 19240, DEEUR 4602 with DNA barcode id. TLMF Lep 23363, K. Nupponen leg., coll. RCTN

ETYMOLOGY. The species is named after the Ural Mountains.

95   *Agonopterix latipennella* (Zerny, 1934)

Agonopterix latipennella Zerny, 1934: 24.

DESCRIPTION. Wingspan 27–28 mm. Head buff to light sandy brown. Labial palp buff, segment 3 unringed. Antenna light brown. Thorax and tegula buff to light sandy brown. Forewing buff to light sandy brown, scattered black scales over whole surface of wing, sometimes organised into lines along veins; normal cell markings absent; fringe pale buff to light brown, with fringe line. Hindwing creamy white to light grey, grey-buff towards apex.

VARIATION. There is slight variation in ground colour and in the extent of black speckling which may be heavy enough to indicate veins in some specimens.

SIMILAR SPECIES. Scattered black scales over whole wing surface and absence of any typical *Agonopterix* markings is characteristic.

MALE GENITALIA. Socii medium-sized, somewhat divergent, nearly contiguous; gnathos lanceolate, about 2.5 times as long as wide, exceeding socii by its full length; valva with costal margin concave, ventral margin gradually curved after middle, distal one-quarter slightly concave, apex obtuse; cuiller from broad base, rapidly tapering to slender distal half, middle part strongly bowed inwards, distal part directed towards middle of costa, apex slightly expanded, rounded, 80–87% of valva width; transtilla slightly thickened; anellus ovate, wider than long, distal margin truncate with median incision; aedeagus tapering beyond wide base, then nearly parallel-sided, gradually curved in its middle part, half as long as valva, cornuti minute, in a dense group, basal process short.

FEMALE GENITALIA. Anterior apophysis about two-thirds length of posterior apophysis; sternite VIII width:length about 5:2, anterior margin steeply inclined outwards, rounded in middle part, posterior margin inclined inwards, emarginated in middle; a pair of short folds from either side of middle of anterior margin, diverging, ostium distal to middle of sternite; ductus bursae narrow, twice as long as elliptical corpus bursae; signum wider than long, narrowly elliptical, with anterior and posterior projections, teeth not numerous, one pair larger than others.

GENITALIA DIAGNOSIS. Male differs from *A. volgensis* as follows: cuiller not reaching costa versus nearly reaching or exceeding costa, socii somewhat elongated versus round, basal process of aedeagus ending concave versus convex. External appearance very different. For further species with similar male genitalia see under *A. bipunctosa*.

In female sternite VIII has anterior margin more steeply inclined outwards than in other species of *pallorella* subgroup.

DISTRIBUTION. Cyprus, Lebanon, Asiatic Turkey, Armenia.

BIONOMICS. Host-plant unknown. Specimens have been taken from May to November, specimens from June to November fresh, those from May rather worn, which makes it likely that adults hibernate.

REMARKS. Two specimens from Cyprus are the only records in our area.

96   *Agonopterix invenustella* Hannemann, 1953

Agonopterix invenustella Hannemann, 1953: 293.

DESCRIPTION. Wingspan 16–20 mm. Head, labial palp and thorax pale straw yellow. Antenna dark fuscous. Forewing pale straw yellow, veins paler than ground colour, a few scattered blackish dots mainly in dorsal half of wing; a black dot in mid-wing at one-third and a larger one at end of cell; terminal dots blackish, distinct: fringe pale straw yellow. Hindwing whitish, greyer towards apex, with several darker grey dashes between veins around margin. Abdomen pale straw-yellow.

VARIATION. Only known from very few specimens, mostly in poor condition. It is not known if the presence of marginal dashes of hindwing is a useful distinguishing character.

SIMILAR SPECIES. Resembles other species of *pallorella* group with very limited black dots. *A. straminella* usually has a subdorsal black dot; *A. squamosa* usually lacks terminal dots.

MALE GENITALIA. Socii medium-sized, less divergent than in related species; gnathos elliptic, about 2.2 times as long as wide, exceeding socii by half its length;

valva with costal margin concave, ventral margin strongly curved at about three-fifths, apex subacute; cuiller stout, paddle-shaped, inner margin straight, outer margin with large bulge from two-fifths to rounded apex, 96–98% of valva width; transtilla slightly thickened in middle; anellus widest in middle, anellus margin with wide V-shaped excavation; aedeagus slender, little tapering, slightly curved, about three-fifths length of valva, cornuti in a dense group, basal process small.

FEMALE GENITALIA. Anterior apophysis about three-fifths length of posterior apophysis; sternite VIII width:length about 10:3, anterior and posterior margins nearly straight; a pair of short folds originating just above anterior margin diverging at wide angle, ostium just distal to middle of sternite; ductus bursae narrow, about equal in length to large elliptical corpus bursae; signum unusually large, wider than long, lateral lobes obtuse, with anterior and posterior projections, with numerous teeth.

GENITALIA DIAGNOSIS. In male paddle-shaped cuiller reaching costa is distinctive. In female nearly straight margins of sternite VIII, folds originating distal to anterior margin and unusually large corpus bursae and signum are characteristic.

DISTRIBUTION. Sardinia, Sicily. Algeria.

BIONOMICS. Host-plant unknown. The first European specimen was collected in March, indicating that it had overwintered as adult.

REMARKS. Described from two males and a female from Algeria. The species remains poorly known. Hannemann's original description compared it with *Exaeretia lepidella*, from which he separated it by characters of the labial palps which are those separating *Agonopterix* from *Exaeretia*. From the male genitalia, we have assumed that it is related to *A. pallorella*, but female genitalia are less similar. The species with closest barcode are *A. carduncelli* and *A. straminella*, both with p-distance of 2.6%.

97   *Agonopterix abditella* Hannemann, 1959

Agonopterix abditella Hannemann, 1959: 40.

DESCRIPTION. Wingspan 20–24 mm. Head straw-yellow. Labial palp pale orange-yellow, segment 2 darkened on outer side. Antenna yellowish brown. Forewing straw-yellow with scattered light brown and orange-brown scales, sometimes forming small patches or short transverse lines, giving the wing an orange-yellow tinge, scattered blackish scales forming similar transverse lines; basal field sometimes defined by a brown partial fascia; narrow orange-brown lines in fold and in mid-line of wing, more diffuse in cell, sometimes also between veins towards apex and termen; a blackish dot at base of costa, outer part of costa with light orange-brown spots, terminal spots absent; a blackish dot in cell at one-third and another at end of cell; a diffuse round greyish blotch between cell and costa at one-half and a greyish band from end of cell towards tornus; fringe concolorous with forewing to pinkish red. Hindwing whitish yellow; fringe whitish yellow with weak line. Abdomen pale straw-yellow.

VARIATION. Some specimens have a partial basal fascia; the narrow orange-brown lines may be restricted to fold and mid-line of wing in terminal third, or may be more extensive giving the wing a reticulate appearance; the degree of development of scattered orange or brown scales is very variable; occasionally the two blackish cell dots are faint or absent. Sometimes the costa and terminal areas are more or less suffused pinkish red.

SIMILAR SPECIES. *A. lidiae, A. volgensis (q.v.)*

MALE GENITALIA. Socii medium-sized, slightly divergent, more or less contiguous; gnathos lanceolate, three times as long as wide, carried beyond socii on long gnathos arms; valva long, costal margin concave, ventral margin slightly curved in middle, apex subacute; sacculus unusually short, cuiller small, straight, directed towards costal base of valva, narrowed at middle to slender distal half, apex slightly expanded, hooked outwards, 65–68% of valva width; transtilla thin; anellus small, about as wide as long, distal margin with broad humps separated by a shallow sinus; aedeagus curved and tapering from middle, less than half as long as valva, cornuti in an inconspicuous group, basal process short.

FEMALE GENITALIA. Not available.

GENITALIA DIAGNOSIS. In male short cuiller ending in a distinct hook in combination with entire length of gnathos exceeding socii is characteristic.

DISTRIBUTION. Russia (Lower Volga). Asia: Russia (Cheliabinsk district, Altai Mts), Daghestan, Armenia (Ivan Richter, pers. comm.).

BIONOMICS. Host-plant unknown. Adults collected in July and August.

REMARKS. The species was described from one male from Daghestan.

98   *Agonopterix lidiae* Buchner, 2020

Agonopterix lidiae Buchner, 2020b: 3.

DESCRIPTION. Wingspan 17–20 mm. Head pale creamy yellow, face brown laterally. Labial palp whitish buff, segment 2 with some orange-brown scales on upper and outer sides. Antenna grey-brown. Thorax and tegula ochreous to pale ochreous-buff, thorax with three longitudinal orange-brown stripes, posterior crest absent. Forewing ochreous to pale ochreous-buff, with abundant scattered orange-brown scales mostly organised into irregular transverse lines, these more numerous towards apex, rarely with some dark grey scales; basal field scarcely differentiated, a brown subdorsal spot, a blackish dot in cell at two-fifths, a diffuse orange spot in cell at one-half, a dark brown dot at end of cell; two diffuse dark grey markings, a blotch between diffuse orange spot and costa and a band running from near end of cell to tornus; a narrow orange-brown line in fold; terminal spots indistinct; fringe concolorous with end of forewing, with a line. Hindwing light grey-buff, slightly darker towards apex; fringe light grey-buff with one line. Forelegs dark grey above, tibia with some cinnamon-brown scales. Abdomen light grey-buff.

VARIATION. There is slight variation in ground colour; the dots in cell at two-fifths and at end of cell are not reliably present and have only been found present together in a single specimen, which may be due to variation or the fact that we have seen no fresh specimens of this species.

SIMILAR SPECIES. *A. abditella* is the most similar species, although due to differences in genitalia and barcode not closely related. Forms with scattered blackish scales mainly organised into irregular transverse lines occur in *A. abditella* but not in *A. lidiae*, but *A. abditella* is very variable and forms are found which are nearly indistinguishable externally; the only fairly constant distinguishing feature is a dark line running through cell, then becoming indistinct before reappearing between 2/3 and 4/5 in *A. abditella*, this line not present in *A. lidiae* (in one specimen we have seen it is present but not interrupted between cell and the distal section). Other species with oblique diffuse band running from near end of cell toward tornus are *A. angelicella* and *A. astrantiae*, but both have a pair of oblique dots and plical dot.

MALE GENITALIA. Socii rather small, overtopped by rounded uncus in Italian specimens, but socii larger and uncus smaller in French specimens; gnathos narrow, parallel-sided in basal half, at least four times as long as wide, exceeding socii by half its length; valva long, costal margin concave, ventral margin slightly curved at around three-fifths, apex obtuse; cuiller long, slender, slightly inclined inwards in basal half then curved outwards, apex subacute, incurved, overlapping costal margin of valva, up to 120% of valva width, but position of cuiller varies considerably according to preparation details; transtilla thin; anellus small, longer than wide, posterior margin with narrow incision; aedeagus slightly tapering in basal half, weakly curved at middle, less than half as long as valva, cornuti in elongate group, basal process rather short.

FEMALE GENITALIA. Anterior apophysis about two-thirds length of posterior apophysis; sternite VIII width:length about 5:2, anterior margin strongly convex, posterior margin inclined inwards, excavated in middle; ostium in middle of sternite; ductus bursae gradually expanding, four times as long as small ovate corpus bursae; signum wider than long, crescent-shaped, with short anterior and posterior projections, teeth not numerous, mainly marginal.

GENITALIA DIAGNOSIS. Rounded uncus overtopping socii or at least reaching their distal end in combination with very long S-curved cuiller distinctive. In female absence of folds in sternite VIII and absence of other distinct features is shared with *A. nanatella* and *A. propinquella*; small differences are found in width:length ratio and shape of anterior margin of sternite VIII and in outline of uncompressed papilla analis in ventral view.

DISTRIBUTION. Friuli-Venezia Giulia province, north-east Italy, where it has been found between 500 and 1300 m. Recently found in two localities in France in Drôme and Hautes-Alpes.

BIONOMICS. Several larvae found on *Laserpitium siler* L. in France. One French larva was found 20 May, moth emerged 6 June. All Italian specimens were collected as adults in August and September.

REMARKS. As the placement in *pallorella* subgroup is questionable, it is also included in the key (pp. 206–209) to *arenella* subgroup. The rearing from *Laserpitium siler* implies a position in group 2A but morphology does not strongly support this.

The earliest record in Italy is from 2003 and it is only known from a limited area, suggesting the possibility that it might be an introduced undescribed species from somewhere outside Europe. However the discovery of two specimens in France in 2022 with barcode difference from the Italian population of 1.4% indicates that these are two populations that have been isolated for a long time.

arenella subgroup (99–109)
Feed on tribe Cynareae. These have more or less typical *Agonopterix* markings; most have posterior crest on thorax.

Female genitalia key to *arenella* subgroup, including *A. lidiae* and species of the small or monotypic subgroups (*A. petasitis, A. cotoneastri, A. multiplicella* and *A. doronicella*)

1a	On either side of ostium a fold forming a narrow triangle, its sharp tip directed posteriorly, often continuing as fine, gently curved line to posterior margin; anterior margin of sternite VIII below ostium with a structure forming a wide "W"; sclerites in ostium strongly diverging (figs. 99a–d)	*A. subpropinquella*
1b	Sternite VIII different	2
2a (1b)	Inside the ostium, lateral to sclerites, a pair of diverging folds, forming a more or less incomplete "V"; between ostium and anterior margin of sternite a distinct lip, which is accompanied on either sides by a short, somewhat diffuse sclerotisation perpendicular to margin of sternite (fig. 112a). Note: the lip may appear as triangle in embedded slides (fig. 112b), because this area tends to produce a fold when compressed dorsoventrally.	*A. multiplicella*

2b	In ostium no such folds, but folds of different form may be present outside the ostium (e.g. fig. 111a)	3
3a (2b)	Ostium just distal to middle of sternite, a curved fold closely proximal to its anterior margin, distant from anterior margin of sternite	*A. arenella*
3b	Folds absent or curved folds or bands close to anterior margin of sternite or central part running along margin	4
4a (3b)	Sternite posterior to ostium distinctly less sclerotised, this area appears as pale "V" or "U" extending to posterior margin of ostium (fig. 111a); ostium outlined by a fold anteriorly and laterally which forms an "U" with width less than 2 × depth; sternite VIII in proximal area often covered by tiny dots	*A. cotoneastri*
4b	Sternite posterior to ostium not less sclerotised (if a pale excavation in middle of distal margin of sternite VIII is present, it does not reach ostium)	5
5a (4b)	Ostium large, occupying almost entire length of sternite, not circular but broadest in lower half, lower lip distinctly outlined (fig. 101a); sternite VIII, apart from this line, without folds, its anterior margin straight to slightly convex	*A. kuznetzovi*
5b	This area different	6
6a (5b)	Signum very small (figs. 110a, c, d)	*A. petasitis*
6b	Signum medium-sized to large	7
7a (6b)	Papilla analis densely covered with very short bristles, medium length bristles very rare or nearly absent, in lower half a few very long, usually upcurved bristles; shape of uncompressed (!) papilla analis in ventral view unusually swollen in lower half, broadest about 20–30 % from base, outer margins concave near middle, then tapering with nearly straight margins toward tip (figs. 106a, b; 107a, b; 108a)	8

7b	Papilla analis with short bristles present, but not remarkably dense, long and also numerous medium length bristles present; shape of uncompressed (!) papilla analis in ventral view: broadest near middle, outline convex throughout, rarely broadest in lower part but this area not markedly swollen and margins toward tip straight In cases of doubt compare *A. nanatella* (figs. 100a–d) and *A. ferocella* (figs. 105a–d), which are keyed out here, but some specimens can give the impression to belong to species group keyed out under 7a	12
8a (7a)	Anterior margin of sternite with median bulge below ostium (figs. 103a, b; 104)	9
8b	Anterior margin of sternite with a more or less distinct excavation below ostium (figs. 106, 107, 108)	10
9a (8a)	Ostium large, sclerites in ostium strongly diverging (figs. 103a, b)	*A. carduella* in part
9b	Ostium of average size, sclerites in ostium directed obliquely upwards	*A. ivinskisi* in part
10a (8b)	Between ostium and anterior margin of sternite a distinct fold, forming a wide "W", margin below it with indistinct excavation, sclerites in ostium strongly diverging (fig. 108a); restricted to Canary Islands	*A. cinerariae*
10b	Between ostium and anterior margin no such fold	11
11a (10b)	Sclerites in ostium directed obliquely upwards, shallow oblong median excavation distinct, its straight central section about 1/5–1/6 of sternite VIII width, ductus bursae with an intestine-like structure near middle (figs. 106a, b)	*A. laterella*
11b	Sclerites in ostium strongly diverging, shallow oblong median excavation distinct, its straight central section about 1/3–1/5 of sternite VIII width (figs. 107a, b), no distinct intestine-like structure in ductus bursae	*A. xeranthemella*
12a (7b)	Sclerites in ostium strongly diverging, anterior margin of sternite with median bulge below ostium, which is large and fills this bulge (figs. 103a, b); sternite without folds in this area	*A. carduella* in part

| 12b | Sclerites in ostium directed obliquely upwards, anterior margin of sternite with or without median bulge, sternite often with folds of different kind | 13 |

| 13a (12b) | Sternite VIII without any kind of folds or bands | 14 |
| 13b | Sternite VIII with folds or bands | 16 |

| 14a (13b) | Anterior margin of sternite VIII straight or evenly convex throughout Female genitalia differences are too small for safe separation, but these 3 species can be separated by external appearance | *A. lidiae* *A. nanatella* *A. propinquella* in part |
| 14b | Anterior margin not straight or evenly convex throughout | 15 |

| 15a (14b) | Anterior margin of sternite VIII with slight median bulge round ostium (fig. 104a) | *A. ivinskisi* in part |
| 15b | Anterior margin inclined outwards, flat in middle one-third, sometimes with a small median hump; external appearance distinct (figs. 113a–c) | *A. doronicella* |

| 16a (13b) | The bent fold below ostium usually running through, not touching anterior margin of sternite VIII, sometimes indistinct; size of corpus bursa and signum below average for *Agonopterix* (figs. 105a, b). Forewing with distinct reddish-brown pattern | *A. ferocella* |
| 16b | Fold on either side anterior to ostium running obliquely into anterior margin and ending there without overtopping margin, usually short (less than half ostium diameter), thin and usually somewhat irregular (fig. 102d). Forewing with large dark grey blotch | *A. propinquella* in part |

99   *Agonopterix subpropinquella* (Stainton, 1849)

Depressaria subpropinquella Stainton, 1849: 156.
   Depressaria intermediella Stainton, 1849: 160.
   Depressaria rhodochrella Herrich-Schäffer, 1854: pl. 59, fig. 425.
   Depressaria himmighofenella Herrich-Schäffer, 1854:

Depressaria thoracica Lederer, 1855: 233.
Depressaria sublutella Staudinger, 1859: 237.
Depressaria variabilis Heinemann, 1870: 152.
Depressaria occaecata Meyrick, 1921a: 391.
Depressaria remota Meyrick, 1921a: 392.
Depressaria crispella Chrétien, 1929: 193.
Depressaria keltella Amsel, 1935b: 217.
Depressaria amilcarella Lucas, 1951: 143.

DESCRIPTION. Wingspan 16–23 mm. Head light ochreous-brown to dark fuscous mixed with mid-brown, face buff. Labial palp buff, segment 2 outer and ventral sides more ochreous with some dark fuscous scales, segment 3 with dark fuscous rings at base and two-thirds, and usually at tip. Antenna ochreous-fuscous, narrowly ringed dark brown. Thorax ochreous-buff to blackish brown, usually with some paler scales. Forewing elongate, typically ochreous-brown, sometimes with slight cinnamon-pink tinge, but coloration ranging from pale straw-yellow or pale buff to mid-brown; basic markings are two oblique blackish dots, a weak dot at end of cell and a fuscous blotch; additional markings often present include a weak subdorsal spot, some costal and terminal spots and an abundant scattering of blackish scales; fringe ochreous to ochreous-brown, with two or more lines. Hindwing whitish or greyish buff, greyer towards termen, with two to five lines. Abdomen ochreous-buff with scattered dark fuscous scales.

VARIATION. Ground colour is extremely variable from pale straw-coloured or pale buff through various shades of ochreous-brown to brown. The head and thorax are sometimes largely dark fuscous; sometimes the forewing has no markings, or the oblique dots are scarcely visible among scattered dark scales; in most specimens there are only the three blackish dots in the cell and the blotch, which can vary in intensity; occasionally there may be additional dark markings at base of dorsum, on costa and around termen; rarely there is a very weakly indicated basal field.

SIMILAR SPECIES. Characterised by the long unicolorous forewing, without basal field and with limited markings. *A. propinquella* has shorter wings, basal field, slightly mottled pattern and strong blotch; *A. scopariella* has right-angled apex and white dots in cell.

MALE GENITALIA. Socii large, well separated, divergent; gnathos ovate to lanceolate, 2–3 times as long as wide, slightly exceeding socii; valva short to long, costal margin concave, ventral margin curved beyond middle, distal part various, apex obtuse or rounded; cuiller tapering from base, straight or curved inwards, apex rounded to subacute, 70–80% of valva width; transtilla slightly thickened; anellus about as long as wide, with a short-stalked U-shaped sclerotisation over whole length of ventral surface; aedeagus tapering, angled at about 45 ° at middle, less than half as long as valva, cornuti in a more or less dense group, basal process rather short, not expanded terminally.

FEMALE GENITALIA. Anterior apophysis about three-fifths length of posterior apophysis; sternite VIII width:length 2:1 to 5:2, anterior margin strongly convex, middle part smoothly rounded to undulate, forming a narrow lip to ostium, with a fold each side diverging at narrow angle, posterior margin concave or inclined inwards, excavated in middle; ostium unusually large, filling proximal part of sternite, antrum lobes spreading to margin of ostium, distally forming folds directed posteriorly; ductus bursae one to two times as long as elliptical corpus bursae; signum wider than long, elliptical, often with anterior and posterior projections, teeth not numerous, one pair larger than others.

GENITALIA DIAGNOSIS. Male generally similar to *A. propinquella*, differs in presence of U-shaped sclerotisation of anellus and details of aedeagus (angled at about 45° at middle, in ventral view distinctly asymmetrical, basal process not wider than aedeagus, ending straight to slightly convex at one-third of total aedeagus length). Female genitalia distinct by combination of these three features: a fold forming a narrow triangle on either side of ostium with sharp tip directed posteriorly, often continuing as fine, gently curved line to posterior margin; anterior margin of sternite VIII below ostium with a structure forming a wide "W"; sclerites in ostium strongly diverging (fig. 96).

DISTRIBUTION. South and middle latitudes of Europe, just extending to southern Fennoscandia; not recorded in a few countries. Asia: Turkey, Lebanon, Israel, Syria, Yemen. North Africa: Morocco.

BIONOMICS. Larvae on *Cirsium vulgare* (Savi) Ten., *C. acaulon* (L.) Scop., *Centaurea jacea* L., *Cyanus segetum* Hill and *Onopordum acanthium* L. (Hannemann, 1953); *Centaurea scabiosa* L. (Palm, 1989); *Arctium lappa* L., *Centaurea nigra* L., *Carduus crispus* L. and *C. tenuiflorus* Curtis (Harper *et al.*, 2002); *Centaurea vallesiaca* (DC.) Jordan, *C. leucophaea* Jordan, *C. stoebe* L. and *Cirsium eriophorum* (L.) Scop. (Sonderegger, *in litt.*); *Cynara cardunculus* L., *Mantisalca salmantica* (L.) Briq. & Cavill., *Centaurea sphaerocephala* L., *Silybum marianum* (L.) Gaertn. and *Cirsium arvense* (L.) Scop. (Corley, 2015); *Galactites tomentosus* Moench, *Cirsium ferox* (L.) DC., *Carduus pycnocephalus* L., *C. defloratus* L., *C. litigiosus* Nocca & Balb., *Centaurea pectinata* L., *C. paniculata* L., *C. aspera* L. (Rymarczyk *et al.*, *in litt.*); *Rhaponticum coniferum* (L.) Greuter (Corley *et al.*, 2020). Emerging in May to July, flying until April of the following year.

REMARKS. Hannemann (1958a) synonymised *keltella* Amsel, 1935b with *subpropinquella*.

Depressaria hamriella Chrétien, 1922: description based on one specimen from Dar-bel-Hamri (Morocco). It is remarkably large (wingspan 25 mm) but further description corresponds with external appearance of *A. subpropinquella*. Specimens from Morocco, High Atlas, stored as *Agonopterix hamriella* in NHMW belong to *A. subpropinquella*. Genitalia of type not checked, therefore the synonymy with *A. subpropinquella* remains unconfirmed.

100   *Agonopterix nanatella* (Stainton, 1849)

Depressaria nanatella Stainton, 1849: 154.
  Depressaria aridella Mann, 1869: 385.

DESCRIPTION. Wingspan 13.5–18 mm. Head whitish buff. Labial palp buff, segment 2 with scattered ferruginous scales on outer side, segment 3 with fuscous ring at four-fifths. Antenna ochreous-buff. Thorax buff with scattered dark ochreous scales, without posterior crest. Forewing buff or ochreous-buff, more or less extensively overlaid ochreous to light brown or olive, sometimes forming an irregularly-shaped median band, narrow on costa and wider towards dorsum, whole wing with scattered fuscous scales; basal field not defined; a dark fuscous subdorsal dot, a blackish dot in cell at two-fifths and another much weaker at end of cell; blotch weak, light fuscous; fringe buff with one fringe line, sometimes absent. Hindwing grey, darker in male than in female, darker towards termen; fringe grey with two or three fringe lines. Abdomen light to dark grey, buff towards lateral margin.

VARIATION. The extent to which the buff ground colour is overlaid by ochreous, light brown or olive is very variable. Oblique dots sometimes both present and distinct, the dot at end of cell is usually small and quite frequently absent. The blotch is usually faint, and may be entirely absent, but sometimes distinct.

SIMILAR SPECIES. Characterised by small size, buff to olive coloration, with weak markings and no basal field or terminal spots. The male hindwing is darker grey than in most species. *A. aspersella* (*q.v.*).

MALE GENITALIA. Socii large, contiguous to small uncus; gnathos ovate to lanceolate, about 2.5 times as long as wide, equalling or exceeding socii by up to half its length; valva with costal margin slightly concave, ventral margin curved at or beyond middle, distal part straight or slightly concave, apex obtuse; cuiller sinuous, straight or slightly curved inwards in basal half, distal part curved outwards, apex rounded to subacute, 60–68% of valva width; transtilla not thickened; anellus about as long as wide, distal margin slightly concave, thickened, overlying a V-shaped incision; aedeagus rather slender, tapering, curved in middle, two-fifths to half as long as valva, cornuti inconspicuous, occupying most of vesica, basal process of medium length.

FEMALE GENITALIA. Anterior apophysis about three-fifths length of posterior apophysis; sternite VIII width:length 5:2 to 3:1, anterior margin slightly convex, posterior margin straight; ostium reaching posterior margin of sternite; ductus bursae three times as long as small ovate corpus bursae; signum wider than long, elliptical, teeth not numerous, and rather small.

GENITALIA DIAGNOSIS. In male cuiller two-thirds of valva width, terminating in a distinct hook in combination with socii overtopped by gnathos not more than half gnathos length is distinctive. Female similar to *A. lidiae* and *A. propinquella*. More details under *A. lidiae*.

DISTRIBUTION. South and middle latitudes of Europe from Portugal and Ireland to Greece and Romania; Crimea (Lvovsky, 2019). Absent from north-east Europe and Mediterranean islands from Sicily eastwards. We have not seen records from Asia.

BIONOMICS. Larva on *Carlina vulgaris* L. (Harper *et al.*, 2002); *C. corymbosa* L. (Carvalho & Corley, 1995); also recorded on *Carduus tenuiflorus* Curtis (Sonderegger, *in litt.*).

REMARKS. Hannemann (1976) treated *Depressaria aridella* Mann, 1869 as a variety of *A. nanatella*.

101   *Agonopterix kuznetzovi* Lvovsky, 1983

Agonopterix kuznetzovi Lvovsky, 1983: 594.

DESCRIPTION. Wingspan 15–22 mm. Head pale ochreous-buff, face pale straw-coloured. Labial palp pale ochreous-buff, segment 3 often with a very weak darker ochreous ring in middle. Antenna pale ochreous-grey, darker in distal half. Thorax pale ochreous-buff. Forewing quite narrow, pale ochreous-buff to ochreous, sometimes with weak pink suffusion or with some darker ochreous suffusion, with numerous scattered pale ochreous-brown scales, sometimes with some scattered blackish scales mainly along veins; basal field not indicated except in more heavily marked specimens; terminal spots grey to blackish; subdorsal spot if present deep brown; oblique dots blackish, another smaller dot at end of cell; blotch faint, greyish; fringe pale buff with one very indistinct fringe line. Hindwing light grey, slightly darker towards termen; fringe very pale greyish buff, with one more or less distinct fringe line, sometimes with other very faint lines towards margin. Abdomen light greyish buff.

VARIATION. Ground colour of forewing varies between pale straw-coloured and light orange-ochreous; the abundance of darker scales varies, sometimes being almost or completely absent or abundant, indicating most of the veins; basal field demarcated or not; the grey blotch is always faint, but may be almost absent. The palest specimens have all markings almost obsolete except the scattered ochreous-buff scales.

SIMILAR SPECIES. *A. kuznetzovi* is too variable to be possible to characterise. When present the pair of oblique dots and the blotch distinguish it from any of the *pallorella* group species. This species should always be confirmed from genitalia.

MALE GENITALIA. Socii large, well separated, divergent; gnathos elliptic, about 2.2 times as long as wide, exceeding socii by about half its length; valva with costal margin concave, ventral margin gradually curved, distal part straight, apex rounded; cuiller stout, expanding above base, widest at two-thirds to three-quarters length, inner margin concave, outer margin convex, apex incurved, acute, outer surface with microsculpture, about 70% of valva width; transtilla thickened in middle; anellus about as long as wide, distal margin slightly concave, narrowly thickened including wide shallow sinus, a broad median projection beyond this; aedeagus stout, tapering, curved in middle, less than half as long as valva, cornuti occupying most of vesica, basal process short.

FEMALE GENITALIA. Anterior apophysis about half length of posterior apophysis; sternite VIII width:length 7:2, anterior margin slightly convex, posterior margin straight; ostium large, occupying almost entire length of sternite, lower lip distinctly outlined; ductus bursae gradually expanding, twice as long as elliptical corpus bursae; signum wider than long, elliptical, sometimes bulging on anterior side, teeth rather small.

GENITALIA DIAGNOSIS. In male shape of cuiller is unique in European *Agonopterix*. Cuiller of *A. carduella* somewhat similar, but anellus is very different. Female distinct in very large ostium, which is not circular but broadest in lower half and with lower lip distinctly outlined (fig. 101a); sternite VIII, apart from this line, without folds.

DISTRIBUTION. South-west England, South Spain, also province Álava in North Spain (Vives & Gastón, 2017), Russia (Orenburg). In Asia known from Turkey, Kyrgystan and Kazakhstan.

BIONOMICS. Larvae on *Serratula tinctoria* L. (Langmaid & Pelham-Clinton, 1984). Moths emerge in June and July and have been found flying in April of the following year.

102   *Agonopterix propinquella* (Treitschke, 1835)

Haemylis propinquella Treitschke, 1835: 184, 280.

DESCRIPTION. Wingspan 16–19 mm. Head light brownish buff, face buff. Labial palp buff, segment 2 with scattered brown scales on outer side, segment 3 sometimes with a few brown scales at base, with a weak to strong greyish brown to blackish ring from half to three-quarters. Antenna grey-brown with narrow dark brown rings. Thorax light brownish buff. Forewing greyish buff, with much of wing suffused light brown, with a light scattering of deep brown scales; costal spots blackish brown; terminal spots grey, rather weak; basal field buff mixed grey; basal fascia dark brown or absent, leaving basal area demarcated by mid-brown shade fading outwardly; oblique dots blackish, the first sometimes absent or the two dots fused to form a short bar; a small dot at end of cell; blotch blackish, large, round, the area between blotch and costa and extending beyond spot to four-fifths shaded with numerous blackish brown scales; fringe light brownish buff with two indistinct fringe lines. Hindwing light greyish buff, slightly darker towards termen; fringe buff with two fringe lines. Abdomen buff.

VARIATION. Forewing coloration varies little, but the intensity of colour of the scattered scales is variable. The usually conspicuous blotch ranges from blackish to pale grey or may be almost absent. The oblique dots and end of cell dot are occasionally associated with one or two pale scales.

SIMILAR SPECIES. Typical specimens have more conspicuous blotch than most other species. The terminal area of forewing tends to be paler than the middle. *A. subpropinquella* lacks the demarcated basal field. *A. arenella* has straw-yellow ground colour.

MALE GENITALIA. Socii large, widely separated, divergent; gnathos elliptical, about 2.2 times as long as wide, equalling or just exceeding socii; valva with costal margin slightly concave, ventral margin gently curved around middle, distal part straight or slightly concave, apex broadly rounded; cuiller of varying thickness, basal part inclined inwards, more slender distal two-fifths curved slightly outwards, apex acute to subacute, 65–70% of valva width; transtilla not thickened; anellus slightly wider than long, distal margin rounded or slightly protruding with variably developed incision or sinus; aedeagus gradually tapering, angled in middle, two-fifths length of valva, cornuti inconspicuous, basal process rather short, expanded terminally.

FEMALE GENITALIA. Anterior apophysis about half to two-thirds length of posterior apophysis; sternite VIII width:length 3:1 to 5:1, anterior margin straight or convex, posterior margin straight or inclined inwards; ostium distal to middle of sternite, often reaching posterior margin; ductus bursae gradually expanding, three times as long as ovate corpus bursae; signum wider than long, elliptical or diamond-shaped, with anterior and posterior projections, teeth variable in number and size.

GENITALIA DIAGNOSIS. Male generally similar to *A. subpropinquella* (*q.v*). Female similar to *A. lidiae* and *A. nanatella*. See under *A. lidiae*.

DISTRIBUTION. Widely distributed in Europe but absent from a few countries and most Mediterranean islands, except Sicily and Malta. Asia: widespread in temperate latitudes, reaching Japan.

BIONOMICS. *Cirsium vulgare* (Savi) Ten., *Arctium lappa* L. and *Serratula* (Hannemann, 1953); *Cirsium arvense* (L.) Scop. (Harper *et al.*, 2002); *C. eriophorum* (L.) Scop. and *Arctium minus* Bernh. (Sonderegger, *in litt.*); *Centaurea* and *Carduus* (Lhomme, 1945). Moths emerging from July to September, rarely June or October, flying until June of the following year.

103 *Agonopterix carduella* (Hübner, 1817)

Tinea carduella Hübner, [1817]: pl. 66, fig. 439.

DESCRIPTION. Wingspan 14–18 mm. Head ochreous-buff, usually tinged cinnamon. Labial palp cinnamon-buff, segment 2 deeper cinnamon ventrally, segment 3 with dark grey ring at two-thirds, sometimes with blackish apex. Antenna pale grey-buff to grey-brown, darker distally. Thorax ochreous-buff to ochreous, usually with cinnamon tinge. Forewing ochreous-buff usually with more or less strong cinnamon tinge all over, with a scattering of light to medium brown and a few blackish scales; terminal spots grey, few and weak; basal field more or less paler than rest of wing; basal fascia mid to deep brown, narrow, often weak; oblique dots blackish, median cell dot usually absent, a blackish dot at end of cell; cell blotch grey, varying from fairly distinct to very faint; the area between first oblique dot and blotch pale due to absence of scattered darker scales, more obvious in dark specimens; fringe ochreous-buff, often tinged cinnamon, with one or two fringe lines. Hindwing whitish buff, pale grey-brown towards termen, fringe whitish buff to pale grey-brown, with two fringe lines. Abdomen buff, sometimes tinged cinnamon.

VARIATION. There is variation in frequency of scattered dark scales, rarely they can be nearly absent, and the wing then appears smooth; median cell dot can be present.

SIMILAR SPECIES. Characterised by small size, usually cinnamon tinge, a scattering of light to medium brown scales and weakly defined pale basal area on forewing. *A. nanatella* has similar patterns, but usually no cinnamon tinge, no pale basal field, second oblique dot larger than first, hindwings darker. *A. laterella* can be extremely similar, but usually wingspan exceeds 20 mm.

MALE GENITALIA. Socii large, contiguous, divergent; gnathos ovate-lanceolate, about 2.3 times as long as wide, exceeding socii by less than half its length; valva with costal margin concave, ventral margin curved from middle to broadly rounded apex; cuiller stout, bowed outwards in basal half, inwards in slightly wider distal half, apex rounded, 68–72% of valva width; transtilla thickened; transtilla lobes narrow; anellus lobes reduced; anellus exceptionally long, reaching transtilla, with two pairs of lateral bulges; aedeagus gradually tapering from stout base, slightly angled in middle, three-fifths length of valva, cornuti in a large dense mass, basal process very short.

FEMALE GENITALIA. Anterior apophysis about two-thirds length of posterior apophysis; sternite VIII width:length 4:1, anterior margin straight with median bulge round ostium, posterior margin inclined inwards with median excavation; ostium large, occupying bulge and extending beyond middle of sternite, strongly delineated on proximal and distal margins; ductus bursae with a slightly expanded section in distal half, twice as long as ovate corpus bursae; signum wider than long, elliptical, teeth variable in number and size.

GENITALIA DIAGNOSIS. In male shape of anellus is unique in European *Agonopterix*. In female distinctive combination of sclerites in ostium strongly diverging and anterior margin of sternite with median bulge below ostium, which is large and fills this bulge (figs. 103a, b).

DISTRIBUTION. Middle latitudes of Europe from Scotland to Poland, extending south to Spain and Albania; absent from several countries and from Mediterranean islands except Sardinia. Russia, South Ural (Lvovsky, 2019).

BIONOMICS. Larva on *Cirsium arvense* (L.) Scop. and *C. acaulon* (L.) Scop. (Hannemann, 1953); *Centaurea nigra* L. and *Carduus* (Palm, 1989); *Arctium* spp., *Cirsium vulgare* (Savi) Ten. and *C. heterophyllum* (L.) Hill (Harper *et al.*, 2002); *Cirsium eriophorum* (L.) Scop., *Carduus personata* (L.) Jacq. and *Cyanus segetum* Hill (Sonderegger, *in litt.*); *Centaurea jacea* L. and *C. uniflora* Turra (Rymarczyk *et al.*, *in litt.*). Emerging June and July, flying until May of the following year.

104   *Agonopterix ivinskisi* Lvovsky, 1992

Agonopterix ivinskisi Lvovsky, 1992: 4.
  Agonopterix flurii Sonderegger, 2013: 3.
  Agonopterix centaureivora Rymarczyk, Dutheil & Nel, 2013: 19.
  Agonopterix ivinskisi subsp. *daghestanica* Lvovsky, 2018: 318, 319 fig. 2.

DESCRIPTION. Wingspan 16.5–20 mm. Head with pale and dark grey scales mixed. Labial palp segment 2 grey buff with some darker brown scales on the outer side, segment 3 half as long as segment 2, pinkish buff with two blackish bands. Antenna fuscous weakly ringed whitish, unringed towards apex. Thorax grey buff with some darker scales. Forewing greyish buff, heavily overlaid with mixture of darker grey and dark brown scales; basal field grey-buff with some dark brown scales admixed, demarcated by an area of deep brown, the basal field extending shortly onto costa; blackish oblique dots at one-third, a median dot and a whitish buff end of cell dot, the oblique and median dots sometimes with associated whitish buff scales; blotch faint, obscure; costa with pale and dark spots, termen with dark grey spots between vein-ends; fringes grey-buff, light or dark grey-brown. Hindwing grey, darker towards apex; fringes grey, with one to three fringe lines.

VARIATION. Forewing coloration varies according to proportions of dark grey and dark brown scales. Oblique dots can be distinct, indistinct or completely absent; rarely all markings nearly invisible.

SIMILAR SPECIES. The rather dark dull grey-brown coloration is characteristic.

MALE GENITALIA. Socii large, nearly contiguous to slightly overlapping, divergent; gnathos lanceolate, about 2.7–3.2 times as long as wide, exceeding socii or not; valva long, almost parallel-sided for most of its length, costal margin concave, ventral margin curved from end of sacculus to broadly rounded apex; cuiller short, erect or inclined outwards, slightly curving outwards at middle, apex subacute, 52–70% of valva width; transtilla thickened in middle; anellus stalk slender, remainder of anellus wider than long, often reaching transtilla, usually with visible V-shaped or linear incision distally; aedeagus variable in length and thickness, gradually tapering from base, slightly angled in middle, one-third to two-fifths length of valva, cornuti in a dense or diffuse mass, basal process short.

FEMALE GENITALIA. Anterior apophysis about half length of posterior apophysis; sternite VIII width:length 7:2, anterior margin straight with slight median bulge round ostium, posterior margin slightly inclined inwards with median excavation; ostium moderately large, in middle of sternite; ductus bursae with distal half slightly expanded, 2.5 times as long as ovate corpus bursae; signum wider than long, narrowly elliptical, constricted in middle, teeth not numerous, largest at margin.

GENITALIA DIAGNOSIS. In male combination of parallel-sided valva and laterally expanded socii with inner edges rather straight and contiguous should be sufficient for determination. Additional distinguishing characters are roundish, wider than long anellus and very broad tegumen. Shape of valva alone is insufficient, it can be found in other species, e.g. *A. lessini* and *A. doronicella*. Female difficult to characterise. Best single feature is structure of papilla analis, which is intermediate between *A. laterella*, *A. cinerariae* and *A. xeranthemella* on one hand and typical *Agonopterix* form on the other hand, i.e. with very long bristles present, medium-length bristles moderately reduced, very short bristles moderately increased, uncompressed papilla analis in ventral view broadest in lower half, but not as much expanded as in *A. laterella*. Additional features are absence of folds in sternite VIII and a slight median bulge at anterior margin.

DISTRIBUTION. Switzerland (Wallis), France (Alpes-Maritimes, Hautes-Alpes, Haute-Savoie), Spain (Cuenca, Teruel), Italy (Cottian Alps) and Crimea. Asia: Turkey, Tajikistan; Daghestan, Armenia (ssp. *daghestanica*).

BIONOMICS. *Centaurea scabiosa* L., *C. uniflora* Turra, *C. leucophaea* Jord. (Rymarczyk *et al.*, 2013); *Cyanus montanus* (L.) Hill (Rymarczyk *et al.*, *in litt.*). Larvae are found July to September, moths emerging predominantly in August and September; moths hibernate and fly until August of the following year (observation from Switzerland, Sonderegger, 2013).

REMARKS. *A. flurii* and *A. centaureivora* were both described in March 2013, the former published on March 8 and the latter on March 20, giving *A. flurii* priority. However both are synonyms of *A. ivinskisi* (Buchner, 2020b). *A. ivinskisi* was described from a single male from Tajikistan, later *A. ivinskisi* ssp. *daghestanica* was described from Daghestan (North Caucasus) and Armenia. European populations cannot be clearly placed in one subspecies or the other, so treatment as two subspecies appears questionable.

105   *Agonopterix ferocella* (Chrétien, 1910)

Depressaria ferocella Chrétien in Spuler, 1910: 340.

DESCRIPTION. Wingspan 16–18 mm. Head orange-brown, scales mostly tipped buff. Labial palp buff, segment 2 tinged cinnamon on outer and ventral sides with a few light fuscous scales on outer side, segment 3 buff tinged cinnamon, with some grey fuscous scales at base and a dark grey ring beyond middle. Antenna light brown. Thorax orange-brown, posteriorly both thorax and tegulae yellowish buff. Forewing pale yellowish buff, overlaid with a matrix of orange-brown scales, mixed with scattered blackish scales, many forming short irregular transverse lines; basal field yellowish buff, confined to a small area in centre of base of wing; a blackish subdorsal spot; dark grey spots along costa and termen; oblique dots small, sometimes hardly distinguishable from scattered blackish scales, a blackish dot at end of cell; blotch at one-half mid-brown to dark grey, slightly irregular and not very conspicuous; a yellowish buff spot between oblique dots and blotch; fringe greyish, tinged orange-brown beyond fringe line. Hindwing light grey, darker towards apex; fringe light grey with two fringe lines. Abdomen light orange-brown.

VARIATION. Some specimens more orange-brown, others more buff.

SIMILAR SPECIES. Characterised by yellowish buff spots in an orange-brown matrix with short transverse blackish lines, and by reduced basal patch. *A. laterella* may have similar pattern but is larger with narrower, more ochreous forewing and basal field larger.

MALE GENITALIA. Socii large, the part of inner margin not attached to tegumen extended far beyond end of tegumen, the two socii slightly overlapping and together forming an obtuse to subacute point, extending well beyond end of gnathos; gnathos elliptical, apex truncate, 1.7–2 times as long as wide; valva with costal margin straight, ventral margin curved from end of sacculus to acute or subacute apex; cuiller stout to quite slender, slightly curved inwards in basal half, more or less curved outwards at middle, apex rounded, 80–88% of valva width; transtilla slightly thickened in middle; anellus slightly wider than long, distally bilobed with V-shaped sinus, beyond this a truncate thickened margin; aedeagus quite slender, hardly tapering after base, curved, about half length of valva, cornuti in an elongate mass, basal process of medium length.

FEMALE GENITALIA. Anterior apophysis approximately half length of posterior apophysis; sternite VIII width:length 5:2 to 3:1, anterior margin slightly to moderately convex, a curved fold between ostium and anterior margin, posterior margin more or less inclined inwards with median excavation; ostium slightly distal to middle of sternite; ductus bursae gradually expanding, 3 times as long as ovate corpus bursae; signum wider than long, with anterior and posterior projections, teeth variable in size and number.

GENITALIA DIAGNOSIS. In male shape of spread socii (together forming an obtuse to subacute point) is distinct. Female difficult to characterise. Structure of papilla analis is similar to that of *A. ivinskisi* (*q.v.*), but narrower in ventral view; a curved fold which nearly touches margin in between ostium and anterior margin of sternite VIII.

DISTRIBUTION. Spain, France, Italy, Switzerland, Austria, Hungary, Slovenia, Bulgaria, Romania, North Macedonia, Greece, Ukraine. Asia: Turkey. Kovács & Kovács (2020) add European Russia and Near East.

BIONOMICS. Larva on *Cirsium ferox* (L.) DC. (Chrétien, 1914); *C. vulgare* (Savi) Ten. (Hannemann, 1953); *C. tuberosum* (L.) All., *Carduus litigiosus* Nocca & Balb., *Echinops ritro* L. and *E. sphaerocephalus* L. (Rymarczyk *et al., in litt.*), Sonderegger (2013) found larvae on *Carlina acanthifolia* All. He also reared larvae on *Arctium lappa* L. Moth hatching in summer, flying until May, rarely June of the next year.

REMARKS. Chrétien's original description (1910) was very brief. Later (Chrétien, 1914) he published a full length description.

106 *Agonopterix laterella* ([Denis & Schiffermüller], 1775)

Tinea laterella [Denis & Schiffermüller], 1775
  Tinea heraclella Hübner, 1813: pl. 62, fig. 417.
  Depressaria incarnatella Zeller, 1854: 266.

DESCRIPTION. Wingspan (19–)20–25 mm. Head ochreous-buff. Labial palp pale buff, segment 2 with a few fuscous scales on outer and ventral sides, and additional light brown scales on ventral side. Segment 3 pale buff with fuscous ring at base and loose dark grey ring beyond middle. Antenna pale buff with narrow fuscous rings. Thorax orange ochreous, buff posteriorly. Forewing often narrow, yellowish buff to light brownish buff, more or less heavily overlaid with patches of orange-ochreous to light brown scales, some weak fuscous or light brown streaks mainly in middle third of wing, a scattering of blackish scales all over wing; basal field buff, bordered by fuscous fascia from dorsum to cell; oblique dots and plical dot black, small, another dot at end of cell; blotch ill-defined to prominent, light brown to dark grey; blackish spots on costa and around termen; fringe greyish buff with distinct fringe line. Hindwing light grey, slightly darker towards apex; fringe pale grey. Abdomen buff.

VARIATION. Mainly in the depth and intensity of the overlying orange-ochreous to orange-brown scales and the fuscous streaks in the forewing. The dark costal and terminal spots can be very weak in some specimens. There are also forms with almost uniform pale straw-yellow to brownish buff coloration, with abundant light brown scattered scales and fewer fuscous scales; basal field and fascia, cell dots, blotch and terminal dots faint or almost obsolete.

SIMILAR SPECIES. *A. arenella* is similar, but the ground colour is less orange and all the blackish markings are stronger and darker. Forms similar to *A. carduella* are not rare, best distinguishable by larger size. Some forms extremely similar to *A. ferocella* including the yellowish buff spot between oblique dots and blotch and absence of orange tinge, also distinguishable by larger size. Pale forms of *laterella* resemble *A. xeranthemella*, but are distinctly larger and always have a black ring on segment 3 of labial palp. Forms with all markings nearly absent are determinable only by dissection.

MALE GENITALIA. Socii large, contiguous or slightly overlapping, divergent; gnathos lanceolate, about three times as long as wide, exceeding socii by less than half its length; valva with costal margin concave, ventral margin curved at about three-fifths, apex variable, mostly acute or subacute; cuiller of medium width to quite slender, slightly inclined inwards in basal half, usually widest at middle, distal half inclined outwards, curved or not, apex rounded or subacute and incurved, 80–85% of valva width; transtilla slightly thickened in middle; anellus slightly wider than long, distal margin thickened, truncate to slightly rounded with V-shaped sinus; aedeagus quite slender, hardly tapering before three-fifths, curved, about half length of valva, cornuti in an elongate mass, basal process of medium length.

FEMALE GENITALIA. Anterior apophysis half to two-thirds length of posterior apophysis; sternite VIII width:length 3:1 to 7:2, anterior margin strongly convex, with a shallow oblong median excavation, posterior margin straight or inclined inwards, median part forming a bridge over quadrate ostium, which is slightly distal to middle of sternite; ductus bursae slightly expanded in middle part, twice as long as ovate to elliptical corpus bursae; signum wider than long, elliptical or diamond-shaped, with or without anterior and posterior projections, teeth variable in size and number.

GENITALIA DIAGNOSIS. In male shape of gnathos (rather narrow, broadest near base, tapering to a remarkably sharp point, outline in distal half slightly convex) is distinctive. Female *A. laterella*, *A. cinerariae* and *A. xeranthemella* differ from other *Agonopterix* species in structure and shape of uncompressed papilla analis, seen in lateral view. For details see key (pp. 206–209).

DISTRIBUTION. Central and north-east Europe, extending to Netherlands and Denmark in north-west, also in a few countries in south Europe from Spain to Greece and Romania; Crimea, Russia eastwards to Ural Mts. (Lvovsky, 2019). Absent from Mediterranean islands except Sardinia. Asia: Turkey.

BIONOMICS. *Cyanus segetum* Hill (Hannemann, 1953); *C. triumfettii* (All.) Á. Löve & D. Löve (Sonderegger, *in litt.*); *C. montanus* (L.) Hill and *Centaurea uniflora* Turra (Rymarczyk *et al.*, *in litt.*). Emerging predominantly in June and July, flying until May, rarely June of the following year.

107   *Agonopterix xeranthemella* Buchner, 2018

Agonopterix xeranthemella Buchner, 2018: 2.

DESCRIPTION. Wingspan 15.5–18 mm. Head yellow-buff. Labial palp segment 2 yellow-buff with a few light brown scales on outer and ventral sides; segment 3 yellow-buff, mixed with a few brownish or blackish scales, without a black ring. Antenna medium brown. Thorax and tegulae yellow-buff. Forewing narrow, pale buff to yellow-buff, irregularly mixed with light brown to orange-brown scales, black scales interspersed in variable numbers; basal field a little paler, weakly bordered; oblique dots and a dot at end of cell black; blotch diffuse, grey-brown, often weakly developed; terminal dots faint or absent; fringes pale yellowish with two distinct darker lines. Hindwing pale grey, pale and translucent at base and becoming darker toward apex, a narrow dark line running along its margin; fringes more greyish than in forewing, with two faint fringe lines. Abdomen yellowish buff.

VARIATION. The amount of scattered black scales on forewing varies from nearly absent (as in both specimens from France) to rather numerous. Also the constant elements of central forewing pattern vary from nearly invisible to clearly visible, partly because they are barely discernible when the scattered black scales are more prevalent.

SIMILAR SPECIES. *A. laterella* is distinctly larger and has a black ring on labial palp segment 3.

MALE GENITALIA. Socii medium-sized, contiguous or slightly overlapping; gnathos triangular with rounded apex, about 1.3 to 2 times as long as wide, exceeding socii by about half its length; valva with costal margin slightly concave, ventral margin somewhat bulging between end of sacculus and four-fifths, apex subacute; cuiller slightly inclined inwards or strongly curved inwards, narrowed at two-thirds, apex narrow but rounded, 82–97% of valva width; transtilla thin; anellus about as wide as long, distally with wide V-shaped sinus between rounded humps, beyond this a triangular finely spinulose membrane; aedeagus nearly parallel-sided to two-fifths, then angled and tapering, about half length of valva, cornuti in an elongate mass, basal process rather short.

FEMALE GENITALIA. Anterior apophysis about three-fifths length of posterior apophysis; sternite VIII width:length 2:1 to 3:1, anterior margin with wide bulge ending in two humps with long shallow excision between, posterior margin slightly inclined inwards; ostium large, closer to proximal edge than to distal edge of sternite; ductus bursae expanded towards rather narrowly elliptical corpus bursae; signum wider than long, elliptical with anterior and posterior projections, teeth mostly rather small, one pair larger than the rest.

GENITALIA DIAGNOSIS. In male shape of gnathos (short, triangular with rounded apex) is distinctive. Shape of cuiller (remarkably thin in distal third) and triangular, spinulose membrane posterior to anellus also characteristic. In female papilla analis as in *A. laterella* and *A. cinerariae*. For details see key (pp. 206–209).

DISTRIBUTION. France, North Macedonia, Greece. Asia: Turkey, Syria.

BIONOMICS. According to label data E. Dattin reared this species from larvae collected on *Xeranthemum* sp. (Asteraceae) from Pyrénées Orientales, France. *Xeranthemum inapertum* (L.) Mill. is present in a place where the moth has been collected by Thiérry Varenne. Moths emerged in middle of July, and a worn specimen has been caught in April, indicating that the species hibernates in the adult stage (Buchner, 2018).

REMARKS. Two specimens in MNHN are labelled 'Agonopterix xeranthemella E. Dattin i.l.', but the species was never described.

108   *Agonopterix cinerariae* (Walsingham, 1908)

Depressaria cinerariae Walsingham, 1908: 955.

DESCRIPTION. Wingspan 17–21 mm. Head pale buff, sometimes tinged ochreous. Labial palp pale buff, segment 3 with blackish ring above middle. Antenna pale buff, becoming grey-brown in outer half. Thorax buff, more or less tinged ochreous anteriorly. Forewing long, apex not rounded; pale straw-yellow to buff, mottled with pale orange-ochreous patches, scattered fuscous to blackish scales, sparse or more dense, then often forming dotted lines on veins, mainly in terminal one-third; basal field pale straw-yellow to buff, weakly delimited by light brown streak from subdorsal spot; a light to dark brown spot at base of costa, a few grey costal spots and a series of light brown to fuscous terminal spots; oblique dots blackish, end of cell dot small, blackish, all dots inconspicuous among scattered blackish scales; blotch usually present, light grey to blackish; fringe concolorous with forewing or tinged fuscous. Hindwing whitish, fringe whitish grey. Abdomen whitish buff.

VARIATION. Variation occurs in the depth of coloration of the forewing, in the abundance of blackish scattered scales and the distinctness of the blotch, which may be very faint in some specimens.

SIMILAR SPECIES. Other *Agonopterix* species on the Canary Islands have well-defined basal field or median cell dot present or both. Beside this, none of these species show combination of pale buff ground colour, absence of white elements associated with central forewing patterns and acute apex.

MALE GENITALIA. Socii medium-sized, contiguous or slightly overlapping; gnathos ovate-lanceolate, twice as long as wide, exceeding socii by less than half its length; valva with costal margin slightly concave, ventral margin curved about three-fifths, apex rounded; cuiller rather stout, straight, slightly inclined outwards, apex rounded to subacute, 75–80% of valva width; transtilla thin; transtilla lobes narrow; anellus slightly wider than long, distally bifid with wide V-shaped sinus with thickened margins, beyond this a wide more or less rounded membrane; aedeagus rather stout, slightly tapering, angled in middle, about half length of valva, cornuti in an elongate mass, basal process of medium length.

FEMALE GENITALIA. Anterior apophysis about three-fifths length of posterior apophysis; sternite VIII width:length about 3:1, anterior margin slightly inclined outwards, with shallow oblong excision in middle, a clearly defined curved fold with a hump in middle close to anterior margin, posterior margin inclined inwards; ostium quadrate, in middle of sternite; ductus bursae of nearly constant width, 2.5 times as long as elliptical corpus bursae; signum wider than long, nearly elliptical but posterior side more bulging than anterior, teeth medium-sized.

GENITALIA DIAGNOSIS. Male with few distinctive features. Aedeagus asymmetrical, angled and notched ventrally in lateral view (*arenella* subgroup feature) in combination with stout, straight cuiller should be sufficient to recognise this species which is restricted to Canary Islands. In female papilla analis as in *A. laterella* and *A. xeranthemella*. For details see key (pp. 206–209).

DISTRIBUTION. Canary Islands.

BIONOMICS. Larvae can be found all year on *Pericallis tussilaginis* (L'Hér.) D. Don and *P. appendiculata* (L. f.) B. Nord. Adults have emerged in June and in December.

REMARKS. The wide divergence of known emergence times remains unexplained.

109 *Agonopterix arenella* ([Denis & Schiffermüller], 1775)

Tinea arenella [Denis & Schiffermüller], 1775: 137.
  Tinea gilvella Hübner, 1796: pl. 14, fig. 96.
  Depressaria immaculana Stephens, 1834: 200.

DESCRIPTION. Wingspan (16–)19–23.5 mm. Head ochreous-buff, face buff. Labial palp ochreous-buff, segment 2 with ferruginous to fuscous scales on outer side and ventrally, segment 3 with fuscous rings at base, sometimes absent and at two-thirds. Antenna light fuscous with narrow deep brown rings. Thorax light brown, darker anteriorly and at posterior margin. Forewing pale buff, extensively overlaid light brown with scattered brown and blackish scales; costal and terminal spots dark grey to black; basal field light buff, with a blackish spot at base of costa; basal fascia blackish, brown outwardly, extending to margin of cell; oblique dots blackish, the second larger, another dot at end of cell; plical dot weak; blotch dark grey, nearly round; fringe pale ochreous, with two fringe lines. Hindwing whitish buff, pale grey towards termen; fringe whitish-buff with two fringe lines. Abdomen ochreous with scattered light fuscous scales.

VARIATION. One of the least variable species of the genus. Ground colour varies little, but the palest parts of forewing, namely first half of costa, a patch following oblique dots and a broad area from middle of dorsum extending between cell and termen and towards costa, have variable amounts of darker scales present; sometimes there are two patches of grey scales near apex.

SIMILAR SPECIES. Characterised by the ochreous coloration and strongly developed dark blotch. *A. petasitis* has paler forewing lacking first oblique dot, with weakly defined basal field, and labial palp with only one black ring.

MALE GENITALIA. Socii rather large, well separated; gnathos elliptical, 2–2.3 times as long as wide, exceeding socii by more than half its length; valva with costal margin slightly concave, ventral margin curved about three-fifths, apex rounded; cuiller stout, slightly inclined inwards, inner margin curved outwards at two-thirds, outer margin abruptly expanded beyond middle, then curving inwards to rounded apex, 75–80% of valva width; transtilla slightly thickened; transtilla lobes narrow; anellus as wide as long, distally truncate with slightly thickened margin; aedeagus tapering above base and in distal one-third, curved in middle part, slightly over half length of valva, cornuti not conspicuous, in an elongate mass, basal process of medium length.

FEMALE GENITALIA. Anterior apophysis about three-fifths length of posterior apophysis; sternite VIII width:length about 2:1, anterior margin strongly but not smoothly convex, posterior margin inclined inwards; ostium rather large, just distal to middle of sternite, a curved fold proximal to its margin; ductus bursae of constant width distally, then expanding, equal in length to large pyriform corpus bursae; signum wider than long, nearly elliptical with projection on posterior side, teeth rather large.

GENITALIA DIAGNOSIS. Male with cuiller terminally expanding and therefore reminiscent of *A. subumbellana* and *A. pallorella*, but stouter at base, constricted in middle, terminal expansion more elliptic than quadrate or triangular. In female best single feature is a curved fold proximal to margin of ostium, much closer to ostium than to anterior margin of sternite VIII.

DISTRIBUTION. Middle and northern Europe, more restricted in the south but extending from Iberian Peninsula to Greece; absent from several countries and from Mediterranean islands except Sardinia. Asia: Turkey, in Russia eastward to Altai Mts. Canada (BOLD).

BIONOMICS. Larvae on *Arctium lappa* L. (Hannemann, 1953); *Cirsium vulgare* (L.) Scop., *Centaurea scabiosa* L., *C. jacea* L. and *C. nigra* L. (Hannemann, 1995); *Serratula tinctoria* L. (Harper *et al.*, 2002); *Centaurea nigrescens* Willd., *Rhaponticum scariosum* Lam., *Cirsium eriophorum* (L.) Scop. and *Carduus personata* (L.) Jacq. (Sonderegger, *in litt.*); *Carlina vulgaris* L., *Cirsium arvense* (L.) Scop. and *Onopordum acanthium* L. (Huisman, 2012); *Arctium minus* (Hill) Bernh., *Cirsium tuberosum* (L.) All., *C. monspessulanum* (L.) Hill and exceptionally on *Achillea millefolium* L. and *Inula montana* L. (Rymarczyk *et al.*, *in litt.*). Palm (1989) also mentions *Sonchus* and *Knautia* but these require confirmation. Emerging from June to August, flying until May, rarely June of following year.

REMARKS. The occurrence of this species in Canada probably represents a recent accidental importation. It was not mentioned for North America in Hodges *et al.* (1983).

petasitis subgroup (110)

110   *Agonopterix petasitis* (Standfuss, 1851)

Depressaria petasitis Standfuss, 1851: 59.

DESCRIPTION. Wingspan 21–25 mm. Head creamy, sometimes tinged ochreous, face concolorous. Labial palp creamy, segment 2 unusually slender with weakly developed ventral tuft and furrow, dark brown on outer side in basal half, segment 3 with a dark brown ring before apex. Antenna light fuscous narrowly ringed dark fuscous. Thorax pale straw-yellow to pale ochreous, posterior crests sometimes tipped mid-brown. Forewing broad, pale straw-yellow to pale ochreous-buff with light ochreous-brown shading in basal one-sixth between dorsum and cell, around blotch at one-half and a broad transverse band in subterminal area; scattered dark fuscous scales in costal third of wing or sometimes more widely; basal field weakly defined by a diffuse brown subdorsal spot, a deep brown dot at base of costa, costal spots small, brown or blackish with a larger spot at three-fifths; terminal spots blackish; first oblique dot usually absent, second blackish, a larger dot at end of cell; blotch blackish, often extending to end of cell dot; fringe pale ochreous-buff. Hindwing with lobe at anal angle almost obsolete; creamy buff, increasingly grey-tinged towards termen; fringe pale creamy buff. Abdomen pale greyish buff, posterior margin of segments pale creamy buff.

VARIATION. There is some variation in the extent of dark fuscous scattered scales, these being quite absent except in the costal third of the wing in some specimens. The first oblique dot is frequently absent.

SIMILAR SPECIES. A distinctive species. *A. arenella* has defined basal area and more abundant dark scattered scales, first oblique dot usually present.

MALE GENITALIA. Socii large, well separated; gnathos elliptical, 1.7–2.2 times as long as wide, shorter than or equalling socii; valva with costal margin slightly concave, ventral margin gently curved about three-fifths, apex rounded to broadly subacute; cuiller rather stout, slightly inclined outwards in basal half, slightly incurved distally, apex rounded, 60–62% of valva width; transtilla thickened; anellus about as wide as long, obpyriform, distally with deep V-shaped sinus; aedeagus weakly tapering above base, angled at two-thirds, slightly over half length of valva, cornuti not conspicuous, in an elongate mass, basal process strongly developed.

FEMALE GENITALIA. Anterior apophysis about half length of posterior apophysis; sternite VIII width:length about 7:2, anterior margin straight with sharply outlined low median bulge, posterior margin inclined inwards; ostium in middle of sternite; ductus bursae slightly expanded in middle part, at least five times as long as small ovate corpus bursae; signum wider than long, approximately elliptical, usually with posterior and anterior projections, teeth variable in size and number.

GENITALIA DIAGNOSIS. In male combination of obpyriform anellus with wide gap to transtilla, short, medium stout, slightly inclined cuiller and gnathos not overtopping socii is distinctive. Female can be characterised by anterior margin of sternite VIII straight with sharply outlined low median bulge in combination with exceptionally small signum.

DISTRIBUTION. Mountain areas of Central Europe from France to Poland and Romania.

BIONOMICS. Larva on *Petasites hybridus* (L.) P. Gaertn. *et al.* (Hannemann, 1953); *P. albus* (L.) P. Gaertn., *P. paradoxus* (Retz.) Baumg. and *Tussilago farfara* L. (Sonderegger, *in litt.*). Moths emerging in May and June, no records from early spring and observation of mating in July (Lepiforum e.V., 2006–2023) indicates that eggs are laid in late summer and autumn and moths do not hibernate.

REMARKS. This species exhibits several characters that are unusual in *Agonopterix*: rather slender segment 2 of labial palp with weak tuft and furrow, broad forewing and hindwing without lobe at anal angle. These characters are shared by *A. furvella* and *A. pupillana*, which we place in group 2B.

cotoneastri subgroup (111–112)
Larvae feed on Asteraceae tribe Senecioneae or tribe Anthemideae.

111   *Agonopterix cotoneastri* (Nickerl, 1864)

Depressaria cotoneastri Nickerl, 1864: 2.
  Depressaria senecionis Nickerl, 1864: 3.
  Depressaria sarracenella Rössler, 1866: 333.
  Depressaria marmotella Frey, 1868: 377.
  Agonopterix seneciovora Fujisawa, 1985: 34, syn. n.

DESCRIPTION. Wingspan 15–19 mm. Head grey-brown, scales tipped buff. Labial palp segment 2 pale buff on inner side, outer side and ventral tuft mixed with lighter and darker grey-brown scales, segment 3 with grey-brown base and dark grey ring above middle. Antenna greyish buff to light fuscous. Thorax greyish buff, anteriorly darker, sometimes tinged orange-brown. Forewing uniformly grey-brown, sometimes tinged light brown, with scattered blackish scales; basal field pale grey-buff sometimes with cinnamon tinge, extending to costa, demarcated by a broad blackish brown fascia reaching costal margin of cell, fading outwardly into ground colour; a blackish dot or dash in basal field on costal edge of cell; oblique dots black, edged or partly edged white, a white median dot and a larger one at end of cell, both more or less edged black and sometimes surrounded by a darkened patch; costal spots dark grey, usually small and weak, terminal spots dark grey; fringe grey with brown tinge. Hindwing grey; fringe light grey.

VARIATION. Forewing colour may be more grey or more brown, white edge to oblique dots sometimes forming a larger white dot between the black dots, white median dot may be very small or absent.

SIMILAR SPECIES. *A. cotoneastri* is characterised by uniform grey-brown forewing, with conspicuous basal field, 2 or 3 white dots in cell and absence of blotch. *A. capreolella* has white dots lying in a slightly darker area of wing, less uniform ground colour and often is smaller, but sizes overlap. It is otherwise very similar and best separated by genitalia.

MALE GENITALIA. Socii rather small, well separated; gnathos narrow, elliptical, about 3.5 times as long as wide, exceeding socii by at least half its length, usually more; valva long with costal margin straight or weakly convex to two thirds, where it is concave, ventral margin gently curved about three-fifths, distal one-third of valva almost parallel-sided or weakly tapering, apex rounded or subacute; cuiller inclined outwards in basal one-third, incurved in distal two-thirds, swollen on outer side at middle and again slightly swollen at rounded apex, crossing costal margin close to base, 104–112% of valva width; transtilla thickened in middle; anellus slightly wider than long, V-shaped sinus reaching nearly to base; aedeagus weakly tapering throughout, almost straight, slightly under half length of valva, cornuti less numerous than in most related species, basal process of medium length.

FEMALE GENITALIA. Anterior apophysis half to three-fifths length of posterior apophysis; sternite VIII width:length 2:1 to 3:1, anterior margin with median bulge of variable length and width, posterior margin slightly inclined inwards; ostium in proximal half of sternite, clearly outlined by fine fold anteriorly and laterally, area posterior to ostium weakly sclerotised; ductus bursae of nearly even width, about three times as long as small narrowly elliptical corpus bursae; signum variable, wider than long, anterior margin longer than posterior, sometimes with posterior bulge or projection; teeth variable in size and number.

GENITALIA DIAGNOSIS. In male shape of cuiller is unique in European *Agonopterix*. Female well characterised by sternite posterior to ostium distinctly less sclerotised, this area appears as pale "V" or "U" extending to posterior margin of ostium (best single feature, but compare *A. cnicella*); ostium outlined by a fold anteriorly and laterally which forms an "U" with width less than twice depth; sternite VIII in proximal area often covered by tiny dots.

DISTRIBUTION. Spain; mountain areas of Central Europe from France to Poland. Asia: Far East Russia, Japan.

BIONOMICS. Larva on *Senecio doronicum* (L.) L. (Hannemann, 1953). Lhomme (1945) also lists *S. cacaliaster* Lam., *S. ovatus* (G. Gaertn. *et al.*) Hoppe, *S. pyrenaicus* L., *S. nemorensis* L., *S. sarracenicus* L. and *S. doria* L. Adults emerge in July and August, flying until April or May after hibernation.

REMARKS. Hannemann (1953) synonymised *sarracenella* Rössler, 1866 with *senecionis*. Hannemann (1958a) established that *cotoneastri* Nickerl, 1864 and *senecionis* Nickerl, 1864 were also synonymous. He used the name *cotoneastri* in preference to *senecionis*. The two names were published in the same work on pages 2 and 3 respectively. The precedence of two (or more) names for the same taxon published in the same work on the same date is determined by the "First Reviser", viz. the person who first published the synonymy between them (ICZN, article 24.2), in this case, Hannemann. However, in Hannemann (1976) he reverted to *senecionis* as the correct name, and followed this in Hannemann (1995). In Hannemann (1996), *cotoneastri* is given as a junior synonym of *senecionis*, which is not the case. In our opinion, Hannemann's decision to reverse his earlier choice was not legitimate and the correct name to use for this species is *cotoneastri*.

We can find no difference in habitus or male genitalia between *S. cotoneastri* and *S. seneciovora* Fujisawa, 1985. Apparent differences in female genitalia seem to be down to Fujisawa's drawing. We therefore place *seneciovora* in synonymy with *cotoneastri*.

112   *Agonopterix multiplicella* (Erschoff, 1877)

Depressaria multiplicella Erschoff, 1877: 344.
 Agonopterix klimeschi Hannemann, 1953: 296.

DESCRIPTION. Wingspan 17–21 mm. Head light grey-brown to orange-brown, scales tipped buff. Labial palp segment 2 pale buff, outer side mixed with a few light brown and grey scales, ventral side of rough scales mainly mid-brown and blackish, segment 3 pale buff with basal third grey to light brown, a ring at about two-thirds and extreme apex grey to black. Antenna buff to light brown, narrowly ringed fuscous. Thorax light brown with scales tipped grey-buff, or grey-buff all over. Forewing light grey-brown, often more or less heavily overlaid with reddish brown scales, some scattered blackish dots; basal field pale buff to light brown, more grey towards costa and not extending along costa, delimited from dorsum to cell by a broad blackish basal fascia; dark grey spots mainly in outer half of costa and terminal spots, often faint; first oblique dot black, outwardly contiguous with clear white dot, second oblique dot very small, usually surrounded by a white dot, white median and end of cell dots; a dark grey to dark reddish brown shade from between white cell dots to tornus; blotch dark grey; fringe concolorous with terminal area, pale grey towards tornus. Hindwing grey, darker towards apex; fringe grey. Abdomen greyish buff.

VARIATION. Ground colour varies from light grey-brown to deep reddish brown according to the amount of overlying red-brown scales; the four white dots are less evident in paler specimens.

SIMILAR SPECIES. The combination of dark shade from end of cell to tornus, four white dots and presence of dark blotch is characteristic for this species. *A. cnicella* has smaller blotch, narrow dark edge to basal patch and basal field with spike along cell margin. Other species with three to four white dots have blotch weak or absent.

MALE GENITALIA. Socii large, widely separated; gnathos lanceolate or elliptical, 2.2–2.7 times as long as wide, shortly exceeding socii; valva long, almost parallel-sided in basal two-thirds, distal part narrower, slightly incurved, apex subacute; cuiller rather stout, inclined slightly outwards in basal one-third, narrowest in middle, swollen on outer side towards rounded apex, 64–80% of valva width; transtilla thin or thickened; anellus longer than wide, oblong or elliptic, posteriorly with narrow median incision, V-shaped sinus reaching nearly to base; aedeagus hardly tapering, weakly curved, about two-thirds length of valva, cornuti less numerous than in most related species, basal process rather small.

FEMALE GENITALIA. Anterior apophysis about half length of posterior apophysis; sternite VIII width:length about 4:1, anterior margin strongly convex, flat in the middle, posterior margin slightly inclined inwards, with rectangular median excavation; ostium in middle of sternite, with distinct lip on anterior margin of sternite and a pair of diverging lateral folds; ductus bursae of nearly even width, 1 to 1.5 times as long as large elliptical corpus bursae; signum wider than long, narrow, with anterior and posterior projections; teeth few and large.

GENITALIA DIAGNOSIS. Male with combination of large, widely separated socii with stout, rather short cuiller slightly expanded terminally should be sufficient for determination. In cases of doubt shape of anellus (longer than wide, oblong) and aedeagus (less curved than in most other *Agonopterix*) should be included. Female well characterised by a pair of diverging folds inside the ostium, lateral to sclerites, forming a more or less incomplete "V" in combination with a distinct lip between ostium and anterior margin of sternite, which is accompanied on either side by a short, somewhat diffuse sclerotisation perpendicular to margin of sternite.

DISTRIBUTION. Central and East Europe from Sweden south to Austria, north Italy and Romania. Asia: Russia extending to Far East, India, China, Japan. Rather local throughout its range.

BIONOMICS. Larva on *Artemisia vulgaris* (Hannemann, 1953). Flying from July until June of following year.

REMARKS. In Far East Russia and Japan, the closely related and very similar *A. yomogiella* Saito, 1980 is found.

doronicella subgroup (113)

113   *Agonopterix doronicella* (Wocke, 1849)

Depressaria doronicella Wocke in Gravenhorst, 1849: 73.
    Depressaria schmidtella Zeller, 1851: 81.
    Depressaria laetella Herrich-Schäffer, 1853: 122.

DESCRIPTION. Wingspan 17–20 mm. Head orange-ochreous on vertex, face buff. Labial palp ochreous-buff flecked blackish fuscous on outer side of segment 2, inner side ochreous-buff, segment 3 dark brown ringed between half and two-thirds, ring less evident on outer side. Antenna orange-buff on anterior side of scape and rings of proximal segments between narrow fuscous rings, gradually becoming entirely fuscous distally. Thorax orange-ochreous anteriorly and on tegulae, posteriorly orange-brown. Forewing straw-yellow to orange-ochreous, with veins towards termen marked orange-brown to fuscous, whole wing with orange-ochreous to orange-brown scattered scales and with greyish fuscous strigulae including occasional black scales, weakly organised in transverse rows, often wider and darker where they meet the costa; costal spots grey, weak; basal field delimited towards base and dorsum by orange-brown to fuscous lines and by an incomplete basal fascia; basal half of wing except towards dorsum, more yellow with two mid-brown lines running through longitudinally; first oblique dot brown to blackish, sometimes very small, in middle of a pale area, second small or absent, lying on a longitudinal line and not obvious, end of cell dot and fuscous blotch obscured by an irregularly shaped diffuse dark brown to fuscous patch around end of cell with extension towards dorsum at one-fifth and sometimes also to four-fifths and to costa at middle; whole distal end of wing sometimes shaded fuscous; fringe yellow, orange-buff or fuscous with two distinct greyish fuscous fringe lines. Hindwing dark grey; fringe light grey with two darker grey fringe lines. Abdomen dark grey to brown.

VARIATION. Central European specimens duller coloured than those from Portugal. There is some variation in ground colour, but more variation in the extent and amount of darker markings and shading of the forewing.

SIMILAR SPECIES. Characterised by yellow-ochreous basal third separated from more orange to fuscous outer part of wing by oblique grey shade. *A. angelicella* has similar colouring, but less intensely orange.

MALE GENITALIA. Socii medium-sized, well separated; gnathos broadly elliptical, about 1.5 times as long as wide, approximately equalling socii; valva parallel-sided for much of its length, strongly curved, hardly narrowed to rounded apex; sacculus unusually long; cuiller tapering from base to middle or beyond, straight, inclined slightly outwards, sharply incurved at subacute apex, 70–75% of valva width; transtilla thickened; anellus wider than long, posteriorly truncate with thickened margin, median incision narrow or indistinct; aedeagus variable, tapering or hardly tapering, rather slender or quite stout, curved at or beyond middle, from well under half length of valva to more than half, cornuti in a more or less compact group, basal process rather long.

FEMALE GENITALIA. Anterior apophysis three-fifths length of posterior apophysis; sternite VIII width:length about 7:2, anterior margin inclined outwards, flat in middle one-third, sometimes with a small median hump, posterior margin inclined inwards, emarginate; ostium in middle of sternite; ductus bursae with median swelling or gradually expanding, three times as long as small elliptical corpus bursae; signum wider than long, more or less elliptical; teeth mostly small.

GENITALIA DIAGNOSIS. In male combination of parallel-sided valva, unusually long sacculus and broad elliptic gnathos not overtopping socii is distinct. Shape of valva alone is insufficient, as it can be found in other species, e.g. *A. lessini* and *A. ivinskisi*. Female characterised by shape of anterior margin of sternite VIII (flat in middle one-third, inclined outwards) in combination with absence of folds or other distinct features.

DISTRIBUTION. Portugal; Middle Europe from France to Poland and Romania, and through Balkan countries to Greece. Russia, Caucasus (Lvovsky, 2019).

BIONOMICS. Larva on *Doronicum austriacum* Jacq. in Central Europe (Hannemann, 1953), on *D. pardalianches* L. in France (Rymarczyk *et al.*, *in litt.*) and on *D. carpetanum* Willk. in Portugal (Corley *et al.*, 2011). Emerging predominantly in June and July, we have not seen moths collected later than September.

REMARKS. Corley *et al.* (2009) reported this species feeding on *Hieracium vulgatum* Fries in northern Portugal, but this proved to be a misidentification of *Doronicum* plants before flowering (Corley *et al.*, 2011).

Group 4 (114–123)
    Species feeding on Fabaceae

Male genitalia key for Fabaceae feeding *Agonopterix* species

1a	Anellus lobes narrow, posterior margin sloping throughout towards centre, separated by wide gap (about half transtilla length)	*A. assimilella*
1b	Anellus lobes touching or with gap not more than 0.2 times transtilla length in between	2
2a	Anellus lobes L-shaped, in standard position (!) widely overlapping	*A. scopariella*
2b	Anellus lobes different	3
3a (1b)	Anellus lobes moderately expanded medially, posterior margin sloping throughout toward centre or slightly bulging section present	4
3b	Anellus lobes strongly bulging medially and here forming an orbicular or elongate elliptic structure, at beginning of expansion a distinctly to strongly rising section present	6 (four species with very similar male genitalia, not possible to determine safely using a simple key)

| 4a (2b) | Anellus lobes moderately expanded medially, but posterior margin sloping throughout towards centre | *A. umbellana* |
| 4b | Anellus lobes moderately expanded medially, but on posterior margin a horizontal or slightly rising section present | 5 |

| 5a (4b) | In standard position, basal sections of costa of valvae forming an angle of 180 °; cuiller usually bent inwards over whole lengh, rather narrow, 75–90% of valva width | *A. conciliatella* |
| 5b | In standard position, basal sections of costa of valvae forming an angle of < 180 °; cuiller usually nearly straight, moderately stout, 90–100% of valva width; socii smaller than in the other species of this group | *A. aspersella* |

| 6a (4b) | Gnathos exceeding socii by up to half of its length; cuiller usually bent outwards, tending to expand in distal half, widest at blunt end | 7 |
| 6b | Gnathos exceeding socii by at least half of its length; cuiller straight or gently curved, of same diameter throughout or slightly wider in distal half, but maximum diameter not at very end | 8 |

| 7a (6a) | Anellus lobes with more or less orbicular expansion centrally, these structures often with indistinct transverse folds | *A. nervosa* |
| 7b | Anellus lobes with elongate elliptic expansion centrally, these structures often with distinct transverse folds | *A. comitella* |

| 8a (6b) | Gap between transtilla lobes usually 0.4–0.5 times transtilla width, cuiller usually slightly S-curved or bent outwards | *A. oinochroa* |
| 8b | Gap between transtilla lobes usually 0.2–0.25 times transtilla width, cuiller usually straight or slightly bent outwards | *A. atomella* |

Female genitalia key for Fabaceae feeding *Agonopterix* species, including *A. perezi*. *A. rutana* and *A. subumbellana*

| 1a | Anterior margin of sternite VIII with a nearly rectangular protrusion, its width about 1/3 of sternite, its margins extending as slightly outward sloping folds on sternite (figs. 116a–d) | *A. scopariella* |
| 1b | Anterior margin of sternite VIII different | 2 |

2a (1b)	Ostium on either side, at level of distal end of sclerites, with small but distinct pale triangles directed laterally (figs. 114a–c)	*A. assimilella*
2b	Ostium without such pale triangles directed laterally	3

3a (2b)	Sternite VIII with a fold on either side in area anterior to ostium, running obliquely into anterior margin. Where meeting anterior margin, the fold may be somewhat swollen, but is not overtopping margin (figs. 115a–c). Note: removing intersegmental skin – connection with sternite VII is a critical step for these structures, because it can happen that parts of the folds are also removed.	*A. atomella*
3b	Sternite VIII with a fold or band between ostium and anterior margin or at anterior margin, sometimes more or less outlining ostium; this structure is usually running through, although sometimes weaker near middle or even interrupted, but not running into anterior margin on either sides and ending there (e.g. figs. 117, 118)	4

4a (3b)	Fold developed as broad band at anterior margin of sternite VIII, more or less parallel-sided in central part, then on either side tapering, proximal margin sometimes continued as a narrow fold running into sternite (figs. 62, 117, 121)	5
4b	Folds and/or bands different	6

5a (4a)	Band at anterior margin of sternite VIII with width:length ratio of about 1:8–1:10, evenly curved, parallel-sided central section 70–80% of band width, then on either side tapering, horizontal expansion of band about 1/3 of segment VIII width, proximal margin continuing as a fold running into sternite on either side, distance of ends of these folds about half segment VIII width (figs. 117a–c). Shape and proportions of the band rather constant within this species and usually not seriously affected by preparation artefacts. Only known from Canary Islands and Madeira.	*A. conciliatella*

5b	Fold or band somewhat similar, but different in details, variable within the species and may be strongly affected by preparation details in embedded slides: In *A. perezi* (fig. 62a) the fold is broader; only known from Canary Islands and Madeira. In *A. oinochroa* (figs. 121a, b), this area can look very similar, but this is a widespread species with clearly different external appearance from presence of ferruginous scales associated with central forewing markings. In *A. umbellana* (figs. 120a, b) the band or fold is usually narrow, but depending on embedding details, sometimes central part forms a triangle with obtuse tip directed anteriorly.	*A. perezi* *A. oinochroa* *A. umbellana* in part
6a (4b)	Fold anterior to ostium forming a bend or "U" with width less than 2.5 × depth (figs. 82, 83)	7
6b	Fold anterior to ostium forming a bend or "U" with width more than 2.5 × depth	9
7a (6a)	Ends of fold on either side of ostium closer to margin of ostium than half of ostium diameter (figs. 82a, b), usually about at level of center of ostium	*A. rutana*
7b	Ends of fold on either side of ostium clearly more distant from margin of ostium than its half diameter	8
8a (7b)	Bend of fold only exceptionally with width less than 2.5 × depth. Fringes at termen reddish, widespread.	*A. nervosa* in part
8b	Bend of fold normally with width less than 2.5 × depth (figs. 83a, b). Fringes at termen not reddish, in Europe only in most eastern parts	*A. subumbellana*
9a (6b)	Horizontal expansion of fold usually exceeding 1/2 of segment VIII width (figs. 118a, b). From Greece eastwards.	*A. comitella*
9b	Horizontal expansion of fold barely reaching 1/2 of segment VIII width (figs. 119, 120, 121). Remark: female genitalia differences between these species are too small for safe determination without further information. Additionally in embedded slides shape of fold in central part of anterior margin of sternite can be seriously affected by preparation details.	*A. aspersella* *A. nervosa* *A.umbellana* in part

114   *Agonopterix assimilella* (Treitschke, 1832)

Haemylis assimilella Treitschke, 1832: 259.
  Depressaria irrorella Stephens, 1834: 203.

DESCRIPTION. Wingspan 16–20 mm. Head and face pale straw-yellow to ochreous-buff. Labial palp buff to light ochreous, segment 2 with variable amount of scattered mid-brown scales, segment 3 often with some grey scaling at base and at two-thirds, more often on inner side and rarely forming a complete ring. Antenna thicker in male than in female, grey-brown, with narrow darker brown rings. Thorax from very pale straw-yellow to ochreous-buff. Forewing straw-yellow to ochreous-buff, often with some light grey-brown suffusion, whole surface of wing often with more or less sparsely scattered grey-brown to dark brown scales, more abundant in outer half of wing, mainly on veins; a dark brown dot at base of costa; terminal dots grey, often faint; basal field sometimes slightly paler, often partially demarcated by a subdorsal brown to dark brown spot; oblique dots black, the second smaller and sometimes absent; a pale orange-brown elongate spot in cell at one-half, enclosing faint whitish buff orange-ringed median and slightly larger end of cell dots; blotch grey, contiguous with median orange spot, often weak or absent; a grey patch extending from end of cell in direction of tornus, but not reaching it, often weak or absent; fringe concolorous with forewing, in darker specimens with two fringe lines. Hindwing pale greyish buff, greyer towards termen, often with some grey spots between veins at termen; fringe whitish buff, with two fringe lines, sometimes very faint. Abdomen pale ochreous to deep ochreous.

VARIATION. Forewing ground colour varies from pale straw-coloured to ochreous-buff, but also in the extent of grey-brown suffusion, which in darker specimens can extend over almost the whole wing, except the basal field and area around the markings in mid-wing. In palest specimens all dark markings are reduced and some may be lost, basal field weakly defined. In the darkest there is a grey blotch on the costal side of the orange-brown mark in mid-wing. The median and end of cell dots are inconspicuous and may be very weak or absent.

SIMILAR SPECIES. The orange-brown mark in the cell at one-half and a weak greyish extension towards tornus, together with weakly expressed median and end of cell dots, are characteristic features of the species. Palest specimens resemble some of the *pallorella*-group species, but these never have both oblique dots present and proximal half of labial palp is not rough-scaled. *A. nervosa*, which also has orange-brown mark between median dot and end of cell dot, can usually be distinguished by wing shape, but in both species forms with straight termen are found, in *A. nervosa* fringes near apex are ferruginous and terminal dots are absent.

MALE GENITALIA. Socii medium-sized, well separated, uncus small but distinct; gnathos lanceolate to elliptical, 2.5–3 times as long as wide, exceeding socii by less than half its length; valva with costal margin slightly concave in basal half, ventral margin curved at about three-fifths, apex subacute; cuiller moderately stout, inclined inwards or not, slightly curved outwards at middle, apex rounded or sometimes hooked outwards, 72–82% of valva width; transtilla slightly thickened in middle; anellus lobes well separated; anellus small, as wide as long, more or less orbicular, with wide V-shaped sinus; aedeagus slightly tapering, weakly curved or angled, half as long as valva, or a little less, cornuti in an elongate group, basal process rather short.

FEMALE GENITALIA. Anterior apophysis half to two-thirds length of posterior apophysis; sternite VIII width:length 5:2 to 3:1, anterior margin convex, posterior margin slightly inclined inwards; ostium distal to middle of sternite; ductus bursae gradually expanding, four times as long as small ovate corpus bursae; signum wider than long, more or less elliptical; teeth variable in size and number.

GENITALIA DIAGNOSIS. In male narrow anellus lobes with wide gap in between are distinct within this species group. In female best single feature is small but distinct pale triangles on either side of ostium, at level of distal end of sclerites, directed laterally (111a–c), but see *A. alstromeriana*.

DISTRIBUTION. Nearly all Europe, but absent from a few countries, extending north-east to Sweden and Latvia and in Mediterranean found on Balearic Islands, Sardinia and Malta. Russia: Ural Mts., eastward to Krasnojansk region (Lvovsky, 2019).

BIONOMICS. Larva feeds in a fork between spun shoots of *Cytisus scoparius* (L.) Link (Hannemann, 1953); *Genista pilosa* L. (Sonderegger, *in litt.*); *G. florida* L. (Corley, 2015); *G. cinerea* (Vill.) DC. and *Cytisus villosus* Pourr. (Rymarczyk *et al.*, *in litt.*). Emerging predominantly in May, flying until autumn, hibernates as half grown larva.

115   *Agonopterix atomella* ([Denis & Schiffermüller], 1775)

Tinea atomella [Denis & Schiffermüller], 1775: 137.
   Tichonia pulverella Hübner, 1825: 412.
   Haemylis respersella Treitschke, 1833: 275.

DESCRIPTION. Wingspan 17–22 mm. Head buff to light brown, sometimes with slight ferruginous tinge, face buff. Labial palp light ochreous-brown on outer side, buff on inner side, segment 2 with admixture of mid-brown scales on outer side and ventrally. Antenna mid to dark brown, with narrow deep brown rings. Thorax buff, with few to many scattered ferruginous or mid-brown scales. Forewing pale buff to pale ochreous-brown, often with some light ferruginous suffusion, dorsal area and area around end of cell with orange-brown scales or suffusion with short projection towards tornus; scattered blackish scales all over wing, in some specimens mainly along veins; basal field buff, sometimes tinged light brown, extending along costa to one-fifth; basal fascia from dark brown subdorsal spot, pinkish brown extending to costal edge of cell; costal dots grey, small and few, terminal dots grey; oblique dots small, blackish; one or two pale buff dots within orange-brown area surrounding cell, the outer at end of cell always more distinct and sometimes partially edged black; blotch sometimes present as a weak grey shade; fringe grey-buff, often strongly tinged pink, the pink tinge starting on costa, but not often reaching tornus, fringe line faint. Hindwing pale grey-buff, becoming grey towards termen; fringe whitish buff to grey, with faint fringe lines. Abdomen grey, buff towards lateral margins.

VARIATION. When newly emerged there is often a distinct tinge of pink to the ferruginous areas, but this fades and is not present in all specimens. Specimens with forewing divided into two parts by colour, ferruginous in dorsal two-thirds and buff in costal one-third, are distinct, but the ferruginous colour is elusive and variable and can be entirely absent or from just a small area at base, in cell between pale buff dots and in fringe, to the other extreme where apart from the basal field and the costa, the whole wing is suffused light ferruginous; the scattered blackish scales also vary greatly and sometimes dominate over the whole wing; the oblique dots are often inconspicuous, poorly marked or absent and the first pale buff dot in cell is often absent.

SIMILAR SPECIES. The ferruginous to pink-tinged forewing fringe also occurs in a few other species, including *A. nervosa* with forewing apex near right-angled and *A. irrorata* (*q.v.*). *A. scopariella* and *A. assimilella* share features with *A. atomella*. *A. scopariella* has squarer wing apex, better defined oblique dots and more whitish dots in cell; *A. assimilella* lacks pale ferruginous-pink coloration, but the pink fades in *atomella* and even slightly worn specimens are difficult to distinguish from *assimilella*, however forms with pale orange-brown mark between median dot and end of cell dot but without ferruginous suffusion in rest of forewing are common in *assimilella* and found only exceptionally in *atomella*. Specimens found in early spring cannot be *A. assimilella* because this species hibernates as half grown larva.

MALE GENITALIA. Socii longer than wide; gnathos elliptical, 2.5–3.5 times as long as wide, exceeding socii by nearly half its length or more; valva with costal margin straight, ventral margin gently curved through most of its length, apex subacute to acute; cuiller of average thickness, slightly inclined outwards or not, very slightly curved outwards at middle, apex rounded, 80–90% of valva width; transtilla slightly thickened; anellus lobes crossing anellus, touching or almost touching each other; anellus slightly wider than long, with wide V-shaped sinus, behind this an extended truncate membrane; aedeagus parallel-sided to three-fifths, then tapering, weakly curved beyond middle, slightly more than half as long as valva, cornuti in an elongate group, basal process short.

FEMALE GENITALIA. Anterior apophysis two-fifths to three-fifths length of posterior apophysis; sternite VIII width:length 2:1, anterior margin convex, an incurved lip below ostium extending into a pair of diverging folds, posterior margin slightly inclined inwards; ostium distal to middle of sternite; ductus bursae gradually expanding, about twice as long as ovate corpus bursae; signum wider than long, more or less elliptical; teeth variable in size and number.

GENITALIA DIAGNOSIS. Male genitalia of *A. atomella*, *A. oinochroa*, *A. nervosa* and *A. comitella* are extremely similar, see key (pp. 232–233) for optional determination features. Female difficult to characterise, best single feature is a fold on either side in area anterior to ostium, usually gently curved and about as long as diameter of ostium, running obliquely into anterior margin. Where meeting anterior margin, the fold may be somewhat swollen but does not form an anteriorly directed bulge (figs. 115a–c).

DISTRIBUTION. Distributed over most of Europe but not recorded from a number of countries particularly around the edges of Europe, being absent from Portugal, Ireland, Norway, Finland and the Baltic States, and in the Mediterranean only present on Sardinia. Asia: Turkey. Syria, Israel, North Africa (Lvovsky *et al.* 2016).

BIONOMICS. Larva on *Genista tinctoria* L. (Hannemann, 1953); *G. pilosa* L., *G. anglica* L. (Palm, 1989); *Genista sagittalis* (L.) P. Gibbs and *Cytisophyllum sessilifolium* (L.) O. Lang (Rymarczyk *et al.*, *in litt.*). Emerging from May to July, flying until May of following year

REMARKS. In Austria and in Great Britain, specimens of *A. scopariella* are found which share barcode with *A. atomella*, therefore if determination as *A. atomella* rests only on barcode, results may be incorrect.

116   *Agonopterix scopariella* (Heinemann, 1870)

Depressaria scopariella Heinemann, 1870: 149.
   Depressaria rubescens Heinemann, 1870: 154.
   Depressaria genistella Walsingham, 1903: 266, syn. n.
   Depressaria cyrniella Rebel, 1929: 45, syn. n.
   Depressaria scopariella subsp. *calycotomella* Amsel, 1958: 73, syn. n.

DESCRIPTION. Wingspan 17–22(–24) mm. Head light brown, through reddish brown to dark brown, face buff. Labial palp whitish buff, segment 2 with scattered light brown scales, segment 3 with fuscous rings at base, between middle and three-quarters and at tip. Antenna buff with narrow dark brown rings. Thorax light brown, reddish-brown, dark brown or blackish, tegulae concolorous except when thorax blackish, then tegulae intermediate between colours of thorax and forewing. Forewing with apex nearly right-angled, buff with slight ochreous tinge, light brown, reddish brown to dark brown, with scattered mid-brown to fuscous scales, the latter when present sometimes in rows following veins; costal and terminal spots dark fuscous, the latter sometimes absent; basal field similar in coloration to rest of forewing, but usually slightly paler; basal fascia often weak, from fuscous subdorsal spot to dorsal margin of cell, where it turns outwards; costa sometimes with deep brown dot at base; oblique dots blackish, often with a few whitish or buff scales beyond the second dot, white or buff median and end of cell dots, often lying in a slightly darker brown or grey area; blotch variable in development, light to dark fuscous, or absent; fringe light brown, often fuscous towards apex and frequently with a cinnamon tinge, with two lines. Hindwing light grey, darker towards termen; fringe whitish to light grey, with several faint lines. Abdomen grey.

VARIATION. Forewing ground colour is quite variable, with some specimens showing a degree of ochreous or pinkish tinge, and some rather darker brown; scattered dark scales very variable in extent, often absent, but when abundant oblique dots are inconspicuous; the basal patch is sometimes very poorly marked; the blotch varies from almost obsolete to well marked; pale dots in cell can be white or buff, sometimes evident, sometimes obscure, varying from two to four in number; terminal dots present or not. Forewing fringes often partially suffused fuscous. Blackish thorax occurs quite frequently.

SIMILAR SPECIES. *A. conciliatella*, *A. atomella* (*q.v.*). Typically the forewings are rather uniformly coloured throughout, although various colours are possible. Other species with three whitish dots in cell do not have the right-angled apex (although this is less evident in worn specimens) and they lack the even coloration of *A. scopariella*. *A. subpropinquella* has longer wings and lacks a basal patch and whitish cell-spots.

MALE GENITALIA. Socii longer than wide, not exceeding short uncus; gnathos fusiform, 4–5 times as long as wide, exceeding socii by nearly half its length; valva with costal margin weakly concave in basal half, ventral margin gently curved at about three-fifths, apex subacute to acute; cuiller of average thickness, straight, inclined outwards, apex acute or subacute, 62–70% of valva width; transtilla thin; anellus lobes L-shaped, free shorter limb either crossing anellus, and generally overlapping or reflexed, depending on preparation; anellus slightly wider than long, with wide V-shaped sinus, behind this an extended truncate membrane; aedeagus parallel-sided to three-quarters, then tapering, gently curved throughout, over two-thirds as long as valva, cornuti in a dense group, basal process of average length.

FEMALE GENITALIA. Anterior apophysis half to two-thirds length of posterior apophysis; sternite VIII width:length about 2:1, anterior margin slightly convex, a wide oblong middle section protruding, its margins forming curved folds on sternite; posterior margin flat or slightly inclined inwards, emarginate; ostium distal to middle of sternite; ductus bursae of nearly even width, about twice as long as elliptical corpus bursae; signum wider than long, more or less elliptical, with short anterior and posterior projections; teeth medium-sized, rather numerous.

GENITALIA DIAGNOSIS. In male details of anellus lobes distinct (L-shaped, overlapping if put into standard position or reflexed if overlap has been avoided). In female anterior margin of sternite VIII with a nearly rectangular protrusion with width about one-third of sternite, its margins extending as slightly outward sloping folds on sternite (figs. 116a–c) is distinctive.

DISTRIBUTION. Canary Islands and Madeira; nearly all European countries and Mediterranean islands east to Sweden, Poland, Romania, Greece and Crimea; the most common *Agonopterix* species in the Mediterranean area. Asia: Turkey, Lebanon, Syria. North Africa: Morocco.

BIONOMICS. *Cytisus scoparius* (L.) Link (Hannemann, 1953); *Calicotome spinosa* (L.) Link (Hannemann, 1995); *Lupinus arboreus* Sims (Harper *et al.*, 2002); *Genista pilosa* L., *G. tinctoria* L., *Laburnum anagyroides* Medik. and *L. alpinum* (Mill.) Bercht. & J. Presl. (Sonderegger, *in litt.*); *Adenocarpus complicatus* (L.) J. Gay, *Cytisus striatus* (Hill) Rothm. and *C. grandiflorus* (Brot.) DC. (Corley, 2015); *Genista scorpius* (L.) DC., *G. monspessulana* (L.) L.A.S. Johnson and *Cytisus oromediterraneus* Rivas Mart. *et al.* (Rymarczyk *et al.*, *in litt.*) and *Cytisophyllum sessilifolium* (L.) O. Lang (reared R. Seliger, 2023).

REMARKS. *A. scopariella* is extremely variable without obvious regional forms. We see no justification for recognition of any subspecies and treat subsp. *calycotomella* Amsel, 1958 as a synonym. We have seen holotypes of both *Depressaria genistella* Walsingham, 1903 from South Spain (NHMUK) and *Depressaria cyrniella* Rebel, 1929 from Corsica (NHMW) and both fall within the variation of *A. scopariella* and are hereby placed in synonymy with *A. scopariella*.

117   *Agonopterix conciliatella* (Rebel, 1892)

Depressaria conciliatella Rebel, 1892: 272.
    Agonopterix mutatella Hannemann, 1989: 391.

DESCRIPTION. Wingspan 18–22(–23) mm. Head variable in colour. Labial palp pale buff, segment 2 outer and ventral sides mixed with light brown and light to dark grey-brown scales, segment 3 with dark grey to black rings at one-third, two-thirds and at tip. Antenna grey-brown. Thorax variable in colour, sometimes nearly black and contrasting with pale tegulae. Forewing elongate, apex approximately right-angled, termen very slightly concave towards apex; coloration and markings very variable, ground colour straw-yellow, buff, various shades of brown or grey-brown, uniformly coloured or with lighter and darker areas, with scattered black scales in small numbers or sometimes following veins; in many specimens veins, except those to costa, heavily marked fuscous to blackish, this coloration sometimes extending to fill almost all of middle part of wing; basal field present except sometimes in darkest specimens, delimited by a subdorsal spot or a blackish line; oblique dots black, often followed by a few white scales, median and end of cell dots white, often edged blackish or reddish; costa not or weakly spotted, terminal spots dark grey-brown, often weak; fringe grey-brown often tinged cinnamon, from middle of termen to apex usually darker. Hindwing light grey-brown; fringe grey-brown, with two weak fringe lines.

VARIATION. Exceptionally variable species, characterised by long wings with right-angled apex and slight concavity of termen, usually with basal field, black oblique dots and usually two creamy cell dots set in a small darker patch. Apart from these features everything else varies. Head from light brown to deep brown, normally with buff-tipped scales, sometimes darker in mid-line; thorax also variable, usually paler posteriorly, tegulae usually concolorous with forewing; forewing ground colour straw-yellow, buff, brownish-buff, orange-brown, reddish-brown, grey-brown, dark brown, sometimes overlaid with orange-brown or darker brown scales particularly on dorsal side following blackish line bordering and near termen; basal field usually paler than ground colour, except where ground colour is also pale, but sometimes not differentiated from ground colour in darker specimens; oblique dots sometimes followed by white scales; cell dots often ringed blackish, but sometimes by orange-brown, occasionally one or other absent; veins often clearly marked dark brown or blackish; sometimes a large area of middle of wing dark fuscous; fringe usually darkened from middle to apex, but sometimes of ground colour.

SIMILAR SPECIES. *A. scopariella* has similar wingshape and likewise often has dark fringe from middle of termen to apex. Only forms of *conciliatella* with uniform forewing coloration are likely to be confused with *scopariella*, but usually the basal field is more defined in *conciliatella*. *A. perezi* which is also very variable and has the same distribution has rounded forewing apex.

MALE GENITALIA. Socii longer than wide; gnathos fusiform, 3–4 times as long as wide, exceeding socii by nearly half its length; valva often rather narrow, with costal margin concave, ventral margin gently curved at about three-fifths, apex subacute; cuiller slender, usually slightly inclined outwards in basal half, gradually curving inwards, or straight or curving outwards from middle, often slightly thickened to rounded apex, 75–90% of valva width; transtilla slightly thickened; transtilla lobes elongate, nearly touching; anellus lobes meeting in middle; anellus wider than long, reaching transtilla lobes; aedeagus parallel-sided to beyond three-quarters, then hardly tapering, gently curved throughout, about three-fifths as long as valva, cornuti in a dense group, basal process short.

FEMALE GENITALIA. Anterior apophysis half to two-thirds length of posterior apophysis; sternite VIII width:length 5:2 to 3:1, anterior margin inclined outwards, a broadly rounded incurved lip below ostium with ends extending onto sternite; posterior margin inclined inwards, slightly excavated in middle; ostium in middle of sternite; ductus bursae gradually expanding, 1.5 times as long as ovate to elliptical corpus bursae; signum narrowly elliptical, with anterior and posterior projections of variable length; teeth rather large, few on plate.

GENITALIA DIAGNOSIS. In male 180 ° angle between basal parts of costa of valvae distinct, see also key (pp. 232–233). In female band at anterior margin of sternite VIII is distinctive, but in *A. oinochroa* and *A. umbellana* it may be somewhat similar, for details see key (pp. 233–235).

DISTRIBUTION. Almost certainly confined to the Canary and Madeira archipelagos. Records from Sicily and Greece are unconfirmed and improbable.

BIONOMICS. A small series was reared from *Adenocarpus* sp. (Fabaceae) by Klimesch, moths emerged 16–26 May. Moths hibernate. No more data on larval phenology, but an extended larval period from early spring to early summer is likely.

REMARKS. Hannemann (1953) placed this species in *Martyrhilda*, based on what he believed to be its male genitalia, later transferring all the *Martyhilda* species to *Depressariodes* and ultimately to *Exaeretia*. The detailed explanation for this error which derived from confusion of dissected abdomens, can be found above, under *Exaeretia thurneri* and also in Buchner (2015b). Later Hannemann (1989) described this polymorphic species from the Canary Islands as *Agonopterix mutatella*, unaware of his earlier error.

118   *Agonopterix comitella* (Lederer, 1855)

Depressaria comitella Lederer, 1855: 232.

DESCRIPTION. Wingspan (22–)23–27 mm. Head creamy buff to light grey, sometimes ochreous-buff, usually contrasting with thorax colour. Labial palp whitish buff, segment 2 with some light brown scales on outer side, segment 3 with grey-brown rings at one-quarter and beyond middle, tipped dark grey. Antenna grey-brown, ringed pale buff. Thorax concolorous with forewing or slightly darker. Forewing long, apex nearly right-angled, termen slightly concave before apex, uniformly ochreous-buff to orange-ochreous, scattered blackish scales sometimes present, sometimes plentiful and following veins; costal and terminal spots not present; basal field not indicated or faintly demarcated with a few blackish scales from subdorsal spot; first oblique dot very small or more often absent, second usually black, conspicuous, usually followed by one or two white scales, median and smaller end of cell dots white, often weakly ringed orange; fringe ochreous-grey tinged cinnamon. Hindwing light grey; fringe light grey. Abdomen buff to ochreous-buff.

VARIATION. Forewing coloration from pale yellowish-buff to a rich orange-ochreous; blackish scales often absent but sometimes forming lines following the veins. Either of the white cell dots may be the larger. Sometimes white scales may be absent leaving an orange dot. A grey blotch is occasionally present on costal margin of cell.

SIMILAR SPECIES. *A. comitella* is characterised by large size, head usually contrasting with thorax colour, forewing apex right-angled, uniform ochreous-buff to orange ground colour, absence of basal field and terminal spots. *A. aspersella* is similar in shape, usually with paler coloration, with paler fringe, and no white dots in forewing. *A. irrorata* has different wing shape, second oblique dot absent, not first, white dot at end of cell larger than median dot and terminal spots present.

MALE GENITALIA. Socii large, longer than wide; gnathos large, fusiform, 3–3.5 times as long as wide, slightly exceeding socii; valva often rather narrow, with costal margin weakly concave, ventral margin gently curved just beyond middle, apex subacute; cuiller tapering from broad base to slender middle part, beyond middle curving outwards and slightly expanded, apex rounded, 85–88% of valva width; transtilla slightly thickened; transtilla lobes elongate, but well separated; anellus lobes meet in middle; anellus much wider than long, two halves widely divergent, separated by sinus with more than 90°, largely concealed by anellus lobes which meet in middle; aedeagus stout, parallel-sided to middle, then tapering, slightly curved beyond middle, about two-thirds as long as valva, cornuti in a dense group, basal process short.

FEMALE GENITALIA. Anterior apophysis half to three-fifths length of posterior apophysis; sternite VIII width:length 5:2 to 3:1, anterior margin convex, a wide incurved lip below ostium extending across sternite as a pair of folds; posterior margin inclined inwards, excavated in middle with a small bulge inside excavation; ostium distal to middle of sternite; ductus bursae gradually expanding, 1.3 to 2 times as long as elliptical corpus bursae; signum wider than long, more or less elliptical, with anterior and posterior projections of variable length; teeth medium-sized.

GENITALIA DIAGNOSIS. For male see remark under *A. atomella* and key (pp. 232–233). In female a wide incurved lip below ostium extending across sternite as a pair of folds is also found in *A. nervosa, A. aspersella* and sometimes in *A. umbellana*. For details see key (pp. 233–235).

DISTRIBUTION. Greece including Crete, Cyprus. Russia: Volgograd Oblast (Lvovsky, 2014). Asia: Turkey, Syria, Israel, Iran.

BIONOMICS. Larvae found on *Anagyris foetida* L. by Klimesch.

119   *Agonopterix nervosa* (Haworth, 1811)

Depressaria nervosa Haworth, 1811: 506
  Depressaria costosa Haworth, 1811: 508.
  Tinea depunctella Hübner, 1813: pl. 56, f. 378.
  Depressaria boicella Freyer, 1836: 120.
  Depressaria dryadoxena Meyrick, 1920: 315.
  Agonopteryx blackmori Busck, 1922: 277.
  Depressaria perstrigella Chrétien, 1925: 259, syn. n.
  Depressaria obscurana Weber, 1945: 373.

DESCRIPTION. Wingspan 16–22(–24) mm. Head ochreous-buff, face pale buff. Labial palp pale buff, segment 2 speckled with ochreous-brown to dark brown scales on outer side, segment 3 with or without a weak basal dark ring and with a greyish ring at two-thirds, tip blackish. Antenna grey-brown. Thorax pale ochreous-buff to ochreous. Forewing with apex weakly falcate due to concave termen, pale yellow-buff, ochreous-buff to orange-ochreous, scattered dark brown scales on veins, few or numerous, often confined to costal and terminal areas; veins marked grey-brown in costal area only or all veins strongly marked or frequently not marked at all; basal field buff, well defined, or absent when brown-marked veins run through it; basal fascia represented by a dark brown subdorsal spot or absent; costa buff to ochreous-buff; a dark brown dot at base of costa; costal spots absent or rarely faint, terminal spots absent; oblique dots small, blackish, the first often reduced or absent, followed by a small area of pale yellow and an orange-brown area extending from position of frequently absent median dot to creamy dot at end of cell; blotch grey, often very faint, outer quarter of wing weakly suffused pink; fringe whitish buff with pink tinge towards tornus, deeper pink around apex with some dark grey tinge, fringe with two lines. Hindwing whitish buff, grey-brown tinged towards termen, sometimes with veins sharply defined; fringe whitish, grey-tinged towards tornus, with two fringe lines, the outer usually faint. Abdomen whitish buff.

VARIATION. Apparently the most variable *Agonopterix* species. Ground colour of forewing can be buff through to an almost orange ochreous. There is much variation in the depth of marking on the veins, which can be completely absent except in costal and terminal areas, or all veins may be heavily marked, running right through to wing margins and usually slightly expanded into small spots at terminal margin. These spots are on the vein endings, unlike those throughout the rest of the genus where the terminal dots are interneural. The orange-brown patch in the end of the cell is sometimes absent, especially in veined forms, but when present may contain not only the creamy end of cell dot, but sometimes a creamy median dot and rarely a third dot between the normal dots. If these dots are absent, the orange-brown is usually more concentrated at the position of the dots. Creamy dots are usually more buff in veined forms and more often absent.

SIMILAR SPECIES. The weakly falcate apex separates this species from nearly all its congeners. Specimens with veins strongly marked resemble *A. umbellana*, but differ in wing shape and position of the terminal dots.

MALE GENITALIA. Socii longer than wide; gnathos large, fusiform, 3–3.5 times as long as wide, exceeding socii; valva with costal margin weakly concave, ventral margin gently curved beyond middle, apex subacute; cuiller narrowest in middle part, where it curves outwards, beyond middle varying from almost parallel-sided to expanded to an obovate knob, twice as wide as middle part, apex rounded or obliquely truncate, 80–87% of valva width; transtilla slightly thickened; anellus lobes almost meet in middle; anellus wider than long, two halves widely divergent with wide-angled sinus, largely concealed by anellus lobes, a rounded membrane may be visible beyond; aedeagus parallel-sided to three-fifths, then tapering, hardly curved, about one-half to three-fifths as long as valva, cornuti in a dense group, basal process short.

FEMALE GENITALIA. Anterior apophysis half to three-fifths length of posterior apophysis; sternite VIII width:length 3:1 to 7:2, anterior margin strongly convex or inclined outwards, a wide incurved lip below ostium extending into a pair of folds; posterior margin inclined inwards, with shallow excavation in middle; ostium distal to middle of sternite; ductus bursae slender, over 3 times as long as ovate to elliptical corpus bursae; signum rather variable, with well-developed anterior and posterior projections of variable length; teeth medium-sized.

GENITALIA DIAGNOSIS. For male see remark under *A. atomella* and key (pp. 232–233). Female has a wide incurved lip below ostium extending across sternite as a pair of folds, a feature shared with *A. comitella*, *A. aspersella* and sometimes in *A. umbellana*. For details see key (pp. 233–235).

DISTRIBUTION. Nearly all Europe, but not recorded from some Balkan countries and in Mediterranean only known from Corsica and Sardinia. In Russia eastwards to Ural Mts. North America (introduced) (BOLD).

BIONOMICS. Larvae on *Cytisus scoparius* (L.) Link and *Laburnum anagyroides* Medik. (Hannemann, 1953); *Ulex* (Benander, 1965); *Lembotropis nigricans* (L.) Griseb. (Palm, 1989); *Genista tinctoria* L., *G. anglica* L. and *Lupinus arboreus* Sims (Harper *et al.*, 2002); *Genista pilosa* L., *G. germanica* L. and *G. radiata* (L.) Scop. (Sonderegger, *in litt.*); *Genista sagittalis* L., *G. cinerea* (Vill.) DC., *Cytisus hirsutus* L., *C. oromediterraneus* Rivas Mart. *et al.* and *Cytisophyllum sessilifolium* (L.) O. Lang (Rymarczyk *et al.*, *in litt.*). Emerging predominantly in June, flying until autumn, we have checked specimens collected up to 9 October (Greece). A female was taken at light in Essex, England on 2 January 1995 (Harper *et al.*, 2002).

REMARKS. This species was known as *costosa* Haworth for many years, but Bradley (1966) showed that the name *nervosa* Haworth which had been applied to the species now known as *Depressaria daucella* actually belonged to this *Agonopterix*. As *nervosa* Haworth has page precedence over *costosa* Haworth, he chose to apply that name to the species. This was not strictly necessary but it would not aid nomenclatural stability to reverse Bradley's choice after so many years.

D. *perstrigella* Chrétien has conspicuously marked veins and no orange-brown mark in cell, but shows no differences from *A. nervosa* with which we synonymise it.

120   *Agonopterix umbellana* (Fabricius, 1794)

Pyralis umbellana Fabricius, 1794: 286.
   Depressaria ulicetella Stainton, 1849: 154
   Depressaria lennigiella Fuchs, 1880: 237.
   Depressaria prostratella Constant, 1884: 215.
   Depressaria knitschkei Predota, 1934: 2.

DESCRIPTION. Wingspan (19–)22–25 mm. Head pale ochreous to ochreous-brown, darker brown in mid-line, face shining light golden brown. Labial palp buff, segment 2 with scattered dark brown scales on outer side and ventrally, segment 3 sometimes ochreous on outer side, blackish at tip. Antenna grey-brown. Thorax and tegula straw-yellow to ochreous-buff, thorax with narrow dark mid-line anteriorly, lateral margins broadly dark brown. Forewing elongate, straw-yellow to ochreous-buff, veins usually marked mid to dark brown, except close to costa and termen, sometimes with some brown suffusion between veins particularly between end of cell and apex, between dorsal margin of cell and fold and in mid-wing in terminal area; terminal spots grey to blackish; subdorsal spot deep brown, sometimes extended to form outwards directed beginning of markings on veins; a small deep brown dot at base of costa; first oblique dot small, blackish, sometimes absent, second large, sometimes elongate; plical dot small, but often larger than first oblique dot; a small blackish end of cell dot usually present; fringe ochreous-buff with one fringe line. Hindwing light greyish buff to grey, grey-brown towards apex with a few dark spots between vein ends around termen; fringe pale buff with one or sometimes two fringe lines. Abdomen ochreous to greyish buff, posterior margins of segments buff.

VARIATION. There is much variation in strength of dark markings on veins of forewing, which may be weak, incomplete or absent.

SIMILAR SPECIES. Characterised by elongate forewing with more or less strongly marked veins not clearly reaching wing margins, when these are absent wingshape can help to separate it from other similar looking species, but dissection is recommended to confirm *A. umbellana* in such cases. Some forms of *A. nervosa* have similarly marked veins, but differ in the strongly darkened fringe around apex and the position of terminal dots which are at ends of veins, not in the usual interneural position that is found in *umbellana*.

MALE GENITALIA. Socii longer than wide; gnathos elliptical, 2–2.5 times as long as wide, exceeding socii by half its length; valva with costal margin weakly concave, ventral margin curved at about three-fifths, apex subacute to acute; cuiller inclined outwards, curved further outwards at middle, tapering towards acute slightly incurved apex or hardly tapering to truncate or obliquely truncate apex, 72–87% of valva width; transtilla thickened; anellus lobes almost meet in middle; anellus wider than long, distal margin hardly sclerotised, rounded or truncate; aedeagus parallel-sided or slightly tapering to three-fifths, then tapering more strongly, slightly curved, about one-half to three-fifths as long as valva, cornuti in a dense elongate group, basal process ending at about 40 % of total aedeagus length, which is (together with *A. oinochroa*) the longest within Fabaceae feeding *Agonopterix* species group.

FEMALE GENITALIA. Anterior apophysis half length of posterior apophysis; sternite VIII width:length 4:1 to 5:1, anterior margin convex, a wide incurved lip below ostium; posterior margin inclined inwards, with shallow excavation in middle; ostium in middle of sternite; ductus bursae slender, 2.5 times as long as ovate to elliptical corpus bursae; signum variable, wider than long, with or without short anterior and posterior projections; teeth numerous, medium-sized.

GENITALIA DIAGNOSIS. Male with anellus lobes expanding centrally, but without a rising section on upper edge, also see key (pp. 232–233). Female with band at anterior margin of sternite VIII similar to that in *A. conciliatella* and *A. oinochroa*, for details see key (pp. 233–235).

DISTRIBUTION. Western Europe east to Denmark, Germany, Italy, Slovenia, Albania. Asia: Turkey. North Africa: Morocco. New Zealand (BOLD).

BIONOMICS. Larvae on *Ulex europaeus* L. (Hannemann, 1953); *Spartium junceum* L. and *Adenocarpus hispanicus* (Lam.) DC. (Hannemann, 1995); *Ulex gallii* Planch., *U. minor* Roth, *Genista pilosa* L. and *G. anglica* L. (Harper *et al.*, 2002); *Genista cinerea* (Vill.) DC., and *G. scorpius* (L.) DC. (Rymarczyk *et al., in litt.*); *Cytisus decumbens* (Durande) Spach. Emerging predominantly in July, flying until May of following year.

REMARKS. Hannemann (1958a) considered *knitschkei* Predota, 1934 to be no more than a form of *umbellana*.

121   *Agonopterix oinochroa* (Turati, 1879)

Depressaria oinochroa Turati, 1879: 200.

DESCRIPTION. Wingspan (14.5–)16–19 mm. Head fuscous, scales tipped buff, face creamy buff. Labial palp with segment 2 pale buff, with scattered light fuscous scales, rough scales fuscous mixed pink, segment 3 pink or buff with blackish rings at base and beyond middle, tip blackish. Antenna fuscous. Thorax mid-brown to fuscous, scales tipped greyish buff, often giving whole thorax greyish buff appearance.

Forewing light brown, mid-brown to fuscous with scales tipped light brown or grey tinged dull pink, with scattered dark fuscous scales mainly in costal third and terminal areas; costal and terminal spots dark fuscous, variable, often small and faint; basal field whitish buff to buff or grey-buff, tinged with ground colour; basal fascia dark brown to blackish, extending to costal margin of cell, fading outwardly into ground colour; costal one-third of wing light brownish buff mixed with greyish buff scales; oblique dots sometimes black, more often ferruginous, joined or not, often associated with some whitish or buff scales; a pair of ferruginous spots, one in cell at one half and one at end of cell, often joined, each enclosing a whitish or buff dot, blotch absent or faint; fringe greyish buff to greyish fuscous, scales sometimes tipped dull pink, with indistinct fringe line. Hindwing grey, dark grey towards termen; fringe dull grey with basal line. Abdomen pale grey, or light to dark fuscous.

VARIATION. Depth of forewing colour varies from light fuscous with buff tinge to dull brownish fuscous with a slight purplish tinge. The oblique dots are very variable in the proportions of black, whitish and ferruginous scales, only the last always present.

SIMILAR SPECIES. *A. oinochroa* is characterised by ferruginous scales surrounding central forewing dots, a feature shared with few other European *Agonopterix* species, namely *A. ocellana*, *A. cnicella*, *A. subtakamukui* and *A. pseudoferulae*. *A. ocellana* is larger and differs in several details, *A. cnicella* has basal field with spike along cell margin, *A. subtakamukui* has subfalcate apex; *A. pseudoferulae* is most similar, but forewing more mottled, blotch and other blackish markings distinct.

MALE GENITALIA. Socii longer than wide; gnathos fusiform, about 3 times as long as wide, exceeding socii by half its length or more; valva with costal margin straight, ventral margin gently curved at about three-fifths, apex subacute; cuiller slightly curved outwards at middle, very slightly wider in distal part, apex subacute to acute, 85–90% of valva width; transtilla more or less thickened; anellus lobes close or touching in middle; anellus largely concealed by anellus lobes, with two halves divergent, V-shaped sinus less than 90°, with a rounded or truncate membrane beyond; aedeagus parallel-sided or slightly tapering to three-fifths, then tapering more strongly, slightly angled at three-fifths, about one-half as long as valva, cornuti in an elongate group, basal process ending at 35–40 % of total aedeagus length, which is (together with *A. umbellana*) the longest within Fabaceae feeding *Agonopterix* species group.

FEMALE GENITALIA. Anterior apophysis half to three-fifths length of posterior apophysis; sternite VIII width:length 7:2, anterior margin slightly convex, a wide incurved lip below ostium; posterior margin inclined inwards; ostium in middle of sternite; ductus bursae slender, 4 times as long as ovate corpus bursae; signum variable in shape, wider than long; teeth numerous, medium-sized.

GENITALIA DIAGNOSIS. For male see remark under *A. atomella* and key (pp. 232–233). In female band at anterior margin of sternite VIII similar to that in *A. conciliatella* and *A. umbellana*, for details see key (pp. 233–235).

DISTRIBUTION. South Europe and southern parts of Central Europe from Spain to Greece, extending north to Czechia. Asia: Turkey.

BIONOMICS. Larvae have been found on *Genista tinctoria* L. and *G. pilosa* L. (Rymarczyk *et al.*, *in litt.*). Emerging predominantly in June, flying until May of following year.

122   *Agonopterix aspersella* (Constant, 1888)

Depressaria aspersella Constant, 1888: 170.
  Depressaria novaspersella Spuler, 1910: 334.
  Depressaria autocnista Meyrick, 1921b: 76.

DESCRIPTION. Wingspan 19–23 mm. Head straw-yellow to light brown. Labial palp pale buff with a few light brown scales on segment 2, segment 3 with blackish ring just beyond middle and apex blackish. Antenna pale buff. Thorax straw-yellow to light brown. Forewing straw-yellow to light brown, with a scattering of light brownish buff scales, sometimes organised into irregular transverse lines, also a few scattered blackish scales, sometimes more abundant, mainly in costal half of wing; basal field not differentiated; costa hardly spotted, terminal dots faint; a dark brown dot at extreme base of costa; first oblique dot obscure or absent, the second blackish, conspicuous, distal pair of cell dots absent, but this area often somewhat paler by absence of scattering of light brownish buff scales, blotch usually developed as hardly visible grey shadow, rarely absent or distinct; fringe pale buff. Hindwing whitish to light grey; fringe light grey, with weak fringe line. Abdomen pale buff.

VARIATION. The first oblique dot is sometimes more strongly marked, but often absent. The scattered blackish scales vary in abundance. One Algarve specimen has darker background colour and abundant blackish scales scattered all over wing, especially on costa and with dashes round termen, light grey blotch present.

SIMILAR SPECIES. Characterised by pale coloration without basal field, most specimens effectively only marked with the second oblique dot. *A. pallorella*-group species have unmarked segment 3 of labial palp. *A. nanatella* also has pale coloration with a scattering of light brownish scales, no basal field and poor markings, and if end of cell dot is completely absent it appears extremely similar, but distinctly smaller, second oblique dot closer to blotch and (at least in males) hindwing usually darker than in *A. aspersella*.

MALE GENITALIA. Socii rather small, longer than wide; gnathos fusiform, 3–3.5 times as long as wide, exceeding socii by more than half its length; valva with distal part narrow, costal margin concave, ventral margin gently curved, apex subacute or acute; cuiller moderately stout, straight or very slightly curved outwards at middle, apex obliquely truncate to acute, 95–100% of valva width; transtilla slightly thickened; anellus lobes close or touching in middle; anellus wider than long, distal margin hardly sclerotised, rounded or truncate; aedeagus parallel-sided or slightly tapering to three-fifths, then tapering more strongly, slightly curving throughout or more curved at three-fifths, about one-half as long as valva, cornuti in an elongate group, basal process very short.

FEMALE GENITALIA. Anterior apophysis about three-fifths length of posterior apophysis; sternite VIII width:length 4:1, anterior margin weakly convex, an incurved lip below ostium; posterior margin inclined inwards; ostium in middle of sternite; ductus bursae slender, 3 times as long as ovate corpus bursae; signum wider than long, in shape of a truncated equilateral triangle, with a short posterior projection; teeth numerous, medium-sized.

GENITALIA DIAGNOSIS. Male with long cuiller (usually reaching costa) in combination with remarkably small socii should be distinct, also see key (pp. 232–233). Female with a wide incurved lip below ostium extending across sternite as a pair of folds, feature is also found in *A. nervosa, A. comitella,* and sometimes in *A. umbellana.* For details see key (pp. 233–235).

DISTRIBUTION. South-west Europe to France; Corsica and Sicily.

BIONOMICS. Larva on *Cytisus hirsutus* L. (Hannemann, 1953); *C. villosus* Pourr. and *Genista monspessulana* (L.) L.A.S. Johnson (Rymarczyk *et al., in litt.*); *Cytisus triflorus* Lam. Adult from June to the following spring.

REMARKS. Meyrick (1921b) proposed the name *autocnista* as he considered *aspersella* to be only a variant of the word *adspersella* and likely to be confused with *adspersella* Kollar. Nevertheless *aspersella* has been in accepted use for many years.

Depressaria Haworth, 1811

Depressaria Haworth, 1811: 505.
   Type species: *Phalaena radiella* Goeze, 1783 (= *Phalaena (Tortrix) heracliana* auct., nec Linnaeus, 1758); subsequent designation.
Piesta Billberg, 1820: 91.
   Type species: *Phalaena heracliana* Linnaeus, 1758; subsequent designation.
Volucrum Berthold, 1827 in Latreille: 484.
   Type species: *Phalaena heracliana* Linnaeus, 1758; monotypy.
Volucra Latreille 1829 in Cuvier: 412.
   Type species: *Phalaena heracliana* Linnaeus, 1758; subsequent designation.
Siganorosis Wallengren, 1881: 94.
   Type species: *Phalaena heracliana* Linnaeus, 1758; subsequent designation.
Schistodepressaria Spuler, 1910: 337; as subgenus.
   Type species: *Tinea depressella* Hübner, 1813; subsequent designation.

Horridopalpus Hannemann, 1953: 320.

Type species: *Haemylis dictamnella* Treitschke, 1835; original designation.

Hasenfussia Fetz, 1994: 249.

Type species *Depressaria hirtipalpis* Zeller, 1854; original designation.

DESCRIPTION. Wingspan 11–37 mm. Head with ocelli present; crown with raised scales, face with appressed scales. Labial palp strongly upcurved, with segment 2 curved, ventrally thickened with spreading scales, usually furrowed or rarely with long tuft; segment 3 one-half to four-fifths length of segment 2, slender, smooth, or with spreading scales below in *hirtipalpis* group. Antenna not ciliate, scape with pecten. Thorax with or without posterior crest. Forewing long, costa almost straight, apex rounded and termen curved, dorsum three-quarters to four-fifths length of costa; veins CuA_1 and CuA_2 arising separately from cell. Hindwing with lobe at anal angle present (except *hirtipalpis* group); veins CuA_1 and M_3 stalked. Abdomen dorsoventrally flattened. Forewing brown or grey, usually with short dark streaks between veins and dark spots around termen and often with a weak angled pale fascia beyond cell.

MALE GENITALIA. Gnathos rounded to elliptical, spinose, in *badiella* group divided into two parts; valva much more variable than in *Agonopterix*, costal margin sometimes with lobes or processes, clavus usually present, sacculus variable, often with terminal process (cuiller), usually crossing valva; anellus lobes usually absent; aedeagus variable, usually slender and slightly curved, with or without basal process on outer side, often with cornuti.

FEMALE GENITALIA. Strongly sclerotised signum usually present, more or less triangular, rhomboid (sometimes appearing triangular in some preparations) or four-pointed star-shaped plate with numerous broad-based triangular thorn-like teeth. In the species descriptions the ratio in standard preparations between the greatest width of segment VIII and its length in the middle on the ventral side (i.e. measured through the ostium) is given. As segment VIII is a ring, the ratio will vary within species depending on how much pressure has been applied.

DISTRIBUTION. The genus has a Holarctic distribution, apparently with few species in other regions. *Depressaria* species are found throughout Europe with substantial numbers of species in all parts of the continent.

BIONOMICS. Larval host-plants of European species belong mainly to families Asteraceae and Apiaceae, with a few species on Rutaceae and one each on Lamiaceae and Rosaceae. According to species larvae feed in spring and summer on leaves, shoots, flowers and developing seeds and internally in stems. The majority of species with known life histories overwinter as adults, but several species in the *douglasella* group do not. It is not clear if these overwinter as eggs or young larvae. For some of the least well-known species, there is just not enough evidence to be sure if they overwinter in the adult stage or in another stage. Pupation is normally in detritus at soil level beneath the host-plant and lasts for only a few weeks. The adults appear to be quite long-lived, even in non-hibernating species. In warmer climates there is probably a period of aestivation which may continue into hibernation or there may be some activity in the autumn. The exact timing of the larval stage in particular is very much dependent on seasonal variation, latitude and altitude.

REMARKS. The type species of *Depressaria* was *Phalaena* (*Tortrix*) *heracliana* Linnaeus, 1758 before the unfortunate lectotypification of Linnaeus's *heracliana* by Bradley (1966) whereby a species of *Agonopterix* was chosen as lectotype of *heracliana*. As a result *heracliana* became a misidentification of the *Depressaria* species and therefore ineligible as type species of the genus. *Phalaena* (*Tinea*) *radiella* Goeze, 1983 is the earliest name available for the species which had been the type species of *Depressaria* and was therefore selected as type species for the genus by Karsholt *et al.* (2006).

Based on external appearance most species are immediately recognisable as belonging to *Depressaria*. However, genitalia show considerable diversity. This resulted in the division of the genus into species groups by Hannemann (1953) and Palm (1989). Hannemann also separated three species into a new genus *Horridopalpus* and later Fetz (1994) separated three more species into genus *Hasenfussia*. We have not recognised these at genus or subgenus level, but we divide the genus into eight species groups, some of which can be further divided into subgroups. This is preferable to division into further genera or subgenera, which would leave a number of species isolated, inevitably resulting in a proliferation of monotypic genera.

Group 1 *Depressaria douglasella* group (species 123–142)

Forewings shorter and broader than in some of the other groups of *Depressaria*. In male genitalia, gnathos short-stalked, round, ovate or shortly elliptic, never elongate; socii usually small, spreading laterally; valva evenly tapering from broad base to narrow rounded apex, slightly curved, sacculus ending beyond middle in a well developed cuiller; clavus well developed (absent in *leucocephala*), hairy or rarely scaly; transtilla with two setose lobes; anellus large, more or less round or wider beyond middle, emarginated at base and apex; saccus short and round or elongate; aedeagus long, curved, ending in narrow point, base usually with a short rounded process, cornuti absent or microscopic. In the female genitalia the anterior apophysis is short, less than half length of posterior apophysis; many species have rounded lobes at the anterior angles of segment VIII, adjacent to the base of the anterior apophyses. The majority of species in this group do not overwinter as adults. Most species with larvae on Apiaceae, but a few on Asteraceae.

This group, particularly the species feeding on Apiaceae and closely related to *D. douglasella*, is taxonomically difficult. External appearance of most species is quite variable with overlap between species. Separation therefore has to be by genitalia and here too there are problems as there are rather few characters available and most of these also show variation. A character that appears to show good separation between one pair of species may be of little use in another instance. Another problem is that of associating females with the right males.

Barcodes should be really useful in resolving taxonomic problems such as this group presents, but there are presently so many misidentified specimens that barcode results appear confused and there is one pair of species that share barcode.

The treatment of the group presented here will not allow the certain identification of every specimen. Reliable separation of some species by genitalia is not possible, especially in the case of females. Barcodes are useful, but there are instances of barcode sharing. Also, the possibility that there are still unrecognised species in the group is a real one.

The group can be further subdivided into subgroups of closely related species:

D. douglasella subgroup (123–131): Host-plant Apiaceae; in male genitalia valva with a field of long setae parallel with costa from base to cuiller. This subgroup can be further split into those species closest to *douglasella*, with cuiller usually forked near apex or at least with the smaller fork represented by a knob, and those species related to *incognitella*, with cuiller not forked at apex.

D. albipunctella subgroup (132–139): Also with host-plant Apiaceae; in male genitalia valva without field of setae parallel with costa.

D. olerella subgroup (140–142): Host-plant Asteraceae.

123   *Depressaria pulcherrimella* Stainton, 1849

Depressaria pulcherrimella Stainton, 1849: 164.
  Depressaria semenovi Krulikovsky, 1903: 181.

DESCRIPTION. Wingspan (14–)16–17(–18.5) mm. Head light reddish brown more or less mixed buff. Labial palp segment 3 four-fifths length of segment 2; segment 2 whitish buff with scattered pale reddish brown scales, mainly ventrally and on outer side, segment 3 light reddish brown with a few whitish buff scales, a broad blackish ring from middle to beyond three-quarters, sometimes with blackish ring at base, tip ochreous. Antenna with scape deep brown, flagellum light grey-brown with fine blackish rings, darker grey-brown beyond middle. Thorax light reddish brown, mixed buff, usually paler posteriorly, sometimes almost white; tegulae reddish brown, sometimes whitish buff posteriorly. Forewing reddish brown with some whitish buff scales mixed in costal and dorsal areas, also forming a weak angled fascia beyond cell and lying between veins around termen before terminal spots. A white or whitish spot at end of cell, and sometimes beneath costa at two-fifths; dark brown to blackish area at base of dorsum, one or two short streaks in fold, at base of cell and on either side of white cell spot, between cell and costa, between veins to costa and termen, on costa between veins beyond mid-wing and forming spots around termen; fringe grey-brown, with one or two darker fringe lines. Hindwing light grey-brown, darker posteriorly; fringe light grey-brown with two fringe lines. Abdomen grey-brown.

VARIATION. The head and thorax vary from light reddish brown to nearly white. Ground colour always has a reddish tinge, but can be very pale with only the basal one-fifth of dorsal half of forewing rich reddish brown; the darkest specimens may have all the whitish markings including the cell spot greatly reduced in extent, although the angled fascia and subterminal spots persist as whitish dots. The dark markings vary from a dull dark brown to black, and can be few and inconspicuous in some specimens.

SIMILAR SPECIES. *D. pulcherrimella* is the smallest species of the subgroup, always under 19 mm wingspan, characterised by the pale head and thorax, and reddish brown ground colour of forewing with rather weak whitish spot at end of cell. *D. douglasella* is similar, but has sombre brown ground colour and more extensive dark marking on segment 3 of labial palp.

MALE GENITALIA. Gnathos round; socii small (smallest in sub-group); apical part of valva narrow, cuiller typically straight, tapering, without angle or step on outer side, angled inwards at costal margin forming a short beak, usually with a small knob on the outer margin at the bend, but there are exceptions to most of these features; clavus just longer than anellus, bushy haired; anellus concave-sided, widest at three-quarters, twice as wide as narrowest part at one-quarter; saccus short, rounded; aedeagus slender, gradually tapering beyond middle to apex.

FEMALE GENITALIA. Anterior apophysis one-third length of posterior apophysis or slightly less; segment VIII width:length 3:1, lateral lobes present; ostium slightly distal to middle of segment VIII, triangular, folds posterior to ostium meeting at a right-angle, folds extending from anterior lip of ostium spreading at very wide angle; ductus bursae sclerotised on one side, sclerotisation about twice as long as anterior apophysis, with a loop before elliptical corpus bursae, signum rhombiform, as wide as long or wider than long, sparsely to densely toothed.

GENITALIA DIAGNOSIS. In male the combination of features of socii (very small), cuiller (stout at base, just exceeding costa, quickly tapering near tip, here incurved with usually indistinct second branch on outer side), valva (remarkably narrow) and aedeagus (slender in lateral view) should be sufficient for certain determination of most European specimens. In female for separation from species with lateral lobes present, the combination of position of ostium (closer to distal edge of VIII sternite), direction of laterally protruding folds (only slightly rising, but steeper in the compared species) and length of sclerotised section of ductus bursae (present, but short) is usually, but not always reliable against *D. sordidatella*.

DISTRIBUTION. Northern half of Europe, extending south to northern Portugal and northern Italy. Georgia and Armenia (Lvovsky, 2004).

BIONOMICS. Larvae from early spring until May or June, in Britain most frequently found on *Conopodium majus* (Gouan) Loret, but also on *Pimpinella saxifraga* L. and *Daucus carota* L., more rarely on *Meum athamanticum* Jacq. and *Seseli libanotis* L. (Harper *et al.*, 2002). Outside Britain the most favoured host-plant is probably *Pimpinella saxifraga*. Also on *Cnidium dubium* (Schkuhr) Thell. (O. Karsholt, pers. comm.). Moths from June to September.

REMARKS. Misidentifications in this subgroup have been so frequent that we have only accepted host-plant records in which we can be really confident. Hannemann (1953) mentions *Bunium bulbocastanum* L. but this report may refer to *D. incognitella* (*q.v.*), a species feeding on this plant, externally similar to *D. pulcherrimella* and still undescribed at that time. Reports of this species feeding on *Valeriana* (Spuler, 1910; Lvovsky, 1981a) must be considered doubtful.

D. pulcherrimella has not been found in Western Asia apart from the Caucasus, but specimens with genitalia close to it suggest the possibility of undescribed species.

124   *Depressaria sordidatella* Tengström, 1848

Depressaria sordidatella Tengström, 1848: 124.
  Depressaria weirella Stainton, 1849: 165.
  Depressaria gudmanni Rebel, 1927: 7.
  Depressaria larseniana Strand, 1927: 282.

DESCRIPTION. Wingspan (18–)19–20.5(–21.5) mm. Head brownish buff. Labial palp segment 3 three-quarters length of segment 2; segment 2 outer side mid-brown, darker at base and apex, with some buff scales in middle dorsally, inner side buff dorsally, ventrally brownish buff; segment 3 cinnamon-brown with blackish brown ring beyond middle, tip buff. Antenna with scape cinnamon-brown, blackish dorsally; flagellum light to dark grey-brown with deep brown rings. Thorax mid-brown, sometimes brownish buff towards anterior margin. Forewing cinnamon-brown, sometimes darker but retaining cinnamon tinge, sometimes overlaid with scattered dark brown scales, particularly in cell and along dorsum to two-fifths; a deep brown slightly oblique streak at base of cell, sometimes extended to end of cell; a few whitish scales in cell beyond short streak and at end of cell forming a dot, sometimes of only one scale; whitish or pale grey scales scattered through ground colour in subcostal area, extending towards cell at two-fifths, and forming a rounded or angled fascia beyond end of cell and extending along fold, and a crescent around termen preceding terminal spots, which are dark grey-brown; fringe grey tinged with cinnamon-brown, usually with two fringe lines, the inner with dark brown tipped scales, the outer grey. Hindwing whitish grey, light grey-brown posteriorly; fringe light grey-brown with two fringe lines. Abdomen grey-brown.

VARIATION. The ground colour varies in intensity of cinnamon-brown colora-tion and extent and depth of dark brown overlying scales. Occasionally dark brown scales are dominant over costal half and terminal quarter of wing, forming dark clouds between costa and cell in between areas with some whitish scales. A black-ish streak is sometimes present throughout length of cell to beyond white dot, and such well-marked specimens may also have short dark streaks between veins to costa in apical third of wing. In some specimens, particularly the darkest, all mark-ings including the white cell spot, pale fascia and dark terminal spots are reduced or obscure.

SIMILAR SPECIES. Characterised by cinnamon-brown ground colour of thorax and forewing, lack of blackish streaks except at base of cell and minute white dot at end of cell, but some specimens with longitudinal blackish streaks distinct (e.g. 121b), in such cases see remarks under *D. nemolella*.

MALE GENITALIA. Gnathos round; socii not as small as in closely related spe-cies; cuiller short (but longer than in *pulcherrimella* or *floridella*, rarely as long as in *beckmanni*), stout, often with step or angle on outer margin near base, meeting costal margin at right-angle, ending in knob or short beak on outer margin and longer inwards directed beak, projecting beyond costal margin; clavus about equal-ling anellus, densely hairy; anellus widest at two-thirds length; saccus short, obtuse; aedeagus less slender than *pulcherrimella*, curved, sometimes by as much as 90 °, often slightly wider beyond middle, tapering in last one-third.

FEMALE GENITALIA. Anterior apophysis about one-third or less length of pos-terior apophysis; segment VIII width:length 3:1, lateral lobes present; ostium in middle of segment VIII, with three slightly concave sides, with lateral folds equal-ling width of ostium on each side; ductus bursae sclerotised on one side in posterior two-fifths, with a loop before elliptical corpus bursae, signum rhombiform, as wide as long, densely toothed.

GENITALIA DIAGNOSIS. In male, for *D. douglasella* and *D. pulcherrimella* (*q.v.*); in *D. floridella* socii larger and aedeagus broader in distal 1/3 with more clearly separated apiculus, in *D. beckmanni* cuiller more slender and often longer, but not always safely discernible from one of these two species. In female, for *D. pulcherri-mella* (*q.v.*), compared with other species of this group anterior apophysis tends to be slightly shorter, but no reliable feature for safe determination exists.

DISTRIBUTION. Northern half of Europe south to France, the Alps and to Romania. Western Asia. Daghestan, southern Siberia, Chita Province, Kamchatka and Sakhalin (Lvovsky, 2004).

BIONOMICS. Larvae on *Anthriscus sylvestris* (L.) Hoffm., less often on *Heracleum sphondylium* L., *Chaerophyllum temulum* L., *Conium maculatum* L. (Harper *et al.*, 2002); *Pastinaca sativa* L. (Hanneman, 1953); *Peucedanum* L., *Angelica* L. and

Aegopodium L. (Palm, 1989), in spun leaflets from spring until May or June, maybe to July depending on local climate, adults June or July to October.

REMARKS. *D. weirella* was synonymised with *sordidatella* by Lvovsky & Jalava (1993). Hannemann (1953) treated *D. gudmanni* Rebel as a good species, but in Hannemann (1995) it is given as a synonym of *D. weirella* (*i.e. D. sordidatella*) with male genitalia illustrated. We have obtained a barcode from the type of *D. gudmanni*; it is 100% similar to *D. sordidatella*, which is a further argument for the synonymy.

125   *Depressaria floridella* Mann, 1864

Depressaria floridella Mann, 1864: 186.

DESCRIPTION. Wingspan (17–)20–22(–23) mm. Head whitish buff, grey or grey-brown. Labial palp segment 3 three-fifths length of segment 2; whitish buff, segment 2 with brown scales mixed in on ventral and outer sides, segment 3 blackish beyond middle nearly to apex. Antenna light brown, narrowly ringed dark brown. Thorax and tegula whitish buff through various shades of grey and grey-brown to reddish brown or blackish, usually paler posteriorly, tegulae often darker than thorax. Forewing light grey-brown, orange-brown to mid-brown or reddish brown, with grey or buff scales particularly near costa, in cell, forming an often weak angled fascia beyond cell and in an arc before blackish terminal dots; a blackish subdorsal spot and sometimes a spot at extreme base of costa; blackish spots on costa at middle and towards apex continuing along termen; a short and a long black streak in cell and a series of black lines between veins to costa and termen; a whitish to buff spot separating black streaks in cell and another at end of cell; fringe various shades of grey or grey-brown with a distinct dark line. Hindwing light grey, darker towards apex; fringe light grey, with distinct darker line. Abdomen light grey to grey-brown.

VARIATION. Extremely variable in ground colour and extent of black lines.

SIMILAR SPECIES. Characteristic pale grey-brown coloration differs from most species in the *douglasella* group, but there are specimens indistinguishable from e.g. *D. douglasella* or *D. beckmanni*.

MALE GENITALIA. Gnathos round; socii small; compared with *D. sordidatella*, valva wider beyond cuiller, cuiller ending in knob on outer margin and longer inwards directed beak, not or scarcely projecting beyond costal margin; clavus about equalling anellus, densely hairy; anellus widest at two-thirds length with wide shallow apical notch; saccus short, obtuse; aedeagus quite broad, curved at about one-third, broadest just before abruptly tapering in last one-quarter to long apiculus.

FEMALE GENITALIA. Anterior apophysis approximately one-third length of posterior apophysis; segment VIII width:length 5:2, lateral lobes present; ostium slightly anterior to middle of segment VIII, anterior margin slightly convex, each end with a straight edge converging to an obtuse angle, lateral folds weakly developed; ductus bursae sclerotised on one side in posterior two-fifths, with a loop before elliptical corpus bursae, signum rhombiform, as wide as long or slightly wider, densely toothed.

GENITALIA DIAGNOSIS. In male combination of features of socii (rather large – only seen if spread well!), cuiller (very stout base, not or only slightly exceeding costa), aedeagus (remarkably broad in distal 1/4 with clearly separated apiculus) and anellus (ratio of width:length 1.1–1.2, larger than in other species) allows a reliable determination of nearly all specimens. Most features similar in *D. nemolella*, but cuiller very different. Female, for *D. pulcherrimella* (q.v.), other species of this group differ in length of sclerotised part of ductus bursae and details in folds around ostium, but insufficient for safe determination of every specimen.

DISTRIBUTION. Scattered through southern Europe from Spain and France to Greece, Bulgaria and Crimea, extending north to Switzerland, Austria, Czechia and Slovakia; not on Mediterranean Islands; Caucasus.

BIONOMICS. Larvae have been found on *Seseli arenarium* M. Bieb. in spun leaves in Crimea (Lvovsky specimens in ZIN; Savchuk & Kajgorodova (2017)), on *Seseli* sp. and *Trinia glauca* (L.) Dumort. in Pyrenées-Orientales (specimens in TLMF), reared from *Seseli annua* L. in Austria, Hundsheim (Sonderegger, one specimen in NMBE). The most abundant *Seseli* sp. in the Austrian range of *D. floridella* is *Seseli hippomarathrum* Jacq., so probably this species is a host-plant also. Larvae from early spring until April or May, adults from May until September or October.

REMARKS. Pupation of larvae reared in Crimea took place in spun leaves and not on the soil, unlike other species close to *D. douglasella*.

Female genitalia of a specimen from France show a different ostium with a wide anterior curve, the ends nearly joined by transverse lines, forming a near semicircle.

126   *Depressaria douglasella* Stainton, 1849

Depressaria douglasella Stainton, 1849: 164.
    Depressaria miserella Herrich-Schäffer, 1854: 119.

DESCRIPTION. Wingspan (15–)18–20(–22) mm. Head white with a few light grey-brown scales. Labial palp segment 3 three-quarters length of segment 2; segment 2 outer side light grey-brown to buff, with dark grey-brown band at one-third and white patch on dorsal side from middle to three-quarters, inner side whitish buff; segment 3 with often poorly delimited or joined deep brown to black rings at base and beyond middle, light grey-brown between, but sometimes nearly joined, tip buff to bright ochreous. Antenna with scape deep brown, flagellum usually light brown with narrow deep brown rings, darker towards apex. Thorax whitish, with admixture of light grey-brown scales of varying extent, but mainly in anterior part, tegulae brownish grey. Forewing grey-brown with admixture of whitish-tipped scales in ill defined areas, mainly beneath costa, in cell, between veins beyond cell indicating an angled fascia and between veins before terminal spots; dark brown-tipped scales mainly in dorsal half and beyond cell; a blackish brown patch at extreme base of wing on dorsal side, followed by a larger area of mid-brown to about one-quarter; blackish brown marks, mainly in the form of short streaks in cell, beneath costa, on costa at one-half and between veins to costa in posterior quarter, occasionally also between veins to termen; a series of dark grey or grey-brown spots around termen extending along costa to three-quarters, sometimes joined to make a border round termen; a white spot at end of cell; fringe grey-brown, slightly paler beyond fringe line. Hindwing light grey-brown, slightly darker posteriorly; fringe light grey-brown with one or sometimes two fringe lines. Abdomen grey-brown, posterior margin of segments whitish to buff.

VARIATION. Ground colour of forewings varies from light grey-brown to quite dark. Darker specimens have fewer or no whitish tipped scales and may have all markings more or less obscure, except whitish spot at end of cell. The extent of brownish scales on the thorax is very variable, from absent to mainly confined to the anterior part, occasionally they may be mixed in all over thorax.

SIMILAR SPECIES. Characteristically darker brown than other species of the subgroup with conspicuous white unringed spot at end of cell, but these features are not invariable, so dissection will often be necessary. *D. pulcherrimella* has reddish forewing ground colour.

MALE GENITALIA. Gnathos round; socii small; compared with *D. sordidatella*, valva slightly broader especially beyond cuiller, cuiller usually without step or angle on outer margin, ending in a fork with outer arm incurved, normally longer than inner arm, slightly projecting beyond costal margin; clavus usually about equalling anellus, densely hairy; anellus widest at three-quarters length, longer than wide with shallow apical notch; saccus short, obtuse; aedeagus slender, curved throughout, gradually tapering in last one-third to long apiculus.

FEMALE GENITALIA. Anterior apophysis about one-third length of posterior apophysis; segment VIII width:length 3:1, lateral lobes present; ostium slightly anterior to middle of segment VIII, anterior lip wide, concave, posterior lip consists of two folds forming an obtuse angle, lateral folds long, directed backwards, straight or curved, all forming a wide W; ductus bursae sclerotised on one side in posterior two-fifths, with a loop before elliptical corpus bursae, signum rhombiform, as wide as long or wider than long, densely toothed.

GENITALIA DIAGNOSIS. In male end of cuiller with two usually unequal branches which together form an arc of a circle, is distinct for this species, although specimens exist where this feature is deviant and not reliably distinguishable from other species with forked cuiller. Female genitalia cannot be safely distinguished from *D. beckmanni*. As these two species share barcode, females can only be safely distinguished if the host-plant is known.

DISTRIBUTION. Widely distributed in Europe from Portugal and Ireland to the Baltic Countries and Turkey, but not confirmed from a number of countries; Cyprus, not on other Mediterranean Islands. Western Asia.

BIONOMICS. Larvae on *Daucus carota* L. (Hannemann, 1953), *D. crinitus* Desf. (Corley, personal observation), *Torilis japonica* (Houtt.) DC. and *Anthriscus sylvestris* (L.) Hoffm. (Harper *et al.*, 2002); also recorded on *Seseli* L. and *Carum* L. (Benander, 1965), among spun leaflets until June, adults June to November.

We have collected larvae in spun leaves of *Daucus carota* and checked many reared specimens, most of them from Central Europe, host-plant was *Daucus carota* throughout.

According to De Prins & Steeman (2003–2023) it is unknown whether the species hibernates in the adult stage or as an egg or early instar caterpillar. According to Sonderegger (*in litt.*) adults hibernate and eggs are laid in early spring, but this rests on a single moth said to have been found in April. According to all other sources moths do not hibernate. Among all specimens we have ever seen none was collected earlier than middle of May and all these early specimens were fresh.

REMARKS. DNA barcodes obtained from around 60 specimens form a tight cluster which includes specimens with male genitalia and biology matching both *D. douglasella* and *D. beckmanni*. In addition there are three more clusters containing specimens that fit *douglasella* and *beckmanni* but also specimens that do not fit either. This could represent infraspecific variation or there may be additional taxa, but it is currently not possible to draw any conclusions.

127   *Depressaria beckmanni* Heinemann, 1870

Depressaria beckmanni Heinemann, 1870: 179.

DESCRIPTION. Wingspan (19–)20–22(–23) mm, largest species of *D. douglasella*-subgroup. Head whitish grey. Labial palp segment 3 three-quarters length of segment 2; segment 2 outer side light grey-brown to buff, with dark grey-brown band in basal third, inner side whitish buff; segment 3 buff with often poorly delimited deep brown rings at base and beyond middle. Antenna grey-brown. Thorax whitish grey, with anterior part brown to a varying extent, tegulae brown, sometimes whitish grey posteriorly. Forewing orange-brown to mid-brown or reddish brown with admixture of whitish grey to grey-buff tipped scales in ill-defined areas, mainly beneath costa, in cell, between veins beyond cell indicating an angled fascia and between veins before terminal spots; a blackish brown patch at extreme base of wing on dorsal side; blackish brown marks, mainly in the form of short streaks in cell, beneath costa, on costa at one-half and between veins to costa in posterior quarter, occasionally also between veins to termen; a series of dark grey or grey-brown spots around termen extending along costa to three-quarters, frequently joined to make a border round termen; a whitish grey to grey-buff spot at end of cell. Inner fringe scales mid-brown, outer light grey-brown. Hindwing light grey-brown, slightly darker posteriorly; fringe light grey-brown with one fringe line. Abdomen pale buff, posterior half often grey.

VARIATION. There is considerable variation in ground colour and the extent of whitish markings and also the development of black streaks. Darkest forms are more like *D. douglasella* while some lighter forms are almost without dark streaks and resemble weakly marked forms of *D. sordidatella*.

SIMILAR SPECIES. The whitish head and thorax appear to be a constant feature of this species, more reliably present than in *D. douglasella*.

MALE GENITALIA. Similar to *D. douglasella*, differing in the cuiller: cuiller usually without step or angle on outer margin, crossing costal margin of valva, often projecting well beyond margin, incurved at tip, sometimes with a knob on outer margin before apex.

FEMALE GENITALIA. There is no reliable character for separation from *D. douglasella*.

GENITALIA DIAGNOSIS. Determination rests on different shape of cuiller. See also remarks under *D. douglasella* and *D. sordidatella*.

DISTRIBUTION. France, Italy, Switzerland, Austria, Albania, Latvia. Likely to be more widely distributed.

BIONOMICS. Larvae on *Pimpinella major* (L.) Huds. between spun or folded leaflets (many reared specimens from Switzerland, Sonderegger, *in litt.*), no other host-plant known. Larvae until June, adults June or July to October.

REMARKS. *D. beckmanni* was described from Gastein, Austria, but the type material is lost. Hannemann (1953) examined specimens from localities that are far from the type locality, with cuiller angled inwards in the middle and ending in a fork with equal arms. It remains unclear how he was able to draw the conclusion it may be *D. beckmanni*, as these characters lie within the variation of *D. douglasella*. All subsequent recognitions of this species based on the features published in Hannemann (1953) & (1995) must therefore be considered unreliable.

However, there is another species of the *douglasella* group which does occur in the Gastein area and has characters that distinguish it from *douglasella*. It is this species that we believe to be Heinemann's *beckmanni*. A neotype has been chosen from the Gastein area which will stabilise usage of the name. Neotype ♂, [Austria], 'Styria, U.[Umgebung] B. [Bad] Gleichen- | berg, Stradner Kg. [Kogel] | 480 m Licht [light trapped] leg. K. Rath | 17.8.1979' (8485, TLMF).

D. beckmanni differs from related *D. douglasella* group species in the shape of the cuiller, particularly in the angle at which it crosses the costa and in the apex. Since the cuiller is fundamental in the definition of species in this group, we recognise *beckmanni* as a good species. This is reinforced by its choice of host-plant, *Pimpinella major* which appears to be the only host-plant of this species in Central

Europe. The problem with recognising this species is that it has barcode identical to that of *D. douglasella* (with quite different cuiller and with main host-plant *Daucus*) and female genitalia that cannot be satisfactorily distinguished owing to variation in this and other species. However, there are other examples of barcode-sharing in Depressariidae and separation of female genitalia in the *douglasella* group is often problematic. *D. beckmanni* is present in several collections but has been consistently misidentified as other species.

128   *Depressaria incognitella* Hannemann, 1990

Depressaria incognitella Hannemann, 1990: 142.

DESCRIPTION. Wingspan 16–18 mm. Head brown to grey-brown with scales tipped grey, face whitish. Labial palp segment 3 three-fifths length of segment 2; segment 2 whitish buff on inner side, outer and ventral sides grey to light fuscous, segment 3 whitish buff with narrow dark grey rings at base and beyond middle. Antenna with scape dark fuscous, flagellum buff, ringed dark fuscous. Thorax brown to grey-brown, scales tipped light grey, anterior margin dark fuscous. Forewing light to dark brown, with extensive admixture of light grey scales, particularly in costal area to one-half, in fold, in cell forming distinct spots at one-third and at end of cell, beyond cell forming a faint angled fascia, and close to termen; a black subdorsal spot at base, a weak dark dot on costa at extreme base; blackish streaks in cell, beneath costa, in fold, beneath costa and beyond angled fascia in subterminal area; a series of blackish dots between veins to termen; fringe grey, pale grey beyond a fuscous fringe line. Hindwing light grey, darker towards costa and apex; fringe pale grey with a distinct fringe line. Abdomen pale grey-buff.

VARIATION. There is some variation in depth of forewing ground colour and the amount of light grey scales.

SIMILAR SPECIES. Smaller than most species in the subgroup, without distinctive characters; the markings are often weak, with faintly marked fascia and obscure end of cell spot.

MALE GENITALIA. Gnathos slightly longer than wide; socii less spreading than in related species, separated by a V or U-shaped sinus; valva with costa slightly convex in basal half, apical part slender, cuiller tapering, slightly angled inwards at two-fifths, more than one-third of its length beyond costal margin, not angled, only extreme apex slightly incurved; clavus hairy, slightly longer than anellus; saccus small, obtusely rounded to broadly acute; aedeagus variable in length, slightly curved, gradually tapering in distal one-third.

FEMALE GENITALIA. Anterior apophysis one-third length of posterior apophysis; segment VIII width:length 3:1, lateral lobes absent, posterior margin with shallow excavation; ostium in middle of segment, triangular, with conspicuous folds extending laterally from its anterior angles; ductus bursae with short lateral sclerotisation distally, towards corpus bursae with a single spiral twist, corpus bursae elliptical, signum rhombiform, sometimes folded and appearing triangular, densely covered with triangular teeth.

GENITALIA DIAGNOSIS. In male combination of features of cuiller (long, straight or nearly straight, rather slender, unforked), outline of field of long setae parallel with costa (rather broad triangular with straight base) and aedeagus (slender) is unique. In female combination of absence of rounded lobes at the anterior angles of segment VIII and presence of sclerotisation in distal part of ductus bursae is unique in this subgroup.

DISTRIBUTION. Spain, Alps of France, Switzerland and Italy, Croatia, North Macedonia and Greece.

BIONOMICS. Larvae on *Bunium bulbocastanum* (L.) Huds. between spun leaflets until May, predominantly on small, non-flowering plants, adults from May to September, earlier in the year than most other species in this group. No other hostplants reported.

REMARKS. Hannemann's (1990) original description was based on one male and one female from the French Alps. Huisman & Sauter (2002) described the species in more detail and pointed out that Hannemann's female actually belonged to *D. ululana*.

129   *Depressaria nemolella* Svensson, 1982

Depressaria nemolella Svensson, 1982: 293.

DESCRIPTION. Wingspan 19–22 mm. Head whitish grey to grey-buff. Labial palp segment 2 with mixed pale buff and fuscous scales beneath, inner side creamy buff, segment 3 reddish brown with blackish ring from middle to near apex, apex creamy buff. Antenna fuscous. Thorax and posterior part of tegulae whitish grey to grey buff, anterior part of tegula fuscous to dark brown. Forewing light brown with scattered whitish grey scales, particularly in costal area, sometimes forming an angled fascia beyond cell; basal one-fifth of wing on dorsal side darker brown; black spot at base of costa absent or very small, spot at base of dorsum small; terminal dots blackish; black lines in cell and between veins; a faint whitish grey dot at end of cell; fringe grey-brown with indistinct lines. Hindwing light grey, darker posteriorly; fringe light grey. Abdomen grey-buff, two basal segments creamy.

VARIATION. Variation is mainly in the amount of whitish scales.

SIMILAR SPECIES. Differs from most closely related species in the tegulae contrasting with thorax colour, forewings evenly coloured with less mottling than in any of the related species; there is a lack or reduction of blackish marks at extreme base of wing and interneural spaces of almost whole wing marked black. Externally resembling forms of *D. sordidatella* with unusually distinct longitudinal blackish streaks. Determination of this extremely local species should never be based solely on external appearance.

MALE GENITALIA. Gnathos nearly round, socii diverging; valva with straight costa, apical part quite wide, sacculus narrow, cuiller tapering, slightly angled inwards at two-fifths, crossing costal margin, pointed tip slightly angled inwards; clavus stout, hairy, shorter than anellus; anellus widest at two-thirds; saccus short, obtuse angled; aedeagus strongly curved, short, stout, abruptly tapering at four-fifths to parallel-sided apex.

FEMALE GENITALIA. Anterior apophysis two-fifths length of posterior apophysis; segment VIII width:length 3:1, lateral lobes absent, posterior margin with broad excavation; ostium slightly anterior to middle of segment, with the folds developed from its ends forming a wide W; ductus bursae without sclerotisation, posteriorly straight, towards corpus bursae with a single spiral twist, corpus bursae elliptical, signum rhombiform, densely covered with triangular teeth.

GENITALIA DIAGNOSIS. In male combination of most features as in *D. floridella* (*q.v.*) but with long, unforked cuiller is unique. In female combination of absence of rounded lobes at the anterior angles of segment VIII with absence of sclerotisation in ductus bursae is unique in this subgroup.

DISTRIBUTION. For many years considered to be endemic to Gotland (Sweden), but now known also from France, Austria and Russia (Ural Mountains and Russian Altai).

BIONOMICS. Larva in rolled lower leaves of *Seseli libanotis* (L.) W.D.J. Koch in June (Svensson, 1982). Adults in July and August.

Based on unpublished observations of Nils Ryrholm in Sweden, larvae live in spinnings on the distal part of the leaf on the lower part of the stem and are fully grown by mid-June.

130   *Depressaria cinderella* Corley, 2002

Depressaria cinderella Corley, 2002: 29.

DESCRIPTION. Wingspan 17.5–19 mm. Head white or whitish with scales of neck and crown brown-based, face white. Labial palp segment 3 four-fifths length of segment 2; segment 2 with two deep grey bands separated by mixed white and grey scales, white at apex, upper edge white, segment 3 with two blackish rings, not always distinct due to mixture of whitish and dark brown scales in middle of segment, tip white. Antenna with scape blackish, flagellum dark grey-brown with narrow deep brown rings. Thorax white or whitish, blackish on anterior margin. Forewing deep grey-brown, with scales more or less tipped blackish; light grey scales scattered over most of wing, more concentrated between cell and costa, at middle of dorsum, a spot near base of costa, another in middle of cell and a larger one at end of cell, an ill-defined curved fascia beyond end of cell, and a series of spots between veins to termen before terminal spots; blackish brown spots at extreme base of costa and dorsum; black streaks at base of cell and in middle between two whitish cell-spots, another between second cell streak and costa, one on costa at middle, one or more streaks between veins to costa, sometimes with narrow streaks between veins to termen, a series of dots between vein-ends from costa at three-quarters around termen; fringe light grey, with deep grey fringe line. Hindwing pale grey, slightly darker posteriorly; fringe light grey with weak fringe line. Abdomen light grey.

VARIATION. Forewings vary in the depth of dark grey-brown coloration and the extent of light grey scales, resulting in some specimens appearing quite light grey and others greyish black. The blackish streaks vary in extent and some may be entirely absent, particularly those between veins towards termen.

SIMILAR SPECIES. *D. cinderella* is characterised by the deep grey-brown colouring with light grey markings. The black streaks are inconspicuous against the dark ground colour. *D. infernella* has slightly less contrasting forewing coloration.

MALE GENITALIA. Gnathos round, socii small, shortly separated, without notch between them; valva narrow, costa slightly convex in basal half, sacculus rather wide, cuiller angled outwards, straight, smooth, tapering evenly, angled inwards where it crosses costal margin, three-tenths of its length beyond costa; clavus short, stout, spinous, without hairs; anellus slightly longer than wide, widest at three-quarters, shallowly notched at apex; saccus as long as wide, broadly triangular, obtuse; aedeagus gradually tapering from base, slightly curved.

FEMALE GENITALIA. Anterior apophysis one-third length of posterior apophysis; segment VIII width:length 2:1, lateral lobes present, posterior margin of sternite excavate; ostium at posterior margin, with broad sclerotised curved lip tapering laterally to a narrow fold; ductus bursae without sclerotisation, posteriorly straight, towards corpus bursae with a single spiral twist; corpus bursae elliptical, signum rhombiform, wider than long, densely covered with triangular teeth.

GENITALIA DIAGNOSIS. In male the clavus without hairs is a unique feature. See also *D. infernella*.

DISTRIBUTION. *D. cinderella* is known from Serra de São Mamede in east-central Portugal and from Jaén and Toledo in Spain (Buchner & Šumpich, 2018).

BIONOMICS. In Portugal larvae were found in April in tubes spun among basal leaves of a plant originally thought to be *Conopodium capillifolium* (Guss.) Boiss. (Corley, 2002), but this identification needs to be confirmed at species level. Altitudinal range from 650 to 800 m. Fresh adults were taken at light on 5 June 1996 and on 14 April 1997. In 1998 and 2000 larvae were found in mid-April, but in early seasons (e.g. 1997) they must be feeding in March. Larvae brought to England refused *Conopodium majus* (Gouan) Loret, but a few were reared successfully on *Anthriscus sylvestris* (L.) Hoffm.

131   *Depressaria infernella* Corley & Buchner, 2019

Depressaria infernella Corley & Buchner *in* Corley, Buchner & Ferreira, 2019: 294.

DESCRIPTION. Wingspan 17–19 mm. Head with brownish-black scales with white tips on neck and crown, face white. Labial palp segment 3 four-fifths length of segment 2; segment 2 whitish with two grey-brown bands, outer side with additional grey-brown scales often obscuring bands, upper edge white, segment 3 with two blackish rings, not always distinct due to mixture of whitish and dark brown scales in middle of segment, tip white to buff. Antenna with scape blackish, flagellum dark grey-brown with narrow deep brown rings. Thorax blackish anteriorly, white posteriorly, position of boundary variable, tegulae with some buff scales. Forewing deep grey-brown, with scales more or less tipped blackish; whitish grey scales thinly scattered over most of wing, absent from area between fold and dorsum before middle, more concentrated between cell and costa, at middle of dorsum, a spot in middle of cell and a larger one at end, an ill-defined curved fascia beyond end of cell, and a series of spots between veins to termen before terminal spots; blackish brown spots at extreme base of costa and dorsum; white spot on costa following basal spot; black streaks at base of cell and in middle between two whitish cell-spots, another between second cell streak and costa, one on costa at middle, interrupted streaks between veins to costa, narrow streaks between veins to termen, spots on margin between vein-ends from costa at three-quarters around termen; fringe light grey, with deep grey fringe line. Hindwing pale grey, slightly darker posteriorly; fringe light grey with two weak fringe lines. Abdomen light grey.

VARIATION. The appearance of the crown of the head varies according to the length of the white tip of the scales. The extent of white on the thorax also varies according to how far the dark scales extend.

SIMILAR SPECIES. *D. infernella* is characterised by the deep grey-brown colouring with whitish grey markings. The black streaks are inconspicuous against the dark ground colour. *D. cinderella* has similar but slightly more contrasting coloration. Externally the species pair *D. cinderella* and *D. infernella* is usually discernible from the rest of *D. douglasella* subgroup by distinct black/white contrast and more prominent white elements especially on head, but specimens remain where these differences are not clear. Identification of both species should be based on genitalia.

MALE GENITALIA. Gnathos from broad base, slightly longer than wide; socii as wide as long, divergent with shallow notch between them; valva with costal margin slightly convex in basal half, sacculus rather wide, cuiller angled outwards, straight, smooth, tapering evenly, curved inwards just before it crosses costal margin, about one-fifth of its length beyond costa, pointed; clavus without hairs at base, otherwise hairy with long hairs exceeding anellus; anellus slightly longer than wide, widest at two-thirds, not notched at apex; saccus with margins slightly concave, apex rounded; aedeagus slightly curved in basal half, very slightly expanded beyond middle, tapering at four-fifths to parallel-sided apex.

FEMALE GENITALIA. Anterior apophysis less than one-third length of posterior apophysis; segment VIII width:length 2:1, lateral lobes present, posterior margin slightly concave, anterior margin of tergite VIII deeply excavated; ostium near posterior margin, with broad sclerotised curved lip tapering laterally to a narrow fold, in lateral view this lip is strongly protruding; ductus bursae without sclerotisation, posteriorly straight, towards corpus bursae with a single spiral twist, corpus bursae elliptical, signum rhombiform, sometimes folded and then appearing triangular, densely covered with triangular teeth.

GENITALIA DIAGNOSIS. In male *D. cinderella* is distinguished by lack of hairs on clavus, *D. infernella* is most similar to *D. incognitella*, safely discernible by combination of differences in cuiller (distinctly bent inward near tip), clavus (hairs longer, reaching near costa, hairless basal part distinctly bent with concave edges on inner side), vinculum (lateral edges concave) and outline of field of long setae parallel with costa (not as clearly triangular as in *D. incognitella*). In female for species pair *D. cinderella* / *infernella*, combination of ostium with broad sclerotised curved lip, presence of lateral lobes and absence of sclerotisation in ductus bursae is unique. Differences between these two species are found in anterior margin of tergite VIII: straight or with shallow excavation in *D. cinderella*, with deep excavation in *D. infernella*. But with only few female specimens of each species available, intraspecific variability is insufficiently known.

DISTRIBUTION. Currently known from the mountains of inland northern Portugal from Serra da Estrela to Serra do Alvão and the adjacent part of Spain: Sierra de Gredos (Avila).

BIONOMICS. Larvae on *Conopodium majus* (Gouan) Loret, feeding from a spinning among stem leaves, flowers or developing seeds. Larvae have been found in late May; adults from late June to September. Altitudinal range from 920 to 1850 m.

REMARKS. *D. infernella* and *D. cinderella* are closely related and occupy nearly contiguous regions. There are constant differences in the male and female genitalia and the larval host-plants and methods of feeding are different. Barcodes show 2.43% difference.

132   *Depressaria indecorella* Rebel, 1917

Depressaria indecorella Rebel *in* Prinz, 1917: 25.

DESCRIPTION. Wingspan 20–22 mm. Head fuscous, face creamy buff. Labial palp segment 3 three-quarters length of segment 2; segment 2 grey-buff on outer and ventral sides, segment 3 whitish with scattered light grey scales. Antenna fuscous. Thorax grey-brown, fuscous anteriorly. Forewing elongate, grey-brown, basal one-sixth and terminal area fuscous; a dark brown subdorsal spot at base, a blackish dot at base of costa; a blackish mark in cell at one-third consisting of two over-lapping contiguous short streaks, a weak blackish mark on costal side of cell at one-third and another in fold also at one-third, some thin blackish lines between veins towards costa, a series of small blackish dots between veins at termen; a buff zigzag mark from costa through middle of cell, sometimes a buff dot at end of cell, a weak angled buff fascia beyond end of cell and a series of dots preceding termi-nal dots; fringe grey-brown to fuscous. Hindwing whitish-grey, darker towards costa and apex; fringe light grey with weak fringe line. Abdomen pale buff.

VARIATION. Besides the form described above, there is a form with light grey-brown ground colour and very weak markings, with the short double streaks at one-third the only black markings.

SIMILAR SPECIES. Externally similar to several *Depressaria* species, e.g. *D. badiella*.

MALE GENITALIA. Gnathos ovate; socii angled outwards; valva broad-based, costal margin straight, ventral side narrowed after end of narrow sacculus, cuiller almost or just reaching costa, meeting it at right angle, expanded in distal one-quar-ter, with apex truncate and slightly notched, finely papillose on outer side in apical part; clavus with appressed hairs, sickle-shaped with point directed towards ventral margin of valva at one-half; anellus wider than long, broadly excised at base and bulging medially at apex; aedeagus base with flange on outer side and triangular process on inner side, long, gradually tapering to slender point, strongly angled at one-quarter and slightly angled at four-fifths.

Female unknown.

GENITALIA DIAGNOSIS. Male with highly distinctive clavus, cuiller, anellus and aedeagus.

DISTRIBUTION. Russia (Orenburg); also in Asia (Kazakhstan).

BIONOMICS. Host-plant unknown. Adults have been collected between 20 May and 26 September.

133   *Depressaria lacticapitella* Klimesch, 1942

Depressaria lacticapitella Klimesch, 1942: 148.

DESCRIPTION. Wingspan 19–24 mm. Head with face and vertex creamy white. Labial palp segment 3 three-fifths length of segment 2, both segments creamy white, segment 3 with some grey-fuscous scales on ventral edge. Antenna dark fuscous. Thorax and tegulae creamy white. Forewing elongate, grey-buff, mainly overlaid fuscous, darkest in basal one-fifth, particularly towards costa and dor-sum, but with additional slightly darker patch beyond end of cell and a broad band before termen; grey-buff ground colour most evident in middle of cell with irregular oblique extension towards costa, a weak spot at end of cell, in fold, a series of dots forming an indistinct angled fascia beyond cell and a series of dots between veins just before termen; a deep brown spot at base of dorsum; blackish dashes in cell at one-third, in fold, beneath costa and between veins within darkened subterminal fuscous band; a series of blackish dots between veins at termen; fringe pale grey with a light fuscous fringe line. Hindwing whitish grey, slightly darkened towards apex; fringe whitish grey without evident fringe line. Abdomen grey-white.

VARIATION. The markings tend to be obscure, except the spots around termen, but there is also a form with head and thorax tinged ochreous and forewing with darker brown ground colour in which the dark markings are obsolete. Thorax, head and labial palp in specimens of type series without dark to blackish scales, but the specimens from Mittewald with blackish scales on outer side of third segment of labial palp.

SIMILAR SPECIES. The creamy white palps, head and thorax including tegulae contrasting with dull fuscous forewings is characteristic. *D. leucocephala* has dark tegulae and extensive fuscous scaling on outer side of labial palps. *D. hofmanni* has more distinct dark streaks in forewing, but determination of this extremely local species should never be based solely on external appearance.

MALE GENITALIA. Gnathos ovate; socii digitate, shorter than gnathos; valva broad, costal side very slightly convex, ventral side narrowed after sacculus, sacculus wide but less than one-third width of valva, cuiller stout, width hardly changing, basal half slightly curving outwards then abruptly bent outwards towards middle between end of sacculus and apex; clavus stout, scaled, evenly curved and tapering; anellus with wide basal part sclerotised, narrowed in middle, distal part narrower, emarginated; saccus broad, rounded; aedeagus base with flange on outer side and triangular process on inner side; with bend at two-fifths, then tapering with slight inwards angle at five-sixths.

FEMALE GENITALIA. Anterior apophysis half as long as posterior apophysis; segment VIII width:length 7:2, lateral lobes absent, posterior margin of sternite with wide excavation, anterior margin sclerotised with shallow median excavation; ostium in middle of segment, triangular, the posterior sides attached to sclerotised collar slightly tapering towards segment margins; posterior part of ductus bursae slightly curved with a small sclerite at anterior end, remainder of ductus with a twist, expanding to elliptical corpus bursae; signum transversely elliptical, covered with triangular teeth, longitudinal axis weakly extended as two triangles, with minute teeth.

GENITALIA DIAGNOSIS. Male very similar to *D. hofmanni*, small differences in size of socii and shape of clavus, cuiller and anellus, compare figures. Female with slightly shorter anterior apophysis, otherwise scarcely different from *D. hofmanni*. With only one female dissected intraspecific variability is unknown and it must remain open if there are reliable differences.

DISTRIBUTION. Austria.

BIONOMICS. Larvae on *Athamanta cretensis* L. (Hannemann, 1953).

REMARKS. Type specimens reared from larvae collected on *Athamanta cretensis* in subalpine zone of "Grosser Pyhrgas" in Upper Austria and emerged end of August, but not found in this area since. On 15 September 2014 two males were light trapped in Mittewald/Drau, East Tyrol, at 865 m in a dry meadow without *Athamanta*, but with abundant *Seseli annuum* L. Here also it has not been found since. It is possible that neither locality represents the "typical" habitat of this species.

134   *Depressaria hofmanni* Stainton, 1861

Depressaria hofmanni Stainton, 1861: 177.

DESCRIPTION. Wingspan 18–23 mm. Head light brown to orange-brown, scales tipped buff, face buff. Labial palp segment 3 three-fifths length of segment 2; segment 2 inner side whitish, outer and ventral sides partly grey-buff to fuscous, segment 3 mainly grey, upper side buff, apex cream. Antenna dark fuscous, underside partly buff. Thorax and tegula light brown to buff. Forewing mid-brown to dark fuscous, basal one-fifth deep brown except near costa, some pale grey-buff scales, in cell, forming an angled fascia beyond end of cell and between veins close to termen; a blackish subdorsal spot close to base, obscure in darker specimens, a dull orange-brown to cinnamon-brown spot at base of costa; black dashes in cell, beneath costa and between veins; a grey-buff dot at end of cell; a series of black dots between veins to termen; fringe light brown to grey-buff with a fringe line. Hindwing light grey, darker towards costa, apex and dorsum; fringe pale grey with a darker grey fringe line. Abdomen light grey to grey buff.

VARIATION. Forewing ground colour varies from a slightly reddish mid-brown to dark fuscous. The thorax coloration also varies, but is always paler than the forewings.

SIMILAR SPECIES. *D. emeritella* is similar, but has black tegulae.

MALE GENITALIA. Gnathos elliptic; socii very small; valva broad with costal side very weakly concave, narrower after end of narrow sacculus, then much reduced, cuiller slightly expanding from base, elbowed outwards in middle, then slightly tapering, directed towards apex of valva, recurved at end; clavus stout, triangular with hairs and coarse scales, reaching end of anellus; anellus basal half sclerotised with wings directed caudad, apical half not sclerotised, weakly notched, extending beyond wings from basal half; saccus short, rounded; aedeagus base with flange on outer side and triangular process on inner side; angled above base and at two-fifths, after bend straight, tapering to slender slightly curved apex.

FEMALE GENITALIA. Anterior apophysis nearly half as long as posterior apophysis; segment VIII width:length 5:2, lateral lobes absent, posterior margin of sternite with wide deep excavation, anterior margin convex, sclerotised; ostium in middle of segment, triangular, the posterior sides attached to sclerotised collar gradually tapering towards segment margins; posterior part of ductus bursae slightly curved with a small sclerite at anterior end, remainder of ductus spirally twisted, corpus bursae pyriform; signum transversely elliptical, covered with triangular teeth, longitudinal axis extended as two triangles, with minute teeth.

GENITALIA DIAGNOSIS. See *D. lacticapitella.*

DISTRIBUTION. Central Europe: Germany, Switzerland, Austria and Slovenia, also in Romania and Russia (South Ural). Siberia (Minusinsk, Yakutsk) (Lvovsky, 2001).

BIONOMICS. Larvae on *Seseli libanotis* (L.) W.D.J. Koch from early spring until April or May, pupation either in the spinning or on the ground, adults from May or June to September, not hibernating (Hannemann, 1953; Buchner, personal observation).

REMARKS. Specimens from North America are listed as *D. hofmanni* in BOLD, but sequence is more distant from European *D. hofmanni* than from several other Western Palaearctic species. We could not check these specimens, it remains unclear if they are really conspecific.

135 *Depressaria albipunctella* ([Denis & Schiffermüller], 1775)

Tinea albipunctella [Denis & Schiffermüller], 1775: 319.
  Depressaria albipuncta Haworth, 1811: 510.
  Depressaria aegopodiella Hübner, 1825: 411.

DESCRIPTION. Wingspan (15.5–) 18–22 mm. Head mid to deep brown. Labial palp segment 3 three-fifths length of segment 2; segment 2 dark grey-brown on outer side with a few buff scales dorsally, inner side buff with some dark grey-brown scales ventrally; segment 3 blackish brown, shortly tipped bright ochreous. Antenna deep brown. Thorax dark brown. Forewing brownish buff, more or less overlaid with dark brown scales in much of proximal half and in two curved bands either side of pale fascia; base deep brown, gradually less deep away from extreme base; a blackish streak in fold and one in cell at one-third sometimes followed by a whitish dot; a whitish spot at end of cell, sometimes with a few blackish scales around it; costa deep grey-brown to three-fifths, often interrupted by a pale or whitish spot at half, further dark grey-brown marks below costa at half, between veins to costa near apex and around termen; whitish buff scales more or less mixed with ground colour forming a curved fascia round end of cell, not reaching costa, and between veins around termen, preceding terminal spots; fringe light grey-brown, with dark tipped scales forming a fringe line. Hindwing light grey, slightly darker and tinged brownish posteriorly; fringe light grey-brown, with two weak fringe lines. Abdomen light grey with slight brownish tinge.

VARIATION. The ground colour may have a weak reddish brown tinge, and the amount and depth of colour of the overlying dark brown scales varies significantly. In the darkest specimens the deep brown colour of the base may extend along the costal half of wing up to about one-third wing length and the markings may be almost obsolete apart from the white cell spot and the blackish curved mark at base of cell. The small white dot following this blackish mark is frequently absent.

SIMILAR SPECIES. *D. albipunctella* is characterised by the dark head and thorax, labial palp with blackish unringed third segment, forewing with blackish curved mark in cell, clear white spot at end of cell and limited development of dark streaks. The curved fascia is better developed than in *D. pulcherrimella* or *D. douglasella*. Pale specimens can be very similar to *D. olerella*, best difference is posterior crest of thorax without pale scales, present in *D. olerella*, but this difference becomes indistinct in worn specimens, especially after hibernation. See also *D. subalbipunctella.*

MALE GENITALIA. Gnathos ovate; socii small; valva wide at base, strongly tapering, apical portion slender, sacculus quite wide, cuiller from stout base contracting in middle portion to parallel-sided, round-ended distal part exceeding costal margin by about two-fifths its length; a scaly mound preceding and adjoining base of clavus, clavus itself broad-based, tapering to fine point, scaled throughout, twice curved outwards with one inwards curve in between, end of clavus close to base of cuiller; anellus about as long as wide, base with deep V-shaped excavation, a pair of curved folds in middle and apex with shallow notch; saccus with obtuse point; aedeagus without basal flange, curved through a right angle at about one-third, tapering to slender point from two-thirds.

FEMALE GENITALIA. Anterior apophysis about two-fifths length of posterior apophysis; segment VIII width:length 2:1, lateral lobes absent, anterior margin of tergite with wide deep excavation; ostium wide, close to posterior margin of segment, adjacent to a narrow sclerotised collar; ductus bursae with a spiral turn; corpus bursae elliptical, signum transversely elliptical covered with large triangular teeth, longitudinal axis slightly extended as two triangles, with minute teeth.

GENITALIA DIAGNOSIS. In male the clavus is characteristic, but compare with *D. subalbipunctella.* In female the sclerotised collar with tiny spicules centrally and an approximately rhombic structure formed by ostium plus central part of collar is only shared with *D. subalbipunctella* (q.v.).

DISTRIBUTION. Widespread from Portugal and England to Norway and Sweden and through central and southern Europe to Romania and Greece; Sardinia. Also in Western Asia and North Africa.

BIONOMICS. Larvae between spun leaflets on *Chaerophyllum bulbosum* L. (Hannemann, 1953); *Anthriscus sylvestris* (L.) Hoffm., *Daucus carota* L., *Chaerophyllum temulum* L., *Torilis japonica* (Houtt.) DC. (Harper *et al.*, 2002); *Conopodium majus* (Gouan) Loret (Corley, personal observation); *Pimpinella* L., *Seseli* L., *Conium maculatum* L. and *Ptychotis heterophylla* (L.) Loret & Barrandon (Sonderegger, 2012); *Peucedanum oreoselinum* (L.) Moench, (Buchner, personal observation), in an extended generation from April to July, adults from late May until May of the following year; eggs are laid in spring, adults often found after hibernation.

136   *Depressaria subalbipunctella* Lvovsky, 1981

Depressaria subalbipunctella Lvovsky, 1981: 78.

DESCRIPTION. Wingspan 17–21 mm. Head chestnut-brown, scales tipped buff. Labial palp segment 3 two-thirds length of segment 2; segment 2 buff tinged red-brown on upper and inner sides, more or less overlaid with brown scales on outer and ventral sides, segment 3 blackish on ventral side, reddish brown dorsally, apex creamy white. Antenna dark fuscous. Thorax fuscous, scales tipped buff, tegulae dark brown anteriorly. Forewing elongate, grey-brown to dark chestnut-brown, mixed with grey-buff and fuscous scales, darker brown at base and in first half of costal area; a weak pale angled fascia beyond cell followed by a broader darker brown angled fascia at four-fifths; a blackish subdorsal spot at base, a small black spot at base of costa; a black line in fold at one-third and an elongated curved black mark with an associated white dot in cell at one-third, a thin subcostal black line reaching to one-third, a few weak black lines between veins mainly within darker fascia at four-fifths; a whitish spot at end of cell; a series of blackish dots between veins at termen; fringe grey-buff, partially mixed fuscous. Hindwing light grey, darker towards apex; fringe whitish grey with indistinct fringe line. Abdomen grey buff.

VARIATION. There is variation in depth of ground coloration; dark marks are obscure in darker specimens; the pale end of cell dot and fascia also vary in distinctness.

SIMILAR SPECIES. Very similar to *D. albipunctella*, differing in wing shape, longer in *subalbipunctella*. The black dashes between the veins and the pale markings tend to be more distinct in *albipunctella*. In its distribution area, for safe determination of species pair *D. albipunctella/subalbipunctella* it is recommended to examine male genitalia, which are clearly different.

MALE GENITALIA. Similar to *D. albipunctella*, but gnathos round; valva with apical portion less slender, cuiller more slender in basal part; scaly mound preceding and adjoining base of clavus massive, clavus itself only slightly curved towards apex, directed towards base of cuiller but not reaching near it; anellus longer than wide, base without excavation, a conspicuous pair of curved folds from base to three-quarters length.

FEMALE GENITALIA. Papilla analis elongate; anterior apophysis about one-half length of posterior apophysis; segment VIII width:length 8:5, lateral lobes absent, anterior margin of sternite strongly convex; ostium wide, close to posterior margin of segment, adjacent to a deep, slightly sclerotised collar; ductus bursae with a spiral turn; corpus bursae elliptical; signum transversely elliptical, covered with large triangular teeth, longitudinal axis slightly extended as two triangles, with minute teeth.

GENITALIA DIAGNOSIS. For male see *D. albipunctella*. According to original description, differences in female genitalia are found in signum and in anterior margin of sternite VIII, which is straight in *D. albipunctella* but concave in the middle in *D. subalbipunctella*, but it is doubtful if this is always clear. The different shape of signum mentioned is probably not reliable as shape of signum often shows remarkable intraspecific variability.

DISTRIBUTION. Greece (mainland and Samos), Crimea. Also known from Turkey, Armenia, Azerbaijan, Georgia and Turkmenistan.

BIONOMICS. Host-plant unknown. Adults have been collected from late May onwards and in March, after hibernation.

137   *Depressaria krasnowodskella* Hannemann, 1953

Depressaria krasnowodskella Hannemann, 1953: 314.

DESCRIPTION. Wingspan 21–24.5(–26) mm. Head cinnamon-brown with scales tipped buff, face creamy buff. Labial palp segment 3 three-quarters length of segment 2; segment 2 pale cinnamon-buff, outer side cinnamon ventrally, with two light grey bands, segment 3 with outer side and dorsal edge cinnamon with a few blackish scales, a blackish ring beyond middle, apical quarter cinnamon-buff. Antenna with scape light to dark brown, flagellum cinnamon-brown to grey-brown with dark grey-brown rings. Thorax cinnamon-buff, scales tipped buff. Forewing quite elongate, with light to mid grey-brown scales, tipped light brownish buff; areas of light brownish buff scales near base of costa, between cell and costa at two-fifths and three-fifths, beyond end of cell, forming an indistinct angled fascia and between veins before terminal spots; a pale buff spot at end of cell; a dark brown spot at extreme base of costa; a larger blackish spot at base of dorsum; short mostly narrow black streaks, at base of cell, between cell and costa, in fold at two-fifths, between most veins to costa and termen, ending some distance from margin; small blackish dots between veins around termen; fringe light grey-brown, with faint cinnamon tinge; no fringe line. Hindwing light brownish grey, darker posteriorly; fringe light grey-buff, with at least two fringe lines.

VARIATION. The depth of light brown coloration of the forewing varies, sometimes with a cinnamon tinge to the paler areas including the spot at end of cell.

SIMILAR SPECIES. *D. krasnowodskella* is characterised by the rather pale forewing coloration, pale buff spot at end of cell, weak fascia and mainly narrow short black streaks; there are no white scales. Pale specimens of *D. radiosquamella* show very similar wing pattern, but usually it is smaller, head and third labial palp segment darker. Apart from rather large pale buff spot at end of cell, a few other species such as *D. badiella* and *D. subnervosa* also can be very similar.

MALE GENITALIA. Gnathos ovate; socii small; valva not very broad, slightly and evenly curved, sacculus narrow, cuiller arising at middle of ventral margin, angled outwards, almost reaching costal margin at three-fifths, slightly curved inwards, gradually tapering to ultimately outcurved apex; clavus slender, angled outwards but curved inwards to reach costal margin at one-sixth, with narrow scales except on outer margin in basal one-third; anellus with wide base shallowly excavated on distal margin, posterior part quadrate divided into two slightly sclerotised rectangles; saccus longer than wide; aedeagus only slightly longer than clavus, without basal flange, after base parallel-sided to five-sixths before tapering to apex.

FEMALE GENITALIA. Anterior apophysis about one-third length of posterior apophysis; segment VIII width:length 5:2, lateral lobes absent, posterior margin of sternite with wide shallow excavation and weakly sclerotised marginal collar-like band, emarginate in middle; ostium wide, towards anterior margin of segment, anterior margin straight, posterior margin formed by two thickened lips meeting at an angle and extending on to emargination in posterior collar; ductus bursae with slight bend before elliptical corpus bursae, signum rhombiform with weak triangular extension on anterior edge, covered with minute triangular teeth and a small number of larger teeth.

GENITALIA DIAGNOSIS. Male unique in European fauna, but similar to *D. assalella* Chrétien, 1915 (North Africa, Near East Asia), which has clavus distinctly exceeding costa. Female genitalia also unique in European fauna, possibly indistinguishable from *D. assalella*.

DISTRIBUTION. Known from Portugal (Algarve and Serra da Arrábida, west of Setúbal) and Spain (Catalonia, Almería and Cadiz); outside Europe only recorded from Turkmenistan and Morocco.

BIONOMICS. Host-plant unknown. The earliest fresh moths were taken at light on 30 April and during May. A worn specimen, obviously after hibernation, has been light trapped 25 March. In Portugal it occurs on hot south-facing limestone slopes at around 300 m. At the Algarve site where adults had been taken in May, a search in March for larvae on the three common Apiaceae of the locality was unsuccessful. Timing of the search may have been wrong, or perhaps the species does not feed on Apiaceae.

REMARKS. *D. krasnowodskella* was described from a single male in the Staudinger collection, taken at Krasnovodsk (Turkmenistan).

D. assalella Chrétien, 1915 described from Algeria is closely related. Male genitalia were illustrated by Hannemann (1958a: 35). Female genitalia cannot be distinguished from *D. krasnowodskella*.

138   *Depressaria tenebricosa* Zeller, 1854

Depressaria tenebricosa Zeller, 1854: 324.
  Depressaria albiocellata Staudinger, 1871: 247.
  Depressaria amblyopa Meyrick, 1921a: 392.

DESCRIPTION. Wingspan (17–)18–22 mm. Head brown to reddish brown. Scales mostly tipped buff; face whitish buff. Labial palp segment 3 three-fifths length of segment 2; segment 2 inner side whitish buff, outer and ventral sides mixed whitish buff, mid-brown and fuscous, segment 3 fuscous at base and beyond middle, whitish buff before middle, apex ochreous-buff, sometimes all blackish except apex. Antenna dark fuscous, buff below. Thorax light brownish buff, mid-brown to dark fuscous, scales mostly paler tipped. Forewing mid- to dark brown or dark fuscous, more or less extensively marked light grey-buff, particularly in a broad costal stripe, in cell and fold, forming an angled fascia beyond end of cell, in subterminal area; a blackish spot at base of dorsum, black lines beneath costa, in cell and in outer part of wing between veins, a series of blackish dots between vein ends towards end of costa and round termen; cell with one to three white dots, that at end of cell largest and often the only such dot; fringe grey with a fuscous fringe line. Hindwing light grey, slightly darker towards apex; fringe pale grey with one distinct fringe line. Abdomen light grey.

VARIATION. There is much variation in the extent of brown to dark brown coloration, some specimens distinctly dark fuscous and then with other markings obscure. Pale grey-buff markings sometimes much reduced; white dots in cell vary from one (at end of cell) to three. Due to this variation and the general similarity of species in the *douglasella* group, this species requires genitalia examination for reliable determination.

SIMILAR SPECIES. *D. radiosquamella* and *D. albipunctella* have third segment of labial palp almost entirely blackish except the apex, but such forms can also be found in *D. tenebricosa*; *D. radiosquamella* is deep brown at the wing base to about one-fifth; *D. albipunctella* and *D. subalbipunctella* have more reduced black markings and reddish brown ground colour. *D. douglasella* is also similar.

MALE GENITALIA. Gnathos ovoid on short stout stalk; socii small; valva with long bulge from near base to beyond middle of costal margin, apical portion slender, ventral margin evenly curved, sacculus ending in cuiller at two-thirds valva length, cuiller short, directed outwards, curving inwards beyond middle with setae on inner margin from middle to apex; clavus without hairs or scales, slender, gradually tapering to very narrow apex, directed outwards, crossing costal margin at about two-fifths, strongly curving inwards to end in middle of tegumen; transtilla exceptionally thickened; transtilla lobes large; anellus with wide basal part, posterior part with two areas of reticulate microsculpture, narrowing to broad nearly round apex; saccus wide, rounded before elongate round-tipped extension; aedeagus slightly tapering from base to two-fifths, where slightly bent, parallel-sided to close to apex where abruptly tapered and slightly curved.

FEMALE GENITALIA. Anterior apophysis less than one-quarter length of posterior apophysis; segment VIII width:length about 4:1, lateral margin straight, sternite with broad, rounded anterior bulge, posterior margin with wide, deep V-shaped excavation with incision at its base extending to ostium which lies in a pouch-like fold close to anterior margin; tergite with broad, shallow V-shaped excavation on anterior margin; antrum weakly sclerotised, ductus bursae with two spiral turns, corpus bursae elliptical, signum three times wider than long, covered in triangular teeth, one pair noticeably larger than all others.

GENITALIA DIAGNOSIS. Male distinctive, but female genitalia are similar to *D. radiosquamella* (*q.v.*). Males of *D. tenebricosa* and *D. radiosquamella* can be determined by genitalia features without dissection, if valvae are spread. In females it is more difficult and usually only possible to separate the species pair from *D. douglasella* subgroup: terminal edge of papilla analis of *D. tenebricosa* and *D. radiosquamella* (figs. 138d, 139c) in lateral view straight, but convex in *D. douglasella* subgroup (figs. 126e, 129b).

DISTRIBUTION. South-east Europe from mainland Italy and Sicily through the Balkans to Bulgaria, Cyprus and Turkey (Asian part).

BIONOMICS. Host-plant apparently unrecorded. Earliest specimen we have seen was caught in May, followed by records throughout the year, latest observation 30 December. Obviously, eggs are laid late autumn to early winter and adults do not hibernate.

REMARKS. Hannemann (1958a) synonymised *D. amblyopa* Meyrick, 1921 and *D. albiocellata* Staudinger, 1871 with *D. tenebricosa*.

139   *Depressaria radiosquamella* Walsingham, 1898

Depressaria radiosquamella Walsingham, 1898: 132.
   Depressaria duplicatella Chrétien, 1915: 343, syn. n.
   Depressaria adustatella Turati, 1927: 339, syn. n.
   Depressaria delphinias Meyrick, 1936a: 623, syn. n.
   Depressaria subtenebricosa Hannemann, 1953: 315, syn. n.

DESCRIPTION. Wingspan (16.5–)18–21 mm. Head with raised mid to dark brown scales on neck and crown, sometimes mixed grey-buff; face smooth, white to cream-white. Labial palp segment 3 three-fifths length of segment 2; segment 2 whitish buff, with two broad grey-brown bands on outer side, segment 3 blackish brown with a few buff scales, particularly on dorsal edge, tip pale buff. Antenna with scape blackish brown, flagellum grey-brown to dark brown, narrowly ringed deep brown. Thorax deep brown with some light grey-buff scales, particularly posteriorly, or largely light grey-buff. Forewing dark grey-brown, deep brown on most of basal one-sixth, if reaching costa sometimes extending along costa to half; some light grey-brown scales especially in fold, a few in middle of cell and forming a very weak angled fascia beyond cell, sometimes a few pale scales between veins around termen, preceding terminal spots; a whitish spot at end of cell; fine blackish streaks in fold and between cell and costa near base, a stout curved streak in base of cell, a shorter streak between this and costa, fine streaks between veins to costa and to termen; grey-brown spots around apex and termen between veins; fringe brownish grey, without obvious fringe line. Hindwing light grey-brown; fringe light grey-brown, with weak fringe line. Abdomen greyish ochreous.

VARIATION. This species shows less variation than most other species in the *douglasella* group. In male genitalia there is considerable variation in width of valva.

SIMILAR SPECIES. Possibly distinguishable from *D. tenebricosa* by the third segment of labial palp which is uniformly dark except at the tip, but this character needs to be verified on a greater range of material. The species needs genitalia examined for certain identification.

MALE GENITALIA. Rather variable, gnathos from shortly ovate to rounded, slightly wider than long, on short stout stalk; socii small; valva sometimes with bulge in basal half of costal margin, apical portion more or less slender, ventral margin evenly curved to beyond end of sacculus, sacculus ending in stout cuiller, directed outwards, strongly curved inwards around middle with narrow scales or hairs except near its base, sometimes apical part crosses costal margin; clavus with narrow scales in apical part, slender, directed outwards, strongly curving inwards after middle, crossing costal margin or not; transtilla strongly thickened; transtilla lobes large; anellus wider than long, U-shaped, in preparations apical portions often lying flat distally; saccus wide, rounded before elongate extension, longer than in *tenebricosa*; aedeagus similar to that of *tenebricosa*, but with longer taper to apex.

FEMALE GENITALIA. Similar to *D. tenebricosa*. Anterior apophysis between one-quarter and one-fifth length of posterior apophysis; segment VIII width:length 4:1, laterally sinuate, sternite with slight, rounded anterior bulge, posterior margin with wide, deep excavation, rounded at base with incision extending to ostium which lies in a pouch-like fold close to anterior margin; tergite with broad, shallow excavation on anterior margin; antrum not sclerotised, ductus bursae with one and a half spiral turns, corpus bursae elliptical, signum three times wider than long, covered in triangular teeth.

GENITALIA DIAGNOSIS. Male distinctive. Female similar to *D. tenebricosa*, but margin of segment VIII is sinuous, posterior excavation of sternite has rounded base and pouch-like fold is deeper, but differences not reliable owing to variation between preparations. For more details see *D. tenebricosa*.

DISTRIBUTION. South-west Europe: Portugal, Spain, France, Italy; Corsica, Sardinia. Also in North Africa

BIONOMICS. Host-plant unknown. Earliest specimen we have seen was caught 25 March, latest observation 19 November, with a remarkable gap when the moths are presumably aestivating between middle July until September. Obviously, eggs are laid late autumn and adults do not hibernate.

REMARKS. *D. radiosquamella* was described from a single female from Corsica, but the specimen in NHMUK is clearly conspecific with *D. adustatella*. *D. duplicatella* was described from two specimens collected in Tunisia in May. Hannemann (1958a) recognised that *D. delphinias* Meyrick, 1936 was an earlier name for his *D. subtenebricosa*.

A sequenced specimen from Croatia is listed as *D. adustatella* (i.e. *radiosquamella*) in BOLD. We consider this specimen misidentified based on its DNA barcode. Among Croatian *D. tenebricosa* we found cases of barcode sharing with *D. radiosquamella*.

140   *Depressaria emeritella* Stainton, 1849

Depressaria emeritella Stainton, 1849: 167.

DESCRIPTION. Wingspan (19–)21–25(–27) mm. Head whitish buff with orange tinge. Labial palp segment 3 three-fifths length of segment 2; segment 2 pale buff on inner side, outer and ventral sides pale buff, mixed light brown and fuscous, segment 3 mainly light to dark fuscous, apex ochreous-buff. Thorax concolorous with head; tegulae deep brown to blackish.
Forewing brown to chestnut-brown, deep brown in basal one-sixth; scattered light grey scales mainly beneath costa in posterior two-thirds, in cell and fold, in dorsal area, sometimes forming an angled fascia beyond cell, and between veins to termen; an interrupted white line in cell, with a distinct white dot at end of cell; blackish lines often interrupted, beneath costa to one-third, in cell, between veins to costa and termen; fringe grey-brown, lighter grey beyond a fuscous fringe line. Hindwing light grey, darker round apex and termen; fringe light grey with indistinct fringe line. Abdomen light grey to grey-buff.

VARIATION. The angled fascia beyond end of cell is generally faint, but may be obsolete.

SIMILAR SPECIES. The creamy orange head and thorax contrasting with the blackish tegulae are distinctive. *D. leucocephala* has similarly contrasting tegulae, but head and thorax are pale cream.

MALE GENITALIA. Gnathos elliptic; socii tongue-like, extending laterally; valva parallel-sided or even expanding to two-thirds, costal margin slightly incurved towards apex, ventral margin curving from end of sacculus, cuiller moderately stout, initially directed outwards up to curve at one-third, then parallel-sided to rounded apex, reaching nearly to costal margin or shortly crossing it; clavus with base swollen on inner side then tapering to point, with scales mainly on inner side, slightly exceeding anellus; transtilla thickened; anellus wide and short, widest at middle, apical margin concave; saccus wide, rounded and very short; aedeagus short and relatively wide, gradually curved, tapering to apex from two-thirds.

FEMALE GENITALIA. Anterior apophysis one-quarter length of posterior apophysis; segment VIII width:length 3:1, tergite broadly and shallowly excavate anteriorly, sternite shorter than tergite with weakly sclerotised slight median excavation on posterior margin, abundant microtrichia around ostium and posterior margin of sternite; ostium close to posterior margin of segment, antrum funnel-shaped, slightly sclerotised; ductus bursae without sclerotisation, with a single turn close to elliptical corpus bursae; signum rhombiform, twice as wide as long, the anterior and posterior angles slightly extended, covered with triangular teeth with two pairs larger than the rest.

GENITALIA DIAGNOSIS. In males *D. emeritella*, *D. leucocephala* and *D. olerella* are distinct, but females are more similar. *D. olerella* differs by sternite with median excavation between a pair of slight humps; in *D. emeritella* and *D. leucocephala* position of ostium and extent of microtrichia are different.

DISTRIBUTION. Locally distributed in Central and Northern Europe: Germany, Austria, Czechia, Slovakia, Slovenia, Hungary, Poland, Fennoscandia, also in Greece, Romania and Russia. In Asia extending to South Siberia (Novosibirsk) (Lvovsky, 2001).

BIONOMICS. Larvae on *Tanacetum vulgare* L., early stages under a web along the midrib near tip of leaf, later in a tube along midrib built of spun leaflets, spring until June or early July, pupation on the soil, adults from July to May, eggs are laid after hibernation (Hannemann, 1953, 1995, Sonderegger, *in litt.*, Lepiforum, Buchner, personal observation).

REMARKS. An old record from England has not been confirmed (Harper *et al.*, 2002).

141   *Depressaria olerella* Zeller, 1854

Depressaria olerella Zeller, 1854: 337.

DESCRIPTION. Wingspan (18–)19.5–22 mm. Head light grey-brown. Labial palp segment 3 four-fifths length of segment 2; segment 2 light grey-brown with some darker scales ventrally, paler on inner side, segment 3 deep brown, buff at tip, outer side with some grey-brown scales mixed in. Antennae dark brown, obscurely ringed paler. Thorax grey-brown with scales tipped greyish buff, posterior crest whitish grey. Forewing ground colour brown or reddish-brown, more rarely grey-brown, always with scales over much of wing tipped whitish buff; extreme base and basal two-fifths of costa sometimes darker than ground colour; markings consist of a variably developed white or whitish spot at end of cell, a short deep brown to blackish curved streak in cell at one-third, and grey-brown, or sometimes dark brown streaks in cell reaching white spot, in fold, more or less interrupted, below costa at one-third and one-half, sometimes between fold and dorsum at one-quarter, and in posterior quarter between veins to costa and termen; grey-brown spots between veins at termen; an obscure pale angled fascia beyond end of cell; fringe grey, light grey towards tips, with one fringe line. Hindwing light grey-brown, slightly darker posteriorly; fringe light grey-brown, with two fringe lines. Abdomen grey-brown, sometimes greyer or tinged ochreous.

VARIATION. The ground colour varies in intensity, and this affects the markings which may be more obscure in darker coloured specimens. In some specimens the markings in the costal half are well marked, but those in dorsal half and terminal area are almost obsolete. Specimens with reddish ground colour are prevalent over much of the distribution area, but those from southern England are grey-brown.

SIMILAR SPECIES. Characterised by the whitish posterior margin of thorax and forewings with white end of cell spot and weak markings, but see remark under *D. albipunctella*.

MALE GENITALIA. Gnathos more or less round, but usually twisted to one side in preparations; socii tongue-like, widely diverging with rounded sinus between; valva tapering to rounded apex, sacculus extending just beyond middle, cuiller broad-based, tapering to apex, not reaching costal margin of valva, outer margin slightly curving outwards before curving in at apex, inner side rough with small spines; clavus with base moderately broad, tapering to slender apex, nearly straight, directed outwards, ending near base of cuiller, with narrow scales except at outer side of base; anellus about as long as wide, base with deep V-shaped cleft, apex truncate, between two long ovate weakly sclerotised areas; saccus wide, curved sides meeting at a near right-angle; aedeagus after broad base, slender, curved, ending in more or less fine point.

FEMALE GENITALIA. Anterior apophysis one-quarter to one-fifth length of posterior apophysis; segment VIII width:length 3:1, tergite with convex anterior margin, sternite with median excavation between a pair of slight humps; ostium close to posterior margin of sternite, antrum broadly funnel-shaped, sclerotised in anterior half; ductus bursae without sclerotisation, with a single spiral close to elliptical corpus bursae; signum cruciform, wider than long, transverse part with large triangular teeth, longitudinal axis with minute teeth, tapering to a fine point at each end.

GENITALIA DIAGNOSIS. See under *D. emeritella*.

DISTRIBUTION. Middle latitudes of Europe from France and England to southern Fennoscandia and Russia, south to Italy and Slovenia; Greece and Romania. Also in Western Asia (Georgia). Azerbaijan, southern Siberia, Amur Province (Lvovsky, 2004).

BIONOMICS. Larvae on *Achillea millefolium* L. (Hannemann, 1953, Sonderegger, *in litt.*), also on *Tanacetum vulgare* L. (Lvovsky, 1981, Lepiforum e.V., 2006–2023) and *Tanacetum corymbosum* (L.) Sch. Bip. (Buchner, personal observation, Lepiforum e.V., 2006–2023) until June, rarely until July. On *Achillea millefolium*, leaves are usually attached to the stem, on *Tanacetum* leaflets are spun into a tube along midrib, pupation between the spun leaves or on the soil, adults from July until May; eggs are laid after hibernation.

REMARKS. In Lepiforum e.V. (2006–2023) a larval spinning on *Senecio* sp. is shown, but it was not reared successfully.

142   *Depressaria leucocephala* Snellen, 1884

Depressaria leucocephala Snellen, 1884: 160.
   Depressaria thomanniella Rebel, 1917a: 23.

DESCRIPTION. Wingspan (17–)19–23 mm. Head creamy white to whitish buff above, face fuscous. Labial palp segment 3 three-fifths length of segment 2; segment 2 whitish buff, outer and ventral sides with variable amount of light brown and fuscous scales, segment 3 sometimes with a few fuscous or light brown scales. Antenna dark fuscous, underside whitish buff. Thorax creamy white to whitish buff, concolorous with head, tegulae deep brown. Forewing light brown, chestnut-brown or deep brown, with light admixture of pale grey scales beneath costa, forming a zigzag line from costa to cell at two-fifths, a weak angled fascia beyond end of cell, sometimes some short dashes between veins in terminal area; ill-defined whitish dashes in cell at two-fifths, in fold and a white dot at end of cell; a blackish spot at base of dorsum, blackish dashes between veins to costa and termen, in cell, in fold and between fold and dorsum; a series of blackish dots between veins around termen; fringe grey with a dark grey fringe line. Hindwing light grey, darker towards termen; fringe light grey with a fringe line. Abdomen at least in proximal part concolorous with head and thorax, distal part tending to light grey-brown.

VARIATION. Scandinavian specimens are smaller and darker than those from the Alps. In the darker specimens, the blackish markings are hardly evident. One specimen from Italy has almost uniform light fuscous forewing with obscure blackish markings.

SIMILAR SPECIES. *D. emeritella* has head and thorax tinged orange, segment 3 of palp dark; underside of antenna not pale.

MALE GENITALIA. Gnathos ovate; socii tongue-like, extending laterally; valva nearly parallel-sided, costal margin weakly convex, ventral margin gently curved throughout, apex broadly rounded, cuiller directed outwards from wide base, tapering unevenly to slightly incurved apex, inner side rough with small spines; clavus absent; anellus wider than long, with a pair of small curved pointed wings at middle of side, and a pair of rounded lobes postero-laterally; saccus short, round; aedeagus rather stout, tapering in distal quarter to relatively broad point.

FEMALE GENITALIA. Anterior apophysis one-half length of posterior apophysis; segment VIII width:length 5:1, posterior margin of sternite with low humps either side of middle, anterior margin concave, microtrichia absent or present in a small area distal to ostium; ostium circular, in middle of segment VIII; antrum funnel-shaped; ductus bursae elongate with two spiral turns; corpus bursae elliptical; signum densely covered in triangular teeth, rhombiform, wider than long with broad, small-toothed extensions on longitudinal axis.

GENITALIA DIAGNOSIS. See under *D. emeritella*.

DISTRIBUTION. Alps: Italy, Switzerland and Austria; North Europe: Norway, Sweden, Finland and Russia, extending to Far East Asia.

BIONOMICS. Larvae on *Artemisia vulgaris* L. (Hannemann, 1953, Sonderegger, *in litt.*, Lepiforum e.V., 2006–2023) until June or mid-July in a large spinning at the shoot tip, pupation in the dry, lower part of the plant (Hannemann, 1995), adults from June to August (checked specimens), until September according to Hannemann (1995).

REMARKS. This species (as *thomanniella*) was placed in a small group with the Asian *D. ruticola* Christoph, 1873 by Hannemann (1953) because of the presence of cuiller and absence of clavus, but these two species have little else in common, even their host-plants are unrelated. Externally *leucocephala* is close to *emeritella* and in male genitalia the cuiller is close to that of *olerella* and the socii resemble those of both these species. Both barcode and host-plant also place it in this subgroup. This array of characters outweighs the absence of clavus. Hannemann (1958a) synonymised *thomanniella* Rebel, 1917 with *D. leucocephala*.

Group 2 *depressana* group (143)

D. depressana has no close relatives in the European fauna. Male genitalia with short-stalked, rounded gnathos, small socii, tegumen narrowed proximally, long saccus, without clavus or cuiller, aedeagus with cornuti; externally the forewing lacks dark streaks. Larval host-plants Apiaceae.

143   *Depressaria depressana* (Fabricius, 1775)

Pyralis depressana Fabricius, 1775: 655.
    Tinea depressella Fabricius, 1798: 485. Unjustified emendation of *Pyralis depressana* Fabricius, 1775.
Depressaria bluntii Curtis, 1828: fol. 221.
Haemylis colarella Zetterstedt, 1839: 999.
Depressaria depressella amasiella Staudinger, 1880: 300.
    Depressaria depressella prangosella Walsingham, 1903: 268.
    Depressaria rhodochlora Meyrick, 1923: 278.

DESCRIPTION. Wingspan (11–)14–20 mm. Head ochreous to pale ochreous. Labial palp ochreous to pale ochreous, segment 3 two-thirds length of segment 2; segment 2 usually with a few light fuscous scales on outer side, segment 3 with a grey-fuscous ring beyond middle. Antenna grey-brown, ringed ochreous grey. Thorax and tegula ochreous to pale ochreous, anterior third of thorax sometimes brown. Forewing dull reddish brown, heavily overlaid dark fuscous except in costal third, a darker patch towards base from centre to dorsum and a small dark brown spot at extreme base of costa, sometimes with short creamy dashes in cell and fold and beyond middle where they can form an indistinct angulate fascia; fringe pale grey with dark grey fringe line, not always distinct. Hindwing light grey, darker posteriorly; fringe light grey. Abdomen grey, tinged ochreous.

VARIATION. The ochreous coloration of head and thorax may be paler or darker and sometimes posterior part of head and anterior part of thorax are reddish brown. Small cream coloured dashes forming a V-shaped fascia and other marks can be completely absent.

SIMILAR SPECIES. The species is characterised by usually small size, ochreous head and thorax and two-tone forewing without blackish markings.

MALE GENITALIA. Gnathos short-stalked, almost circular to wider than long; socii digitate, only reaching middle of gnathos; tegumen narrow, especially proximally; valva nearly parallel-sided, costal side slightly convex, ending in broad rounded hump, end of valva concave, ventral side with submarginal reinforcement before apex, sacculus reaching one-half ending in a process that is hooked towards costa; saccus elongate, rounded at apex; aedeagus slender, slightly tapering, a group of about 8 cornuti ranging from long to short.

FEMALE GENITALIA. Anterior apophysis two-fifths length of posterior apophysis; segment VIII width:length 5:4, posterior margin of sternite with shallow excavation; ostium in middle of sternite, semicircular, antrum parallel-sided, slightly sclerotised; ductus bursae with a spiral turn before elliptical corpus bursae; signum small, six times as wide as long, covered with small triangular teeth.

GENITALIA DIAGNOSIS. Both sexes unique in European fauna, closest species *D. falkovitshi* Lvovsky, 1990 (Central Asia, aedeagus with short terminal spine, antrum with deep V-shaped excavation).

DISTRIBUTION. Widespread in Europe from Portugal and England (extinct since 1890 (Harper *et al.*, 2002)) to Fennoscandia and south to Greece; Cyprus. In Asia extending at least to Altai Mountains. Far East Russia and North Africa (Lvovsky, 2001). Also recorded from North America, where it may be introduced.

BIONOMICS. Larvae on flowers of *Daucus carota* L. (Hannemann, 1953); *Pimpinella saxifraga* L. (Harper *et al.*, 2002); *Coriandrum* L., *Anethum* L. (Lvovsky, 2001); *Seseli libanotis* (L.) W.D.J. Koch, *Sium latifolium* L., *Laserpitium halleri* Crantz, *Peucedanum palustre* (L.) Moench (Sonderegger, *in litt.*); *Silaum silaus* (L.) Schinz & Thell., *Laserpitium gallicum* L. (Lepiforum e.V., 2006–2023); *Pastinaca sativa* L. (De Prins & Steeman, 2003–2023); *Carum carvi* L. (M. Mutanen, pers. comm.), from May to July or August, occasionally to September in spun umbels, pupation in the spinning or on the soil. Adults emerge from July to September, flying until June (July?) of the following year. Lhomme (1945) lists other host-plants: *Angelica sylvestris* L., *Crithmum maritimum* L., *Falcaria vulgaris* L., *Ferula communis* L., *Foeniculum vulgare* Miller, *Heracleum sphondylium* L., *Orlaya grandiflora* (L.) Hoffm., *Peucedanum oreoselinum* (L.) Moench, *Seseli tortuosum* L. Confirmation of these is needed. REMARKS. Palm (1989) states that the species has two generations, which fly from August to May and in June respectively, but Ole Karsholt (pers. comm.) does not accept that this idea matches the observed phenology in Denmark. We agree that two generations are very unlikely in north Europe, but in the south at least a partial second generation can occur, according to an observation in France by Ruben Meert: reared moths emerging on 8 July mated immediately (Lepiforum e.V., 2006–2023). We have seen both fresh and worn specimens collected in July.

Group 3 *artemisiae* group (144–159)
   This grouping, proposed by Hannemann (1953), is characterised by male genitalia with clavus absent or weakly developed, cuiller usually absent; aedeagus with well-developed cornuti. We recognise three subgroups: *artemisiae* subgroup associated with Asteraceae, *zelleri* and *chaerophylli* subgroups associated with Apiaceae.

D. artemisiae subgroup (144–149): With ovate gnathos on short stalk, socii mostly stalked, valva with upturned point at end of costal margin, transtilla very large with microsculpture. Two species have small clavus. Larvae on Asteraceae.

D. zelleri subgroup (150–151): With gnathos not elongate, not on long stalk, sacculus less than half width of valva, clavus present. Larvae on Apiaceae.

D. chaerophylli subgroup (152–159): With elongate gnathos on long stalk, socii variable, sacculus broad, occupying half or more of width of valva. Only *D. ululana* and *D. peregrinella* have a clavus. Placement of *D. peregrinella* is tentative – sacculus is narrower and the transtilla has microsculpture, but it does not otherwise agree with *artemisiae* subgroup. Larvae on Apiaceae.

144   *Depressaria absynthiella* Herrich-Schäffer, 1865

Depressaria absynthiella Herrich-Schäffer, 1865: 110.
   Depressaria absinthivora Frey, 1880: 355.

DESCRIPTION. Wingspan (14–)16–19 mm. Head light brown, reddish brown to fuscous, scales tipped buff, face light fuscous. Labial palp segment 3 three-fifths length of segment 2; segment 2 whitish buff, outer and ventral sides mixed with light and dark fuscous scales, segment 3 whitish buff with grey rings at base and at three-quarters. Antenna light fuscous, narrowly ringed dark fuscous, underside whitish buff. Thorax whitish buff, browner anteriorly, sometimes fuscous to dark fuscous, only pale posteriorly. Forewing from orange-brown to light or dark fuscous with scattered whitish grey scales mainly along veins, in cell and in fold, sometimes forming a weak angled fascia beyond end of cell; a blackish subdorsal spot at base; interrupted black lines below costa, in cell and fold and between veins towards termen; a stronger elongate black spot in cell at one-third; a series of rather small dark grey dots between veins at termen; fringe grey-brown, with fuscous fringe line. Hindwing often narrowed towards apex which is relatively pointed compared with most allied species, light grey, darker towards costa and apex; fringe whitish grey, with weak fringe line. Abdomen light grey-buff.

VARIATION. Colour of head, thorax and forewings is variable, the fascia beyond end of cell is not often evident. Occasionally there is a broad pale grey band along costa at least to middle. The amount of light grey scaling is much reduced in darker specimens.

SIMILAR SPECIES. *D. artemisiae* is similar, with a similar range of variation. It has broader, less pointed hindwing and forewing costa usually has a blackish spot at middle. *D. tenerifae* (*q.v.*).

MALE GENITALIA. Gnathos narrowly ovate, equalling or exceeding socii; socii stalked, paddle-shaped with oblique end; valva with rounded apex and a short point at end of costa, sacculus simple, sometimes with a small hump on dorsal margin; transtilla expanded medially, distally scobinate; anellus broadly pyriform, apex truncate; saccus short, wide, rounded; aedeagus slightly curved, nearly parallel-sided after bulbous base, single cornutus as long as parallel-sided part.

FEMALE GENITALIA. Anterior apophysis about three-fifths length of posterior apophysis; segment VIII width:length 3:1; ostium on anterior margin of segment, circular; antrum short, cylindrical, sclerotised laterally; ductus bursae with posterior one-third slightly expanded, weakly sclerotised, remaining two-thirds not sclerotised, more slender, gradually expanding to narrowly elliptic corpus bursae; signum small, rhombiform, wider than long, covered with small triangular teeth.

GENITALIA DIAGNOSIS. In male all species with upturned point at end of costal margin of valva appear similar. Species without clavus, namely *D. absynthiella*, *D. tenerifae*, *D. artemisiae*, *D. fuscovirgatella* and *D. atrostrigella* show differences predominantly in aedeagus: *D. absynthiella* and *D. tenerifae* with a very long, single cornutus, *D. artemisiae* with a small group of short cornuti, *D. fuscovirgatella* with a compact group of about 10 small cornuti and one separated longer cornutus. In *D. atrostrigella* Clarke, 1941, external appearance and aedeagus are similar to that of *D. fuscovirgatella*, but transtilla is clearly different. *D. zelleri* and *D. pyrenaella* differ by presence of clavus. Female genitalia most similar to *D. tenerifae* (*q.v.*).

DISTRIBUTION. Spain, France, Italy, Germany, Switzerland, Austria, Czechia, Bulgaria and Russia. In Asia extending at least to Altai Mountains; Far East Russia (Primorskii kraj) (Lvovsky, 2001).

BIONOMICS. Larvae on *Artemisia absinthium* L. (Hannemann, 1953); *A. frigida* Willd. (Lvovsky, 2001); *Artemisia chamaemelifolia* Vill. and *A. vallesiaca* All. (Sonderegger, *in litt.*), between spun leaflets (or on *Artemisia vallesiaca* usually two side branches attached to main stem, Sonderegger, *in litt.*) from April to June or rarely July, pupation in the spinning or on the soil, emerging predominantly in June (rarely in May or July), adults June to October, usually not hibernating.

REMARKS. *Depressaria tenerifae* Walsingham, 1908, described from the Canary Islands has male genitalia very similar to *D. absynthiella* and has in consequence been treated as a subspecies of *D. absynthiella*, but significant differences in female genitalia and barcode justify retention as a separate species.

145   *Depressaria tenerifae* Walsingham, 1908 stat. rev.

Depressaria tenerifae Walsingham, 1908: 958.

DESCRIPTION. Wingspan (15–)16–19.5 mm. Head light fuscous, scales tipped buff, face light fuscous. Labial palp segment 3 three-fifths length of segment 2; segment 2 whitish buff, outer and ventral sides mixed with light and dark fuscous scales, segment 3 whitish buff with grey ring at three-quarters. Antenna light fuscous, narrowly ringed dark fuscous. Thorax light fuscous, scales tipped buff. Forewing light fuscous with slightly paler markings mainly along veins, in cell and in fold, sometimes forming a weak angled fascia beyond end of cell; a large blackish subdorsal spot at base; a slightly elongate black spot in cell at one-third and another at end of cell; black lines largely absent, sometimes one beneath costa at one-fifth, a small one in fold at one-third and a few weak dashes between veins to termen; fringe grey-brown, with fuscous fringe line. Hindwing light grey, darker towards costa and apex; fringe whitish grey, with weak fringe line. Abdomen light grey-buff.

VARIATION. The black lines on forewings vary from absent to weakly developed.

SIMILAR SPECIES. *D. absynthiella* is more variable in coloration, forewing markings more contrasted with well-developed black lines and scattered whitish or light grey scales more evident. Labial palp segment 3 with two dark grey rings.

MALE GENITALIA. Similar to *D. absynthiella* but gnathos more broadly ovate, not reaching end of socii.

FEMALE GENITALIA. Anterior apophysis about three-fifths length of posterior apophysis; segment VIII width:length 5:2, sternite with wide V-shaped anterior extension accommodating circular ostium, posterior margin incised in middle down to ostium, tergite with slightly convex posterior margin; antrum short, sclerotised; ductus bursae with three sections, posteriorly expanded, smoothly cylindrical, sclerotised, middle part not sclerotised, initially narrow then expanded following a bend, abruptly contracted to anterior section gradually expanding into narrowly elliptic corpus bursae; signum wider than long with many small and a few larger triangular teeth.

GENITALIA DIAGNOSIS. *D. absynthiella* and *D. tenerifae* are nowhere found together. Male genitalia scarcely distinguishable. In female ductus bursae differs from *D. absinthiella* in having an expanded section in middle.

DISTRIBUTION. Canary Islands: Tenerife, Gran Canaria and El Gomera.

BIONOMICS. Reared from *Artemisia canariensis* (Besser) Less. (Klimesch, larvae collected beginning of March). Flight period not quite clear: among checked specimens two light-trapped 3 and 5 March, the second rather worn, but this is no proof for hibernation. First clearly fresh light-trapped specimen 19 April, latest in the year 11 November, in rather good condition.

REMARKS. *D. tenerifae* was synonymised with *absynthiella* by Hannemann (1958a). Sometimes treated as a subspecies of *D. absynthiella*, due to only small differences in male genitalia, but differences in barcode and female genitalia support treatment as a separate species.

146   *Depressaria artemisiae* Nickerl, 1864

Depressaria artemisiae Nickerl, 1864: 4.

DESCRIPTION. Wingspan (13–)15–19 mm. Head pale buff to mid-brown or orange-brown, with scales tipped buff, face light fuscous. Labial palp segment 3 two-thirds length of segment 2; segment 2 pale buff, outer side with variable amount of fuscous to dark grey scales, particularly proximally, ventral side with dark scales intermixed, segment 3 with grey or blackish ring at base and another beyond middle almost to tip which is light ochreous-buff. Antenna with scape fuscous, flagellum pale buff with narrow fuscous rings. Thorax pale buff, orange-brown or light brown, sometimes fuscous, usually slightly darker anteriorly. Forewing orange-brown to fuscous; light grey-buff in a costal band reaching one-half, dashes in cell and in fold and in an arc forming a weak fascia beyond end of cell; a blackish subdorsal spot at base, a dot at base of costa, elongated blackish spots on costa at one half, beneath this at margin of cell, in cell at one-third and at end of cell, often another between these and in fold at one-third; blackish dots around apex and termen between veins, fine black dashes between veins to termen and sometimes additional dashes in fold; fringe grey with a fuscous fringe line. Hindwing light grey, darker towards apex; fringe light grey with one or two fringe lines. Abdomen pale buff.

VARIATION. A very variable species in ground colour and extent of markings, the light grey-buff markings including the costal band can be more or less obsolete. The curved fascia beyond middle of cell is usually indistinct and often absent. The black dot in middle of cell is sometimes absent. In some specimens pale buff lines follow the veins, sometimes heavily speckled with black.

SIMILAR SPECIES. *A. absynthiella* can be similar in some of its forms, but a blackish spot on costa at one-half is often present in *artemisiae*, only exceptionally in *absynthiella*.

MALE GENITALIA. Gnathos ovate; socii short-stalked, narrowly elliptic; valva with rounded apex, costa concave ending in broad hook, sacculus with angled hump near base; transtilla very large, distally scobinate; anellus small; saccus short, wide, rounded; aedeagus slightly curved, nearly parallel-sided after bulbous base tapering from two-thirds to very slender apex, about six small cornuti in a tight group.

FEMALE GENITALIA. Anterior apophysis one half to five-eighths length of posterior apophysis; segment VIII width:length 5:2, sternite with wide convex anterior extension accommodating circular ostium; antrum short, sclerotised, wider than long, posterior margin longer than anterior; ductus bursae short, slightly longer than anterior apophysis, wide, scobinate except at posterior end, densely so from middle to three-quarters towards broadly pyriform corpus bursae; signum isodiametric with a mixture of small and large triangular teeth.

GENITALIA DIAGNOSIS. Male, see remarks under *D. absynthiella*. In female the short and wide ductus bursae is distinct.

DISTRIBUTION. Alpine region from France eastwards extending to Hungary and Poland; Fennoscandia, Russia. Widespread in Asia, extending to far east. Also recorded from North America (Hodges, 1974).

BIONOMICS. Larvae usually on *Artemisia campestris* L. (Hannemann, 1953) between spun leaflets, one record on *Artemisia borealis* Pall. (Swiss Alps at 2500 m, Jürg Schmid, pers. comm.), Nearctic host-plant is *Artemisia dracunculus* L. (Robinson *et al.*, 2010). Larvae from early spring to May or June, rarely until July, adults from June to October, not hibernating.

147   *Depressaria fuscovirgatella* Hannemann, 1967

Depressaria fuscovirgatella Hannemann, 1967: 166.
    Depressaria pagmanella Amsel, 1972: 140.

DESCRIPTION. Wingspan (18–)19–25 mm. Head grey-buff. Labial palp grey-buff, segment 2 with blackish scales on ventral and outer sides, segment 3 buff. Antenna pale buff narrowly ringed fuscous throughout. Thorax and tegula grey-buff. Forewing elongate, grey-buff, flushed light brown except in costal third of wing; a blackish subdorsal spot, a series of black dots around apex and termen; a thin black line in fold from base to one-third, a black line in cell and a series of black lines between veins to outer end of costa and to termen; fringe grey. Hindwing rather narrow, apex acute; pale grey, darker towards apex; fringe pale grey, with one line. Abdomen grey tinged yellow-brown.

VARIATION. Only a few specimens examined, but most showing little variation. One specimen with black lines reduced to one near base in fold, one in cell and short lines between veins to termen and with terminal area greyer than rest of wing.

SIMILAR SPECIES. Coloration and markings similar to *D. zelleri*, but this lacks black lines in basal quarter of wing, has distinct angled fascia, relatively broader forewing and hindwing and quite different male and female genitalia. *D. pyrenaella* also has broader fore and hindwings. *D. atrostrigella* (*q.v.*). Further species with distinct longitudinal blackish streaks, especially *D. sarahae*, can appear very similar at first glance.

MALE GENITALIA. Gnathos ovate; socii digitate, slightly expanded towards apex; valva with rounded apex, costa concave ending in broad hook, sacculus simple; saccus short, wide, rounded; aedeagus curved, gradually tapering to very acute apex, about three long straight cornuti and a group of small cornuti.

FEMALE GENITALIA. Anterior apophysis half to five-eighths length of posterior apophysis; segment VIII width:length 2:1, sternite with wide semicircular anterior bulge accommodating large circular ostium; antrum short, weakly sclerotised; ductus bursae long and rather wide, especially in posterior half which contains a sclerotised section on one side; corpus bursae shortly ovoid; signum small to medium-sized, more or less rhombiform with small triangular teeth, larger around margin.

GENITALIA DIAGNOSIS. Male, see remarks under *D. absynthiella*. In female general character similar to *D. silesiaca* and *D. zelleri* in presence of pair of distinct triangular sclerites in ostium, posterior third of ductus bursae with a sclerotised section, corpus bursae rather small. Ostium of *D. fuscovirgatella* much larger than that of *D. silesiaca*. *D. zelleri* differs from both by unsclerotised area posterior of ostium extending to posterior margin of sternite VIII, anterior margin of sternite VIII not bulged and sclerotised part of ductus bursae narrow. *D. atrostrigella* has scelerotised colliculum.

DISTRIBUTION. Russia (Orenburg oblast); in Asia to Mongolia.

BIONOMICS. Host-plant unknown. Adults have been light-trapped from June to September, the autumn specimens also in good condition.

148   *Depressaria atrostrigella* Clarke, 1941

Depressaria atrostrigella Clarke, 1941: 168.

DESCRIPTION. Wingspan 22–25 mm. Head fuscous. Labial palp grey-buff, segment 2 with dark fuscous scales on ventral and outer sides, segment 3 buff. Antenna fuscous.

Thorax and tegula grey-buff, fuscous anteriorly. Forewing elongate, grey-buff, browner towards base and apex; a large blackish subdorsal spot, a series of black dots around apex and termen; a black line in fold from base to two-fifths, a thick black line in cell, distinctly wider proximally and a series of black lines between veins to outer end of costa and to termen; fringe grey. Hindwing whitish buff, greyer towards apex and termen; fringe pale grey, with one line. Abdomen fuscous.

VARIATION. In some specimens many of the black lines are not or weakly marked, but line from base to fold and thick line in cell are always distinct.

SIMILAR SPECIES. Similar to *D. fuscovirgatella*, but the black line in cell is not thickened in that species.

MALE GENITALIA. Similar to *D. fuscovirgatella*. Transtilla thickened, with a pair of posteriorly-directed bulges; anellus longer; aedeagus with apex less slender, with a group of about three straight cornuti, shorter than in *fuscovirgatella*, and a shorter group of small cornuti in middle with another straight cornutus.

FEMALE GENITALIA. Apophyses as in *D. fuscovirgatella*; segment VIII width:length 5:2, sternite with wide anterior extension, flat for half width of sternite, ostium reaching anterior margin of sternite; antrum short, moderately sclerotised; a strongly sclerotised colliculum at posterior end of ductus bursae; signum larger than in *fuscovirgatella*.

GENITALIA DIAGNOSIS. Male, for distinction from *D. fuscovirgatella* see above. In female the colliculum is an unusual structure in *Depressaria*; a similar structure in *D. absinthiae* is heavily microsculptured.

DISTRIBUTION. Russia (Bashkiria). Asia: Altai, Irkutsk, Yakutia, Tajikistan, Mongolia (Lvovsky, 2001). North America: Manitoba, Colorado (Hodges, 1974).

BIONOMICS. Host-plant unknown. Adults have been light-trapped from July to October, the autumn specimens in poor condition. It has been recorded at 3400 m altitude in Tajikistan.

149   *Depressaria silesiaca* Heinemann, 1870

Depressaria silesiaca Heinemann, 1870: 184.
  Depressaria millefoliella Chrétien, 1908a: 186.
  Schistodepressaria freyi Hering, 1924: 80.

DESCRIPTION. Wingspan (13)15–18.5 mm. Head greyish white mixed with pinkish scales. Labial palp segment 3 four-fifths length of segment 2; outer side of segment 2 dull grey-brown with slight pinkish tinge, segment 3 dark brown, with some pinkish buff scales in middle, giving indistinct rings at base and beyond middle, pink at apex; inner side of segment 2 pinkish buff, segment 3 similar to outer side, but paler in middle. Antenna dark brown, narrowly ringed buff. Thorax greyish white mixed with pinkish scales. Forewing light greyish brown, suffused dull orange-brown with a tinge of pink when fresh, particularly in basal one-fifth, away from costa and in centre of wing; small area of dark brown scales at dorsal side of base; poorly defined short blackish brown streaks in fold, in cell, sometimes one between cell streak and costa, and scattered scales in posterior third of wing extending along dorsum, sometimes organised into streaks between veins near termen; a dark grey-brown spot at end of cell and weak spots between end of veins on termen or a poorly defined line around termen; whitish or whitish grey scales forming a broad stripe along costa to one quarter, sometimes much further, but becoming mixed with pinkish brown scales, whitish scales also present beyond middle of fold, in cell before dark spot, in terminal area and sometimes forming an angled fascia beyond end of cell spot; fringe grey-brown, with pink tinge on costa and dorsum. Hindwing very light grey, slightly darker posteriorly; fringe light-grey with grey fringe line, terminally greyish buff. Abdomen dull grey-brown.

VARIATION. The balance between pinkish and whitish scales on the forewing is variable, particularly along the costa, and the angled fascia may be obsolete. Blackish brown streaks in the apical area are evident in some specimens, but not in others, where there are only scattered scales.

SIMILAR SPECIES. The species is characterised by small size, dull pinkish brown colouring, whitish costa at least towards base, dark spot at end of cell and usually weak development of dark streaks between veins.

MALE GENITALIA. Gnathos lanceolate; socii extending laterally, clavate; valva with long process, recurved at apex, on costal side at one-half, sacculus with C-shaped sclerotised process near base and a long-haired cuiller reaching half-way across valva; transtilla not enlarged, not scobinate; saccus short, wide, rounded; aedeagus slightly curved, tapering from bulbous base to acute apex, about 4 long straight cornuti and a compact group of small cornuti.

FEMALE GENITALIA. Anterior apophysis one half length of posterior apophysis; segment VIII width:length 3:1, sternite with convex anterior extension accommodating transversely widened ostium; antrum funnel-shaped, sclerotised; ductus bursae with posterior two-fifths swollen, scobinate on one side and containing a sclerite; corpus bursae ovoid; signum twice as wide as long, bearing triangular teeth.

GENITALIA DIAGNOSIS. In male long process on costal margin of valva is very characteristic. For female see under *D. fuscovirgatella*.

DISTRIBUTION. Alps: France, Italy, Switzerland, Austria; Czechia, Poland, Scotland and Fennoscandia. Karelia, Leningrad Province, northern Caucasus (Teberdinskii Nature Reserve), Kamchatka (Lvovsky, 2004).

BIONOMICS. Host-plants *Achillea millefolium* L., *Tanacetum vulgare* L. (Harper *et al.*, 2002); *Achillea setacea* Waldst. & Kit., rarely on *Artemisia vulgaris* L. (Sonderegger, *in litt.*; Deutsch, 2016), from May to July, rarely to August at higher altitudes, adults from July onwards, probably to May of the following year.

According to Hannemann (1995) adults fly from June to September, but according to Harper *et al.* (2002) adults hibernate; Sonderegger (*in litt.*) was unable to attract moths to light, even in well known localities. This may be the reason that adults are not found in spring, although they hibernate. M. Mutanen (pers. comm.) has taken a few at light.

REMARKS. The synonymy of *D. millefoliella* Chrétien and *D. silesiaca* was established by Rymarczyk *et al.* (2015b).

150   *Depressaria zelleri* Staudinger, 1879

Depressaria zelleri Staudinger, 1879: 300.

DESCRIPTION. Wingspan 21–24 mm. Head fuscous, scales tipped grey-buff, face whitish. Labial palp whitish buff, segment 3 two-thirds length of segment 2, spreading ventral scales of segment 2 long, some scales mid-brown to dark fuscous. Antenna grey-buff narrowly ringed fuscous. Thorax light grey-buff, tinged fuscous anteriorly. Forewing light grey-buff, especially near costa and in dorsal area either side of fold, elsewhere lightly tinged fuscous, particularly in dorsal half of wing and in subterminal area, sometimes whole wing tinged fuscous or various shades of brown; a black subdorsal spot at base, a small dot at base of costa, a long black line from cell at one-third extending to termen, often broken beyond cell by angled fascia; additional black lines in fold, and between veins to apex and termen; a series of grey-brown dots between veins around termen; fringe grey-buff. Hindwing light grey-brown to grey, darker towards costa and apex; fringe pale grey with a darker fringe line. Abdomen grey-buff.

VARIATION. There is variation in ground colour from palest almost silvery buff through to quite dark brown, also in the continuity of the black lines, which may be interrupted by a pale angled fascia.

SIMILAR SPECIES. *D. pyrenaella* (*q.v.*), *D. fuscovirgatella*. The ventral scales of segment 2 of labial palp are longer than in related species.

MALE GENITALIA. Gnathos short-stalked, ovate to elliptic; socii digitate, reaching middle of gnathos; valva with low bulge on costa, slightly concave on ventral margin at three-fifths, apex rounded, a slightly hooked process at end of costa before apex, sacculus with small process at end, directed towards costa; clavus hairy, about as long as half-width of valva near base; transtilla expanded medially, distally finely scobinate; anellus broader than long, apex shallowly emarginate; saccus short, wide; aedeagus curved before middle, apex slender, a group of cornuti of variable length, the longest about one-third length of aedeagus.

FEMALE GENITALIA. Anterior apophysis about one-half length of posterior apophysis; segment VIII width:length 5:1, sternite sclerotised with convex anterior margin and concave posterior margin, tergite with anterior margin concave; ostium adjacent to anterior margin of sternite, semicircular, posterior margin extending to posterior margin of sternite; antrum short, weakly sclerotised; ductus bursae long with a curved section containing a sclerite at one-third length from ostium; corpus bursae ovoid; signum deltoid, covered with triangular teeth.

GENITALIA DIAGNOSIS. Male most similar to *D. pyrenaella* (*q.v.*). See also remarks under *D. absynthiella*. For female see under *D. fuscovirgatella*.

DISTRIBUTION. France (Gard) (Grange *et al.*, 2011), Italy, North Macedonia, Greece. Outside Europe in Caucasus.

BIONOMICS. The species was reared from *Trinia glauca* (L.) Dumort. from Italy by Burmann. Larvae always solitary, feeding on flowers or leaves, which are connected via a silken tube to the ground where this tube continues about 10–15 cm within detritus, here the larva rests or hides if disturbed. Klimesch (1953) gives a detailed description of larva and its behaviour, found in June in Trento, Italy on *Trinia glauca*, but misidentified as *Depressaria cervicella* (cited uncorrected in Hannemann, 1995, correction of determination as *D. zelleri* in Lvovsky, 2001: 536). The few adults we could check were light-trapped in July and August.

REMARKS. Female genitalia have not been described or illustrated previously.

151   *Depressaria pyrenaella* Šumpich, 2013

Depressaria pyrenaella Šumpich, 2013: 114

DESCRIPTION. Wingspan 29–30 mm. Head light brown, face creamy. Labial palp creamy, mixed brown on segments 2 and 3. Antenna brown, weakly ringed fuscous. Thorax mid-brown with deep brown median stripe. Forewing medium brown to dark fuscous; a blackish subdorsal spot; long black lines between veins running into terminal dots around apex and most of termen, similar lines in cell, two or three lines crossing lines in cell at a narrow angle, the first following fold to a variable distance; fringe concolorous, with or without a weak fringe line. Hindwing light grey, grey-brown towards apex, with a fine darker grey line around margin; fringe light grey-brown, with one more or less faint fringe line.

VARIATION. The length and strength of the black lines is variable, particularly the first line crossing cell at narrow angle which may continue far along fold,

SIMILAR SPECIES. The almost uninterrupted black lines running into the terminal dots and the black lines crossing the cell at a narrow angle are characteristic, but *D. cervicella* and *D. fuscovirgatella* share these features. The long line in *D. zelleri* normally begins at about one-third from base.

MALE GENITALIA. Similar to *D. zelleri*, gnathos shorter, ovate, valva with a triangular point near middle of costa and slightly outward-curving apical process; aedeagus only slightly curved in middle, with a small group of cornuti pointing towards apex and beyond these a group of smaller cornuti directed across aedeagus.

FEMALE GENITALIA. Unknown.

GENITALIA DIAGNOSIS. The triangular point on costa of valva distinguishes this species from *D. zelleri*.

DISTRIBUTION. Pyrenees of France and Spain.

BIONOMICS. Host-plant unknown but likely to be Apiaceae. Adults have been taken in August.

152   *Depressaria marcella* Rebel, 1901

Depressaria marcella Rebel, 1901: 174.
  Depressaria cuprinella Walsingham, 1907: 214.
  Platyedra cruenta Meyrick, 1920: 298.
  Depressaria chneouriella Lucas, 1940: 228.

DESCRIPTION. Wingspan 13.5–17.5 mm. Head greyish buff, face pale buff. Labial palp segment 3 three-fifths length of segment 2; segment 2 whitish buff, dark grey-brown ventrally and at base of outer side, segment 3 dark grey-brown with a few buff scales, tip buff. Antenna with scape deep brown, flagellum light to dark grey-brown, weakly ringed darker. Thorax narrowly deep brown anteriorly, otherwise pale buff. Forewing light grey-buff overlaid with darker grey-brown, particularly along costa; small deep brown patch at base of costa, larger subdorsal patch, small elongate patch in base of cell, similar but smaller elongate spots in cell at mid-wing, at end of cell and towards end of fold, a cloud between cell and costa at two-fifths, a broad curved band at three-quarters and a terminal band; light brownish buff scales beneath costa at one-fifth, subdorsally at one-quarter, in cell, a curved fascia preceding dark band at three-quarters and a few dots beyond the dark band; fine black streaks along veins in terminal quarter of wing; fringe dull grey-brown, without fringe line. Hindwing light greyish-brown, slightly darker posteriorly; fringe light grey-brown with an indistinct fringe line. Abdomen light grey-brown.

VARIATION. The depth of the ground colour varies as does the extent of light buff scales on the forewing, which can give the impression of light and dark bands across the wing, but are sometimes more or less obsolete, especially in darker specimens.

SIMILAR SPECIES. *D. marcella* is characterised by small size, light buff thorax and face and dull grey-brown forewings with hint of transverse banding, markings are mostly diffuse.

MALE GENITALIA. Gnathos on long stalk, elliptic; end of tegumen broad, shallowly lobed, socii small, circular; valva with costal margin straight, ventral margin curved, sacculus half width of valva, extending to apex; transtilla papillose; anellus with rounded lobes, deeply emarginate; saccus short, wide, rounded; aedeagus curved beyond middle, four or five small cornuti in a tight group.

FEMALE GENITALIA. Anterior apophysis three-quarters length of posterior apophysis; segment VIII width:length 2:1, sternite with anterior margin convex; ostium wide, on anterior margin of sternite; antrum not distinct; ductus bursae wide, posterior two-fifths sclerotised, initially straight then with double bend, remaining part thin-walled, gradually expanding to ovoid corpus bursae; signum absent.

GENITALIA DIAGNOSIS. Male genitalia unique in European fauna. In female absence of signum and of triangular sclerites in ostium in combination with short and broad ductus bursae sclerotised in distal part is unique.

DISTRIBUTION. Southern Europe from Portugal to Greece, Bulgaria and Crimea; Slovakia, Hungary; Sicily, Madeira, Azores. North Africa (Buchner & Karsholt, 2019). Asia: Turkey, Armenia. Georgia, Israel, Iran (Lvovsky, 2004). Likely to be more widespread.

BIONOMICS. Larva on flowers of *Daucus carota* L. (Lvovsky, 2004) and *D. muricatus* (L.) L. (Corley, personal observation), spring to June, rarely to July, adults emerging May to July, flying until April (rarely May?) of next year.

REMARKS. Hannemann (1958a) synonymised *D. cuprinella* Walsingham, 1907 with *D. marcella. D. marcella* has been considered a pest of carrot seed crops in Italy (Celli, 1970).

153   *Depressaria peregrinella* Hannemann, 1967

Depressaria peregrinella Hannemann, 1967: 167.

DESCRIPTION. Wingspan (12–)15–18 mm. Head dark fuscous. Labial palp segment 3 half as long as segment 2; dark brown, inner side of segment 2 buff with some fuscous scales, tip of segment 3 cream. Antenna dark fuscous. Thorax fuscous, darker anteriorly. Forewing elongated, fuscous to dark brown, dark fuscous in basal one-third of costal area, with rather few scattered grey-buff scales, mainly near costa and dorsum, in cell and fold and between veins; base on costal side blackish; thin blackish often interrupted lines beneath costa, in cell and fold and between veins particularly to costa and apex; a dark fuscous dot at end of cell and small dots between veins around termen; fringe with fuscous fringe line. Hindwing pale grey, darker towards costa and apex; fringe pale grey, with distinct fringe line. Abdomen grey-buff to fuscous.

VARIATION. Ground colour varies from fuscous to dark brown, darkest specimens with very few scattered grey-buff scales; the blackish lines are often obscure.

SIMILAR SPECIES. Characterised by uniformly fuscous forewing with obscure markings and dark dot at end of cell, without pale fascia or other whitish markings.

MALE GENITALIA. Gnathos long-stalked, elliptic; socii elliptic-ovate, more convex on outer margin than inner; valva broad at base, rather abruptly narrowed at one-third and two-thirds due to curves on the ventral margin, costal margin straight to two-thirds, ending in a triangular projection with rounded apex, sacculus ending at three-fifths in broad digitate cuiller directed initially towards apex then towards costa; transtilla with honeycomb pattern; anellus wide and shallow, apical processes extended laterally across base of valva; saccus short, wide, rounded; aedeagus slightly curved, stout, about six well-spaced medium-sized thorn-like cornuti with broad bases.

FEMALE GENITALIA. Papilla analis strongly elongate; anterior apophysis one-half length of posterior apophysis; segment VIII width:length 7:2; ostium wide, on anterior margin of sternite; antrum cup-shaped, margin sclerotised, produced at angles; ductus bursae thin-walled; corpus bursae elliptical; signum small, weakly sclerotised, three times as wide as long, with marginal triangular teeth.

GENITALIA DIAGNOSIS. Male characteristic, especially combination of narrowed distal third of valva and thorn-like cornuti. In female the cup-shaped antrum at proximal margin of sternite VIII in combination with ductus bursae unsclerotised but thickened and longitudinally streaked in central part is shared with *D. heydenii*. In *D. peregrinella* ostium without triangular sclerites and without microtrichia (only a field of tiny dots close to distal edges of antrum), thickened part of ductus bursae without microsculptures. In *D. heydenii* ostium with pair of triangular sclerites and with microtrichia, thickened part of ductus bursae with microsculptures in variable number.

Note: in standard genitalia slides, i.e. dorsoventrally compressed, the cup-shaped antrum is susceptible to preparation artefacts, so details of shape of antrum in such slides must be taken with care.

DISTRIBUTION. In Europe only known from Greece and North Macedonia. Turkey, Armenia, Iran.

BIONOMICS. Klimesch reared this species from a larva collected from an undetermined Apiaceae from Peloponnese, Greece, the moth emerged 11 July. Adults hibernate, we have seen worn specimens caught in spring up to 26 April.

154   *Depressaria heydenii* Zeller, 1854

Depressaria heydenii Zeller, 1854: 296.

DESCRIPTION. Wingspan 15–20 mm. Head with face buff, vertex brown, scales tipped buff. Labial palp segment 3 two-thirds length of segment 2; both segments pale buff, segment 3 with some fuscous scales at base and towards apex. Antennal scape dark fuscous with a few buff and mid-brown scales, flagellum buff, narrowly ringed fuscous. Thorax dull chestnut-brown, more or less mixed grey, especially posteriorly. Forewing light grey, partially to largely overlaid dull chestnut-brown or sometimes darker brown, particularly in basal one-fifth except near costa, and with similarly coloured longitudinal stripes and scattered scales over most of wing, least evident from middle of cell to costa; a blackish subdorsal spot at base; thick short black streaks in cell at one-third, in fold, beneath costa at one-third and at middle, a spot at end of cell, narrower often interrupted black lines at base beneath costa and in central line, near dorsum and between veins to termen; a series of blackish dots between veins around termen; fringe mixed grey and light chestnut-brown, with or without distinct fringe line. Hindwing pale grey, darker grey towards costa and around apex; fringe pale grey with distinct fringe line. Abdomen grey-buff.

VARIATION. The proportion of grey and chestnut-brown scales varies, sometimes the brown areas are heavily shaded with fuscous.

SIMILAR SPECIES. *D. heydenii* has no diagnostic external features separating it from other *Depressaria* species, but it has no white spot in cell and the pale fascia beyond end of cell is absent or weakly marked.

MALE GENITALIA. Gnathos unusually large and long-stalked, lanceolate, the expanded part without spinules proximally; socii digitate, divergent, quite long but not reaching base of gnathos; tegumen short; valva nearly parallel-sided, at end gradually rounded to subacute apex, sacculus more than half width of valva, ending at two-thirds in a small rounded hump; anellus wider than long, expanding to posterior angle, with a turret-like posterior median extension; saccus short, wide, rounded; aedeagus stout, with broad base, abruptly tapering to acute point, 10 or more long cornuti forming a broad band with longest three-quarters length of aedeagus.

FEMALE GENITALIA. Anterior apophysis one-third length of posterior apophysis; segment VIII width:length 5:2, sternite with rectangular anterior extension accommodating transversely widened ostium; antrum wider than long, widening posteriorly, sclerotised; ductus bursae thin-walled, somewhat swollen in anterior half with minute scobination, mainly on one side; corpus bursae ovoid; signum as wide as long, rhombiform, with longitudinal angles diffuse, covered in triangular teeth.

GENITALIA DIAGNOSIS. Male characteristic, especially large long-stalked gnathos. For female see *D. peregrinella*.

DISTRIBUTION. Alpine region from France and Italy to Germany, Switzerland, Austria, Slovakia and Romania.

BIONOMICS. Larvae on *Chaerophyllum hirsutum* L., *C. villarsii* W.D.J. Koch, *Carum carvi* L., *Pimpinella major* (L.) Huds., *P. saxifraga* L., *Laserpitium latifolium* L., *L. halleri* Crantz, *L. krapfii* Crantz, *L. siler* L., *Heracleum austriacum* L., *H. sphondylium* L., *H. juranum* Rapin, *Peucedanum venetum* (Spreng.) W.D.J. Koch, *P. ostruthium* (L.) W.D.J. Koch, *Meum athamanticum* Jacq. and *Ligusticum mutellina* (L.) Crantz (Sonderegger, *in litt.*); usually in spun umbels, or in spun leaves if flowers are not available, from June to August, rarely to September. Pupation usually on the soil, occasionally also in spinnings of host-plant. Adults from August to June of the following year.

One of the *Depressaria* species with the widest spectrum of host-plants.

155 *Depressaria fuscipedella* Chrétien, 1915

Depressaria fuscipedella Chrétien, 1915: 343.

DESCRIPTION. Wingspan 20–22 mm. Head with face buff, vertex dark fuscous, scales tipped buff. Labial palp segment 3 two-thirds length of segment 2; segment 2 pale buff with numerous dark fuscous scales on outer side, mainly in middle, segment 3 dark fuscous with some buff scales in middle, mainly on inner side, apex creamy buff. Antenna dark fuscous, narrowly ringed buff. Thorax dark fuscous, mixed grey-buff posteriorly or almost all over. Forewing dark fuscous, darkest at base; an elongate black spot in mid-line at one-sixth, a black triangular spot at one-third in cell, various short black dashes between veins to costa and in terminal area, terminal dots black; rather few scattered grey-buff scales over wing surface, more numerous along costa, more extensive light grey-buff scales in cell after triangular dot and beyond cell forming a broken angled fascia; fringe dark fuscous without distinct fringe line. Hindwing with dark fuscous scales, more concentrated around margins and in terminal area; fringe basally dark fuscous, fuscous beyond distinct fringe line. Tarsi almost black, ringed pale buff at end of segments. Abdomen dark fuscous.

VARIATION. The thorax and tegulae vary from almost all dark fuscous with buff scales near to posterior margin, to nearly whole of thorax pale buff, only darker anteriorly.

SIMILAR SPECIES. *D. fuscipedella* is characterised by dark fuscous coloration, which includes fore and hindwing, most of segment 3 of palps, tarsi and abdomen. It could easily be mistaken for a dark form of *D. badiella*.

MALE GENITALIA. Gnathos elongate, exceeding digitate socii; tegumen tapering to apex; valva with costal margin convex in middle, apically strongly incurved to fine acumen, ventral margin gradually curving, towards apex more strongly curved to join apiculus, sacculus two-fifths width of valva, ending in a rounded hump; transtilla almost semicircular, transtilla lobes large; anellus wider than long, expanding from base to two-thirds length, distal margin with two obtuse teeth near middle; saccus short, wide, rounded; aedeagus slender, slightly tapering to acute point, long cornuti forming a dense group less than half length of aedeagus.

FEMALE GENITALIA. Anterior apophysis three-fifths length of posterior apophysis; segment VIII width:length 2:1, sternite shorter than tergite; ostium at anterior margin of sternite, with sclerotised crescent-shaped lip, antrum short, goblet-shaped, weakly sclerotised,; ductus bursae long, thin-walled except a short sclerotised section towards posterior end, very gradually expanding into narrow club-shaped corpus bursae; signum small, wider than long with three lobes, lateral lobes roughly circular, with one or more large teeth, central lobe with some smaller teeth near middle of plate.

GENITALIA DIAGNOSIS. Male similar to *D. heydenii*, but with smaller gnathos and fine incurved acumen of valva. For female see description and remarks.

DISTRIBUTION. Described from Algeria, also apparently in south Italy, see remarks, below.

BIONOMICS. Host-plant unknown. Adult in July in Algeria; Italian specimens taken in August.

REMARKS. Chrétien (1915) described *D. fuscipedella* from one male taken near Oran, Algeria. It is characterised by the very dark coloration of fore- and hindwings, abdomen, labial palps and legs. In August 2020 F. Graf collected two females fitting the description of *D. fuscipedella* in Puglia, Italy. If these are truly *D. fuscipedella* the species is new for Europe. However, females of *fuscipedella* from Algeria are unknown and no males of the Italian *Depressaria* were collected. The conspecificity of Italian and Algerian specimens, although probable, cannot be proved unless a barcode can be obtained from the type or fresh males can be found. The description above is based on the Italian females, apart from male genitalia, based on the type specimen.

156   *Depressaria ululana* Rössler, 1866

Depressaria ululana Rössler, 1866: 334.

DESCRIPTION. Wingspan 18–21 mm. Head light brown, face light brownish buff to whitish. Labial palp segment 3 three-fifths length of segment 2; segment 2 creamy buff with a few dark brown scales and a blackish band at three-quarters, segment 3 deep brown with a few buff scales, tipped buff, inner side with more buff scales in middle, leaving deep brown bands at base and beyond middle. Antenna with scape deep brown, basal third of flagellum buff with narrow dark brown rings, posterior two-thirds dark brown. Thorax and tegulae brown anteriorly, middle part of thorax brown to whitish with scattered grey-brown scales, posterior crests buff, tegulae whitish, buff to light brown. Forewing mid-brown to deep reddish brown; deep brown or blackish spot at extreme base of costa and at base of wing on dorsal side, and a larger patch in basal area from fold to one-sixth, these deep brown areas nearly surrounding a whitish or pale buff subtriangular spot from which a pale streak of the same colour runs along costa to one-quarter or one-half; additional whitish buff scales between veins forming an indistinct angled fascia beyond cell, and an irregular-shaped partial fascia from costa at one-half to middle of wing, reappearing in fold; a blackish dot at end of cell, short blackish lines between veins, a few in cell and edging costal streak, and a series between veins to costa, followed by whitish buff spots before terminal blackish spots; fringe pale greyish brown, with two darker fringe lines, the inner line the darkest. Hindwing grey, posteriorly darker; fringe with two indistinct fringe lines. Abdomen grey-buff.

VARIATION. The development of a whitish costal streak is very variable and may be obsolete, but the white triangular mark at base of costa is always evident. Some specimens have pink scales on costa and in fold. Male genitalia show considerable variation in shape of socii and anellus and number and arrangement of cornuti.

SIMILAR SPECIES. Characterised by whitish tegula and triangle at base of costa running into narrow whitish subcostal streak, a black spot at end of cell.

MALE GENITALIA. Gnathos long-stalked, narrowly elliptic; socii short-rounded, reaching base of gnathos; valva broad, costa straight, ventral margin with bulge at three-quarters, apex a narrow triangle, sacculus half width of valva, extending into extreme apex, at three-fifths with digitate cuiller just crossing costal margin; clavus present, two-thirds length of anellus, hairy; anellus large, from broad base narrowest beyond middle, widening distally, apex bifid with semicircular or V-shaped sinus; saccus short, wide, rounded; aedeagus stout, curved, tapering from broad base, apex obtuse, about 10–20 cornuti of varying length.

FEMALE GENITALIA. Papilla analis elongate; anterior apophysis three-fifths length of posterior apophysis; segment VIII width:length 2:1, posterior margin of sternite extended in a wide arc, anterior margin of tergite convex; ostium wide, in anterior half of sternite; antrum shortly cylindrical, with slight curve, weakly sclerotised; ductus bursae arising asymmetrically from antrum, posterior third weakly sclerotised, twisted, remaining part thin-walled, eventually expanding to elliptical corpus bursae; signum variable, sometimes absent.

GENITALIA DIAGNOSIS. Male similar to *D. chaerophylli*, best difference in clavus: present in *D. ululana*, absent in *D. chaerophylli*, and small socii of *D. ululana*. Note: socii can be turned inward or outward in slides, this causes rather different appearance, but has no diagnostic significance. Female characterised by ductus bursae originating asymmetrically from antrum, leaving distinct bulge on right side (this bulge sometimes small or nearly absent). See also Remarks below. *D. manglisiella* very similar, but antrum much shorter, ostium closer to proximal edge of sternite VIII.

DISTRIBUTION. Scattered in Southern Europe from Portugal to Greece; Switzerland, Germany and Poland. In Asia known from Turkey, Armenia and Azerbaijan.

BIONOMICS. Larvae on flowers and developing seeds of *Bunium bulbocastanum* L. (Hannemann, 1953), *Conopodium majus* (Gouan) Loret (Corley, personal observation), *Carum verticillatum* (L.) Koch and *Scaligeria napiformis* (Spreng.) Grande (Sonderegger, *in litt.*) in spun umbels, from May to end of June, rarely to July; adults from July to April (rarely May) of the following year. *Ptychotis saxifraga* (L.) Loret & Barrandon is mentioned by Lhomme (1945) quoting Chrétien. Also recorded from *Nigella arvensis* L. (Fazekas & Schreurs, 2013), but see Remarks below.

REMARKS. Hannemann (1953) placed *D. ululana* in the *douglasella* group because it has clavus and cuiller, but it differs from that group in most other characters of male genitalia. It is much better placed close to *D. chaerophylli*.

We have seen a single female from Armenia with barcode 100% same as *ululana*, but ductus bursae with posterior half of its length sclerotised and without bulge, corpus bursae with signum. This needs further investigation.

Fazekas & Schreurs (2013) added *D. ululana* to the fauna of Hungary, based on larvae found on *Bunium persicum* (Boiss.) B. Fedtsch. and *Nigella arvensis* L. The figured male genitalia belong to *D. chaerophylli*, but the figured larva is undoubtedly *D. ululana*, although the host-plant of that larva is not specified. In view of these uncertainties we have not included Hungary in the distribution of *D. ululana*. *Bunium persicum*, more often placed in genus *Elwendia* as *E. persica* (Boiss.) Pimenov & Kljuykov is an oriental herb cultivated for its seeds, apparently also cultivated in Hungary. From its appearance it can easily be mistaken for *Nigella sativa* L. (both are called Black Cumin). This probably explains the inclusion of *N. arvensis* as a host-plant for *D. ululana*. As a member of the Ranunculaceae, it is a most improbable host-plant for any Depressariidae.

157   *Depressaria manglisiella* Lvovsky, 1981

Depressaria manglisiella Lvovsky, 1981: 78.
   Depressaria venustella Hannemann, 1990: 138, syn. n.

DESCRIPTION. Wingspan 17–20 mm. Head with cinnamon-brown to fuscous scales, often tipped ochreous-brown. Labial palp segment 3 two-thirds length of segment 2; segment 2 pinkish buff on inner side, outer and ventral sides with mixed cinnamon and blackish scales, segment 3 nearly all black with pale cinnamon tip. Antenna dark fuscous. Thorax cinnamon-brown or chestnut-brown, paler posteriorly. Forewing grey with admixture of light brown or cinnamon-brown scales; a blackish subdorsal patch, a small blackish dot at extreme base of costa outwardly broadly edged creamy buff to cinnamon-buff, black longitudinal streaks, in middle of base, in cell, beneath costa, shorter streaks in terminal area between veins to termen and blackish dots between veins around termen and onto costa; creamy buff to cinnamon-buff streaks or spots preceding, following or edging some of the black streaks, in middle of cell and forming an angled fascia beyond cell and a row of dots before terminal dots; fringe light grey-brown, with faint cinnamon tinge; no fringe line. Hindwing light grey, darker posteriorly; fringe light grey-buff, with a fringe line.

VARIATION. Fresh specimens have a cinnamon tinge to the head and forewing costa; the pale markings on wing are characteristically creamy buff but may be more or less tinged cinnamon.

SIMILAR SPECIES. *D. manglisiella* is characterised by creamy or creamy buff markings associated with the black streaks and the basal costal spot.

MALE GENITALIA. Gnathos short-stalked, narrowly elliptic; socii nearly reniform but narrower in apical third, almost contiguous distally; tegumen conspicuously bulging at level of base of gnathos arms; valva gradually tapering to blunt apex, ventral margin with a bulge at three-quarters opposite a hump in sacculus margin, sacculus wide, clavus curving in direction of apex of valva, untidily hairy; anellus approximately X-shaped, proximal limbs wider than distal which are more erect and heavily sclerotised on inner margin; saccus a broad obtuse triangle; aedeagus nearly straight, cylindrical, ending in a stout recurved point, 9 to 11 medium-sized cornuti.

FEMALE GENITALIA. Papilla analis elongate; anterior apophysis three-fifths length of posterior apophysis; sternite VIII width:length 3:1, posterior margin of sternite slightly concave, anterior margin of tergite slightly convex with slight additional bulge around wide ostium; antrum sclerotised, with bulge on right side immediately below ostium, then tapering to slender ductus bursae which expands slightly to ovate corpus bursae; signum a weakly sclerotised polygon with somewhat irregularly toothed margins.

GENITALIA DIAGNOSIS. Male characteristic, especially the terminal hook of aedeagus. For female see under *D. ululana*.

DISTRIBUTION. Described from Georgia, now known from Sicily, Greece, Serbia and Croatia.

BIONOMICS. Host-plant unknown. Adults collected in June and early July were worn, those collected end of August and September were fresh, which makes it likely that larvae develop in summer and adults hibernate.

REMARKS. The relationships of *D. manglisiella* were obscure, but following the discovery of a female which shows considerable similarity in genitalia to *D. ululana*, similarities with *D. ululana* in male genitalia are also apparent.

D. venustella Hannemann, 1990 was described from four males collected on Mount Etna, Sicily. From Hannemann's drawing of male genitalia, it is clear that it is synonymous with *manglisiella*.

158   *Depressaria chaerophylli* Zeller, 1839

Depressaria chaerophylli Zeller, 1839: 196.

DESCRIPTION. Wingspan (16–)19–22.5 mm. Head mid-brown, face paler brown. Labial palp segment 3 three-fifths length of segment 2; segment 2 outer side dark grey-brown with a few light grey-buff scales, ochreous at apex, inner side grey-buff with some dark grey-brown scales; segment 3 deep brown with apical one-quarter pale ochreous. Antenna dark grey-brown. Thorax always mid to dark brown anteriorly, greyish white posteriorly, sometimes extending towards front of thorax, posterior tufts fuscous. Forewing reddish brown; a deep brown spot at base of costa and a larger extended spot at base of dorsum; basal quarter of ground colour with a blackish streak through middle, remainder of wing heavily marked with short blackish lines and streaks between veins; a series of blackish spots between veins around end of costa and termen; light grey-buff scales scattered in costal and terminal areas, especially across middle of wing, preceding blackish spots between vein ends around termen and forming an angled fascia beyond end of cell; fringe grey-brown, tipped grey-buff, with two more or less distinct fringe lines. Hindwing light grey-brown, darker posteriorly; fringe light grey-brown with two distinct fringe lines. Abdomen light brown to mid-brown.

VARIATION. The ground colour may have a hint of ochreous coloration, and the extent of the reddish brown scales varies; the blackish streaks in posterior part of wing may be absent. Rarely the reddish basal part of costa is overlaid with fuscous scales.

SIMILAR SPECIES. The species is characterised by the reddish brown base of wing in costal half extending to one-quarter. *D. radiella* has this area similarly without markings, but with quite different coloration.

MALE GENITALIA. Gnathos long-stalked, narrowly elliptic; socii very large, curved, extended laterally; tegumen conspicuously bulging at level of base of gnathos arms; valva broad, short, apex truncate, but with a sclerotised claw extending through apex on costal side, sacculus more than half width of valva, ending in a triangular point with sclerotised tip just reaching costa; anellus large, from broad base narrowing to two limbs; saccus short, wide, rounded; aedeagus slightly curved, tapering from broad base, apex obtuse, about eight smaller cornuti and sometimes a longer one. There is some variation, particularly in the shape of end of valva and in the cornuti.

FEMALE GENITALIA. Papilla analis elongate; anterior apophysis one-half length of posterior apophysis; segment VIII width:length 2:1, anterior margin of tergite with deep rounded excavation; ostium wide, at one-third sternite length from anterior margin; antrum not distinct; ductus bursae with posterior two-fifths cylindrical, weakly sclerotised, remaining part thin-walled, slightly swollen between two-fifths and four-fifths; corpus bursae ovoid; signum small, rhombiform, about as wide as long, bearing triangular teeth.

GENITALIA DIAGNOSIS. For male see *D. ululana*. Female recognisable by concave anterior margin of tergite VIII in combination with ductus bursae starting with shallow V or U slightly distal of anterior margin of sternite VIII, only posterior two-fifths weakly sclerotised and cylindrical.

DISTRIBUTION. Middle latitudes of Europe from France and Wales to southern Sweden and Finland and east to Romania, also in Asia: Turkey and Georgia. Armenia, Palestine, Libya (Cyrenaica) (Lvovsky, 2001).

BIONOMICS. Larvae at first on leaves then on flowers and developing seeds of *Chaerophyllum temulum* L. (Harper *et al.*, 2002), *C. aureum* L. (Sonderegger, *in litt.*), *C. bulbosum* L. (Buchner, personal observation), exceptionally on *Anthriscus sylvestris* (L.) Hoffm., (Sonderegger, *in litt.*, Lepiforum e.V., 2006–2023), May to July; adults from July to June of the following year. Lhomme (1945) also mentions *Athamanta* L. as host-plant, but this needs confirmation.

159   *Depressaria longipennella* Lvovsky, 1981

Depressaria longipennella Lvovsky, 1981b: 75.

DESCRIPTION. Male wingspan 30–37 mm, female brachypterous, wingspan 22 mm. Head brown, scales tipped buff. Labial palp segment 3 three-fifths length of segment 2; light grey-brown, with a scattering of fuscous scales. Antenna fuscous. Thorax and tegula brown, scales tipped buff. Forewing extremely elongated, gradually widening from narrow base to four-fifths, grey-buff, more or less overlaid mid-brown, particularly a streak between costa and cell and in middle of wing around three-quarters, buff scales numerous, beneath costa, in dorsal one-third of wing, in cell and among veins to termen; costa narrowly fuscous; a blackish subdorsal spot and series of terminal dots; a thin black line from wing base along fold to one-fifth, a short black mark in mid-wing at one-quarter, a line along costal margin of cell from one-quarter to two-thirds, two lines between veins to apex, short lines between veins to termen reaching terminal dots, these lines longer towards dorsum; fringe light to mid-brown mixed buff. Hindwing light grey, darker towards apex; fringe very pale yellowish grey, darker towards apex, with up to four lines. Abdomen grey-brown.

VARIATION. Ground colour varies from grey-buff to mid-brown; the dark lines are sometimes represented by a scattering of black scales

SIMILAR SPECIES. The male of this species has the largest wingspan of any European *Depressaria*, but the female is brachypterous. Its sister species *D. zapryagaevi* Lvovsky, 2001 from Tajikistan is even larger, up to 44 mm wingspan.

MALE GENITALIA. Gnathos lanceolate; socii clavate; valva broad, costal margin highly convex, ending in hairy digitate process, ventral margin slightly convex, near end with large rounded hump equalling end of costal margin and separated from it by a deep rounded sinus, sacculus wide; transtilla with pair of broad, round-ended projections distally, covered with minute erect scales; anellus pyriform, emarginate at apex; saccus short, wide, rounded; aedeagus slightly narrowed beyond middle, two rows of about 15 small cornuti.

FEMALE GENITALIA. Anterior apophysis about one half length of posterior apophysis; segment VIII width:length 3:1, sternite with an anterior median bulge partly accommodating large, circular ostium; antrum short, weakly sclerotised; ductus bursae gradually expanding, not clearly separated from corpus bursae; signum small, transversely elliptical with triangular teeth, longitudinal axis with triangular extensions with small teeth.

GENITALIA DIAGNOSIS. Male unique in European fauna, but very similar to *D. zapryagaevi* with transtilla broad throughout, without pair of projections.

DISTRIBUTION. Russia (Orenburg oblast). In Asia known from Afghanistan, Kazakhstan, Uzbekistan and Turkmenistan.

BIONOMICS. Host-plant unknown. Lvovsky (2004) refers to records from Turkmenistan of this species coming to light from November to late January, then again in the spring. We have seen a worn specimen taken in April and fresh specimens taken from July onwards.

REMARKS. We have not seen female genitalia of this species. The description above relies on the drawing by Lvovsky (2004) which is not clear in all details. Lvovsky (2004) does not give an upper limit to the wingspan of the brachypterous females.

Group 4 *radiella* group (160–170)

Clavus well developed; no distinct cuiller; aedeagus with cornuti. Larval host-plants Apiaceae. We recognise three subgroups.

daucella subgroup (160–162): Socii well exceeding gnathos, valva with costal margin entire, saccus acute or acuminate.

radiella subgroup (163–168): Gnathos short to long stalked, ovate to lanceolate, socii digitate, costal margin of valva with specific features, clavus elongate, appearing to consist of several long scales, transtilla with characteristic hooded lateral lobes. In female genitalia ductus bursae after a narrow beginning with a swollen, thick-walled and usually longitudinally streaked, but unsclerotised section at about one-quarter, not found in that form in any other *Depressaria* spp.

bupleurella subgroup (169–170): Gnathos elongate ovate, exceeding small socii, costal margin of valva with hump and sinus, short clavus.

160   *Depressaria daucella* ([Denis & Schiffermüller], 1775)

Tinea daucella [Denis & Schiffermüller], 1775: 137.
　　Tinea rubricella [Denis & Schiffermüller], 1775: 142.
　　Tinea apiella Hübner, 1796: pl. 14, fig. 94.
　　Depressaria nervosa Stephens, 1834: 198

DESCRIPTION. Wingspan (18–)20–26 mm. Head light brown. Labial palp segment 3 four-fifths length of segment 2; segment 2 outer side mid-brown, buff at apex, inner side buff, darker grey-brown ventrally, segment 3 dull grey-brown on outer side, ringed dark grey-brown beyond middle, apex buff, inner side brownish buff in middle. Antenna with scape dark brown, flagellum light grey-brown, sometimes with reddish tinge, narrowly ringed deep brown. Thorax light brown. Forewing light grey-brown, usually with reddish tinge, with varying amounts of whitish buff scales in centre, particularly in cell and forming a weak sharply angled fascia around and beyond end of cell; costa to three-fifths and dorsum to two-fifths often darkened with dark grey-brown scales; a deep brown streak in cell from one-fifth to one-third, and another shorter, between this and costa; brown streaks on either side of cell, between veins to costa, termen and dorsum; dark grey spots at margin between veins around apex, often absent; fringe light grey-brown, often reddish tinged, with two darker fringe lines. Hindwing light brownish grey, slightly darker on dorsum and posteriorly; fringe light brownish grey with two fringe lines. Abdomen greyish buff.

VARIATION. Fresh specimens generally have a reddish tinge which can fade with age. The dark streaks are much more evident in some specimens than others. Whitish buff scales vary much in extent and are almost lacking in some specimens which thus have no detectable fascia. Rarely the whole forewing is dark fuscous without reddish coloration.

SIMILAR SPECIES. *D. daucella* is characterised by its relatively large size, reddish tinge, rather weak dark streaks, absence of white spot and weak development of sharply angled fascia.

MALE GENITALIA. Gnathos short-stalked, narrowly elliptical; socii tapering from broad base to rounded apex, far exceeding gnathos; valva boat-shaped, with straight costa, ventral margin straight then gradually curving round to meet costa at a right-angle, sacculus simple, about half width of valva, extending to two-thirds valva length; clavus as long as sacculus, curved outwards, with an additional curve close to apex (inwards or outwards according to preparation), with a fibrous core from near base to apex where exposed like cut end of a rope; anellus wider than long, very broad-based, rapidly tapering to turret-shaped posterior end; saccus short, wide, ending in blunt triangle; aedeagus slender, tapering, with about three small to medium-sized cornuti.

FEMALE GENITALIA. Papilla analis elongate; anterior apophysis about one-third length of posterior apophysis; segment VIII width:length 3:1, anterior margin of sternite with low median bulge accommodating ostium, posterior margin concave; ostium adjacent to anterior margin of sternite, wide; antrum long, sclerotised with slight contraction in middle, expanding to ostium; ductus bursae with posterior one-third straight, sclerotised, middle third thin-walled with a spiral turn, posterior third expanding to small ovoid corpus bursae; signum roughly triangular, outline with triangular teeth.

GENITALIA DIAGNOSIS. Male characteristic, especially the very long clavus. Female characterised by the very narrow elliptic ostium just at anterior margin of sternite VIII and long antrum. *D. marcella* appears similar only at cursory glance.

DISTRIBUTION. Almost all continental Europe; Britain, Ireland, Sardinia, Madeira and Canary Islands. Also recorded from western North America, where it may be introduced (Hodges, 1974).

BIONOMICS. Larvae on leaves and flowers of several species of wet habitats like *Cicuta virosa* L. (Hannemann, 1953); *Oenanthe crocata* L., *O. aquatica* (L.) Poir., *O. pimpinelloides* L., *O. fistulosa* L., *Carum verticillatum* (L.) W.D.J. Koch, rarely on

Sison amomum L. (Harper *et al.*, 2002); *Helosciadium nodiflorum* (L.) W.D.J. Koch (Corley, personal observation). It can be a pest on cultivated *Carum carvi* L. in Finland, where it is less confined to wet habitats (M. Mutanen, pers. comm.). Introduced in the Nearctic, where also found on *Cicuta douglasii* (DC.) Coult. & Rose and *Oenanthe sarmentosa* C. Presl. ex DC. (McKenna & Berenbaum, 2003). Further host-plants are mentioned in Lvovsky (2001): *Pastinaca* L., *Daucus* L., *Sium* L., *Chaerophyllum* L., *Petroselinum* Hill. Apart from *Sium* these appear doubtful, either they are used only exceptionally if they grow close to wet habitats or refer to other *Depressaria* species. Larvae April to July, adults from July to May (rarely June) of following year.

REMARKS. For many years this species passed under the name *D. nervosa* Stephens, 1834, but Bradley (1966) observed that the name is a junior homonym of *nervosa* Haworth, 1811, now placed in *Agonopterix*. Stephens' name was anyway junior to *daucella* [Denis & Schiffermüller], 1775 and *rubricella* [Denis & Schiffermüller], 1775. Bradley (1966) chose *daucella* as the replacement name for this species.

161   *Depressaria ultimella* Stainton, 1849

Depressaria ultimella Stainton, 1849: 166.

DESCRIPTION. Wingspan (15–)17–21 mm. Head mid-brown, slightly darker on face. Labial palp with segment 3 four-fifths length segment 2; light brown on outer side, segment 2 with broad deep brown band before apex, segment 3 deep brown at base and at three-quarters, with some deep brown scales between these bands, light brown at apex; inner side buff. Antenna deep brown. Thorax mid-brown, sometimes paler. Forewing light brown, often with slight reddish tinge, deep brown at base on costal and dorsal sides, blackish brown streaks in mid-line of wing and lying between veins in costal half of wing and beyond end of cell in terminal area; some dark brown scales in fold; pale buff scales, usually mixed with a few brown scales, forming an elongate patch in cell often between blackish brown streaks, likewise along or within the streak on the subcostal vein and often interrupting the streaks between veins from cell to costa; some additional buff scales along streaks to termen; outer half of costa with small areas of pinkish buff scales between vein ends; fringe light grey-brown, paler, sometimes faintly pinkish apically beyond a weak grey fringe line. Hindwing light grey, light grey-brown posteriorly; fringe light grey-brown, pale grey-buff beyond second greyish fringe line. Abdomen light brown.

VARIATION. Ground colour of forewing varies from slightly reddish brown through shades of mid-brown occasionally to dull brown. The strength and extent of the blackish brown streaks can extend from mid-wing to costa, or not appear till some distance from mid-line, similarly the amount of buff scaling varies. When most developed there is a faint hint of the angled fascia that is present in many other *Depressaria* species. In the lightest coloured specimens the markings are weakly developed.

SIMILAR SPECIES. *D. halophilella* (q.v.). *D. daucella* is larger with wings more elongate.

MALE GENITALIA. Gnathos circular or short ovate; socii short-stalked, straight on outer margin, bulging on inner margin; valva slightly wider at base than apex, costa slightly convex, apex truncate, ventral margin slightly curved at one half, sacculus reaching one-half; clavus equalling anellus in length, widest at one-third, tapering to apex, inner margin with projecting scales; anellus as wide as long, distal margin with shallow excavation or emargination; saccus nearly triangular, sides slightly to strongly concave, apex broadly or narrowly pointed; aedeagus at base wide, tapering, then parallel-sided, with 3–8 medium sized cornuti in a loose group.

FEMALE GENITALIA. Anterior apophysis short, about one-tenth length of posterior apophysis; segment VIII width:length 5:2, posterior margin of sternite with shallow excavation, anterior margin sclerotised, convex; ostium in middle of segment, with strongly sclerotised ring; antrum cylindrical, sclerotised, ending obliquely; posterior part of ductus bursae swollen on one side, scobinate, contracted to slender thin-walled section then gradually widening to long ovoid corpus bursae; signum longer than wide, subquadrate with longitudinal extensions covered in small triangular teeth, larger teeth across middle.

GENITALIA DIAGNOSIS. See *D. halophilella*.

DISTRIBUTION. Widespread in middle latitudes of Europe, in south recorded from Portugal, Spain, Sicily, Croatia and Greece, not recorded from mainland Italy and Switzerland. Also in Asia: Kyrgyzstan. Azerbaijan, western Kazakhstan, western Mongolia and north-western China (Lvovsky, 2004).

BIONOMICS. Larvae in the hollow stems of *Cicuta virosa* (Hannemann, 1953), *Helosciadium nodiflorum* (L.) W.D.J. Koch (Heckford, 1983), *Oenanthe aquatica* (L.) Poir., *O. crocata* L., *Sium latifolium* L. (Harper *et al.*, 2002) and *Berula* Besser ex W.D.J. Koch (Lvovsky, 2004), where it may be submerged. When feeding finishes it makes a hole in the side of the stem which is covered with silk and frass making it watertight, through which the adult emerges. Larvae May to September, adults from July to June (rarely July) of following year.

REMARKS. The record from Madeira belongs to *D. halophilella* (Buchner & Karsholt, 2019).

For a possible record of this species from Canary Islands see under *D. halophilella*.

162   *Depressaria halophilella* Chrétien, 1908

Depressaria halophilella Chrétien, 1908: 60.

DESCRIPTION. Wingspan 15–18.5 mm. Head light brown, scales tipped buff, face whitish buff. Labial palp segment 3 three-quarters length of segment 2; segment 2 whitish buff, outer side with a few grey scales above base and below apex, segment 3 whitish buff with a few grey scales, mainly above middle. Antenna fuscous, underside mainly whitish buff. Thorax mid-brown, slightly darker towards head. Forewing light brown, sometimes buff, with a few scattered whitish buff scales, particularly below costa and between veins towards termen; a dark brown subdorsal spot at base, another at base of costa; a black line in cell from one-third to two-thirds, an interrupted line parallel with costa, a series of lines between veins in outer one-third of wing, sometimes widening towards termen, more or less joined to series of black dots between vein ends; fringe light brown with cinnamon-tinge, with a weak fringe line. Hindwing light grey, darker towards apex; fringe light grey with faint fringe line. Abdomen light grey-buff.

VARIATION. Ground colour can be light to mid-brown or buff. Black line in cell sometimes thickened at outer end; an angled pale fascia beyond cell is sometimes faintly indicated.

SIMILAR SPECIES. *D. halophilella* lacks any black lines in dorsal half of wing before two-thirds. *D. ultimella* is very closely related. It has more white scales in cell, the angled fascia is more often present, although not strong and the black lines are mostly shorter.

MALE GENITALIA. Similar to those of *D. ultimella*. Valva obliquely truncate; cornuti 8–20.

FEMALE GENITALIA. Very similar to those of *D. ultimella*.

GENITALIA DIAGNOSIS. Male extremely similar to *D. ultimella* and not always distinguishable.

Typical *D. halophilella* has angle between costa and truncate tip of valva acute, lateral outline of vinculum moderately to markedly convex and usually 8–20 cornuti whereas *D. ultimella* has valva right-angled at apex, vinculum straight-sided or slightly convex and 3–8 cornuti. Note: these features are variable and sometimes overlapping. Females not distinguishable.

DISTRIBUTION. On rocky coasts of Madeira, Spain, France, Italy, Sardinia, Sicily, Croatia and Greece.

BIONOMICS. Larva in stems of *Crithmum maritimum* L. (Chrétien, 1908), from November to March. Pupa in the stem. Adults appear in March, latest emergence in July, latest collection date of checked specimens middle of October.

REMARKS. In areas where *D. halophilella* is lacking but *D. ultimella* and *Crithmum maritimum* are present, (e.g. south coast of Great Britain), *D. ultimella* has never been found on this plant. This underlines that they are close, but clearly distinct species. Barcode difference from *D. ultimella* is 1.23%.

A single female which could belong to either *D. halophilella* or *D. ultimella* (Canary Islands: Gran Canaria, Virgen-Moya, 400m, 20.vii.1984, leg. Olsen, Skule and Stadel, ZMUC SL2485) has been studied but a certain determination is not possible. In view of the presence of *D. halophilella* on Madeira it is perhaps more likely to be this species.

163 *Depressaria radiella* (Goeze, 1783)

Phalaena (*Tinea*) *radiella* Goeze, 1783: 162.
  Phalaena (*Tinea*) *heracliana* auct., nec Linnaeus, 1758
  Phalaena (*Tinea*) *heraclei* auct., nec Retzius, 1783
  Tinea radiata Geoffroy *in* Fourcroy, 1785: 320.
  Haemilis pastinacella Duponchel, 1838: 153.
  Depressaria sphondiliella Bruand, 1851: 73.
  Depressaria ontariella Bethune, 1870: 3.
  Depressaria caucasica Christoph, 1877: 293.

DESCRIPTION. Wingspan (22–)25–28(–31) mm. Head light brownish buff, face buff. Labial palp segment 3 three-fifths of segment 2; segment 2 buff, outer side with a few mid-brown scales, ventral side mid-brown, segment 3 buff, deep brown on ventral edge and from middle to near tip, tip pale buff. Antenna with scape brownish buff, flagellum buff with grey-brown rings. Thorax buff. Forewing light brownish buff, with a light dusting of dark brown scales, particularly on dorsum, near base of costa and in terminal quarter of wing; light buff scales beneath costa at one-fifth and two-fifths, in cell, in fold at distal end, forming a sharply angled fascia beyond end of cell and round termen; dark brown spots at extreme base of costa and at base of dorsum, narrowly bordered buff; dark brown broad streaks in base of fold, in base of cell and a spot at end of cell, further fine streaks in pairs in costa and cell and singly between veins to costa, termen and dorsum, interrupted by angled fascia; grey-brown spots between vein ends around apex and termen; fringe greyish buff with indistinct fringe lines. Hindwing light grey-brown, slightly darker posteriorly; fringe pale greyish buff, with two indistinct fringe lines. Abdomen buff.

VARIATION. The species varies less than most, but there is nevertheless some variation in intensity of ground colour, strength of dark markings and visibility of the pale fascia. There is a rare form with whole wing dark fuscous somewhat obscuring the dark markings, but the angled pale fascia is evident.

SIMILAR SPECIES. *D. radiella* is characterised by its large size and buff ground colour. Some specimens of *D. daucella* may appear similar, but have narrower wings and more sharply angled fascia.

MALE GENITALIA. Gnathos short-stalked, ovate; socii curved, digitate, about equalling gnathos; valva short and wide, costa with median hump followed by a sclerotised sinus and another hump before rounded apex, ventral margin nearly straight, sacculus one-third to one-quarter width of valva, ending at three-quarters; clavus about as long as width of valva, consisting of several long pointed scales; anellus flask-shaped; aedeagus curved, with slender pointed apex, about nine cornuti of various lengths

FEMALE GENITALIA. Anterior apophysis about one-half length of posterior apophysis; segment VIII width:length 10:3, sternite with pair of unsclerotised triangles lateral to ostium, anterior margin of tergite with shallow excavation; ostium in middle of sternite, with pair of distinct triangular sclerites; ductus bursae thin-walled throughout, with swollen section at one-quarter and sometimes also beyond middle; corpus bursae long ovoid; signum with four concave sides, as long as or longer than wide, often not symmetrical, longitudinal axis with minute teeth, laterally with large teeth.

GENITALIA DIAGNOSIS. Male extremely similar to *D. pimpinellae*, where tip of valva is less blunt, no further reliable differences (easier to separate these two species by size and external appearance). Female very similar to *D. pimpinellae*, where pair of unsclerotised triangles lateral to ostium are broader, swelling in ductus bursae is wider than long, but longer than wide in *D. radiella*.

DISTRIBUTION. Mainly in middle latitudes of Europe, extending north to central Finland and south to Spain, Sardinia, Croatia, North Macedonia and Romania. Southern Siberia, Primorskii kraj, Sakhalin and Kunashir (Lvovsky, 2004).

BIONOMICS. Larvae feed gregariously in flower heads of *Pastinaca sativa* L. (Hannemann, 1953) and *Heracleum sphondylium* L., where they produce abundant silk and frass, more rarely on *Helosciadium nodiflorum* (L.) W.D.J. Koch. (Harper *et al.*, 2002; De Prins & Steeman, 2003–2023). Further host-plants are mentioned in literature including *Oenanthe* L., *Carum* L., and *Anethum* L. (Lvovsky, 2004), but it is doubtful if they are more than exceptions or they may be based on incorrect determinations. Larvae from May to August, pupation takes place inside the stem of *Heracleum* (sometimes several pupae in one stem) but on the ground beneath *Pastinaca* (Sonderegger, *in litt.*). Adults from August to June of the following year.

REMARKS. After the lectotypification of *P. (T.) heracliana* Linnaeus by Bradley (1966) the name *pastinacella* was used for this species, although *heraclei* was also sometimes used, but is ineligible because it has the same type specimen as *heracliana*. Karsholt *et al.* (2006) consider the nomenclature and synonymy of *D. radiella* in detail.

In the 19th century *pastinacella* was wrongly applied to both *D. badiella* and *D. discipunctella* by various authors, giving rise to confusion over host-plants of these species.

Depressaria caucasica Cristoph, 1877 was placed in synonymy with *pastinacella* by Lvovsky (1998b).

164   *Depressaria libanotidella* Schläger, 1849

Depressaria libanotidella Schläger, 1849: 44.
   Depressaria laserpitii Nickerl, 1864: 1.
   Depressaria daucivorella Ragonot, 1889: 106.
   Depressaria mesopotamica Amsel, 1949: 314.

DESCRIPTION. Wingspan (21–)23–28.5(–30) mm. Head orange-brown, scales tipped buff, face whitish buff. Labial palp segment 3 three-quarters length of segment 2;

segment 2 whitish buff on inner side, outer and ventral sides grey-fuscous, segment 3 whitish buff, grey-fuscous at base, along ventral edge and in a partial band at three-quarters, tip pale ochreous. Antenna dark fuscous above, pale buff below. Thorax dull orange-brown, ochreous-buff or creamy buff, sometimes fuscous anteriorly. Forewing dull orange-brown to grey-brown mixed buff in costal one-third, abruptly changing on dorsal side of blackish line in cell to fuscous or dark brown in central and dorsal areas, extending to termen; scattered whitish grey or grey-buff scales sometimes more numerous forming a stripe in middle of cell, a weak angled fascia beyond end of cell and lines between veins in terminal area; a blackish subdorsal spot at base and a dot at base of costa; blackish lines in fold, beneath costa, in cell and between veins to termen; a series of blackish dots between veins around termen; fringe grey-brown, outwardly grey beyond a fringe line. Hindwing light grey, darker towards apex; fringe whitish grey, with or without fringe line. Abdomen grey-buff.

VARIATION. Some specimens show marked contrast between more or less orange-brown costal area and remainder of wing, with black lines well-marked, but others are grey-fuscous with little contrast between costal area and remainder of wing, longitudinal blackish lines few and more or less indistinct.

SIMILAR SPECIES. *D. pimpinellae*, *D. villosae*, *D. bantiella*, and *D. velox* are externally too similar to separate without genitalia examination.

MALE GENITALIA. Gnathos lanceolate, far exceeding the digitate socii, shorter than anellus; valva almost parallel-sided, apex rounded, but forming a near right-angle with slight apiculus at end of costa, costal margin with weak median hump; sacculus ending at two-thirds valva length in a broad lobe directed towards costa; clavus narrowly lanceolate with bristles on the inner side, of variable length; transtilla posteriorly with a pair of large hood-like lobes; anellus flask-shaped, reaching base of tegumen; aedeagus slender to quite stout, with about eight cornuti present in a tight group.

FEMALE GENITALIA. Anterior apophysis one-half length of posterior apophysis; segment VIII width:length 3:1; ostium in middle of segment with narrow, acutely pointed lateral lobes overlapping at base and extending beyond antrum by less than width of antrum; antrum short, parallel-sided, weakly sclerotised; ductus bursae thin-walled, with a swelling at one-quarter length which is wider than long; corpus bursae ovoid; signum small, wider than long, with large triangular teeth.

The description of female genitalia above is from a specimen from Regensburg, Germany, not very distant from Jena, the type locality of *D. libanotidella*. Other female genitalia examined show considerable divergence: segment VIII width:length 2:1; ostium with a continuous sclerotised lip of varying width and length, extending beyond antrum by more than width of antrum or not; antrum sclerotised, concave-sided; part of ductus bursae anterior to swelling widened at varying distance from swelling, not clearly distinct from elliptical corpus bursae;

signum small or large, wider than long or rhomboidal, longer than wide, sometimes with all sides concave.

GENITALIA DIAGNOSIS. Dissection can safely exclude *D. pimpinellae* and *D. villosae*, but in *D. libanotidella*, *D. bantiella* and *D. velox* some doubtful specimens remain. Male genitalia of *D. velox* are distinct in gnathos not exceeding socii in combination with hook at end of costa. Differences between *D. libanotidella* and *D. bantiella* are found in relative length of gnathos and lateral outline of aedeagus, but without knowledge of biology some specimens will remain indeterminable. Female genitalia of all three species are similar and not always safely determinable. In *D. libanotidella* most distinct detail is shape of ductus swelling, it appears more as lateral appendix than swollen section in its course, ostium/antrum region variable, but usually different from typical *D. bantiella* and *D. velox*. In these two species the ductus swelling is longer than wide and ostium with broad rim, tapering rapidly to narrowly pointed upturned lateral lobes, this rim rather narrow and usually straight in *D. bantiella*, very broad and convex in *D. velox*.

DISTRIBUTION. From France and Italy through Central Europe, extending north to Sweden and Finland and south to Albania and Romania; Sardinia and Sicily. Not reported from several countries within this range. In Asia extending at least to Altai Mountains. Far East Asia (Lvovsky, 2004).

BIONOMICS. Larvae on *Seseli libanotis* (L.) W.D.J. Koch (Hannemann, 1953). Rarely on *Laserpitium latifolium* L. (De Prins & Steeman, 2003–2023). At higher altitudes on *Laserpitium halleri* Crantz (Sonderegger, *in litt.*), *Ligusticum lucidum* Mill. and *L. ferulaceum* All. (Rymarczyk *et al.*, 2015a). Lhomme (1945) adds *Athamanta cretensis* L., *Peucedanum cervaria* (L.) Lapeyr. and *Endressia pyrenaica* (DC.) J. Gay, but these are unconfirmed. Reports from *Pimpinella anisum* L. refer to *Depressaria bantiella*. Larvae on *Daucus carota* L. (Hannemann, 1953), were referred to *D. daucivorella*, which we treat as a synonym of *D. libanotidella*. Larvae May to August, rarely to September. Adults from July, but may emerge as late as September, to June (rarely July) of the following year.

REMARKS. *D. daucivorella* Ragonot, 1889 (recorded from France, Switzerland and Italy) has long been treated as a separate species from *D. libanotidella*. Hannemann (1995, 1996) continued to treat it as a good species differing from *libanotidella* in more yellow-grey coloration and in host-plant. However in his 1995 work he remarked that male genitalia show no difference from *libanotidella*. In consequence we have treated it as synonymous with *libanotidella*, but its status deserves further investigation.

Based on male and female genitalia, which show considerable variation and on barcodes, which indicate at least two distinct groups under the name *libanotidella*, it appears that *D. libanotidella* is a species complex. Externally specimens look very similar, as also do *D. bantiella*, *D. villosae*, *D. pimpinellae* and some specimens of *D. velox*. Complete resolution of the taxonomy of this complex is not possible in the short term because there is insufficient material available for study from the full geographical range. We therefore treat *libanotidella* as a single variable species with synonyms *laserpitii* Nickerl, 1864 and *daucivorella* Ragonot, 1889. We have not examined *Depressaria mesopotamica* Amsel, 1949 which according to Hannemann (1958a), is a synonym of *daucivorella*.

165   *Depressaria bantiella* (Rocci, 1934)

Schistodepressaria libanotidella f. *bantiella* Rocci, 1934: 222.

DESCRIPTION. Wingspan (20–)22–28 mm. Head orange-brown to dark fuscous, scales tipped buff in paler forms, face whitish buff. Labial palp segment 3 three-quarters length of segment 2; segment 2 with outer and ventral sides deep brown, upper and inner sides buff, segment 3 yellowish, with a blackish ring near base and another before apex which is pale ochreous. Antenna dark fuscous. Thorax orange-brown, sometimes fuscous anteriorly or entirely fuscous. Forewing dull orange-brown to grey-brown mixed buff in costal one-third, abruptly changing on dorsal side of blackish line in cell to fuscous or dark brown in central and dorsal areas, extending to termen; scattered whitish grey or grey-buff scales sometimes more numerous to form a stripe in middle of cell, a weak angled fascia beyond end of cell and lines between veins in terminal area; a blackish subdorsal spot at base and a dot at base of costa; blackish lines in fold, beneath costa, in cell and between veins to termen; a series of blackish dots between veins around termen; fringe grey-brown, outwardly grey beyond a fringe line. Hindwing light grey, darker towards apex; fringe basally dark grey, whitish grey terminally. Abdomen grey-buff.

VARIATION. French specimens closely resemble *D. libanotidella* in coloration, but south-east European specimens are darker, with less pale and orange-brown scales on palps, head, thorax and forewing. In the darkest specimens the black streaks on the forewing are obscure, the angled fascia is absent and the paler costal area is suffused fuscous and does not reach costal margin. Such forms also have fuscous hindwing.

SIMILAR SPECIES. Externally similar to *D. libanotidella* (*q.v.*) etc.

MALE GENITALIA. Similar to *D. libanotidella*, but gnathos longer, about as long as anellus.

FEMALE GENITALIA. Anterior apophysis three-fifths length of posterior apophysis; segment VIII width:length 4:1; ostium near anterior margin of sternite with acutely pointed lateral lobes extending beyond antrum by less than width of antrum; antrum short, weakly sclerotised, somewhat funnel-shaped; ductus bursae thin-walled with a twisted swelling at one-quarter length which is longer than wide, then somewhat expanded again beyond middle; corpus bursae narrowly ovoid; signum small, wider than long, with large triangular teeth.

GENITALIA DIAGNOSIS. See under *D. libanotidella*.

DISTRIBUTION. Italy, France, North Macedonia, Greece, Cyprus.

BIONOMICS. Host-plants *Pimpinella anisum* L. (Rocci, 1934) and *Pimpinella peregrina* L. (Rymarczyk *et al.*, 2015a).

REMARKS. *Depressaria bantiella* was described from Italy as a form of *D. libanotidella* based on adults reared from *Pimpinella anisum* L. on which the larvae were a pest. Larvae have been found in France on *Pimpinella peregrina* L. by Rymarczyk and Dutheil which differ from those of *D. libanotidella* in having the anal plate black, not green (Rymarczyk *et al.*, 2015a). These authors raised *bantiella* to species and illustrated differences in male and female genitalia. *D. bantiella* has barcodes in three groups. Species status for *D. bantiella* is accepted here, but it should be included in any future investigation of the *D. libanotidella* complex.

166   *Depressaria velox* Staudinger, 1859

Depressaria velox Staudinger, 1859: 237.
   Depressaria tortuosella Chrétien, 1908a: 186.

DESCRIPTION. Wingspan (20–)22–27 mm. Head mid-brown to dark fuscous, scales tipped buff; face creamy white. Labial palp segment 3 three-fifths length of segment 2; segment 2 whitish buff on inner side, outer side cinnamon-brown with a few blackish scales, segment 3 whitish buff, a few dark grey scales at base and a dark grey ring beyond middle, extreme tip black. Antenna dark fuscous. Thorax light brown through dull chestnut-brown to fuscous or deep brown, usually paler posteriorly. Forewing light brown through dull chestnut-brown and reddish-brown to deep brown, when reddish or chestnut-brown, this coloration particularly in costal one-third of wing; scattered whitish scales, mainly near costa, in fold, in cell, usually forming a weak to conspicuous angled fascia beyond end of cell, some spots between veins close to termen; a black subdorsal spot at base, often obscured by general dark coloration of basal area, a black dot at base of costa; a broad black dash in cell from one-third through to end of cell sometimes reaching nearly to wing base, interrupted by one or two groups of whitish scales; a blackish line in fold and a series of black lines between veins to termen; a series of black dots between veins at termen; fringe deep brown, outer ends grey beyond fringe line. Hindwing pale grey, darker towards apex; fringe with broad dark grey basal line, light grey beyond, sometimes with additional faint lines. Abdomen pale grey to grey-buff.

VARIATION. The black lines can be very obscure in the darkest specimens. An angled pale fascia beyond the end of cell is sometimes present, sometimes obsolete or very faint. In different regions, the populations of *D. velox* appear very different. On the eastern coast of the Adriatic Sea and in France, specimens have at least the costal one-third of wing width dull chestnut-brown, this colour tending to penetrate other parts of the wing also; the pale fascia is faint. In Portugal, specimens are very dark brown over the whole wing surface, with rather few white scales, no angled fascia and obscure black markings. The type of *D. tortuosella* Chrétien, 1908 is pale brown mixed buff, with whitish grey marks in cell and angled fascia.

SIMILAR SPECIES. Specimens with costa chestnut-brown to reddish-brown externally resemble *D. libanotidella*, *D. bantiella*, *D. villosae* or *D. pimpinellae*.

MALE GENITALIA. Gnathos ovate, not exceeding curved digitate socii; valva parallel-sided, costal margin with small hump in middle and a small hook before end; sacculus ending in a short lobe directed towards costa; clavus short, lanceolate with bristles on the inner side; transtilla lobes large, not hood-like; anellus straight-sided; aedeagus quite stout, numerous cornuti in a group half as long as aedeagus.

FEMALE GENITALIA. Anterior apophysis one-half length of posterior apophysis; sternite VIII width:length 3:1; ostium slightly anterior to middle of sternite, with broad rim, tapering rapidly to narrowly pointed lateral lobes, extending beyond antrum to about width of antrum; antrum short, funnel-shaped, sclerotised posteriorly; ductus bursae thin-walled, with swelling at one-fifth length which is slightly longer than wide, with an appendix-like diverticulum near posterior end; corpus bursae elliptic; signum small, wider than long, with triangular teeth.

GENITALIA DIAGNOSIS. See *D. libanotidella*.

DISTRIBUTION. Appears to be strictly coastal in Portugal, Spain and probably France, with larvae on *Seseli tortuosum*. Further east from Italy through the Balkans to Ukraine it occurs mainly in the mountains; in Turkey, Kazakhstan and Iran it ascends to 3000 m. *S. tortuosum sensu stricto* is not present outside south-west Europe.

BIONOMICS. Larvae on *Seseli tortuosum* L. (Chrétien, 1908). Staudinger (1859) gives *Ferula* sp. as the host-plant, but this may be a misidentification of *S. tortuosum*. It has also been reared from *Ligusticum lucidum* Mill. (Sonderegger, *in litt.*). In the eastern part of its range the use of other host-plants is likely. Larvae spring to early summer, adults from June to April (rarely May) of the following year.

REMARKS. In view of the difference in habitat and host-plant between western and eastern populations, *D. velox* should also be included in any future investigation of the taxonomy of this group.

167   *Depressaria pimpinellae* Zeller, 1839

Depressaria pimpinellae Zeller, 1839: 195.
  Haemylis pulverella Eversmann, 1844: 568.
  Depressaria reichlini Heinemann, 1870: 173.
  Depressaria reichlini subsp. *hungarica* Szent-Ivány, 1943: 99.

DESCRIPTION. Wingspan 17–21 mm. Head mid-brown. Labial palp segment 3 five-sixths length of segment 2; segment 2 dark brown on outer side, inner side light brown, segment 3 buff with deep brown base and band between half and two-thirds. Antenna deep brown. Thorax buff to light brown, sometimes with reddish tinge, always darker towards head. Forewing light brown with slight reddish tinge; dark brown spot at base of costa often weak, a dark brown area at base of wing towards dorsal side; deep brown streak in mid-line of wing from one quarter to three-quarters, a double streak along fold, a short narrow streak beneath costa at one quarter, similar streaks originating beyond cell, between veins in posterior third of wing; grey-brown dots between vein ends around apex and along termen; scattered greyish buff scales most concentrated in middle of cell, in fold and beyond cell in mid-line, less concentrated in costal area, at base of streaks running to termen and costa and around vein ends; fringe grey-brown, paler towards tips. Hindwing light grey-brown, darker posteriorly; fringe grey-brown, paler terminally, with one grey fringe line. Abdomen dull grey-brown.

VARIATION. Ground colour sometimes darker brown. The dark streaks in the outer third of the forewing are sometimes almost obsolete, or may be thicker and longer. The greyish buff scales also vary considerably, sometimes almost absent, in other specimens with more numerous and larger concentrations.

SIMILAR SPECIES. The female is characterised by slightly reddish brown costal third of forewing almost without dark streaks, but compare *D. libanotidella*, *D. bantiella*, *D. villosae* and *D. velox*.

MALE GENITALIA. Gnathos short-stalked, elliptic, equalling or slightly longer than socii; socii digitate, angled close to base; valva with evenly rounded apex, costal margin with well-developed median hump, at end a strongly sclerotised nearly circular protuberance; sacculus ending at three-quarters valva length in a small lobe directed towards costa; clavus narrowly lanceolate; transtilla posteriorly with a pair of large hood-like lobes; anellus flask-shaped, reaching base of tegumen; aedeagus quite stout, with about eight elongate cornuti in an extended group.

FEMALE GENITALIA. Anterior apophysis two-fifths length of posterior apophysis; sternite VIII width:length 3:1; sternite with pair of unsclerotised triangles lateral to ostium, ostium in anterior half of sternite, ovate with pair of sclerites, in addition there is a pair of narrow plates on ventral margin of ostium, not always visibly distinct from sclerites; antrum not strongly sclerotised; ductus bursae thin-walled, with wide swelling on left side near posterior end, then shortly constricted before long expanded section, not clearly differentiated from corpus bursae; signum wider than long, with large triangular teeth, but with small-toothed anterior and posterior extensions.

GENITALIA DIAGNOSIS. Both sexes most similar to *D. radiella* (*q.v.*).

DISTRIBUTION. Most of Europe, but not in the south-west or Mediterranean islands; in the Balkans only recorded from North Macedonia and Greece; Asia: Turkey, Russia. Siberia, Turkmenistan, Kazakhstan (Lvovsky, 2004).

BIONOMICS. Larvae in April or May between two spun leaflets, from June onwards mainly on flowers of *Pimpinella saxifraga* L., *P. major* (L.) Huds. (Hannemann, 1953) and *P. nigra* Mill. (Sonderegger, *in litt.*). Adults emerging from June to September with a peak in August, flying to May (rarely June) of the following year.

REMARKS. Barcodes of *D. pimpinellae* from Austria and *D. villosae* from Portugal are identical.

168   *Depressaria villosae* Corley & Buchner, 2018

Depressaria villosae Corley & Buchner, 2018: 105.

DESCRIPTION. Wingspan 20–24 mm. Head red-brown. Labial palp segment 3 three-quarters length of segment 2; segment 2 deep brown with or without reddish tinge on outer side and on inner side ventrally, rest of inner side whitish to whitish buff, segment 3 deep brown on outer side except tip, inner side light grey-brown with whitish buff band before middle and tip whitish buff. Antenna dark grey-brown with indistinct deep brown rings. Thorax and tegula sometimes reddish brown, more often a mixture of various proportions of light grey-buff, grey-brown, deep brown and reddish brown or reddish brown-tipped scales. Forewing with costal third dull reddish brown, middle and dorsal thirds deep grey-brown or dark grey; blackish brown at extreme base of costa, beyond which some dark grey-brown scales usually mixed in at least to middle; scattered whitish scales in fold, in cell, beyond cell at three-quarters, around termen before terminal spots and in costal third; indistinct short blackish streaks in cell at one-third and two-thirds, 3–4 longer and narrower streaks between veins to termen before subterminal white spots; fringe light grey-brown, sometimes tinged reddish brown near apex and on dorsum, with more or less distinct fringe line. Hindwing grey, darker posteriorly; fringe light grey with darker basal fringe line. Abdomen grey.

VARIATION. Darkest specimens have reddish brown costal third reduced and tinged grey, and dark markings almost invisible on dark ground colour; paler specimens may have reddish brown tipped scales in areas between cell and fold and between fold and dorsum.

SIMILAR SPECIES. *D. libanotidella*, *D. bantiella*, *D. pimpinellae* and *D. velox* are all externally similar.

MALE GENITALIA. Gnathos elliptic, equalling or slightly longer than socii; socii digitate, angled close to base; valva almost parallel-sided, apex rounded, costal margin with low median hump, ending in a small hardly sclerotised, slightly curved digitate process, just exceeding margin before apex; sacculus ending at two-thirds valva length in a lobe directed towards costa; clavus narrowly lanceolate with bristles on the inner side, about same length as width of valva before apex; transtilla expanded in middle, posteriorly with a pair of large hood-like lobes; anellus flask-shaped, overlapping base of tegumen; aedeagus slender, with about six elongate cornuti present in a tight group.

FEMALE GENITALIA. Anterior apophysis just over half length of posterior apophysis; sternite VIII width:length 5:2; ostium in middle of sternite, lip narrow, acutely pointed laterally extending beyond antrum by about half width of antrum;

antrum short, funnel-shaped, slightly sclerotised; ductus bursae thin-walled, with wide swelling at one-sixth length, gradually widening to elliptic corpus bursae; signum medium-sized, wider than long, almost triangular with triangular teeth.

GENITALIA DIAGNOSIS. In male the digitate process before apex of valva is distinct. In female ductus bursae wider and the swelling in its posterior part less distinct than in rest of *D. radiella* subgroup.

DISTRIBUTION. Portugal (Trás-os-Montes and Beira Litoral), Spain (Toledo), Sicily and Greece.

BIONOMICS. Larvae spin the basal leaves of *Pimpinella villosa* Schousb. in May in Portugal; in Sicily and Greece the host-plant must be different as *P. villosa* is not present there (Corley & Buchner, 2018). Adults emerge in June, and have been taken in poor condition in October and November. Probably they overwinter in this stage. Iberian records have altitudinal range from 620 m to 990 m; in Greece it was collected at 1200 m.

REMARKS. Only males were reared from larvae prior to the description of this species. At the time an unidentified female from Sicily was considered as a possibility for the female of *villosae*, but that could not be proved. Subsequently females have been reared from larvae from the type locality in Portugal, allowing description of female genitalia and confirming the presence of *villosae* in Sicily.

169   *Depressaria bupleurella* Heinemann, 1870

Depressaria bupleurella Heinemann, 1870: 171.

DESCRIPTION. Wingspan 18–24 mm. Head dull brown to dull chestnut-brown, scales tipped buff. Labial palp segment 3 three-fifths length of segment 2; segment 2 pale buff on inner side, outer and ventral sides pale buff mixed light brown, segment 3 whitish, mixed with light brown and fuscous scales in basal one-third and beyond middle, not forming distinct rings, apex pale ochreous. Antenna fuscous with some buff scales mainly on lower side. Thorax deep brown, posteriorly tending to grey, tegulae more extensively grey posteriorly. Forewing elongate, dull chestnut-brown to dark fuscous with weak chestnut tinge, darkest in basal one-sixth, obscuring subdorsal spot at base; scattered grey-buff scales over most of wing except base, particularly near costa and dorsum, in middle of cell and in terminal area, usually forming a weak angled fascia beyond end of cell; dark brown to blackish streak in fold and three short lines or spots in cell; dots between veins around termen usually weak and obscure, sometimes better marked; fringe grey, more or less tinged chestnut-brown, sometimes with fuscous fringe line. Hindwing light grey, darker towards apex; fringe pale grey with one fringe line. Abdomen grey-buff.

VARIATION. Ground colour of forewing variable from chestnut-brown to dull dark brown, but usually retaining some chestnut coloration. Extent of grey-buff scattered scales variable, occasionally a hint of an angled fascia present.

SIMILAR SPECIES. *D. bupleurella* tends to be rather weakly marked, with costal half of wing more red-brown than dorsal half. *D. depressana* can appear very similar at cursory glance, but it is smaller and thorax usually yellowish, contrasting against darker forewings.

MALE GENITALIA. Gnathos short-stalked, ovate; socii reduced to small pads on posterior angles of tegumen; valva short and broad, costa with small rounded sinus slightly deeper than wide in distal three-fifth, ventral margin nearly straight, from two-thirds gradually curving into apex, sacculus one-third of valva width, ending in a short digitate process; clavus short, triangular, hairy; anellus flask-shaped; saccus short, rounded; aedeagus broad-based, gradually tapering to slender apex, one small and one larger cornutus.

FEMALE GENITALIA. Anterior apophysis slightly less than half length of posterior apophysis; segment VIII width:length 5:3; ostium in middle of segment; antrum cup-shaped, sclerotised posteriorly; ductus bursae long, mainly thin-walled, with ribbed swelling from half to three-quarters length; corpus bursae ovoid; signum subquadrate to nearly circular, with large triangular teeth.

GENITALIA DIAGNOSIS. In male the small, rounded sinus on costa is distinct, in *D. sarahae* this sinus is also present but it is much larger. In female the cup-shaped antrum in combination with long, narrow ductus bursae with longitudinally ribbed swelling near middle is distinct.

DISTRIBUTION. Mountain areas of Central Europe from France through to Romania and Ukraine. Russia (Chita Province), Turkey, Iran, Mongolia (Lvovsky, 2004).

BIONOMICS. Larvae on *Bupleurum falcatum* L. (Hannemann, 1953), *Bupleurum stellatum* L., *Bupleurum ranunculoides* L. (Sonderegger, *in litt.*), also on *Bupleurum rotundifolium* L., *Bupleurum angulosum* L. and *Bupleurum petraeum* L. (Lhomme, 1945). On *Bupleurum falcatum* resting in a tube formed by a longitudinally folded leaf (Lepiforum e.V., 2006–2023), on *Bupleurum stellatum* found between spun leaves and flowers (Schmid, 2019), April to July or August. According to Sonderegger (*in litt.*) larvae grow more slowly than any other *Depressaria* larvae. Adults from July onwards to April (rarely May) of the following year.

170   *Depressaria sarahae* Gastón & Vives, 2017

Depressaria sarahae Gastón & Vives, 2017: 352.
  Depressaria saharae [*sic*] subsp. *tabelli* Buchner, 2017: 143.

DESCRIPTION. Wingspan 18–23 mm. Head greyish brown, scales tipped buff. Labial palp segment 3 three-fifths length of segment 2; segment 2 pale buff on inner side, outer and ventral sides pale buff heavily mixed dark fuscous, segment 3 pale buff with some dark grey scales near base. Antenna dark fuscous with some buff scales mainly on lower side. Thorax and tegula dark brown near anterior margin, otherwise grey-buff, thorax with median and lateral longitudinal dark lines, tegulae more extensively grey posteriorly. Forewing elongate, grey-brown, with some admixture of light grey-buff scales particularly at base on dorsal side and in terminal third of wing, scattered grey-buff scales forming indistinct streaks near costa and in middle of wing and a V-shaped fascia beyond end of cell; dark grey to blackish spot at base of costa and lines between veins over most of wing; terminal dots blackish, distinct; fringe yellowish buff, mixed with dark grey scales, no clear fringe line. Hindwing light grey, darker towards costa and apex; fringe pale grey with one fringe line. Abdomen grey-buff.

VARIATION. In mainland Spain few specimens have been seen, all males. They belong to the nominotypical subspecies. In the Canary Islands, subspecies *tabelli* Buchner, 2017 occurs with different external appearance, see description below:

DESCRIPTION. Subspecies *tabelli* differs from subspecies *sarahae* in slightly larger size, wingspan 22–24 mm; head reddish brown mixed yellowish buff; labial palp segment 2 buff, outer and ventral sides with dark rusty brown scales, segment 3 with basal half light brown mixed dark grey, distal half blackish grey, extreme tip ochreous; thorax orange-brown strongly mixed yellowish buff, without longitudinal dark lines; forewing orange-brown, with a scattering of light grey-buff scales, and overlaid with fuscous scales; blackish markings and pale fascia similar to nominotypical subspecies, but blackish markings and scattered fuscous scales almost absent from much of costal one-third. Hindwings paler than in subsp. *sarahae*.

SIMILAR SPECIES. Subsp. *sarahae* is similar to several grey species with distinct longitudinal streaks on forewings, but in these species the streaks are either upcurved distally in costal half (e.g. *D. cervicella*, *D. fuscovirgatella*) or larger areas are unstreaked (e.g. *D. zelleri*); subsp. *tabelli* is similar to other warm brown species, determination should be based on genitalia.

MALE GENITALIA. Similar to *D. bupleurella*, but socii even more reduced, costa with rounded lobe larger and sinus in distal three-fifth larger and wider than deep; ventral margin concave at three-fifths, sacculus less than one-third of valva width, ending in a short digitate process, larger than that of *bupleurella*; clavus short, triangular, spiny; aedeagus broad-based nearly parallel-sided from above base to apex, one cornutus.

FEMALE GENITALIA. Subspecies *tabelli*. Anterior apophysis three-fifths length of posterior apophysis; segment VIII width:length 3:1, middle section of posterior margin parallel to anterior margin at base of wide excavation; ostium slightly beyond middle of segment; antrum narrowly funnel-shaped, sclerotised laterally; ductus bursae long, mainly thin-walled, swollen in middle, swelling slightly sclerotised on one side; corpus bursae small, elliptical; signum three times as wide as long, with large triangular teeth, smaller in middle.

GENITALIA DIAGNOSIS. Male distinct, but compare *D. bupleurella*. In female (subspecies *tabelli*) narrowly funnel-shaped antrum in combination with long ductus bursae with swelling in the middle slightly sclerotised on one side is distinct.

DISTRIBUTION. Subsp. *sarahae* is only known from mainland Spain. Subsp. *tabelli* is so far only known from Tenerife, Canary Islands.

BIONOMICS. Larvae of subspecies *tabelli* on *Bupleurum salicifolium* subsp. *aciphyllum* (Webb & Berthel.) Sunding & G. Kunkel. Larvae of subsp. *sarahae* are unknown, but the host-plant is likely to be a *Bupleurum*. Fresh adults of subspecies *sarahae* have been collected from June to October; we have not seen worn specimens. A rather worn specimen of subspecies *tabelli* was taken in January, a very fresh one 23 April.

REMARKS. In Buchner *et al.*, 2017, *D. sarahae* was misspelled as *saharae* throughout.

Subsp. *tabelli* has male genitalia extremely similar to subspecies *sarahae*. Female genitalia of the typical subspecies remain unknown.

Group 5 *badiella* group (171–173)

Male genitalia with gnathos divided into two parts; valva ending in three lobes, clavus and sacculus present; aedeagus with cornuti. Female genitalia with anterior apophysis broadly triangular; sternite VIII largely membranous.

171   *Depressaria badiella* (Hübner, 1796)

Tinea badiella Hübner, 1796: pl. 14, fig. 92.
   Depressaria corticinella Zeller, 1854: 328, syn. n.
   Depressaria brunneella Ragonot, 1874: 2585.
   Depressaria aurantiella Tutt, 1893: 241.
   Depressaria uhrykella Fuchs, 1903: 244.
   Depressaria frigidella Turati, 1919: 335.
   Depressaria badiella ssp. *frustratella* Rebel, 1936a: 97.

DESCRIPTION. Wingspan 19–26 mm. Head mid-brown. Labial palp segment 3 three-fifths length segment 2; segment 2 outer side with mixture of whitish buff and mid to dark brown scales, inner side whitish buff, mid to dark brown ventrally, segment 3 mid-brown, dark brown base on inner side and ring beyond middle, tip ochreous buff. Antenna light greyish buff with narrow grey-brown rings. Thorax mid-brown with a few buff scales, posteriorly mainly buff, especially on tegulae. Forewing mid-brown, more or less heavily overlaid with deep brown scales, less so beneath costa and between fold and dorsum from one-sixth to one-third; small blackish brown spot at base of costa and a larger subdorsal spot; short broad blackish lines at base of cell, above this beneath costa, at end of cell and beneath costa at half; fine blackish streaks between most veins, terminating in grey-brown spots at margin from costa at three-quarters around termen; a few scattered whitish buff scales mainly in cell and forming an angled fascia beyond the end of cell; fringe dark grey-brown without obvious fringe line. Hindwing light brownish grey, darker posteriorly; fringe grey-brown or light grey-brown with weak fringe line. Abdomen dark grey-brown.

VARIATION. Forewing ground colour varies from mid-brown to deep brown; sometimes with some pinkish scales on costa and fringe with a pinkish tinge. A form with a broad pale band mixed with ground colour lying along the costa is moderately frequent. Dark specimens have inconspicuous dark streaks. The pale angled fascia is almost obsolete in some specimens, moderately evident in others.

SIMILAR SPECIES. *D. badiella* is characterised by its fairly large size, dark brown forewings with weak pale fascia and often obscure dark streaks. *D. pseudobadiella* (*q.v.*) is not distinguishable externally. *D. subnervosa* (*q.v.*) is also similar. Due to external variability it is recommended to check genitalia for safe determination.

MALE GENITALIA. Gnathos divided into a pair of subspherical spinulose lobes; socii digitate; valva broad, ending in three lobes, a high hump on costal side, a long beak-like process with parallel-sides at right-angle or greater angle to costal hump, and a rounded hump at end of ventral margin, the sinus between this and the beak-like process often parallel-sided; clavus stout, reaching three-fifths to two-thirds length of valva; anellus distally with two small triangles separated by wide shallow V-shaped sinus; aedeagus with a group of 5–14 long cornuti.

FEMALE GENITALIA. Posterior apophysis normal, two to two and a half times length of papilla analis; anterior apophysis broad, triangular; sternite VIII width:length variable, from 2:1 to nearly 4:1, with concave anterior margin; ostium on anterior margin of sternite, a sclerotised ring with short lateral extension or none; antrum funnel-shaped or cylindrical, not or weakly sclerotised; ductus bursae thin-walled, with little change in width; corpus bursae elliptical; signum large, about twice as long as wide, rhomboidal with sides concave, with triangular teeth largest across mid-line.

GENITALIA DIAGNOSIS. Male unique for species pair *D. badiella/pseudobadiella*, but these two species are not distinguishable. In female combination of broad triangular anterior apophysis and large signum about twice as long as wide is unique.

DISTRIBUTION. All Europe. Russia (southern Siberia, Sakhalin), Turkey, Georgia, Armenia, Azerbaijan, Kazakhstan, Mongolia; Libya (Lvovsky, 2004). Some of these records may belong to *D. pseudobadiella*, then undescribed.

BIONOMICS. Host-plants various Asteraceae tribe Cichorieae including *Hypochaeris radicata* L., *Sonchus arvensis* L., *Taraxacum* F.H. Wigg. (Meyrick, 1927); also on *Leontodon* L. and *Hieracium* L. (Lhomme, 1945). Host-plants in the Apiaceae have been repeatedly mentioned in the literature, but see Remarks below. According to Harper *et al.* (2002), the first instars spin two leaves together and eat small blotches in these leaves. Later the larva enters the soil and constructs a silken tube adjacent to the root of the host-plant on which it feeds. Larva probably from early spring to June or July. Pupation inside the silken tube. We have never seen the larvae ourselves. Adults from April to December.

REMARKS. *D. aurantiella* Tutt, 1893 is a dark form with orange palps. We suspect that this was a unique aberration and of no taxonomic value. *D. corticinella* Zeller has genitalia within the variation of *D. badiella*, so is placed in synonymy here. Hannemann (1976a) placed *frigidella* Turati, 1919 in synonymy with *badiella*. In the same work he treated *D. uhrykella* Fuchs, *D. brunneella* Ragonot and *D. frustratella* Rebel as subspecies of *badiella*. *D. badiella* shows considerable variation wherever it occurs, so we can find no justification for the recognition of subspecies.

In the 19th century the name *D. pastinacella* was wrongly applied to *D. badiella* by some authors, resulting in *Pastinaca sativa* L. and *Heracleum sphondylium* L. being given as host-plants for *badiella*.

172   *Depressaria pseudobadiella* Nel, 2011

Depressaria pseudobadiella Nel, 2011: 4.

DESCRIPTION. Wingspan 23–25 mm (Nel, 2011), which is in the upper range for *D. badiella*, otherwise the species is indistinguishable externally from *D. badiella*. Few specimens known so far, so smaller specimens may eventually be found.

VARIATION. Quite probably this species is just as variable externally as *D. badiella*. There is some variation in the genitalia.

SIMILAR SPECIES. *D. badiella*, *D. subnervosa* (*q.v.*). Due to external variability it is recommended to check genitalia for safe determination.

MALE GENITALIA. Nel (2011) indicates several differences in male genitalia between *badiella* and *pseudobadiella*, but there is considerable variation in the characters mentioned. Possibly there is a tendency for *pseudobadiella* to have more cornuti (10–12 against 5–10) and a longer aedeagus. Shape of anellus and of the end lobes of the valva is too variable to be useful. Possibly the gnathos lobes are different in *pseudobadiella*, being elliptical rather than round, but this needs confirmation.

FEMALE GENITALIA. As in *badiella* except the signum which is rhomboidal in both species, around five times as long as wide in *pseudobadiella* but only twice as long as wide in *badiella*, but there is some variation in both species. In specimens we have examined the anterior margin of sternite VIII is broadly convex, but this character is not figured by Nel (2011) so may be variation within the small sample examined.

GENITALIA DIAGNOSIS. Male characters unique for species pair *D. badiella*/ *pseudobadiella*, but these two species are not distinguishable. In female, combination of broad triangular anterior apophysis and very large signum, about five times as long as wide is unique.

DISTRIBUTION. Present in Spain, Portugal, France including Corsica, Italy including Sardinia. Also found in Turkey (Asian part).

BIONOMICS. Host-plant unknown. Flight period of checked specimens falls within that of *D. badiella*.

REMARKS. *Depressaria pseudobadiella* Nel, 2011 is indistinguishable externally and by male genitalia from *D. badiella*. In the female genitalia the strongly elongated signum is distinct, but the differences in the male genitalia figured by Nel (2011) are insufficient for reliable identification. The signum is frequently a variable structure in *Depressaria*, but it is the sole morphological means of distinguishing *pseudobadiella*, so justification for treating this as a distinct species is weak. This form of signum has been found in specimens from several departments in the south of France and also from Portugal, Spain, Italy and Turkey. Barcode difference of 3.43% from *D. badiella* has been found, which would suggest that *D. pseudobadiella* is a valid species, but two specimens (from France and Italy) with signum of the *pseudobadiella* form share barcode with *badiella*. Further evidence is needed to establish the status of this taxon, which we provisionally treat as a species separate from *D. badiella*.

173   *Depressaria subnervosa* Oberthür, 1888

Depressaria subnervosa Oberthür, 1888: 42.
   Depressaria fusconigerella Hannemann, 1990: 140, syn. n.

DESCRIPTION. Wingspan 20–24 mm. Head with face yellowish brown, vertex grey-buff to dark brown. Labial palp segment 2 grey-brown, segment 3 yellowish-grey, with ill-defined blackish ring before apex. Antenna blackish brown. Thorax and tegula grey-buff to dark brown, yellowish brown posteriorly. Forewing mid-brown to grey-brown with a few buff scales, particularly in cell and adjacent to fold, also forming an indistinct angled fascia beyond end of cell, sometimes forming a broad grey-buff band along costa with extensions to cell between black lines, then also with a series of spots from outer third of costa, round apex and termen; a blackish brown subdorsal spot; blackish lines in cell and fold and between veins towards end of costa and around termen; terminal dots weakly developed; fringe concolorous with forewing. Hindwing light grey, darker posteriorly; fringe light grey-brown with distinct basal fringe line. Abdomen grey-brown.

VARIATION. Ground colour grey-brown without contrasting pale markings, or mid-brown with strongly contrasting pale markings, particularly along costa.

SIMILAR SPECIES. Externally not distinguishable from *D. badiella* and *D. pseudobadiella*.

MALE GENITALIA. Gnathos divided into a pair of subspherical spinulose lobes on stout stalks; socii much reduced, on posterior angles of tegumen; valva with ventral margin almost straight, ending in a blunt triangular lobe, on costal side a high hump followed by a digitate lobe almost parallel with lobe at end of ventral margin, separated from it by a deep sinus; clavus absent; transtilla a flattened V-shape, pointing posteriorly with a long hairy triangular structure from its middle reaching base of gnathos arms; anellus subrectangular, wider than long, with a large parabolic incision on posterior side. Aedeagus long, tapering from broad base, without cornuti.

FEMALE GENITALIA. Posterior apophysis normal, two and a half to four times as long as papilla analis; anterior apophysis broad, triangular; sternite VIII from slightly longer than wide to slightly wider than long, anterior margin slightly extended; ostium on anterior margin of sternite, bowl-shaped with pair of triangular processes on posterior margin; ductus bursae thin-walled, gradually enlarged to ovate or elliptic corpus bursae; signum small, wider than long, rhomboidal or triangular with triangular teeth.

GENITALIA DIAGNOSIS. Male unique. In female combination of broad triangular anterior apophysis and small signum not longer than wide is unique.

DISTRIBUTION. Spain (Almería). Originally described from Oran (Algeria).

BIONOMICS. Host-plant unknown. Checked specimens were collected from April to September.

REMARKS. The characteristic shape of the valva is similar to that of *D. badiella* which also has divided gnathos, otherwise male genitalia show little similarity. However female genitalia place this species close to *D. badiella*.

Group 6 *veneficella* group (174–185)

Large species with long forewings. Hindwing with large dorsal lobe and quite acute apex. Scales unusually long on forewings, thorax and face. Male genitalia: gnathos lanceolate, long-stalked (except *D. altaica*); socii small, club-shaped; costal margin of valva convex in basal part, sometimes with additional low lobe or lobes, apex incurved, more or less hooked, sacculus wide (half as wide as valva) and long, with one or two lobes, clavus absent; saccus usually elongate; aedeagus elongate with single, often long cornutus. All known larvae on Apiaceae. This is a group containing many closely related species which cannot be divided into convenient subgroups.

Key to male and female genitalia of *Depressaria veneficella* group.
The inserted figures are not intended to replace comparison of genitalia on the plates.

Male genitalia

| 1a | Valva more or less sharply pointed | 2 |
| 1b | Valva broad in distal part, blunt (not yet reported from Europe) | 14 |

2a (1a)	Saccus short, triangular (length about equalling width at base)	3
2b	Saccus clearly longer than width at base	5

3a (2a)	Gnathos short (length about equalling width), valva with subapical sharp point	*D. altaica*
3b	Gnathos long (length at least 2 × width)	4

4a (3b)	Valva short, broad	*D. cervicella*

4b	Valva long, narrow	*D. gallicella*

5a (2b)	Valva gradually tapering to a sharp tip, but often slightly more strongly bent inward near tip	6
5b	Outer edge of the valva with a strong bend shortly before tip ("subapical hump") then running straight to the tip, this part running downwards in standard preparation	11

6a (5a)	Cornutus longer than half of aedeagus length	7
6b	Cornutus shorter than half of aedeagus length	9

7a
(6a)
Distal part of valva, beyond median bulge, slender, length to width ratio of this part 3:1 or more

D. eryngiella

7b
Distal part of valva, beyond median bulge, not so slender, length to width ratio of this part clearly below 3:1

8

8a
(7b)
Saccus about 3 times as long as wide at base, valva only slightly more strongly bent near tip

D. deverrella

8b
Saccus about 2 times as long as wide at base, valva distinctly more strongly bent near tip

D. hansjoachimi sp.n.

9a
(6b)
Cornutus short, shorter than one-third of aedeagus

D. veneficella

9b
Cornutus longer than one-third of aedeagus

10

10a
(9b)
Saccus less than half as long as valva

D. albarracinella

| 10b | Saccus longer, more than half as long as valva | *D. discipunctella* |

| 11a (5b) | Outline of valva from tip to tegumen rather straight, without a strong bulge | 12 |
| 11b | Outline of valva from tip to tegumen with a strong bulge, forming a deep semicircular excavation between tip and bulge | 13 |

| 12a (11a) | Valva rather slender in distal part, cornutus half of aedeagus length or slightly longer | *D. pentheri* |

| 12b | Valva more broad in distal part, cornutus less than half of aedeagus length | *D. junnilaineni* |

| 13a (11b) | Cornutus very short (about one fifth of aedeagus length), valva very broad in distal part | *D. erzurumella* |

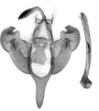

| 13b | Cornutus one-third of aedeagus length or slightly longer, valva more slender in distal part | *D. hannemanniana* |

| 14a (1b) | Gnathos short, length 1–1.5 × width, saccus long, parallel-sided | *D. kailai* (Turkey, Armenia, Kyrgyzstan, Tajikistan) |  |

| 14b | Gnathos longer, length 1.5–3 × width, saccus more or less triangular | 15 | |

| 15a (14b) | Saccus short triangular (shorter than width at base), one stout cornutus less than one third aedeagus length | *D. rjabovi* (Asia: Turkmenistan) |  |

| 15b | Saccus longer, cornutus about one half aedeagus length | 16 | |

| 16a (15b) | Saccus slightly longer than width at base, one rather slender cornutus | *D. almatinka* (Asia: Kazakhstan) |  |

| 16b | Saccus clearly longer than width at base, one stout cornutus | *D. ivinskisi* (Asia: Tajikistan, Afghanistan) | |

Female genitalia (females of *D. altaica*, *D. erzurumella*, *D. rjabovi* and *D. almatinka* unknown)

| 1a | Ductus bursae gradually expanding to corpus bursae, without distinct structures there | 2 | 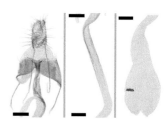 |

1b   Ductus bursae narrow and more or less straight to slightly bent over most of its length, shortly before it meets bursa expanded and strongly bent with distinctly thickened wall on concave side, then usually again constricted until it abruptly widens to corpus bursae   5

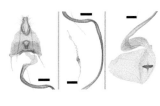

2a (1a)   Anterior margin of sternite with distinct median sinus   *D. albarracinella*

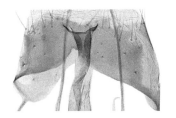

2b   Anterior margin of sternite slightly concave, straight or slightly convex   3

3a (2b)   Ostium in middle of sternite, sternite distal of ostium evenly sclerotised, anterior margin of sternite slightly convex   *D. gallicella*

3b   Ostium slightly distal of middle of sternite, sternite distal of ostium less sclerotised or excavated   4

4a (3b)   Sternite distal of ostium less sclerotised, ductus bursae with swelling at two-fifths from antrum   *D. cervicella*

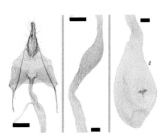

4b	Sternite distal of ostium with rectangular or trapezoid excavation, ductus bursae without swelling	*D. kailai* (not in Europe)	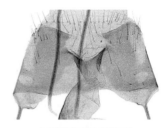
5a (1b)	Anterior margin of sternite strongly sclerotised with conical projections on either side of antrum	*D. veneficella*	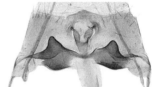
5b	Anterior margin without such projections	6	
6a (5b)	Ostium near posterior margin of sternite but well separated from margin (if signum exceptionally small, cf. *D. hansjoachimi*)	7	
6b	Ostium open to posterior margin of sternite	8	
7a (6a)	Margin of sternite with distinct apiculus posterior of ostium	*D. hannemanniana*	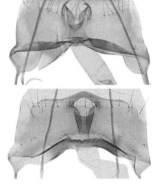
7b	Margin of sternite with slight bulge or nearly straight posterior of ostium	*D. discipunctella, junnilaineni, pentheri, deverrella*	

Four species show this feature, they are not safely distinguishable by female genitalia, although in *D. junnilaineni* and

D. pentheri the slight bulge posterior of ostium tends to be more distinct than in *D. discipunctella* and *D. deverrella*; ductus bursae tends to be shorter in *D. junnilaineni* than in the other compared species.

8a (6b)	Anterior margin of sternite with median excavation and pair of distinct folds directed towards antrum, ostium close to posterior margin of sternite, always open	*D. eryngiella*	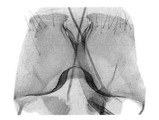
8b	Anterior margin of sternite with median excavation or straight, folds indistinct or absent	9	

9a (8b)	Ostium close to posterior margin of sternite, open or (rarely) closed by a narrow scletotised band, anterior margin of sternite with distinct median excavation, signum exceptionally small	*D. hansjoachimi* sp.n.	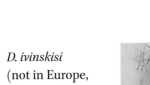
9b	Ostium distant from posterior margin of sternite, but sternite distal of ostium unsclerotised, anterior margin of sternite straight, signum of average size for this group	*D. ivinskisi* (not in Europe, only one female known so far)	

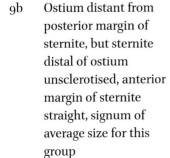

174   *Depressaria cervicella* Herrich-Schäffer, 1854

Depressaria cervicella Herrich-Schäffer, 1854: 130.

DESCRIPTION. Wingspan 18–26 mm. Head fuscous to dark fuscous. Labial palp segment 3 just over half as long as segment 2; segment 2 with inner side and apex pale pinkish buff, outer and ventral sides dark fuscous, segment 3 pale pinkish buff, base and a few scales at apex grey-fuscous. Antenna ringed light and dark fuscous. Thorax fuscous to dark fuscous. Forewing fuscous to fuscous-buff, a blackish spot towards dorsal side at base, long black lines between veins running into terminal dots around apex and most of termen, similar lines in cell, two or three lines crossing lines in cell at a narrow angle, the first following fold to a variable distance; fringe fuscous, with or without a weak fringe line. Hindwing light grey, grey-brown towards apex, with a fine darker grey line around margin; fringe light grey-brown, with one more or less faint fringe line. Abdomen fuscous-buff.

VARIATION. The length and strength of the black lines is variable, particularly the first line crossing cell at narrow angle which may continue far along fold.

SIMILAR SPECIES. The almost uninterrupted black lines running into the terminal dots and the black lines crossing the cell at a narrow angle are characteristic features, but shared with *D. pyrenaella*, which usually has paler ground colour.

MALE GENITALIA. Valva weakly convex in basal half of costal margin, distal half broad, curved inwards, strongly tapering to narrow apex in distal quarter, sacculus more than half width of valva, an asymmetrical hump in basal half, a terminal process just reaching costal margin; anellus broadly pear-shaped; saccus short and wide, apex obtuse; aedeagus straight, almost as long as valva, with single cornutus about quarter of its length.

FEMALE GENITALIA. Posterior apophysis about twice as long as anterior apophysis; sternite VIII width:length 4:1; ostium in middle of sternite, with lips on lateral edges, antrum heavily sclerotised, funnel-shaped; long, thin-walled ductus bursae with a swelling at two-fifths from antrum, swollen part with minute papillae; corpus bursae ovate; signum small, 'moth-shaped' with a body of diffuse minute teeth and a pair of wings with larger triangular teeth.

GENITALIA DIAGNOSIS. See key on pp. 328–335.

DISTRIBUTION. Austria (apparently extinct), Hungary (no recent records) and Czechia (not seen since 1973, J. Správce, pers. comm.). Turkey, Iran, Mongolia (Lvovsky, 2004).

BIONOMICS. Host-plant unknown, but likely to be an Apiaceae. Adults hibernate, one checked specimen was collected end of April.

REMARKS. The description of larva and larval host-plant given in Hannemann (1995) were based on a misidentification of *D. zelleri* by Klimesch (1953) corrected by Lvovsky (2001). This may also be the source of the record from France in various works. We have not been able to confirm its presence in France or Spain; the latter may be based on *D. pyrenaella* before it was described.

175   *Depressaria gallicella* Chrétien, 1908

Depressaria gallicella Chrétien, 1908: 127.
   Depressaria quintana Weber, 1945: 374.

DESCRIPTION. Wingspan 20–26 mm. Head pale grey-buff. Labial palp segment 3 two-thirds to three-quarters length of segment 2; segment 2 entirely pale flesh-coloured to pale greyish buff with outer and ventral sides more or less grey-fuscous, segment 3 flesh-coloured or greyish buff, with a grey-fuscous ring beyond middle, apex ochreous-buff. Antenna pale grey, narrowly ringed dark fuscous. Thorax buff to light grey-buff. Forewing light grey-fuscous, costa broadly light grey-buff to about middle; a blackish spot at base of dorsum, a blackish dot at base of costa, black dots in cell at one-third and at end of cell, a short dash in fold at two-fifths, sometimes extended or with another faint dash beyond, sometimes some faint blackish lines between veins to termen, dark grey dots from end of costa around termen; indistinct streaks paler than ground colour in cell and fold and forming an angled fascia beyond end of cell; fringe pale grey-buff without fringe line. Hindwing pale grey, more or less darkened towards costa and termen; fringe whitish grey with weak fringe line. Abdomen grey-buff.

VARIATION. Most variation is in the strength of the dark dots in cell and fold, with the median cell dot often obsolete. Some specimens have a weakly defined grey-fuscous spot obliquely placed before the first cell dot.

SIMILAR SPECIES. *D. pentheri* is externally extremely similar, but has different distribution. The cell and plical dots are reminiscent of the markings of some *Agonopterix* species.

MALE GENITALIA. Valva slightly convex in basal half of costal margin, distal half narrow, curved inwards from one-half to three-quarters, final quarter nearly straight, sacculus wide, basal third recurved, a pointed hump at one-half, then extending nearly to apex; anellus wider than long, widest near base, distal margin with wide sinus; saccus as wide as long, margins slightly concave, apex pointed; aedeagus lightly curved above base, slender, about half as long as valva, with single cornutus about half its length.

FEMALE GENITALIA. Posterior apophysis about twice as long as anterior apophysis; sternite VIII width:length 3:1; ostium in middle of sternite, antrum heavily sclerotised, tapering to long thin-walled ductus bursae; corpus bursae long ovate; signum very small, wider than long, with a few triangular teeth.

GENITALIA DIAGNOSIS. See key on pp. 328–335.

DISTRIBUTION. France, Italy and Switzerland. Also in Asia (Armenia, Kazakhstan). Uzbekistan (Lvovsky, 2004).

BIONOMICS. Larvae on *Laserpitium gallicum* L. (Chrétien, 1908), *L. prutenicum* L. (Hannemann, 1953) and *L. siler* L. (Sonderegger, *in litt.*) in loosely spun leaves or umbels, May to June, rarely July, adults from July to May of the following year.

176   *Depressaria altaica* Zeller, 1854

Depressaria altaica Zeller, 1854: 309.

DESCRIPTION. Wingspan 22–26 mm. Head grey-buff. Labial palp buff, segment 3 with dark grey ring beyond middle, apex buff. Antenna fuscous. Thorax and tegula grey-buff. Forewing elongate, light grey-brown, dusted with light fuscous over whole wing except a broad stripe along costa to one-half; a dark brown subdorsal spot; a thin blackish line from middle of base to one-sixth, two short lines in cell and one between end of cell and costa, other lines between veins and dots around termen very weakly expressed; fringe grey-buff. Hindwing whitish, grey towards apex; fringe whitish to light grey with darker line. Abdomen grey-buff.

VARIATION. We have seen very little material of this species, so there may be more variation than described. Ground colour can be light grey-brown or mid-brown, the costa has a more conspicuous pale band in basal third in greyer specimens. There is no angled fascia beyond cell.

SIMILAR SPECIES. *D. discipunctella* has a series of short dark dashes, usually four, along mid-line of cell.

MALE GENITALIA. Gnathos short-stalked, rounded, slightly longer than wide; socii small, incurved; valva with long hump on costal margin followed by a short hump at one-half, distal part of valva narrow, incurved, ending in large spine directed towards end of tegumen, sacculus wide, running into apex, almost as wide as valva in distal half, with a hump at one-third; anellus wider than long, widest just above base, narrowed to obtuse, slightly emarginated apex; saccus short, triangular, bluntly pointed; aedeagus lightly curved, two-thirds length of valva, slender with a single cornutus two-thirds its length.

FEMALE GENITALIA. Unknown.

GENITALIA DIAGNOSIS. See key on pp. 328–332.

DISTRIBUTION. Russia (Orenburg). Kazakhstan.

BIONOMICS. A very fresh specimen was collected end of September, the fresh lectotype in December, and a moderately worn specimen in May, indicating this species hibernates in the adult stage. Host-plant unknown, but likely to be an Apiaceae, see also Remarks.

REMARKS. According to the original description, the type series consists of two males and one female. One male is lost, the other male was selected as lectotype (Hannemann, 1953). The female was dissected during preparation of this book and found to belong to *D. schaidurovi* Lvovsky, 1981. Other females stored in different collections as *D. altaica* have been found to belong to *D. eryngiella*. One male without abdomen and without corresponding genitalia slide, with label data "Altai, Mongolia, 3. 13" is stored in NHMUK. Based on external appearance it might be *D. altaica* but also e.g. *D. rjabovi* Lvovsky, 1990 or *D. ivinskisi* Lvovsky, 1990, so correctness of determination must remain open. In total we have found only three confirmed specimens: one is the lectotype, collected in Kazakh Altai and two others collected by K. Nupponen in Russia, Orenburg district, Schibendy Valley.

Lvovsky (2001) reported *D. altaica* from Tajikistan, with additional information "The only moth examined by the author was reared ..... from larva collected in the Kondara Gorge (Tajikistan) ..... Larva fed on *Prangos pabularia* Lindl ..." Male genitalia of this specimen are depicted, but do not belong to *D. altaica*. External appearance of the moth was checked during a visit to ZIN, confirming that it is not *D. altaica*, but determination was not possible.

177   *Depressaria albarracinella* Corley, 2017

Depressaria albarracinella Corley in Buchner, Corley & Junnilainen, 2017: 121.

DESCRIPTION. Wingspan 22–26 mm. Head cinnamon-brown on neck and crown; face light brownish buff. Labial palp with segment 3 two-thirds length of segment 2; segment 2 buff with tufted scales on ventral side cinnamon; segment 3 cinnamon with dark grey ring beyond middle, tip cinnamon-buff. Antenna light grey-brown, narrowly ringed dark brown. Thorax light brownish buff. Forewing light brown, often with slight cinnamon tinge, subdorsal spot absent or weak, very weakly marked but sometimes with more or less faint grey-brown interrupted streaks in cell, in fold, beyond cell, between veins to costa and between veins to termen; equally indistinct grey-brown spots between vein-ends at termen; fringe light brown, without obvious fringe line. Hindwing light grey, slightly darker posteriorly, with narrow grey-brown line around terminal and dorsal margins; fringe light grey-brown at apex to almost white at dorsal base, with a fine darker fringe line. Abdomen light grey-brown.

VARIATION. The forewing markings vary from almost completely obsolete to present but generally faint compared with other *Depressaria* species.

SIMILAR SPECIES. *D. albarracinella* is characterised by the uniformly coloured forewings usually with weak or obsolete markings; the basal dark spots which occur so widely in *Depressaria* are absent or the subdorsal spot may be present but weak. There is usually no angled fascia beyond cell, only rarely developed but faint. Specimens with stronger markings can be undistinguishable from further species of this group, e.g. *D. eryngiella* and *D. veneficella*.

MALE GENITALIA. Socii as long as stalk of gnathos; valva with broad base, costal margin with a weak hump in basal half and a larger one in outer half, apex incurved, sacculus with two lobes, the inner broadly triangular, the second longer and narrower, slightly incurved, sacculus running on into apex; anellus broadly pear-shaped; saccus triangular, longer than wide; aedeagus lightly curved, slender, cornutus about two-fifths of its length.

FEMALE GENITALIA. Posterior apophysis about twice as long as anterior apophysis; sternite VIII width:length 5:1, anterior margin of sternite with wide median sinus; ostium near posterior margin of sternite, narrow curved margin extending beyond edges of antrum, antrum heavily sclerotised, slightly tapering to ductus bursae with fine longitudinal folding and abundant minute papillae; corpus bursae ovate; signum small, wider than long, with triangular teeth.

GENITALIA DIAGNOSIS. See key on pp. 328–335.

DISTRIBUTION. Spain, France, Greece, Turkey (Asian part).

BIONOMICS. Host-plant unknown. Adult moths have been taken from May to July and in October, and also in March after hibernation.

178   *Depressaria eryngiella* Millière, 1881

Depressaria eryngiella Millière, 1881: 7.
    Depressaria campestrella Chrétien, 1896: 104.
    Depressaria deliciosella Turati, 1924: 174.
    Depressaria obolucha Meyrick, 1936b: 51.

DESCRIPTION. Wingspan 18–24(–26) mm. Head pale brown. Labial palp segment 3 one-half length of segment 2; segment 2 pale buff, outer side with a few light brown scales, segment 3 pale buff with a blackish ring in middle. Antenna light brown, narrowly ringed fuscous. Thorax light brownish-buff. Forewing light to mid-brown, sometimes with broad slightly paler stripe along costa to middle, most scales tipped darker than ground colour; base with blackish subdorsal spot, base of costa with blackish dot, a weak blackish spot on costa at middle, a blackish dash on anterior margin of cell at one-third, another below this in cell followed by a series of dashes, end of cell with a black dot, similar weak black dashes in proximal half of fold, between veins to termen and a series of weak blackish dots around termen; narrow buff streaks in cell on costal side of black markings, and among veins in terminal area, an angled fascia beyond cell sometimes indicated; fringe light grey, without distinct fringe line. Hindwing pale grey, darker towards costa and termen, fringe light grey to whitish-grey on dorsum, with weak fringe line.

VARIATION. Costal stripe often not evident; black dashes in cell, fold and ter-minal area narrow, rather irregular, sometimes forming a continuous line in cell or fold, in other specimens very weak; wings often not symmetrically marked.

SIMILAR SPECIES. *D. discipunctella* is very similar, but usually the line of black markings in the cell is stronger with larger dots, but occasionally the reverse is true.

MALE GENITALIA. Valva evenly tapering from broad base to slender incurved apex, costal margin with low hump, sacculus over half as wide as valva, with triangu-lar hump followed by process crossing costal margin; anellus broadly pear-shaped; saccus about twice as long as wide, slightly tapering to rounded end; aedeagus very long, lightly curved from broad base, slender, with slender cornutus more than half its length (in 13 of 18 checked specimens cornutus length between 55 and 66 %, 1 with 53 % and 3 between 67 and 74 % of aedeagus length).

FEMALE GENITALIA. Posterior apophysis about twice as long as anterior apo-physis; sternite VIII width:length 3:1, anterior margin of sternite with wide shallow excavation and pair of folds directed towards antrum; ostium open to posterior margin of sternite; antrum sclerotised, slightly tapering; ductus bursae with fine longitudinal folding and abundant minute papillae, particularly in expanded ante-rior one-quarter; corpus bursae ovate; signum medium-sized, wider than long, slightly extended posteriorly, more so anteriorly, with triangular teeth.

GENITALIA DIAGNOSIS. See key on pp. 328–335.

DISTRIBUTION. Southern Europe: Spain, France, Sardinia, Croatia, Bulgaria, Greece, Ukraine. Asia: Turkey, Armenia, Uzbekistan, Tajikistan, Kyrgyzstan, Iran, Turkmenistan, Kazakhstan, Iraq (Lvovsky, 2004)

BIONOMICS. Larvae on *Eryngium campestre* L. (Hannemann, 1953). Adult moths have been taken from May to September, and both fresh and worn specimens in May, presumably the worn had hibernated.

REMARKS. Hannemann (1958a) synonymised *Depressaria obolucha* Meyrick, 1936 with *D. campestrella* Chrétien, 1896, and subsequently (Hannemann, 1976b) he also synonymised *D. deliciosella* Turati, 1924 with *campestrella*. Later, (Hannemann, 1983) he synonymised *campestrella* with *eryngiella* Millière, 1881.

179   *Depressaria veneficella* Zeller, 1847

Depressaria veneficella Zeller, 1847: 842.

DESCRIPTION. Wingspan 23–28 mm. Head light-brown with scales tipped buff on neck and crown, face buff. Labial palp segment 3 four-fifths length of seg-ment 2; segment 2 buff, outer side much mixed dark brown, ventral tuft mid-brown, segment 3 buff with two deep brown bands and with tip ochreous buff. Antenna light brownish buff, narrowly ringed dark grey-brown. Thorax light brownish buff. Forewing light brown to mid-brown, often appearing greyer due to scales tipped grey-brown, darker grey-brown in basal one-sixth except towards costa; a small dark brown spot at extreme base of costa and a larger one at base of dorsum, finely bordered buff; cell more or less filled with grey scales with some buff and light grey-brown scales mixed in; fine dark grey streaks between veins to costa and termen, interrupted by pale angled fascia, grey-brown spots between vein-ends on costa in apical quarter and around termen; fringe grey-brown, without fringe line. Hindwing light grey, slightly darker posteriorly, with narrow grey-brown line around terminal and dorsal margins; fringe light grey-brown at apex to almost white at dorsal base, with a fine darker fringe line. Abdomen ochreous-brown to fuscous.

VARIATION. There is some variation in depth of ground colour and distinctness of black markings.

SIMILAR SPECIES. *D. discipunctella* is similar in size, colour and wing shape, but differs in the cell markings which are blacker and clearly divided into four cell spots. Some specimens show a broad stripe along costa paler than rest of wing, fading into ground colour beyond middle, which is typical feature of *D. junnilaineni*, so external appearance of *D. veneficella* with rich markings is overlapping with poorly marked specimens of *D. junnilaineni*.

MALE GENITALIA. Valva with broad base, valva with slight hump on costa, slightly concave on ventral margin at three-fifths, in distal quarter sharply turned into slender pointed apex, sacculus half as wide as valva, with triangular hump followed by broad process not crossing costal margin; anellus as long as wide with deep incision in distal margin; saccus about twice as long as wide, concave-sided near base then straight-sided, tapering to broadly acute end; aedeagus twice as long as saccus, lightly curved from broad base, slender, with cornutus about one-third its length.

FEMALE GENITALIA. Posterior apophysis about twice as long as anterior apophysis; sternite VIII width:length 5:2, anterior margin of sternite strongly sclerotised with conical projections on either side of antrum; ostium near posterior margin of sternite which bulges slightly around margin of ostium; antrum hardly sclerotised, tapering; ductus bursae with fine longitudinal folding and abundant minute papillae; corpus bursae ovate; signum medium-sized, rhombiform, wider than long, with numerous teeth.

GENITALIA DIAGNOSIS. See key on pp. 328–335.

DISTRIBUTION. Canary Islands, Spain, Portugal, Croatia and Greece; also on Balearic Islands, Sardinia, Sicily and Crete. North Africa: Morocco. Asia: Turkey, Azerbaijan, Turkmenistan, Kazakhstan, Israel, Mongolia (Lvovsky, 2004). We have not seen specimens from Asia, and due to numerous misidentifications in *D. veneficella* group these records must be treated with caution.

BIONOMICS. The larva eats the flower buds within the leaf-sheaths of *Thapsia garganica* L., later continuing to feed on the flowers as they expand, in spring up to June, adults start to emerge in May (Zeller, 1847). A further confirmed host-plant is *Thapsia villosa* L. (several reared specimens). Also reported to have been reared from *Ferula* (Klimesch, 1985, Baez, 1998). The earliest fresh specimens we have checked were taken at beginning of April (Morocco), adults hibernate and fly until April of following year.

REMARKS. *Depressaria riadella* Amsel, 1972 was treated as a subspecies of *D. veneficella* by Hannemann (1976), but is actually synonymous with the non-European *D. deverrella* Chrétien, 1915.

The record from Malta (Sammut, 1984) needs confirmation as several species near *D. veneficella* have been described since 1984.

Thapsia is not known from the Canary Islands, but according to Klimesch (1985) the species was reared from *Ferula communis* L. This was presumably a misdentification of *F. linkii* Webb as *F. communis* is not known from Canary Islands.

The host-plant *Thapsia* should be handled with caution as it can cause severe phytophotosensitive reactions in some people. Zeller (1847) gives a harrowing account of his experiences in rearing this species.

180   *Depressaria discipunctella* Herrich-Schäffer, 1854

Depressaria discipunctella Herrich-Schäffer, 1854: 128.
  Depressaria pastinacella Stainton, 1849 nec Duponchel, 1838.

DESCRIPTION. Wingspan 20–27 mm. Head with neck and crown mid-brown, face light brownish buff. Labial palp segment 3 three-fifths length of segment 2; segment 2 light brown to cinnamon-brown, upper edge whitish buff or rarely deep brown, segment 3 light or cinnamon-brown with blackish rings at base and at two-thirds, tip ochreous-buff. Antenna with scape pale brown, flagellum pale brown, narrowly ringed deep brown. Thorax light brown with scattered buff scales, particularly on tegulae. Forewing light brownish buff to mid-brown or mid-grey brown, sometimes with cinnamon tinge, often with most scales tipped mid-brown; light buff to brownish buff scales present in middle of cell, usually in fold, beyond cell forming a weak angled fascia and between veins close to termen; a small deep brown spot at base of costa and a larger subdorsal spot; blackish grey spots at base and end of cell with two black spots or streaks between them; further fine black streaks may be present in basal part of fold, between veins to costa and veins to termen; grey brown spots between veins at termen weakly developed; fringe mid-brown, without fringe line. Hindwing light grey-brown, slightly darker posteriorly; fringe with two weak fringe lines.

VARIATION. The ground colour of forewing varies from light greyish buff through to mid grey-brown, occasionally there is a cinnamon tinge. The black streaks in fold and between veins can be obsolete or very weakly developed.

SIMILAR SPECIES. *D. discipunctella* is characterised by the main dark markings being a series (usually four) of short blackish marks along length of cell. *D. eryngiella* has similar but weaker marks.

MALE GENITALIA. Valva with broad base, valva with slight hump on costa, slightly concave on ventral margin at three-fifths, in distal quarter strongly incurved, edges curved on both sides, rapidly tapering to a pointed but not hooked apex directed towards base of socii in standard preparation, sacculus more than half as wide as valva, with triangular hump followed by narrow process not crossing costal margin; anellus slightly longer than wide, widest near base, lateral margins slightly concave, with small incision in distal margin; saccus about twice as long as width of base, concave sided near base then sides straight and parallel to rounded apex; aedeagus with broad base, twice as long as saccus, lightly curved in distal half, slender, with stout cornutus two-fifths to half its length (in 20 of 21 checked specimens cornutus length between 40 and 50 %, one with 54 % of aedeagus length).

FEMALE GENITALIA. Posterior apophysis about twice as long as anterior apophysis; sternite VIII width:length about 5:2, anterior margin of sternite slightly excavated, forming wide flattened W-shape with a sclerotised fold; ostium extending into slight bulge in posterior margin of sternite; antrum sclerotised, tapering; ductus bursae long, thin, initially unsclerotised to about one-tenth of its length, seven-tenths with fine longitudinal folding and abundant minute papillae, anterior one-fifth papillose, including sudden expansion leading to ovate corpus bursae; signum medium-sized, longer than wide, four pointed, transverse parts with large teeth, longitudinal parts with small teeth.

GENITALIA DIAGNOSIS. See key on pp. 328–335.

DISTRIBUTION. Southern Europe, extending north to England (not seen since 1924), Netherlands, Germany and Slovakia. North Africa. Asia: we have seen specimens from Turkey, Jordan, Yemen, Armenia, Azerbaijan, Lebanon, Syria, Iran, Turkmenistan and Tajikistan. Lvovsky (2004) adds Mongolia.

BIONOMICS. The larva and its host-plant appear to be unknown, see Remarks below. Adults hibernate and fly up to May of the following year.

REMARKS. According to the literature larvae feed on *Ferula* (Hannemann, 1953); among flowers and developing seeds of *Heracleum sphondylium* L., *Pastinaca sativa* L., *Angelica sylvestris* (L.) Hoffm. (Harper *et al.*, 2002), and on *Peucedanum* L. and *Coriandrum* L. (Lvovsky, 2001). While these host-plants are not unlikely, we have never seen a reared specimen, we have not traced any published record of an actual rearing and we do not know anyone who has seen the larva. If it really feeds among spun flowers and seeds, it should be easy to find. In the 19th century the name *D. pastinacella* was wrongly applied to *D. discipunctella* by some authors which may explain the assertion that *discipunctella* feeds on *Heracleum* and *Pastinaca*. While the other mentioned host-plants remain possible, confirmation is needed. In the absence of convincing evidence, we have to regard the host-plant of this species as unknown.

Recorded from Canary Islands (Baez, 1998; Vives, 2014) but we have seen no specimens from the islands. In view of past confusion in this group and the known presence of *D. veneficella* we treat *D. discipunctella* as unconfirmed.

181   *Depressaria hansjoachimi sp. n.*

Type locality: Greece, Evritania, Karpenision

DESCRIPTION. Wingspan (19–)22–25 mm. Head with face light grey, vertex mid-brown, scales buff-tipped. Labial palp grey-brown, segment 2 with mixture of blackish scales, segment 3 with blackish ring of varying length beyond middle. Antenna dark brown. Thorax and tegula light to mid-brown. Forewing greyish buff to mid-brown, with a yellowish buff line along middle of cell and an angled fascia beyond end of cell indicated by short yellowish buff dashes, scattered yellowish buff scales, mainly in fold and towards apex on costal side; a broad band along costa without markings to beyond middle; a large blackish subdorsal spot, blackish spots around apex and termen, often weak; central line of cell with three blackish dashes and a dot, the first dash nearer costa than the others, usually with additional short blackish markings in fold and between veins to costa and termen; fringe grey. Hindwing almost translucent, posteriorly light grey; fringe light grey with distinct fringe line. Abdomen light grey to buff.

VARIATION. Forewing varies from a pale buff-brown to mid-brown. There is considerable variation in extent of black lines.

SIMILAR SPECIES. *D. discipunctella* does not have pale costal band.

MALE GENITALIA. In general similar to *D. discipunctella*, but with several differences (see figs. 181a–d): end of valva with short, inwardly directed hook, the subapical section less strongly tapering than in *D. discipunctella*, inner edge nearly parallel with longitudinal axis; saccus triangular, 1.0–1.4 × longer than wide at base, lateral edges straight or rarely weakly concave in basal 1/3; aedeagus slightly curved throughout, cornutus at least two-thirds length of aedeagus (in 13 of 17 checked specimens cornutus length between 70 and 75 %, one with 68 % and three between 77 and 81 % of aedeagus length).

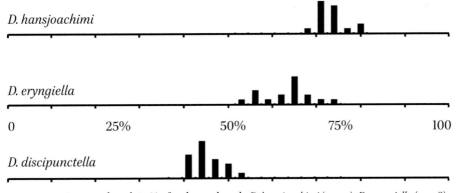

FIGURE 16   Cornutus length in % of aedeagus length, *D. hansjoachimi* (n = 17), *D. eryngiella* (n = 18) and *D. discipunctella* (n = 21) compared.

FEMALE GENITALIA. In general similar to *D. discipunctella* and *D. eryngiella*. Posterior half of sternite VIII and ostium intermediate between these species, anterior margin with deep U-shaped sinus of about half length and one quarter width of sternite VIII, also in the same area, a sclerotised fold of intersegmental membrane similar to that in *D. discipunctella* but less distinct (this fold susceptible to being lost during preparation); signum exceptionally small, smaller than in either compared species.

GENITALIA DIAGNOSIS. See key on pp. 328–335.

DISTRIBUTION. Croatia, Greece, Crete. Also in Turkey (Asian part), Armenia.

BIONOMICS. Host-plant unknown, but likely to be an Apiaceae. Fresh specimens from June onwards, worn specimens have been taken in May, which makes it likely phenology is as in *D. discipunctella*.

MATERIAL EXAMINED. Holotype ♂: Greece, Evritania, Karpenision, 1850m, 28.57 ° N, 21.44 ° E, 24.ix.1986, gen. prep. DEEUR 3090 with DNA barcode sample id TLMF Lep 26281 (658bp), H.P. Schreier leg., ex. coll. Knud Larsen, to be deposited in TLMF

Paratypes (in order of collection date):

1 ♀, Croatia, Gravosa near Dubrovnik, vi.1937, gen. prep. DEEUR 6835, H. Fabigan leg. (ZSM).

1 ♂, Greece, Peloponnes, Chelmos, 2100 m, 10.vii.1963, gen. prep. DEEUR 6837, J. Klimesch leg. (ZSM).

1 ♂, Greece, Evritania, Karpenision, 1850m, 24.ix.1986, gen. prep. DEEUR 3089, H.P. Schreier leg., coll. Knud Larsen.

1 ♂, Greece, Peloponnes, Parnoros, 14 km S Aq. Andreas, 20 m, 4.v.1987, gen. prep. DEEUR 7493, F. Schepler leg. (ZMUC).

1 ♂, Greece, Crete, Ida-Andron, 6.xi.1991, gen. prep. DEEUR 7481 (in glycerol), M. Fibiger leg. (ZMUC).

1 ♀, Turkey, Ankara, Camlidere, 1300 m, 6.–10.v.1996, gen. prep. DEEUR 5404, F. Schepler leg. (ZMUC).

1 ♂, Croatia, Pirovac, 12.vi.2001, gen. prep. DEEUR 2217 with DNA barcode sample id TLMF Lep 26339 (658bp), L. Srnka leg. (RCLS).

1 ♂, Greece, Loutra, Kilinis south of Patras, 0 m, 28.vi.2007, gen. prep. DEEUR 5993, J. Viehmann leg. (RCWS).

1 ♂, Greece, Parnass, Itea-Desfina, 1.vii.2007, gen. prep. DEEUR 6007 with DNA barcode sample id TLMF Lep 23285 (658bp), J. Viehmann leg. (RCWS).

1 ♂, Greece, Achaia, Mt. Chmelos, 2100 m, 1.vii.2011, gen. prep. DEEUR 4683 with DNA barcode sample id NGS-29134-C09 (658bp), T. Nupponen leg. (RCTN).

1 ♀, Greece, Crete, Heraklion, Damasta, 35.35 ° N, 24.94 ° E, 21.ii.2014, gen. prep. DEEUR 1891, F. Grünwald leg. (RCFGr).

1 ♂, Greece, Crete, Omalos Plateau, 1040 m, 35.34 ° N, 23.90 ° E, 14.–20.vi.2014, gen. prep. DEEUR 5377, C. Hviid, O. Karsholt & F. Vilhelmsen leg. (ZMUC).

3 ♂, 1 ♀, Greece, Timfristos Mts., 1800 m, 11.vi.2017, gen. prep. ♂ DEEUR 6001 with DNA barcode sample id TLMF Lep 23289 (658bp), 6003 with DNA barcode sample id TLMF Lep 23291 (658bp), 6004 with DNA barcode sample id TLMF Lep 23292 (658bp), ♀ DEEUR 6002 with DNA barcode sample id TLMF Lep 23290 (658bp), J. Viehmann leg. (RCWS).

1 ♀, Greece, Falakron Mountains, 1800 m, 25.vii.2017, gen. prep. DEEUR 6023, J. Viehmann leg. (RCWS).

1 ♀, Greece, Central Greece, Genimakia, 10 km SE of Itea, 38.39 ° N, 22.52 ° E, 10.v.2018, gen. prep. DEEUR 10352 (in glycerol), M. Dvorak leg. (RCAM).

1 ♂, 3 ♀, Armenia, Areni, Novarank road, 1240 m, 39.74 ° N 45.59 ° E, 2.iv.2022, gen. prep. ♂ DEEUR 10421 (in glycerol), ♀ DEEUR 10422, 10423, 10424 (all in glycerol), A. Saldaitis leg. (RCAM).

7 ♂, Armenia, Vayots Dzor province, Jermuk, Apra River Valley, 1430 m, 39.74 ° N 45.60 ° E, 3.iv.2022, gen. prep. DEEUR 10400, 10403, 10409–10413 (10409, 10410, 10413 in glycerol), A. Saldaitis leg. (RCAM).

5 ♂, 1 ♀, Armenia, Selim pass road near Aghnjadzor, 1800 m, 39.92 ° N 45.23 ° E, 19.–21.iii.2023, gen. prep. ♂ DEEUR 9904–9908, ♀ DEEUR 9897 (in glycerol) with DNA barcode sample id TLMF_Lep_38444 (658bp), J. Duda & A. Saldaitis leg. (RCAM).

3 ♂, 1 ♀, Armenia, old Jermuk road, 1340 m, 39.70 ° N 45.60 ° E, 22.&25.iii.2023, gen. prep. ♂ DEEUR 9909, 9910 with DNA barcode sample id TLMF_Lep_38437 (565[3n]bp), 9916, ♀ DEEUR 9915, J. Duda & A. Saldaitis leg. (RCAM).

3 ♂, 6 ♀, Armenia, Areni, Novarank road, 1240 m, 39.74 ° N 45.59 ° E, 19.–26.iii.2023, gen. prep. ♂ DEEUR 9918, 9936, 9939, ♀ DEEUR 9920, 9921, 9924, 9926, 9937, 9938, J. Duda & A. Saldaitis leg. (RCAM).

1 ♂, Armenia, road to Zedea, 10 km SE Jeghegnadzor, 1250 m, 39.71° N 45.43 ° E, 22.iii.2023, gen. prep. DEEUR 9944, J. Duda & A. Saldaitis leg. (RCAM).

1 ♂, Armenia, Areni, Novarank road, 1480 m, 39.68 ° N 45.24 ° E, 23.iii.2023, DEEUR 10394, J. Duda & A. Saldaitis leg. (RCAM).

Some males from Armenia were determined by brushing away scales from the tip of the abdomen rather than from a genitalia preparation.

ETYMOLOGY. The species is named after Hans-Joachim Hannemann (1925–2010) whose many papers on Palaearctic Depressariidae are the foundation on which more recent work is built.

REMARKS. External and some genitalia features are intermediate between *D. discipunctella* and *D. eryngiella*. Possibly this taxon may have originated as a hybrid between these two species which subsequently developed into a distinct species. This is only a hypothesis at the moment, karyogenetic investigation might bring an answer. More important is the fact, that several genitalia features are not intermediate and remarkably constant throughout its distribution range, which contradicts the possibility that the specimens are simply hybrids.

Barcodes were obtained from 17 specimens of *D. discipunctella*, found in two clusters with about 3.4% distance, one cluster (n = 10) with specimens from Austria, Croatia, Greece, Jordan, Iran, Russia and Tajikistan, within this cluster also 11 specimens of *D. hansjoachimi* from Croatia and Greece. The second cluster contains specimens from Spain, Italy, Tunisia and Iran, without *D. hansjoachimi*. 24 specimens of *D. eryngiella* were sequenced, found in three clusters, the two main clusters with about 2.9 % distance, one (n = 16) with specimen from Spain, France, Croatia, Greece, Turkey, Armenia, Iran, Tajikistan and Uzbekistan, the second (n = 7) with specimens from Turkey, Armenia and Iran, and a third cluster with only one *D. eryngiella* from Armenia and one *D. hansjoachimi* from Armenia, distance about 6.2 % from first and 6.5 % from second *D. eryngiella* cluster.

The *D. eryngiella* barcode in the Armenian specimen may be the result of introgression. Prerequisite for introgression are cases of hybidisation (here *D. hansjoachimi* x *D. eryngiella*). Among all checked *D. hansjoachimi* specimens (n = 51) we found one specimen, a female, with genitalia features different from typical *D. hansjoachimi*, intermediate between *D. hansjoachimi* and *D. discipunctella*. If this specimen is in fact a hybrid or the features fall within the intraspecific variability must remain open, but with the barcode background, it cannot be excluded that this specimen is a *D. hansjoachimi* x *D. discipunctella* hybrid. Anyway, the possibility of hybrids of *D. hansjoachimi* with *D. discipunctella* or *D. eryngiella* must be kept in mind.

182   *Depressaria junnilaineni* Buchner, 2017

Depressaria junnilaineni Buchner, 2017b: 3.

DESCRIPTION. Wingspan 20–26 mm. Head with face light grey, vertex mid-brown, scales buff-tipped. Labial palp grey-brown, segment 2 with mixture of blackish scales, segment 3 with blackish rings at base and above middle. Antenna dark brown. Thorax and tegula light to mid-brown. Forewing light grey-brown to mid-brown, with some light buff scales forming short lines in cell and an angled fascia beyond end of cell, a broad stripe along costa buff to light brown, paler than rest of wing, fading into ground colour beyond middle, often including a narrow black line parallel to costa; a blackish subdorsal spot, black spots around apex and termen; a series of black lines of varying length between pale costal streak and mid-line of wing, the first slightly nearer costa than the rest, blackish lines between veins in terminal one-third of wing; fringe grey-brown. Hindwing almost translucent, posteriorly light grey; fringe light grey with distinct fringe line. Abdomen light grey to buff.

VARIATION. Forewing colour varies slightly from light grey-brown to mid-brown.

SIMILAR SPECIES. The pale costa and extensive black markings distinguish *junnilaineni* from other species of the *veneficella* group, except *D. hannemanniana*. The species are best separated by male genitalia; see remark on female genitalia under *D. discipunctella*.

MALE GENITALIA. Similar to *D. discipunctella*, end of valva broader with hooked apex less sharply pointed; anellus wider, posteriorly with shallow notch; aedeagus slightly curved throughout, less slender, cornutus more slender. Also similar to *D. pentheri*, see key (pp. 328–332).

FEMALE GENITALIA. Posterior apophysis about three times as long as anterior apophysis; sternite VIII width:length about 3:1, anterior margin of tergite with wide median sinus of varying depth; ostium near posterior margin of sternite, but not bulging, antrum short, weakly sclerotised, slightly tapering; ductus bursae usually shorter than in related species, initially thin-walled then with fine longitudinal folding and abundant minute papillae; corpus bursae ovate; signum large, four-pointed with sectors of equal length, broad-based, transverse sectors with large teeth, longitudinal sectors finely toothed, wider than long, with triangular teeth.

GENITALIA DIAGNOSIS. See key on pp. 328–335.

DISTRIBUTION. Spain, North Macedonia, Greece. Also in Morocco, Armenia, Turkey and Syria.

BIONOMICS. Host-plant unknown, but likely to be an Apiaceae. Moths have been captured in spring, summer and autumn, indicating that it overwinters in the adult stage.

183   *Depressaria pentheri* Rebel, 1904

Depressaria pentheri Rebel, 1904: 360.

DESCRIPTION. Wingspan 22–25 mm. Head grey-buff. Labial palp segment 3 three-fifths length of segment 2; pale grey-buff, segment 2 mainly brown on outer side, segment 3 with blackish ring at base and beyond middle. Antenna pale grey, narrowly ringed dark fuscous. Thorax grey-buff to dark grey, tegula whitish posteriorly. Forewing grey, whitish grey in basal part of costa; a blackish subdorsal spot and terminal spots, one or two oblique dots at one-third, a plical dot, a black dot at end of cell, sometimes one in middle of cell, faint blackish lines between the veins particularly in terminal one-third; some grey-buff scales in middle of cell and in terminal one-third, angled fascia beyond cell weakly indicated; fringe pale brownish. Hindwing pale grey, posteriorly dark grey; fringe brownish. Abdomen grey.

VARIATION. Main variation is in the development of the black dots in the cell. The plical dot and dot in middle of cell may be reduced or absent.

SIMILAR SPECIES. *D. gallicella* is externally very similar, but has different distribution.

MALE GENITALIA. Similar to *D. discipunctella*, end of valva broader with hooked apex more curved round, less sharply pointed, process at end of sacculus with wide base; anellus evenly tapering, distally with wide shallow notch; saccus stouter; aedeagus slightly curved in distal half, less slender, cornutus more slender, nearly half its length. Also very similar to *D. junnilaineni* (see key on pp. 328–336).

FEMALE GENITALIA. Posterior apophysis about three times as long as anterior apophysis; sternite VIII width:length 5:2, anterior margin of sternite nearly straight; ostium extending into slight bulge in posterior margin of sternite, antrum sclerotised, tapering; ductus bursae initially thin-walled, most of its length with fine longitudinal folding and abundant minute papillae, abruptly expanded before ovate corpus bursae; signum medium-sized, rhomboidal, wider than long, with larger teeth on transverse part than on longitudinal part.

GENITALIA DIAGNOSIS. See key on pp. 328–335.

DISTRIBUTION. Croatia, Bosnia Herzegovina, Albania.

BIONOMICS. Host-plant unknown, but likely to be an Apiaceae. Moths have been collected from June to October and again in March, indicating that it hibernates in the adult stage.

184   *Depressaria hannemanniana* Lvovsky, 1990

Depressaria hannemanniana Lvovsky, 1990: 647.

DESCRIPTION. Wingspan 19–25 mm. Head mid-brown, scales buff-tipped, face light grey. Labial palp grey-brown, segment 2 with mixture of blackish scales, segment 3 with blackish band from one-third to four-fifths. Antenna dark brown. Thorax and tegula light to mid-brown. Forewing light grey-brown to mid-brown, with a yellowish buff line along middle of cell and an angled fascia beyond end of cell indicated by short yellowish buff dashes, scattered yellowish buff scales, mainly in fold and towards apex on costal side; darker specimens sometimes with a broad buff stripe along costa fading into ground colour beyond middle, sometimes including a narrow black line parallel to costa; a blackish subdorsal spot, a narrow stripe from middle of base to one-fifth, black spots around apex and termen; buff line in cell interrupted by blackish spots or short dashes, sometimes some additional short blackish markings in fold and between veins to costa and termen; fringe grey. Hindwing almost translucent, posteriorly light grey; fringe light grey with distinct fringe line. Abdomen light grey to buff.

VARIATION. There is variation in ground colour, with some specimens relatively dark brown; dark markings also vary in extent, but typically mainly consisting of interrupted line down middle of cell, but sometimes more extensive.

SIMILAR SPECIES. The most heavily marked specimens resemble *D. junnilaineni*. Safe identification requires genitalia examination.

MALE GENITALIA. Similar to *D. discipunctella*, hump in middle of costa of valva high, rounded, end of valva broader with hooked apex more curved round, less sharply pointed, directed towards top of hump, process in middle of sacculus very acute, that at end with wide base; anellus with lateral margin slightly concave, distally with wide shallow notch; saccus stouter, gradually tapering to rounded apex; aedeagus slightly curved throughout, less slender, cornutus nearly half its length.

FEMALE GENITALIA. Posterior apophysis about three times as long as anterior apophysis; sternite VIII width:length 5:2, posterior margin of sternite with median bulge; anterior margin of tergite with median excavation of variable size; ostium near posterior margin of sternite but well separated from margin, with posterior apiculus; antrum sclerotised, very short; ductus bursae with fine longitudinal folding and abundant minute papillae; corpus bursae shortly ovoid, signum a four pointed star, teeth absent in longitudinal midline, largest teeth on lateral branches.

GENITALIA DIAGNOSIS. See key on pp. 328–335.

DISTRIBUTION. Russia: Orenburg oblast. Widespread in Asia: Turkey, Azerbaijan, Russia, Kazakhstan, Tajikistan, Mongolia.

BIONOMICS. Larva recorded feeding on buds and flowers of *Ferula litwinowiana* K.-Pol. (Lvovsky, 1990). We have not seen specimens taken in early spring, but hibernation in the adult stage is likely.

185   *Depressaria erzurumella* Lvovsky, 1996

Depressaria erzurumella Lvovsky, 1996: 421.

DESCRIPTION. Wingspan 25–27 mm. Head grey-brown, scales tipped buff, face light grey. Labial palp creamy buff, segment 2 outer and ventral sides mixed grey and blackish, segment 3 with two blackish rings. Antenna fuscous. Forewing dark grey, tinged brown; subdorsal spot blackish, terminal dots blackish; short dark lines at base in middle of wing, on costal side of cell, in outer half of wing before and after a pale buff angled fascia; three very short pale buff lines among dark lines beside cell, and additional pale buff lines in terminal one-quarter of wing; fringe concolorous with wing. Hindwing whitish-grey, translucent, posterior parts of wing light grey-brown; fringe light grey with one line.

VARIATION. There is slight variation in depth of ground colour and in extent of black lines

SIMILAR SPECIES. Externally similar to *D. hannemanniana* and several other species in the *eryngiella* group.

MALE GENITALIA. Similar to *D. hannemanniana*, costal margin of valva with rounded hump, end of valva broader with hooked apex less sharply pointed; anellus broadly pear-shaped, distally with shallow notch, saccus less than twice as long as wide, sides concave near base, hardly tapering before broadly acute apex; aedeagus slightly curved throughout, cornutus about one-sixth length of aedeagus.

FEMALE: Unknown.

GENITALIA DIAGNOSIS. See key on pp. 328–332.

DISTRIBUTION. Greece. Described from a single male from Turkey (Erzurum).

BIONOMICS. Host-plant unknown, but likely to be an Apiaceae. Adult recorded in June and July.

REMARKS. Description based on Lvovsky (1996) and one specimen from Greece. These are the only known specimens.

Group 7 *dictamnella* group (186–187)
Gnathos elongate, exceeding curved digitate socii; valva simple, cuiller curved outwards, saccus rounded; aedeagus without cornuti, but with terminal thorn in *D. moranella*. Larvae on Rutaceae.

Hannemann (1953) separated these two species together with *D. hystricella* as a new genus *Horridopalpus*, based on the long tuft of scales projecting from segment 2 of labial palp, but this character is not strongly expressed in *D. hystricella*. *D. dictamnella* and *D. moranella* share several features in the male genitalia, but *D. hystricella* shows significant differences in the sacculus, anellus and aedeagus. There are equally significant differences in the female genitalia with the unusual projection from which the anterior apophysis arises and the very large signum of *D. hystricella*. The host-plants are also quite different. Furthermore, barcode places *D. hystricella* far from *D. dictamnella* and *D. moranella*. We conclude that *D. hystricella* belongs to a different group. It has features in common with the non-European *D. taciturna*, so we place it in a group with that species and its close allies. Barcodes support this relationship.

We do not find *Horridopalpus* to be more distinct than some other groups within *Depressaria*, therefore the genus is not recognised here. Lvovsky (2001) downgraded *Horridopalpus* to subgenus of *Depressaria*. We prefer to treat it as a group within *Depressaria*.

186   *Depressaria dictamnella* (Treitschke, 1835)

Haemylis dictamnella Treitschke, 1835: 181.

DESCRIPTION. Wingspan 21–28 mm. Head light grey-buff. Labial palp segment 2 with long projecting scales on underside, exceeding length of segment 3; segment 2 pale cinnamon buff, outer side fuscous to beyond middle of segment and of tuft, segment 3 with a weak fuscous ring in lower half. Antenna grey-buff, indistinctly ringed fuscous. Thorax light grey-buff to dark brown, tegulae dark brown or fuscous, posterior tufts usually grey. Forewing light brown to deep reddish brown from dorsum beyond base, through middle of wing to beyond cell, darkest between veins; light grey-buff across extreme base and along costa, occupying one-quarter to one-fifth of wing width with an expansion just beyond base, base adjacent to grey-buff area dark fuscous fading into ground colour dorsally and towards cell; veins on all outer parts of wing paler than ground colour, separated by streaks with mixture of grey-buff and fuscous scales; a narrow blackish line in cell and dark grey spots between veins around termen; sometimes a pale spot at end of cell; fringe grey-buff with one or two weak fringe lines. Hindwing light grey, darker towards apex; fringe light grey with one or more weak fringe lines. Abdomen grey-buff.

VARIATION. There are two main forms, one with ground colour light grey buff and with dark brown coloration of wing base extending outwards through cell and between veins to costa, gradually becoming paler; the other form has richer deep chestnut markings, which may sometimes extend between veins in dorsal half of wing.

SIMILAR SPECIES. *D. dictamnella* is characterised by the pale greyish wing base and costa including a slight expansion just beyond base, together with the light brown markings on veins and near absence of blackish streaks. One form of *D. moranella* is similar (*q.v.*).

MALE GENITALIA. Gnathos long-stalked, narrowly lanceolate; socii digitate, long, reaching base of gnathos; valva widest at one-third, slightly tapering to broad truncate apex, sacculus widest at base, ending in outwards directed cuiller, broadest at base, tapering to slightly incurved rounded apex; clavus absent; anellus wide at base, with lateral bulges at middle, narrowed beyond before widening to apex, distal margin shallowly excavated; saccus short, rounded; aedeagus curved, most strongly after middle where it is widest and abruptly contracts to slender apical one-third.

FEMALE GENITALIA. Posterior apophysis about one and a half times as long as anterior apophysis; sternite VIII width:length 3:1, posterior margin with wide excavation; anterior margin of sternite with median expansion, shallowly emarginated in middle; tergite with wide excavation between anterior apophyses; ostium and short antrum within anterior bulge of sternite; ductus bursae initially expanded, thin-walled with small sclerite, then slender before expanding in anterior three-fifths, not clearly differentiated from narrowly pyriform corpus bursae, without signum.

GENITALIA DIAGNOSIS. Male unique with several differences from *D. moranella*. Female similar to *D. moranella* but differing in shape of bulge on anterior margin of sternite VIII and absence of signum.

DISTRIBUTION. South-central and south-east Europe from France to Ukraine, but absent from a number of countries, extinct in Germany and Austria. Moldavia, Georgia, Asian Russia (Amur and Primorskii region) (Lvovsky, 2001).

BIONOMICS. Larva on *Dictamnus albus* L. on leaves, later on flowers (Hannemann, 1953), in spring to June. Adult from late June to September, not hibernating.

187   *Depressaria moranella* Chrétien, 1907

Depressaria moranella Chrétien, 1907: 277.
    Depressaria arabica Amsel, 1972: 138.

DESCRIPTION. Wingspan 16–19 mm. Head grey-buff. Labial palp segment 2 with long projecting scales on underside, approximately equalling length of segment 3;

segment 2 dark fuscous in proximal half, including part of tuft, distal part and segment 3 pale ochreous to grey-buff, sometimes with scattered brown scales. Antenna grey-buff, indistinctly ringed fuscous. Thorax light ochreous to grey-buff, tegulae light ochreous to fuscous. Forewing light ochreous to mid-brown; a costal stripe occupying about one-quarter wing width from middle of base to two-thirds often light grey or pale ochreous, sometimes demarcated by deep brown line, deep brown coloration then spreading into basal area; mid-brown, fuscous or blackish lines between veins in and near cell and in subterminal area; greyish fuscous dots between veins around termen weak or obsolete; sometimes with a weak pale dot at end of cell; fringe light grey-buff, sometimes with weak fringe line. Hindwing light grey, darker grey around margin; fringe pale grey with one or more distinct fringe lines. Abdomen pale grey.

VARIATION. Coloration of head, thorax and forewings, varies from pale ochreous through to mid-brown; costal streak strongly differentiated in browner forms, but almost obsolete in more ochreous forms; dark lines very variable in quantity, often only a few present.

SIMILAR SPECIES. Close to *D. dictamnella*, but smaller and with slightly shorter tuft of scales on labial palp; forewings with or without smaller area of dark brown in basal part.

MALE GENITALIA. Gnathos ovate with medium length stalk; socii digitate, divergent, incurved beyond middle, reaching base of gnathos; valva narrowed at base on costal side, apex broadly rounded, sacculus nearly half width of valva, cuiller short, directed outwards at base, angled outwards in middle, pointed; clavus absent; anellus lobes large, fused into a broad rectangle with rounded distal excavation; anellus with wide base, narrow in middle, distal one-fifth abruptly expanded to subrectangle with shallow excavation in distal margin; aedeagus slightly narrowing from base to 70° bend in middle, narrower in distal half, with beak-like apiculus.

FEMALE GENITALIA. Posterior apophysis about twice as long as anterior apophysis; sternite VIII width:length 5:2, posterior margin concave, anterior margin with semicircular median bulge around ostium; antrum short, sclerotised; ductus bursae thin-walled, initially expanded with bulge on left side enclosing heavily papillose sclerite, then constricted at one-third, before gradually expanding, not clearly differentiated from narrowly pyriform corpus bursae; signum small, weakly sclerotised, nearly circular with about 12 triangular teeth.

GENITALIA DIAGNOSIS. See under *D. dictamnella*.

DISTRIBUTION. Balkans: North Macedonia, Albania, Greece, Bulgaria. North Africa: Morocco, Tunisia, Lybia. Asia: Turkey, Afghanistan.

BIONOMICS. Chrétien (1907) reared this species from *Haplophyllum tuberculatum* (Forssk.) A. Juss. in Tunisia. Moths have been taken from April to September, those from September still in good condition.

REMARKS. *Haplophyllum tuberculatum* is not present in Europe, but other species of *Haplophyllum* are available.

Group 8 *hystricella* group (188)

Gnathos elongate, exceeding uncus which forms a hood around gnathos stalk; socii not distinct from uncus; valva long, simple, cuiller arising at one-quarter, sickle-shaped, incurved over anellus; aedeagus short, without cornuti. Larva on Rosaceae.

188   *Depressaria hystricella* Möschler, 1860

Depressaria hystricella Möschler, 1860: 275.

DESCRIPTION. Wingspan 28–29 mm. Head whitish grey, most scales tipped light brown. Labial palp segment 2 with short tuft of projecting scales below, proximal two-thirds dark fuscous, distal one-third white with scales tipped light brown, segment 3 with some whitish grey scales but mainly light fuscous, apex blackish. Antenna dark fuscous, narrowly ringed pale grey. Thorax light fuscous, tegulae pale grey mixed with some light brown scales. Forewing whitish grey, overlaid over most of wing with light brown, except for extreme base and a broad stripe along costa to mid-wing or beyond; a black dot at base of costa, a black streak in cell from near base to two-fifths where it forks around a whitish spot, the dorsal fork short, the costal fork extends beyond the whitish spot to one-half; a small faint white spot at end of cell; thin or dotted black lines in fold, beneath costa and between veins to termen; a rather weak series of greyish fuscous spots around termen; fringe pale grey mixed light brown, with two weak fringe lines. Hindwing whitish grey, darker towards costa and termen; fringe whitish to whitish grey, with about two weak fringe lines.

VARIATION. Very limited material seen. The length of the dorsal fork of the black median streak is variable, sometimes only just crossing the white spot, in other specimens curving round to extend beyond it. The whitish dot at end of cell is weak or sometimes absent.

SIMILAR SPECIES. *D. hystricella* is characterised by rather weak tuft on second segment of labial palp, the segment dark brown to two-thirds sharply contrasting with whitish distal one-third; forewing median black line diverted around both sides of conspicuous white spot at two-fifths, but only the costal fork extending well beyond the white spot.

MALE GENITALIA. Gnathos elliptic, exceeding truncate uncus; socii not developed; valva long, costal margin convex in first one-third, slightly incurved in distal one-third, sacculus narrow and short, ending before one-third, cuiller C-shaped curving in towards anellus, basal one-third tapering, with numerous fine straight setae, beyond one-third hardly tapering, ending in rounded apex with minute teeth; anellus small, base with lateral processes, middle and distal part broadly Y-shaped; saccus short and rounded; aedeagus small, short and wide, tapering, curved beyond middle, without cornuti.

FEMALE GENITALIA. Posterior apophysis about twice as long as free section of anterior apophysis; ventral and posterior margins of segment VIII laterally forming rectangular extension with rod-like margins, the ventral extended as anterior apophysis; sternite VIII width:length 2:1, posterior margin concave, anterior margin with longer than wide median bulge around pyriform ostium; antrum not sclerotised; ductus bursae thin-walled, gradually expanding, not clearly differentiated from elliptic corpus bursae; signum almost linear, wider in middle part, five times as long as its greatest width, mid-line with small teeth, margins irregular.

GENITALIA DIAGNOSIS. Absence of socii in combination with cuiller C-shaped curving in towards anellus is unique. In female segment VIII laterally with rectangular extension in combination with longer than wide median bulge around pyriform ostium is unique.

DISTRIBUTION. Spain, Slovakia, Russia. Asia: Armenia, Kazakhstan, Asian Russia (Cheliabinsk district). Uzbekistan, Asian Russia extending to Amur Province and Khabarovsk Territory (Lvovsky, 2004).

BIONOMICS. Larva on *Spiraea media* Schmidt (Lvovsky, 2004). Larva has been found in June, moth emerging in July. A fresh moth was collected in Spain on 26 January (Lepiforum e.V., 2006–2023).

REMARKS. Reasons for excluding this species from the *dictamnella* group are given under the introduction to that group.

Group 9 *hirtipalpis* group (189–192)

Labial palp segment 2 with spreading scales ventrally but not furrowed, segment 3 also with some spreading scales on ventral side. Pecten with long scales, almost equalling diameter of eye. Forewing with raised scales on costal margin, except in *peniculatella*. Forewing markings consisting of various dark spots often of indefinite shape but usually at least one triangular with one side concave. Hindwing without anal bulge. Male genitalia with socii forming a partial hood over small rounded gnathos with stout stalk; clavus absent, cuiller stout, usually angled in middle, heavily sclerotised; aedeagus with mixture of small and large cornuti. Larvae feed internally in stems. *D. hirtipalpis* is the only *Depressaria* known to feed on Lamiaceae. Fetz (1994) separated these three species into a new genus *Hasenfussia*, but we prefer to retain it within *Depressaria*.

189   *Depressaria hirtipalpis* Zeller, 1854

Depressaria hirtipalpis Zeller, 1854: 342.

DESCRIPTION. Wingspan 18–22 mm. Head grey-brown to brown. Labial palp brown with spreading scales on underside on segments 2 and 3. Antenna dark fuscous. Thorax and tegula brown to grey-brown. Forewing broad at base, almost parallel-sided, costal margin with raised scales, uniformly coloured brown to grey-brown; a blackish brown subdorsal spot, a spot at base of costa and series of spots from outer part of costa around termen; markings rather poorly defined, a roughly triangular spot at one-fifth closer to costa than dorsum, a larger triangular spot with outer margin concave in mid-line at two-fifths, a crescent-shaped mark with open side facing wing base at end of cell; a plical spot of varying shape at two-fifths; fringe grey to grey-brown. Hindwing without anal bulge, light grey, slightly darker towards apex; fringe light grey with one or two lines. Abdomen grey-brown.

VARIATION. Ground colour varies through various shades of dull brown and grey-brown. The forewing markings show some variation in shape and definition, the crescent at end of cell and plical spot can be reduced, faint or absent, sometimes all markings are absent.

SIMILAR SPECIES. *D. hirtipalpis* is characterised by conspicuous spreading scales along costa, further parts of forewing and on third segment of labial palp, but these can be lost. *D. peniculatella* shows this feature more weakly or not at all, but wing patterns are very similar. Although there are tendencies towards more dark ground colour and less elongated spot in middle of forewing in *D. hirtipalpis*, for safe determination dissection may be necessary. For differences from *D. erinaceella* (*q.v.*)

MALE GENITALIA. Gnathos on short, stout stalk, wider than long; socii rounded, fused to whole lateral margin of uncus, widely separated; tegumen slightly widened towards gnathos; valva quite small, costal margin straight, ventral margin with low hump between sacculus and apex, a small incision at apex, sacculus increasingly sclerotised on inner half, this forming first part of cuiller, eventually bent at right angles and expanded to apex shaped like bell of a trumpet; clavus absent; anellus wider than long, widest beyond middle, distal margin slightly emarginate; saccus 1.5 times as long as wide, apex rounded; aedeagus curved beyond middle, with about 40 small cornuti and a single stout apical spine.

FEMALE GENITALIA. Posterior apophysis about twice as long as anterior apophysis; sternite VIII width:length about 2:1, posterior margin with small emargination, anterior margin strongly convex around very wide ostium; antrum sclerotised, wide, tapering; ductus bursae thin-walled, quite wide, initially looped behind antrum, anteriorly with another spiral turn; corpus bursae broadly pyriform; signum large, rhombiform, densely covered with teeth, largest across middle.

GENITALIA DIAGNOSIS. Male safely distinguishable from *D. peniculatella* and *D. erinaceella* by clearly longer than wide saccus and aedeagus with a single stout apical spine and a cluster of about 40 sharp cornuti of about 0.1 mm length. In female the very wide ostium and antrum are characteristic.

DISTRIBUTION. Spain, Balearic Islands, Croatia, Montenegro, North Macedonia, Greece, European and Asian Turkey. Azerbaijan (Lvovsky, 2004).

BIONOMICS. Larva on *Salvia officinalis* L. (Hannemann, 1953) feeding in the base of the stem. Reared from *Salvia fruticosa* Mill. in Greece and *Salvia lavandulifolia* Vahl in Spain. Moths have been taken from May to October, stage of hibernation not clear.

190   *Depressaria erinaceella* Staudinger, 1870

Depressaria erinaceella Staudinger, 1870: 247.
  Depressaria sardoniella Rebel, 1936a: 97.

DESCRIPTION. Wingspan 24–26 mm. Head dark grey-brown, face smooth, pale buff. Labial palp segment 2 with coarse rough scales of irregular length ventrally, segment 3 rough-scaled in basal two-fifths, then with slightly raised scales to smooth tip, three-quarters length of segment 2; segment 2 buff on upper side, grey-brown ventrally or entirely grey-brown, segment 3 grey-brown in basal two-fifths or more, buff beyond to tip. Antenna with scape light grey-brown, apically buff; flagellum buff to grey-brown, with or without narrow darker brown rings. Thorax buff with some dark brown scales, particularly anteriorly. Forewing ochreous-buff to pale brown, with thinly scattered dark brown scales all over and overlaid fuscous from base of fold along dorsal edge of cell to one-half, a patch on costal edge at end of cell and in terminal one-fifth of wing, extending into dorsal area; costa narrowly darker grey-brown to middle; a deep brown spot at base of costa, and another larger and more diffuse near base of dorsum and a fuscous spot on costa at one-half; three deep brown spots in mid-line of wing, the first largest and of indefinite shape at one-fifth, the second small, at one-third, and the third small at end of cell; some indistinct grey-brown spots between vein-ends around termen; fringe grey-buff, without fringe line. Hindwing very pale greyish buff with scattered grey-brown scales near margins; fringe buff, with two weak fringe lines. Abdomen buff.

VARIATION. Some specimens appear darker than others through differences in ground colour.

SIMILAR SPECIES. *D. hirtipalpis* and *D. peniculatella*. *D. erinaceella* differs by larger size, indefinite shape of the three spots in mid-line of wing and presence of several diffuse fuscous areas.

MALE GENITALIA. Gnathos short-stalked, varying from slightly wider than long to distinctly longer than wide; socii rounded, fused to lateral margin of uncus, shortly separated; tegumen more or less parallel-sided; valva tapering throughout, with costal margin straight, ventral margin weakly sinuate beyond end of sacculus, apex rounded; sacculus weakly sclerotised, cuiller broad, perpendicular to costal margin, apical part stout, boot-like, directed towards apex, heavily sclerotised; clavus absent; anellus triangular, widest posteriorly with long or digitate lateral processes overlapping base of valvae, distal margin with broad U or V-shaped sinus; saccus short, broadly rounded; aedeagus curved, with a small apical tooth and a field of up to about 40 thorn-like cornuti connected to a plate.

FEMALE GENITALIA. Posterior apophysis about twice as long as anterior apophysis; sternite VIII width:length about 2:1, posterior margin with wide shallow flat excavation, anterior margin slightly convex; ostium less than half as wide as sternite; antrum not sclerotised, expanding towards ductus, abruptly contracted at junction with ductus bursae; ductus bursae thin-walled, wrinkled, with slight double bend;

corpus bursae elongate, contracted in middle, longer than combined antrum and ductus; signum long, more or less elliptic with margins irregular, densely covered with teeth, but without the larger teeth of related species.

GENITALIA DIAGNOSIS. In male combination of rounded, nearly not separated socii with aedeagus ending in one or a few short spines and up to about 40 thorn-like cornuti connected to a plate is unique. Female differs from *D. hirtipalpis* in smaller ostium and from *D. peniculatella* in weakly sclerotised antrum.

DISTRIBUTION. Portugal, Spain, France, Sardinia, Sicily, Malta, North Macedonia.

BIONOMICS. The species is recognised as a pest of artichoke (*Cynara scolymus* L.) (Robinson *et al.*, 2010). Presumably also on other species of *Cynara* L. According to Roberti (1968) larvae start feeding in the midrib of the leaves in October and November, later moving to the developing flower heads and pupating in the soil in April. Reared adults emerged in May and June (Iglesias *et al.*, 2002). In the field adults reappear in worn condition in late September and October.

191   *Depressaria peniculatella* Turati, 1922

Depressaria peniculatella Turati *in* Turati & Zanoni, 1922: 177.
    Depressaria rungsiella sensu Hannemann, 1976: 241.

DESCRIPTION. Wingspan 17–21 mm. Head grey-brown with slight cinnamon tinge; face creamy buff with some slightly reddish brown scales. Labial palp segment 3 with slightly spreading scales, three-quarters length of segment 2, which has course rough scales ventrally, forming an irregular furrow; segment 2 buff with pale cinnamon tinge with a mixture of light to mid-brown scales; segment 3 pale cinnamon-buff, sometimes grey banded on basal half. Antenna with scape light grey-brown, apically pale cinnamon-buff, pecten unusually long; flagellum light grey-brown with mid-brown rings. Thorax light brownish-buff to mid-brown with some dark brown scales, particularly anteriorly, posterior half of tegulae pale brown to cinnamon-buff. Forewing light brownish-buff to cinnamon-buff overlaid with dark grey-brown tipped scales, but these sometimes fewer or lacking beneath costa to half, in middle of cell and in an arc beyond cell reaching costa but not dorsum; a deep brown spot at base of costa and a larger often faint subdorsal patch; blackish marks in cell consisting of varying scratchy marks at one-fifth, from two-fifths to half a large triangular spot with outer margin concave, and a dot at end of cell, a dot in fold; about four dark grey-brown spots on costa in outer half, similar spots on termen very obscure; fringe light brownish buff, without fringe line. Hindwing pale grey-brown; fringe brownish buff, with one fringe line.

VARIATION. Ground colour varies from brownish buff to a slightly rusty or cinnamon tinged mid-brown. Black marks vary somewhat, particularly those nearest base, but also the exact shape of the mid-cell triangular spot.

SIMILAR SPECIES. *D. hirtipalpis* (*q.v.*), less similar to *D. erinaceella* (*q.v.*).

MALE GENITALIA. Gnathos on short, stout stalk, wider than long; socii rounded, widely separated, extending well beyond undifferentiated uncus; tegumen slightly widened distally; valva gradually narrowing, more abruptly after end of sacculus, costal margin straight, apical part narrow, sacculus increasingly sclerotised on costal side, this forming first part of cuiller, eventually bent at 45 ° and slightly expanded to hoof-like apex, crossing costal margin of valva or not, partly depending on preparation; clavus absent; transtilla lobes large, rounded, sclerotised, directed outwards; anellus wider than long, widest beyond middle with lateral margin forming stout process directed posteriorly, distal margin somewhat variable; saccus short, wide; aedeagus hardly curved beyond middle, with a strongly sclerotised hook at apex, about nine cornuti in a row, the last one much longer than the others.

FEMALE GENITALIA. Posterior apophysis about twice as long as anterior apophysis; sternite VIII width:length about 2:1, anterior margin with large median bulge around ostium; antrum sclerotised, twice as long as wide with a diagonal fold, strongly angled to left; ductus bursae thin-walled, less than half as long as elliptic corpus bursae; signum large, rhombiform, more or less elongated longitudinally densely covered with small teeth, a few large teeth across middle.

GENITALIA DIAGNOSIS. In male aedeagus characteristic with a blunt, strongly sclerotised hook at apex and few (up to about 10) cornuti of different length attached to a plate. In female clearly distinguished from other *D. hirtipalpis* group species by strongly asymmetric, sclerotised antrum.

DISTRIBUTION. Portugal, Spain, Greece. North Africa: Morocco, Algeria, Tunisia, Libya.

BIONOMICS. Host-plant unknown. Adults have been taken at light from May to end of November, the last still in good condition, so probably overwinters as an adult.

REMARKS. See under *D. rungsiella*.

192   *Depressaria rungsiella* Hannemann, 1953, stat. rest.

Depressaria rungsiella Hannemann, 1953: 317

DESCRIPTION. Similar to *D. peniculatella*, differing mainly in the cornuti and shape of socii.

DISTRIBUTION. Morocco.

BIONOMICS. Host-plant unknown.

REMARKS. *Depressaria rungsiella* Hannemann, 1953 was treated as synonym of *D. peniculatella* by Hannemann (1976) in spite of the difference in size and number of the cornuti and different shape of socii. Other apparent differences in male genitalia were attributed to differences in the preparations of the relevant type specimens. This may be the case, at least in part, but we consider the differences in the cornuti of the two taxa to be significant so we treat them as separate species. While both species are known from North Africa, as yet only *D. peniculatella* has been recorded in Europe.

Excluded Species

Agonopterix agyrella (Rebel, 1917)
Depressaria agyrella Rebel, 1917b: 193.

Romania: Crişana (Căpuşe and Kovács 1987: 50), as *argyrella* (Rebel), the record was based on two specimens collected in Ineu and identified by L. Diószeghy. Both were examined (Kovács & Kovács, 2020) and they proved to be *A. propinquella* (Treitschke, 1935) therefore excluded from fauna of Romania. This was the only European record.

Agonopterix dumitrescui Georgesco. 1965
Agonopterix dumitrescui Georgesco, 1965: 111.

Two males, Romania: Hunedoara, Hatzeg, Şesul Leordei cave, 2.xi.1963, M. Dumitrescu leg.
We have not located a specimen, nor can we establish the identity of the species from the description and male genitalia figure, but we consider it is likely to be one of the known European species.

Agonopterix rimulella (Caradja, 1920)
Depressaria rimulella Caradja, 1920: 128.

Far East Russian species erroneously listed for European Russia in Hannemann (1996).

© PETER BUCHNER AND MARTIN CORLEY, 2025 | DOI:10.1163/9789004713116_011

CHAPTER 11

Distribution Table

The countries included here are those generally treated as part of Europe, including Russia north of the Caucasus and west of the Ural Mountains and the part of Turkey west of the Bosphorus. Also included are the Macaronesian archipelagos of Madeira, Azores and Canary Islands. In the Mediterranean in addition to the main European Islands, we also include Cyprus.

The boundaries of countries are those existing at the end of 2021, but for purely practical reasons we have recognised some geographical areas that do not coincide with current political boundaries, especially where separate records are not readily available. This has no intended political significance. Ireland includes Northern Ireland and the Republic of Ireland; Gibraltar is included with Spain; Kosovo is included with Serbia; Crimea is treated as a separate entity. The main islands and archipelagos are treated separately from the countries of which they are constituents (Madeira, Azores, Canary Islands, Balearic Islands, Corsica, Sardinia, Sicily and Crete). We have no records from Azores or Iceland, so these are omitted from the table. We have found no representative list for Andorra or Liechtenstein so these are also omitted.

The distribution table is based on more than 10,000 specimens from public and private collections which were checked during preparation of this volume. Our aim is to show the distribution range of every species, therefore data has also been taken from important publications dealing with the distribution of Depressariidae, but it is outside the scope of this book to check and use all available literature. Even from sources considered to be mainly reliable, data were taken with care, and cases of doubt were omitted, if no specimen or an image of the moth or its genitalia could be checked to confirm the determination. Reasons for omission include frequent misidentification or misinterpretation of the species in the past or any other reason for doubt such as a report of a species clearly outside its known range and habitat. Some cases are mentioned under "Distribution" and the reason for omission is given in detail. In spite of this cautious approach we cannot be entirely certain that there are no incorrect reports.

In published papers dealing with species, distribution data are usually presented, even if this is not the main topic of a publication. Such papers are listed in References. The following papers, books and websites were checked especially to find distribution data:

All Europe (BOLD, 2007–2023; Lepiforum e.V. 2006–2023); Madeira (Aguiar & Karsholt, 2006); Canary Islands ((Klimesch, 1985; Baez, 1998); Spain (Vives Moreno, 2014; Catalonia and Balearic Islands Ylla *et al.*, 2020); Malta (Sammut, 1984); Serbia

(Jakšić, 2016); Romania (Kovács & Kovács, 2020); Netherlands (Huisman, 2012); Czech Republic (Laštůvka & Liška, 2011; Šumpich & Liška, 2018); Fennoscandia (Palm, 1989; with Baltic countries Aarvik et al., 2017; 2021); Russia (Lvovsky, 2019); Turkey (Koçak & Kemal, 2009).

In the table X means that the species is extinct in the territory. X? means that the species is probably extinct.

Abbreviations for the Distribution Catalogue

AL	Albania	IT	Italy
AT	Austria	KRI	Greece: Crete
AZO	Portugal: Azores Islands	LT	Lithuania
BA	Bosnia-Herzegovina	LU	Luxembourg
BAL	Spain: Balearic Islans	LV	Latvia
BG	Bulgaria	MA	Malta
BL	Belgium	MD	Moldova
BY	Belarus	MDR	Portugal: Madeira Islands
CAN	Spain: Canary Islands	MK	North Macedonia
CH	Switzerland	MN	Montenegro
COR	France: Corsica	NL	The Netherlands
CRI	Crimea	NO	Norway
CY	Cyprus	PL	Poland
CZ	Czechia	PT	Portugal
DK	Denmark	RO	Romania
EE	Estonia	RU	Russia
ES	Spain	SAR	Italy: Sardinia
FI	Finland	SB	Serbia
FR	France	SE	Sweden
GB	Great Britain	SL	Slovenia
GE	Germany	SIC	Italy: Sicily
GR	Greece	SK	Slovakia
HR	Croatia	TUR	Turkey (European part)
HU	Hungary	UA	Ukraine
IR	Ireland		

	CAN	MDR	AZO	PT	ES	BAL	FR	COR	IT	SAR	SIC	MA	SL	HR	MN	SB	BA	MK	AL	GR	KRI
S. avellanella							FR		IT					HR		SB					
S. oculella							FR		IT												
S. steinkellneriana					ES				IT		SIC			HR			BA				
S. strigulana							FR		IT								BA				
L. lobella									IT					HR				MK			
L. orientella																		MK	AL	GR	
L. osthelderi																					
E. hepatariella							FR		IT												
E. allisella																					
E. ciniflonella									IT												
E. preisseckeri									IT					HR							
E. lutosella				PT	ES	BAL	FR	COR	IT	SAR	SIC		SL	HR			BA			GR	
E. thurneri																	BA	MK		GR	
E. ledereri					ES									HR				MK		GR	
E. praeustella																					
E. nebulosella																					
E. lepidella																					
E. buvati							FR														
E. mongolicella																					
E. lvovskyi																					
E. stramentella								X?							MN			MK			
E. niviferella																					
E. indubitatella																					
A. impurella							FR		IT												
A. liturosa				PT	ES		FR		IT											GR	
A. conterminella							FR		IT												
A. arctica																					
A. ocellana				PT	ES		FR		IT					HR		SB				GR	
A. fruticosella				PT	ES					SAR											
A. rigidella				PT	ES		FR														
A. olusatri	CAN			PT	ES		FR			SAR	SIC	MA								GR	KRI
A. leucadensis																				GR	KRI
A. adspersella					ES		FR		IT	SAR	SIC			HR	MN			MK		GR	KRI
A. thapsiella				PT	ES		FR		IT	SAR	SIC		SL	HR						GR	KRI

CY	BG	RO	MD	IR	GB	NL	BL	LU	GE	CH	AT	CZ	SK	HU	PL	DK	NO	SE	FI	EE	LT	LV	BY	UA	RU	CRI	TUR
		RO		IR	GB				GE	CH	AT	CZ	SK	HU		DK	NO		FI	EE	LT	LV		UA	RU		
		RO							GE	CH	AT	CZ	SK	HU		DK	NO	SE	FI	EE	LT	LV		UA	RU		
		RO		IR	GB				GE	CH	AT	CZ	SK	HU	PL	DK	NO	SE	FI	EE	LT	LV			RU		
		RO							GE	CH	AT	CZ	SK	HU		DK	NO	SE	FI	EE	LT	LV			RU		
	BG	RO			GB				GE		AT	CZ	SK	HU		DK	NO	SE			LT	LV			RU		
																									RU		
		RO			GB	NL				CH	AT	CZ	SK		PL	DK	NO	SE	FI	EE	LT	LV					
		RO		IR	GB		BL		GE	CH		CZ	SK		PL	DK	NO	SE	FI	EE	LT	LV			RU		
					GB				GE	CH	AT	CZ		PL		NO	SE	FI	EE		LV			RU		
	BG	RO									AT	CZ		HU													
																										CRI	
	BG	RO																									
CY	BG	RO																								CRI	
		RO													PL			SE	FI	EE	LT	LV		UA	RU		
																									RU		
																									RU		
															PL						LT				RU		
																									RU		
									X?		X?	CZ	SK	HU	X?										RU	CRI	
																								UA	RU		
																									RU		
		RO							GE	CH	AT	CZ	SK	HU	PL		NO	SE	FI	EE	LT	LV			RU		
	BG	RO		IR	GB	NL			GE	CH	AT	CZ	SK	HU	PL		NO	SE	FI	EE	LT	LV			RU		
		RO		IR	GB	NL			GE	CH	AT	CZ	SK	HU		DK	NO	SE	FI	EE	LT	LV					
																	NO	SE	FI						RU		
		RO		IR	GB	NL	BL	LU	GE	CH	AT	CZ	SK	HU		DK	NO	SE	FI	EE	LT	LV			RU		
CY																											
CY	BG								GE	CH	AT	CZ	SK	HU													
	BG																										

	CAN	MDR	AZO	PT	ES	BAL	FR	COR	IT	SAR	SIC	MA	SL	HR	MN	SB	BA	MK	AL	GR	KRI
A. chironiella							FR		IT					HR						GR	
A. cervariella							FR	X?						X?							
A. paracervariella									IT				SL								
A. cadurciella					ES		FR		IT		SIC			HR	MN			MK	AL		
A. nodiflorella	CAN			PT	ES		FR	COR	IT	SAR	SIC			HR						GR	KRI
A. rotundella				PT	ES		FR		IT	SAR	SIC			HR	MN					GR	KRI
A. purpurea				PT	ES		FR		IT	SAR	SIC			HR		SB			AL	GR	KRI
A. curvipunctosa				PT	ES		FR		IT	SAR				HR		SB				GR	KRI
A. vendettella	CAN	MDR		PT	ES		FR	COR	IT	SAR											
A. alpigena							FR		IT				SL	HR				MK			
A. richteri																				GR	
A. coenosella																					
A. kayseriensis																				GR	
A. cachritis				PT	X?																
A. ferulae				PT	ES		FR	COR	IT	SAR	SIC			HR				MK		GR	
A. galicicensis																		MK			
A. langmaidi											SIC										KRI
A. lessini							FR		IT				SL	HR						GR	KRI
A. selini					ES		FR		IT				SL	HR				MK		GR	
A. ordubadensis																					
A. socerbi									IT				SL								
A. angelicella							FR		IT					HR							
A. paraselini							FR						SL								
A. parilella							FR		IT				SL	HR							
A. ciliella				PT	ES		FR		IT					HR							
A. orophilella							FR														
A. heracliana				PT	ES		FR		IT				SL	HR		SB				GR	
A. perezi	CAN	MDR																			
A. putridella							FR								MN					GR	
A. putridella ssp.																					
A. quadripunctata																					
A. hippomarathri					ES		FR	COR	IT	SAR			SL	HR	MN					GR	
A. astrantiae					ES		FR		IT				SL	HR							
A. cnicella				PT	ES		FR		IT	SAR	SIC		SL	HR		SB		MK	AL	GR	KRI
A. melancholica																					

CY	BG	RO	MD	IR	GB	NL	BL	LU	GE	CH	AT	CZ	SK	HU	PL	DK	NO	SE	FI	EE	LT	LV	BY	UA	RU	CRI	TUR
									X	X	AT			HU													
									CH																		
														HU													
CY																											
				IR	GB		BL		GE	CH	AT	CZ	SK	HU													
	BG	RO		IR	GB	NL			GE	CH	AT	CZ	SK	HU	PL	DK	NO	SE	FI	EE	LT	LV					
CY	BG	RO			GB	NL			GE		AT	CZ	SK	HU		DK	NO	SE							RU		
	BG								GE	CH	AT																
	BG																										
		RO																							RU		
		RO																									TUR
CY																											
		RO																								CRI	
		RO							GE		AT	CZ	SK	HU		DK	NO	SE	FI	EE	LT	LV					
																									RU		
		RO		IR	GB	NL			GE	CH	AT	CZ	SK	HU			NO	SE	FI						RU		
									GE	CH	AT	CZ	SK		PL												
		RO							GE	CH	AT	CZ	SK	HU				SE			LT	LV			RU		
		RO		IR	GB	NL			GE	CH	AT	CZ	SK	HU		DK	NO	SE	FI	EE	LT	LV			RU		
	BG	RO		IR	GB	NL			GE	CH	AT	CZ	SK	HU		DK	NO	SE	FI	EE	LT	LV					
		RO			GB				GE		AT	CZ		HU										UA	RU		
																	NO	SE	FI								
		RO											SK		PL			SE	FI	EE							
		RO								CH	AT	CZ	SK	HU										UA			
		RO		IR	GB				GE	CH	AT	CZ	SK	HU		DK	NO	SE	FI	EE	LT	LV					
CY	BG	RO			GB	NL	BL		GE	CH	AT	CZ	SK	HU	PL	DK	NO										
		RO																							RU		

	CAN	MDR	AZO	PT	ES	BAL	FR	COR	IT	SAR	SIC	MA	SL	HR	MN	SB	BA	MK	AL	GR	KRI
A. alstromeriana				PT	ES		FR		IT	SAR	SIC			HR		SB				GR	KRI
A. capreolella				PT	ES		FR		IT	SAR				HR		SB		MK	AL		
A. yeatiana				PT	ES		FR		IT	SAR	SIC	MA	SL	HR		SB			AL	GR	KRI
A. silerella							FR		IT				SL		MN						
A. ligusticella							FR		IT				SL	HR						GR	
A. medelichensis									IT				SL	HR					AL	GR	
A. irrorata							FR	COR	IT		SIC			HR	MN			MK	AL	GR	KRI
A. graecella									IT						MN			MK	AL	GR	
A. pseudoferulae									IT	SAR	SIC								AL	GR	
A. subtakamukui									IT												
A. guanchella	CAN																				
A. furvella							FR	COR	IT					HR		SB		MK			
A. pupillana							FR		IT										AL		
A. rutana				PT	ES		FR	COR	IT	SAR	SIC	MA	SL	HR				MK	AL	GR	
A. subumbellana																					
A. pallorella				PT	ES		FR		IT		SIC		SL	HR	MN	SB		MK	AL	GR	
A. straminella				PT	ES		FR													GR	KRI
A. carduncelli				PT	ES		FR			SAR	SIC									GR	
A. bipunctosa				PT					IT				SL	HR							
A. kyzyltashensis																					
A. squamosa					ES		FR		IT	SAR	SIC			HR				MK	AL	GR	
A. tschorbadjiewi																					
A. volgensis																					
A. kaekeritziana				PT	ES		FR		IT		SIC			HR				MK	AL	GR	
A. broennoeensis																					
A. uralensis																					
A. latipennella																					
A. invenustella										SAR	SIC										
A. abditella																					
A. lidiae							FR		IT												
A. subpropinquella				PT	ES	BAL	FR		IT	SAR	SIC	MA		HR				MK	AL	GR	KRI
A. nanatella				PT	ES		FR		IT	SAR			SL	HR		SB			AL		
A. kuznetzovi					ES																
A. propinquella					ES		FR		IT		SIC	MA		HR		SB					
A. carduella					ES		FR		IT	SAR			SL	HR		SB			AL		

CY	BG	RO	MD	IR	GB	NL	BL	LU	GE	CH	AT	CZ	SK	HU	PL	DK	NO	SE	FI	EE	LT	LV	BY	UA	RU	CRI	TUR
	BG	RO		IR	GB	NL			GE	CH	AT	CZ	SK	HU		DK	NO	SE	FI	EE	LT	LV					
		RO		IR	GB				GE	CH	AT	CZ	SK	HU		DK	NO	SE	FI	EE	LT	LV			RU		
	BG	RO		IR	GB	NL			GE	CH	AT	CZ	SK	HU		DK		SE	FI	EE	LT	LV			RU		
									GE	CH	AT																
		RO																									
									X			CZ	SK	HU													
CY	BG	RO																						UA		CRI	
									GE		AT			HU													
	BG	RO							GE	CH	AT	CZ	SK	HU											RU		
	BG									CH																	
																									RU		
	BG	RO		IR	GB	NL			GE	CH	AT	CZ	SK	HU		DK	NO	SE	FI	EE	LT	LV			RU		
CY																											
		RO			GB							CZ		HU	PL	DK		SW	FI			LV			RU		
																									RU		
	BG																										
	BG																										
																									RU		
	BG	RO		IR	GB	NL			GE	CH	AT	CZ	SK	HU		DK	NO	SW	FI	EE	LT	LV			RU		
																	NO	SW	FI						RU		
																									RU		
CY																											
																									RU		
CY	BG			IR	GB	NL			GE	CH	AT	CZ		HU		DK	NO	SE	FI		LT	LV					
		RO		IR	GB	NL			GE	CH	AT	CZ	SK	HU													
					GB																				RU		
		RO		IR	GB	NL			GE	CH	AT	CZ	SK	HU	PL	DK	NO	SE	FI		LT	LV			RU		
					GB				GE	CH	AT	CZ	SK	HU											RU		

	CAN	MDR	AZO	PT	ES	BAL	FR	COR	IT	SAR	SIC	MA	SL	HR	MN	SB	BA	MK	AL	GR	KRI
A. ivinskisi					ES		FR		IT												
A. ferocella					ES		FR		IT				SL					MK		GR	
A. laterella					ES		FR		IT	SAR				HR		SB				GR	
A. xeranthemella							FR											MK		GR	
A. cinerariae	CAN																				
A. arenella				PT	ES		FR		IT	SAR						SB					
A. petasitis							FR		IT												
A. cotoneastri					ES		FR		IT				SL								
A. multiplicella									IT												
A. doronicella				PT			FR		IT							SB				GR	
A. assimilella				PT	ES		FR		IT	SAR		MA	SL	HR				MK		GR	
A. atomella					ES		FR		IT	SAR				HR		SB					
A. scopariella	CAN	MDR		PT	ES		FR	COR	IT	SAR			SL	HR	MN			MK	AL	GR	KRI
A. conciliatella	CAN	MDR																			
A. comitella																				GR	KRI
A. nervosa				PT	ES		FR	COR	IT	SAR				HR		SB		MK		GR	
A. umbellana				PT	ES		FR		IT				SL						AL		
A. oinochroa					ES		FR		IT				SL			SB			AL	GR	
A. aspersella				PT	ES		FR	COR		SAR	SIC										
D. pulcherrimella				PT	ES				IT					HR							
D. sordidatella							FR														
D. floridella					ES		FR		IT							SB			AL	GR	KRI
D. douglasella				PT	ES		FR		IT				SL	HR		SB		MK	AL	GR	KRI
D. beckmanni							FR		IT										AL		
D. incognitella					ES		FR		IT	SAR				HR				MK		GR	
D. nemolella							FR														
D. cinderella				PT	ES																
D. infernella				PT	ES																
D. indecorella																					
D. lacticapitella													SL								
D. hofmanni					ES		FR		IT		SIC		SL							GR	
D. albipunctella				PT	ES		FR	COR	IT	SAR	SIC		SL	HR			BA	MK		GR	KRI
D. subalbipunctella																				GR	
D. krasnowodskella				PT	ES																
D. tenebricosa									IT		SIC			HR				MK	AL	GR	KRI

CY	BG	RO	MD	IR	GB	NL	BL	LU	GE	CH	AT	CZ	SK	HU	PL	DK	NO	SE	FI	EE	LT	LV	BY	UA	RU	CRI	TUR
										CH																CRI	
	BG	RO								CH	AT			HU												CRI	
	BG	RO				NL			GE		AT	CZ	SK	HU	PL	DK		SE	FI	EE	LT	LV					
	BG	RO		IR	GB	NL			GE	CH	AT	CZ	SK	HU		DK	NO	SE	FI	EE	LT	LV			RU		
		RO							GE	CH	AT	CZ	SK	HU	PL												
									GE	CH	AT	CZ	SK	HU	PL												
		RO									AT		SK		PL	DK		SE	FI	EE	LT	LV			RU		
		RO							GE		AT	CZ	SK	HU	PL												
	BG	RO		IR	GB	NL			GE	CH	AT	CZ	SK	HU		DK	NO	SE			LT	LV			RU	CRI	
	BG	RO			GB				GE	CH	AT	CZ	SK	HU		DK		SE									
CY	BG	RO			GB	NL			GE	CH		CZ	SK	HU		DK	NO	SE									
CY																									RU		
	BG	RO		IR	GB	NL			GE	CH	AT	CZ	SK	HU		DK	NO	SE	FI	EE	LT	LV			RU		
				IR	GB	NL			GE			CZ				DK											
									GE	CH	AT	CZ	SK	HU													
				IR	GB				GE	CH	AT	CZ	SK	HU	PL	DK	NO	SE	FI	EE	LT	LV					
		RO			GB				GE		AT	CZ	SK			DK	NO	SE	FI	EE	LT	LV			RU		
	BG									CH	AT	CZ	SK													CRI	
CY	BG	RO		IR	GB				GE	CH	AT	CZ	SK	HU		DK		SE	FI	EE	LT	LV					TUR
										CH	AT											LV					
										CH																	
											AT							SE							RU		
																									RU		
											AT																
		RO							GE	CH	AT	CZ	SK		PL										RU		TUR
	BG	RO			GB		BL	LU	GE	CH	AT	CZ	SK	HU	PL	DK	NO	SE	FI		LT			UA			
																								UA	RU	CRI	
CY	BG																										TUR

	CAN	MDR	AZO	PT	ES	BAL	FR	COR	IT	SAR	SIC	MA	SL	HR	MN	SB	BA	MK	AL	GR	KRI
D. radiosquamella				PT	ES		FR	COR	IT	SAR											
D. emeritella							FR		IT											GR	
D. olerella					ES		FR		IT				SL	HR						GR	
D. leucocephala									IT												
D. depressana				PT	ES		FR	COR	IT		SIC	MA	SL	HR				MK	AL	GR	
D. absynthiella					ES		FR		IT									MK	AL	GR	
D. tenerifae	CAN																				
D. artemisiae							FR		IT		SIC					SB					
D. fuscovirgatella																					
D. atrostrigella																					
D. silesiaca							FR		IT												
D. zelleri									IT									MK		GR	
D. pyrenaella					ES		FR														
D. marcella		MDR	AZO	PT	ES		FR		IT	SAR	SIC	MA		HR				MK	AL	GR	
D. peregrinella																		MK		GR	
D. heydenii							FR		IT												
D. fuscipedella									IT												
D. ululana				PT	ES		FR		IT									MK	AL	GR	KRI
D. manglisiella											SIC			HR		SB				GR	
D. chaerophylli					ES		FR		IT					HR		SB	BA	MK		GR	
D. longipennella																					
D. daucella	CAN			PT	ES		FR		IT	SAR			SL	HR		SB			AL	GR	
D. ultimella				PT	ES		FR				SIC			HR						GR	KRI
D. halophilella	?	MDR			ES	BAL	FR		IT	SAR				HR						GR	KRI
D. radiella					ES		FR		IT	SAR				HR				MK			
D. libanotidella					ES		FR		IT	SAR	SIC			HR				MK	AL		
D. bantiella							FR		IT	SAR					MN		BA	MK	AL	GR	KRI
D. velox				PT	ES		FR		IT				SL					MK	AL	GR	KRI
D. pimpinellae							FR		IT				SL			SB		MK		GR	
D. villosae				PT	ES						SIC									GR	
D. bupleurella							FR		IT												
D. sarahae	CAN				ES																
D. badiella				PT	ES		FR	COR	IT	SAR	SIC			HR	MN	SB		MK	AL	GR	KRI
D. pseudobadiella				PT	ES		FR	COR		SAR											
D. subnervosa					ES																

CY	BG	RO	MD	IR	GB	NL	BL	LU	GE	CH	AT	CZ	SK	HU	PL	DK	NO	SE	FI	EE	LT	LV	BY	UA	RU	CRI	TUR
		RO				NL	BL		GE	CH	AT	CZ	SK	HU	PL	DK	NO	SE	FI	EE	LT	LV	BY	UA	RU		
		RO			GB		BL	LU	GE	CH	AT	CZ	SK	HU	PL	DK	NO	SE	FI	EE	LT	LV	BY	UA	RU		
										CH	AT						NO	SE	FI	EE	LT	LV	BY		RU		
CY	BG	RO			X	NL	BL		GE	CH	AT	CZ	SK	HU	PL	DK	NO	SE	FI	EE	LT	LV	BY	UA	RU		TUR
	BG								GE	CH	AT	CZ		HU										UA	RU		
								LU	GE	CH	AT	CZ	SK	HU	PL	DK	NO	SE	FI	EE	LT	LV		UA	RU		
																									RU		
																									RU		
					GB					CH	AT	CZ			PL		NO	SE	FI	EE		LV			RU		
																									RU		
	BG	RO											SK	HU													TUR
		RO							GE	CH	AT		SK		PL												
									GE	CH					PL												
		RO			GB	NL	BL	LU	GE	CH	AT	CZ	SK	HU	PL			SE	FI	EE	LT	LV	BY	UA			TUR
																									RU		
	BG	RO		IR	GB	NL	BL	LU	GE	CH	AT	CZ	SK	HU	PL	DK	NO	SE	FI	EE	LT	LV	BY				
		RO	MD	IR	GB	NL	BL		GE		AT	CZ	SK	HU	PL	DK		SE		EE	LT	LV	BY	UA	RU		
		RO		IR	GB	NL		LU	GE	CH	AT	CZ	SK	HU	PL	DK	NO	SE	FI	EE	LT	LV	BY	UA	RU		
		RO						LU	GE	CH	AT		SK					SE	FI	EE		LV	BY	UA	RU		
CY																											
	BG																							UA			
		RO			GB	NL	BL		GE	CH	AT	CZ	SK	HU	PL	DK	NO	SE	FI	EE	LT	LV	BY	UA	RU		
		RO							GE	CH	AT	CZ	SK		PL									UA			
CY	BG	RO		IR	GB	NL	BL		GE	CH	AT	CZ	SK	HU	PL	DK	NO	SE	FI	EE	LT	LV		UA	RU		

	CAN	MDR	AZO	PT	ES	BAL	FR	COR	IT	SAR	SIC	MA	SL	HR	MN	SB	BA	MK	AL	GR	KRI
D. cervicella														X?							
D. gallicella							FR		IT												
D. altaica																					
D. albarracinella					ES		FR													GR	
D. eryngiella					ES		FR		IT	SAR				HR						GR	
D. veneficella	CAN			PT	ES	BAL				SAR	SIC			HR						GR	KRI
D. discipunctella				PT	ES		FR		IT		SIC			HR				MK		GR	KRI
D. hansjoachimi											SIC			HR						GR	KRI
D. junnilaineni					ES													MK		GR	
D. pentheri														HR			BA		AL		
D. hannemanniana																					
D. erzurumella																				GR	
D. dictamnella							FR		IT					HR				MK			
D. moranella																		MK	AL	GR	
D. hystricella					ES																
D. hirtipalpis					ES	BAL								HR	MN			MK		GR	
D. erinaceella				PT	ES		FR			SAR	SIC	MA						MK			
D. peniculatella				PT	ES															GR	KRI
	CAN	MDR	AZO	PT	ES	BAL	FR	COR	IT	SAR	SIC	MA	SL	HR	MN	SB	BA	MK	AL	GR	KRI

CY	BG	RO	MD	IR	GB	NL	BL	LU	GE	CH	AT	CZ	SK	HU	PL	DK	NO	SE	FI	EE	LT	LV	BY	UA	RU	CRI	TUR
									X		X	X		HU													
										CH																	
																									RU		
	BG																							UA			
																											TUR
CY	BG	RO			X	NL	BL		GE		AT		SK											UA		CRI	
																									RU		
		MD							X		X			HU										UA			
	BG																										
													SK												RU		
																											TUR
CY	BG	RO	MD	IR	GB	NL	BL	LU	GE	CH	AT	CZ	SK	HU	PL	DK	NO	SE	FI	EE	LT	LV	BY	UA	RU	CRI	TUR

CHAPTER 12

Colour Plates

Colour-Plate 1

Remark on number before collection: this is the "DEEUR [Depressariinae of Europe]-number", a unique number which PB had pinned to every specimen of Depressariidae ever worked with, both in all public and all private collections. It helps to locate exactly the specimen which is depicted, if a researcher wants to check a distinct specimen, therefore it is essential to add this information here.

1. *Semioscopis avellanella* (Hübner, 1793) –
   a) ♀, Austria, Nordtirol, Umhausen, 24.iii. 1957, K. Burmann leg. (0394, TLMF)
   b) ♂, Austria, Nordtirol, Umhausen, 17.iii. 1957, K. Burmann leg. (0396, TLMF)
   c) ♂, labial palp in lateral view, Austria, Lower Austria, Schwarzau am Steinfeld, 25.iii.2003, P. Buchner leg. (8499, RCPB)
2. *Semioscopis oculella* (Thunberg, 1794) –
   a) ♀, Austria, Bisamberg, 11.iii.1911 (0071, NHMW)
   b) ♀, Austria, Nordtirol, Umhausen, 17.iii. 1957, K. Burmann leg. (0397, TLMF)
   c) ♀, labial palp in lateral view, Austria, Bisamberg, 11.iii.1911 (0071, NHMW)
3. *Semioscopis steinkellneriana* ([Denis & Schiffermüller], 1775) –
   a) ♀, Austria, Nordtirol, Fließ, 24.v.1987, K. Burmann & P. Huemer leg. (0364, TLMF)
   b) ♀, Austria, Niederösterreich, Wienerwald, 4.iv.1916, K. Predota leg. (0218, NHMW)
   c) ♀, labial palp in lateral view, Switzerland, Neuchatel, 4.iv.2005, P. Sonderegger leg. (0429, NMBE)
4. *Semioscopis strigulana* (Fabricius, 1787) –
   a) ♀, Austria, Niederösterreich, Wienerwald, 11.ii.1912 (0217, NHMW)
   b) ♂, Austria, Langenzersdorf, 27.iii.1908, Spitz leg. (0216, NHMW)
   c) ♀, labial palp in lateral view, Switzerland, Murten, 12.iii.2011, P. Sonderegger leg. (0431, NMBE)
5. *Luquetia lobella* ([Denis & Schiffermüller], 1775) –
   a) ♀, Austria, Niederösterreich, Ternitz, ex l., 24.v.2010, P. Buchner leg. (0028, RCPB)

   b) ♂, labial palp in lateral view, Austria, Lower Austria, Schwarzau am Steinfeld, 25.v.2003, P. Buchner leg. (0342, RCPB)
   c) ♀, labial palp in lateral view, Austria, Niederösterreich, Ternitz, ex l., 24.v. 2010, P. Buchner leg. (0028, RCPB)
6. *Luquetia orientella* (Rebel, 1893) –
   a) ♀, Greece, Peloponnes, Mt. Chelmos, 1700 m, 4.vi.1989, D. Stengel leg. (1923, RCHB)
   b) ♀, labial palp in lateral view, North Macedonia, Skopje, 14.v.1984, A. Mayr leg. (1899, RCAM)
7. *Luquetia osthelderi* (Rebel, 1936) –
   a) Syntype ♀, Turkey, Taurus, Kahramanmaraş, 18.v.1928, L. Osthelder leg. (5107, NHMW)
   b) Labial palp in lateral view, same specimen as a),
8. *Exaeretia hepatariella* (Lienig & Zeller, 1846) –
   a) ♀, France, Alpes Maritimes, Marguareis, 2150 m, 21.viii.2001, A. Mayr leg. (1773, RCAM)
   b) ♂, Austria, Nordtirol, Rettenbachtal, 2600 m, 2.viii.2002, M. Dvorak leg. (1232, NMPC)
9. *Exaeretia allisella* Stainton, 1849 –
   a) ♂, Danmark, Jylland, Glatved, 3.viii. 1973, E.S. Nielsen leg. (2101, TLMF)
   b) ♀, England, year 1868, without further data (0053, NHMW)
10. *Exaeretia ciniflonella* (Lienig & Zeller, 1846) –
    a) ♀, Austria, Nordtirol, Patsch, 5.ix.1942, K. Burmann leg. (0897, ZSM)
    b) ♂, Finland, Otava, Mikkeli, 6.ix.1935, W. Brandt leg. (0898, ZSM)

1a   1c (3x)   1b

2a   2c (3x)   2b

3a   3c (2x)   3b

4a   4c (3x)   4b

5a   5b (3x)   5c (3x)   6a

7a   6b (3x)   8a

8b   7b (3x)   9a

9b   10a   10b

Colour-Plate 2

11. *Exaeretia preisseckeri* (Rebel, 1937) –
    a) ♂, Austria, Niederösterreich, Glaslauterriegel, 18.vi.1960, F. Kasy leg. (0252, NHMW)
    b) ♂, Hungary, Vértes hegyseg, Csákvár, 2.vi.1984, L. Ronkay leg. (5901, HNHM)
12. *Exaeretia lutosella* (Herrich-Schäffer, 1854) –
    a) ♂, Spain, Aragon, Albarracin, 8.vii.1924, H. Zerny leg. (0153, NHMW)
    b) ♀, Portugal, Algarve, Vila Real, ex l., 15.v.1993, M. Corley leg. (0753, RCMC)
    c) ♂, Greece, Litochron, 22.vi.1957, J. Klimesch leg. (0892, ZSM)
13. *Exaeretia thurneri* (Rebel, 1941) –
    a) ♂, North Macedonia, Ochrid, 31.viii.1936, J. Wolfschläger leg. (0906, ZSM)
    b) ♀, Bulgaria, Tuzlata, 29.ix.2011, Ig. Richter leg. (1149, RCIR)
14. *Exaeretia ledereri* (Zeller, 1854) –
    a) ♂, Turkey, Mardin, without date, ex coll. C.S. Larsen (2490, ZMUC)
    b) ♂, Turkey, Pontus, Trabzon, 500 m, 7.vi.1969, E. Arenberger leg. (1671, ZSM)
    c) ♂, Lebanon, Beyrouth, without date, ex coll. C.S. Larsen (3837, ZMUC)

    d) ♂, Iran, Mountains east of Kasri Schirin, 24.v.1963, F. Kasy leg. (0253, NHMW)
15. *Exaeretia praeustella* (Rebel, 1917) –
    a) ♀, Sweden, Öland, Vickleby, ex l., 15.vii.1959, I. Svensson leg. (0890, ZSM)
    b) ♂, Sweden, Öland, Vickleby, ex l., 15.vii.1959, I. Svensson leg. (0891, ZSM)
16. *Exaeretia nebulosella* (Caradja, 1920) –
    a) Paralectotype ♂, Kazakhstan, Uralsk, 6.v.1907, Bartel leg. (6241, MGAB)
    b) ♂, Russia, Cheliabinskskaya Oblast, Moskovo, 26.v.1998, K. & T. Nupponen leg. (4599, RCTN)
    c) ♂, Russia, Astrachanskaya Oblast, Bogdinsko Baskunchakskiy Reserve, 3.v.2004, S. Nedoschivina leg. (6546, ZIN)
    d) ♂, Russia, Saratovskaya Oblast, Algaiskii District, 3.v.2004, V. Anikin leg. (7131, ZIN)
17. *Exaeretia lepidella* (Christoph, 1872) –
    a) ♂, Russia, Cheliabinskskaya Oblast, Arkaim Reserve, 15.vi.1986, K. Nupponen leg. (4581, RCTN)
18. *Exaeretia buvati* Nel & Grange, 2014 –
    a) ♂, France, Pyrénées Orientales, Col de Puymorens, 1950 m, 30.vi.2003, J. Junnilainen leg. (4571, RCJJ)

11a

11b

12a

12b

12c

13a

13b

14a

14b

14c

14d

15a

15b

16c

16a

16b

16d

17a

18a

Colour-Plate 3

19. *Exaeretia mongolicella* (Christoph, 1882) –
    a) ♂, Russia, Primorskaya Oblast, Nio-
       bodka, vii.1994, Kuznecov leg. (1929,
       RCHB)
    b) ♂, Russia, Altai Republic, Aktasch,
       1300 m, 15.vi.2012, B. Schacht leg.
       (2004, RCHB)
    c) ♀, Russia, Primorskaya Oblast, Nio-
       bodka, vii.1994, V. Kuznecov leg. (1926,
       RCHB)
    d) ♂, Russia, Orenburgskaya Oblast, 12
       km S Kuvandyk, 14.vi.1998, K. & T. Nup-
       ponen leg. (4608, RCTN)

20. *Exaeretia lvovskyi* Buchner, Junnilainen &
    Nupponen, 2019 –
    a) Paratype ♂, Russia, Altai mts., Kurais-
       kaja steppe, 1700 m, 25.vi.2000, K. &
       T. Nupponen leg. (4609, RCTN)

21. *Exaeretia stramentella* (Eversmann, 1844) –
    a) ♀, Austria, Burgenland, Schieferberg,
       ex l., 25.vi.1960, F. Kasy leg. (0249,
       NHMW)

22. *Exaeretia niviferella* (Christoph, 1872) –
    a) ♀, Turkey, 10 km E Eskisehir, 28.v.1964,
       J. Klimesch leg. (0843, ZSM)

23. *Exaeretia indubitatella* (Hannemann,
    1971) –
    a) ♂, Russia, Altai mts., 56 km SE Belyashi,
       2400 m, 26.vii.2017, J. Šumpich leg.
       (7876, NMPC)
    b) ♂, Russia, Altai mts., 25 km SE Belyashi,
       2400 m, 28.vii.2017, J. Šumpich leg.
       (7879, NMPC)

    c) ♂, Russia, Cheliabinskaya Oblast,
       Arkaim Reserve, 18.vi.1996 K. & T. Nup-
       ponen leg. (4557, RCTN)

24. *Agonopterix impurella* (Treitschke, 1835) –
    a) ♂, Austria, Nordtirol, Innsbruck, 20.vii.
       1953, K. Burmann leg. (0909, ZSM)

25. *Agonopterix liturosa* (Haworth, 1811) –
    a) ♂, Austria, Oberösterreich, Wegscheid,
       ex l., 16.vi.1931, J. Klimesch leg. (0910,
       ZSM)
    b) ♂, Austria, Niederösterreich, Raabs,
       480 m, 20.vi.2016, W. Stark leg. (4933,
       RCWS)

26. *Agonopterix conterminella* (Zeller, 1839) –
    a) ♂, Germany, Marbach, ex l., 6.vi.1965,
       L. Süssner leg. (0382, TLMF)
    b) ♀, Sweden, Öland, ex l., 15.vi.1980,
       B.H. Thomsen leg. (3798, ZMUC)
    c) ♂, Germany, Marbach, ex l., 24.vi.1965,
       L. Süssner leg. (0381, TLMF)

27. *Agonopterix arctica* Strand, 1902 –
    a) ♂, Sweden, Funäsdalen, 2.viii.1979,
       B.Å. Bengtsson leg. (0209, NHMW)
    b) ♀, Norway, Finnmark, Børselv, ex l.,
       14.vii.2012, W. Wittland leg. (3931,
       RCWW)

28. *Agonopterix ocellana* (Fabricius, 1775) –
    a) ♂, Austria, Nordtirol, Sillschlucht, ex
       l., 8.vii.1994, S. Erlebach leg. (0397,
       TLMF)

29. *Agonopterix fruticosella* (Walsingham,
    1903) –
    a) ♀, Portugal, Algarve, Serra de Mon-
       chique, ex l., 29.v.1998, M. Corley leg.
       (0784, RCMC)
    b) ♀, Spain, Aragon, Borau, 1190 m,
       26.vii.2006, A. Mayr leg. (1730, RCAM)

19a   19b   19c

19d   20a   21a

22a   23a

23b   23c

24a   25a   25b

26a   26b   26c

27a   27b   29a

28a   29b

Colour-Plate 4

30. *Agonopterix rigidella* (Chrétien, 1907) –

a) ♂, Portugal, Algarve, Sta. Bárbara de Nexe, Gorjões, ex l., 13.vi.2003, M. Corley leg. (0773, RCMC)

b) ♂, France, Mas de Londre, 180 m, ex l., 28.vi.2013, F. Rymarczyk & M. Dutheil leg. (7341, RCRD)

c) ♀, Portugal, Algarve, Sta. Bárbara de Nexe, Gorjões, ex l., 20.vi.2002, M. Corley leg. (0780, RCMC)

31. *Agonopterix olusatri* Corley & Buchner, 2019 –

a) Paratype ♀, Portugal, Algarve, Serra de Monte Figo, ex l., 2.v.2011, M. Corley leg. (0801, RCMC)

b) Paratype ♂, Greece, Rhodos, Faliraki, ex l., 25.v.1988, J. Klimesch leg. (4216, ZSM)

c) Detail of labial palp, paratype ♂, Portugal, Algarve, Sao Romao, ex l., 7.v.2011, M. Corley leg. (0807, RCMC)

d) Paratype ♀, Spain, Malaga, Casares, ex l., 26.v.2011, P. Hale leg. (0754, RCMC)

e) Paratype ♂, France, Sète, ex l., 19.v.2008, P. Sonderegger leg. (7705, NMBE)

f) Paratype ♂, France, Pyrenees Orientales, Vernet, vi.1924, K. Predota leg. (0098, NHMW)

32. *Agonopterix leucadensis* (Rebel, 1932) –

a) Detail of labial palp, Greece, Kalavryta, 1.vi.2009, P. Sonderegger leg. (0486, NMBE)

b) ♂, Greece, Lakonia, Githion, 100 m, 30.v.1994, O. Karsholt leg. (3839, ZMUC.

c) ♂, Greece, Peloponnes, Leonidio, 190 m, 18.v.2009, A. Mayr leg. (1797, RCAM)

d) ♀, Greece, Peloponnes, Zachlorou, 30.vi.1958, J. Klimesch leg. (0869, ZSM)

33. *Agonopterix adspersella* (Kollar, 1832) –

a) ♀, Austria, Oberösterreich, Gr. Pyhrgas, 1600 m, ex l., 13.viii.1940, J. Klimesch leg. (5165, TLMF)

b) ♂, Austria, Niederösterreich, Türkensturz, ex l., 25.v.2008, P. Buchner leg. (0004, RCPB)

c) Detail of labial palp, same specimen as b)

d) ♀, Switzerland, Bern, Crémines, 890 m, ex l., 22.v.2007, P. Sonderegger leg. (7710, NMBE)

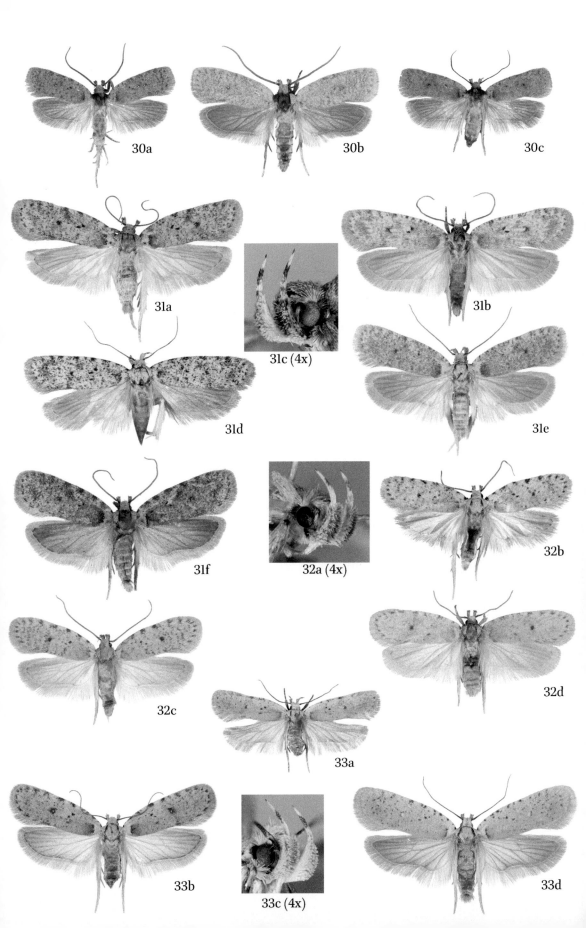

30a 30b 30c

31a 31c (4x) 31b

31d 31e

31f 32a (4x) 32b

32c 32d

33a

33b 33c (4x) 33d

Colour-Plate 5

33. *Agonopterix adspersella* (Kollar, 1832) –
     e) ♂, Austria, Niederösterreich, Dürn-
         stein, ex l., 14.vi.1963, J. Klimesch leg.
         (0895, ZSM)
     f) ♂, Switzerland, Graubünden, Poschi-
         avo Laghet d'Ur, 2400 m, ex l.,
         5.viii.2007, P. Sonderegger leg. (7711,
         NMBE)
33.1a. *Agonopterix adspersella* subsp. *pavida*
     (Meyrick, 2013) **stat. rev.**–
     a) ♂, Turkey, Gaziantep, Kadirli, 700m,
         10.vii.1987, M. Fibiger leg. (2716, RCKL)
     b) ♀, Turkey, Mersin, Arslanköy, 1300m,
         11.vii.1987, M. Fibiger leg. (3120, RCKL)
     c) Detail of labial palp, same specimen as
         b)
     d) *A. pavida* – Holotype ♀, Turkey, Taurus
         (NHMUK010293014, NHMUK)

34. *Agonopterix thapsiella* (Zeller, 1847) –
     a) ♂, Portugal, Algarve, São Romão, ex l.,
         22.iv.2011, M. Corley leg. (0802, RCMC)
     b) ♂, Portugal, Algarve, São Romão, ex l.,
         28.iv.2011, M. Corley leg. (0803, RCMC)
     c) ♂, Italy, Puglia, Gargano, Vieste, ex l.,
         27.iv.2016, P. Sonderegger leg. (7698,
         NMBE)
     d) ♀, Greece, Crete, Petres, Askifou, ex
         l., 1.v.2012, P. Sonderegger leg. (5189,
         NMBE)
35. *Agonopterix chironiella* (Constant, 1893) –
     a) ♀, Italy, Mt. Vulture, Val d'Ofanto, ex l.,
         30.v.1966, J. Klimesch leg. (4213, ZSM)
     b) ♀, France, Alpes Maritimes, 1898,
         A. Constant leg. (5729, ZMHB)
     c) ♂, Italy, Puglia, Gargano, Mt. St.
         Angelo, 400 m, ex l., 7.v.2016, P. Sonde-
         regger leg. (7695, NMBE)
     d) ♀, Italy, Mt. Vulture, Laghi di Montic-
         chio, 1150 m, ex l., 20.vi.1967, F. Hartig
         leg. (4214, ZSM)

33e

33f

33.1a

33.1c (4x)

33.1b

33.1d

34a

34b

34c

34d

35a

35b

35c

35d

Colour-Plate 6

36. *Agonopterix cervariella* (Constant, 1884) –
    a) ♀, Austria, Wien, Leopoldsberg, ex l., 22.vi.1927, F. Preißecker leg. (0101, NHMW)
    b) detail of labial palp, Austria, Niederösterreich, Gumpoldskirchen, ex l., P. Buchner leg. (3408, RCPB)
    c) ♀, Italy, Trento sopra Sasso, ex l., 13.vi.1949, J. Klimesch leg. (0911, ZSM)
    d) ♀, Austria, Niederösterreich, Gumpoldskirchen, ex l., Ronninger leg. (0912, ZSM)
37. *Agonopterix paracervariella* **sp. n.** –
    a) Paratype ♀, Italy, Alpi Giulle, Altiplano Montasio, ex l., 18.vii.1950, J. Klimesch leg. (5152, ZSM)
    b) Holotype ♂, Switzerland, Tessin, Sonvico, 1400 m, ex l., 15.vi.2011, P. Sonderegger leg. (0601, NMBE)
    c) Detail of labial palp, same specimen as b)
    d) Paratype ♂, Slovenia, Kamniske Alpe, Jezersko, 1000 m, 18.ix.2012, J. Skyva leg. (6358, NMPC)
38. *Agonopterix cadurciella* (Chrétien, 1914) –
    a) ♂, Hungary, without further data (4242, NHMW)
    b) Detail of labial palp, Italy, Gran Sasso, 1750 m, 15.vii.2010, P. Huemer leg. (4742, TLMF)

    c) ♀, Croatia, Karlobag, 5.vi.2015, J. Junnilainen leg. (4575, RCJJ)
    d) ♀, France, Alpes Maritimes, Fontan, Maurion, 800 m, ex l., 12.vii.2013, P. Sonderegger leg. (7708, NMBE)
    e) ♂, Hungary, Bucka-hegy, Csákberény, ex l., 7.vi.2008, Iv. Richter leg. (1318, RCIvR)
39. *Agonopterix nodiflorella* (Millière, 1866) –
    a) ♀, Canary Islands, Teneriffa, Orotava, ex l., 30.iv.1967, F. Kasy leg. (0157, NHMW)
    b) ♂, Canary Islands, Teneriffa, Orotava, ex l., 30.iv.1967, F. Kasy leg. (0246, NHMW)
    c) ♀, Croatia, Greece, Rhodos, Mt. Smith, ex l., 2.v.1986, J. Klimesch leg. (0884, ZSM)
40. *Agonopterix rotundella* (Douglas, 1846) –
    a) ♂, Germany, Markgröningen, ex l., 25.viii.1965, L. Süssner leg. (0366, TLMF)
    b) ♂, Switzerland, Valais, Fully, 880 m, ex l., e.p. 28.vi.2010, P. Sonderegger leg. (0620, NMBE)
41. *Agonopterix purpurea* (Haworth, 1811) –
    a) ♀, Germany, Oberstenfeld, ex l., 28.viii.1968, L. Süssner leg. (0370, TLMF)
    b) ♀, Austria, Niederösterreich, Bisamberg, 26.iv.2010, W. Stark leg. (0273, RCWS)

36a

36b (4x)

36c

36d

37a

37b

37c (4x)

37d

38a

38b (4x)

38c

38d

38e

39a

39b

39c

40a

40b

41a

41b

Colour-Plate 7

42. *Agonopterix curvipunctosa* (Haworth, 1811) –
    a) ♂, Austria, Niederösterreich, Gumpoldskirchen, 16.ix.1934, A. Ortner leg. (0907, ZSM)
    b) ♂, Greece, Crete, Krasi, 600 m, 5.vi.2001, R. Mörtter leg. (2018, RCRM)
    c) ♀, England, Somerset, Berrow, ex l., 2010, R. Heckford leg. (10207, RCRH, photo R. Heckford)
    d) ♂, Austria, Niederösterreich, Hundsheim, 26.vii.2008, P. Buchner leg. (7288, RCPB)
    e) ♂, Croatia, ex l., 2023, R. Heckford leg. (10206, RCRH, photo R. Heckford)

43. *Agonopterix vendettella* (Chrétien, 1908) –
    a) ♀, Portugal, Algarve, Loulé, Fonte d'Apra, ex l., 30.v.1993, M. Corley leg. (0796, RCMC)
    b) ♀, Italy, Sardinia, Aritzo, 10.ix.1934, K. Predota leg. (*A. iliensis*-type, 0090, NHMW)
    c) ♂, Madeira, Faja da Nogueira, 500 m, ex l., ex p. 20.vi.1993, O. Karsholt leg. (3717, ZMUC)
    d) ♀, Canary Islands, La Palma, 25.i.1962, R. Pinker leg. (0415, TLMF)

44. *Agonopterix alpigena* (Frey, 1870) –
    a) ♀, Austria, Niederösterreich, Gumpoldskirchen, ex l., 14.vi.1977, F. Kasy leg. (0236, NHMW)
    b) ♀, Italy, Pesaro e Urbino, Marche, Mt. Catria, 1400 m, 14.vii.2018, G. Govi leg. (6995, RCGFi)

c) ♂, Switzerland, Wallis, Vouangnoz, ex l., 16.vii.2009, P. Sonderegger leg. (0602, NMBE)
d) ♀, Italy, Trentino, Vezzano, 600 m, ex l., 15.vi.1983, K. Burmann leg. (1644, TLMF)
e) ♂, Switzerland, Graubünden, Ardez, ex l., 28.vi.2005, P. Sonderegger leg. (0603, NMBE)
f) ♂, France, Vaucluse, ex l., 10.v.2011, P. Sonderegger leg. (0605, NMBE)

45. *Agonopterix richteri* **sp. n.**
    a) Holotype ♀, Bulgaria, Struma Valley, Ilindentci, 500 m, 17.vii.2014, J. Junnilainen leg. (4461, RCJJ)
    b) Paratype ♀, Bulgaria, Sandanski, Ploski, 270 m, 24.ix.2011, Ig. Richter leg. (1157, RCIR)
    c) Holotype, head and labial palp details, 10× natural size

46. *Agonopterix coenosella* (Zerny, 1940) –
    a) ♂, Russia, Orenburgskaya Oblast, 40 km W Orsk, 25.vi.2003, K. Nupponen leg. (4615, RCTN)
    b) ♂, Iran, Khorasan, Golestan NP, Almeh Valley, 23.v.2001, P. Huemer leg. (8429, TLMF)
    c) ♂, Turkey, Gümüşhane Province, Kop geçidi, 31.vii.1978, W. Thomas leg. (1963, museum Dessau)
    d) ♀, Romania, Banat, Dubova, 300 m, 23.ix.2019, S. & Z. Kovacs leg. (8149, RCZK)
    e) ♂, Turkey, Nevşehir Province, Ürgüp, 1300m, 2.vii.1987, M. Fibiger leg., detail of head and thorax, 10 × natural size (5502, ZMUC)

Colour-Plate 8

47. *Agonopterix kayseriensis* Buchner, 2020 –
    a) Paratype ♀, Turkey, Çanakkale Province, Kurudağı geçidi, 300 m, 1.vii.1997, G. Baisch leg. (4376, SMNK)
    b) ♀, Greece, Parnassos, Delphi, 700m, ex l., 11.vi.2001, G. Baisch leg. (4374, SMNK)
    c) Paratype ♂, Turkey, Erzurum Province, 10 km S İspir, 26.vii.2001, B. Schacht leg. (5767, ZMHB)
    d) ♂, Turkey, Nevşehir Province, Ürgüp, 1300m, 2.vii.1987, M. Fibiger leg., detail of head and thorax, 10 × natural size (5495, ZMUC)

48. *Agonopterix cachritis* (Staudinger, 1859) –
    a) Paratype ♂, Spain, "Paratypoid 13", without further data, O. Staudinger leg. (5807, ZMHB)
    b) ♂, Spain, 1874, without further data (0931, NHMW)

49. *Agonopterix ferulae* (Zeller, 1847) –
    a) ♀, Portugal, Alto Alentejo, Portalegre, Monte Paleiro, ex l., 24.v.1996, M. Corley leg. (0794, RCMC)
    b) ♂, France, Var, Bandol, Kreisel, ex l., 13.v.2007, P. Sonderegger leg. (1999, NMBE)
    c) ♀, Italy, Sardinia, Gennargentu, Belvi, 800 m, 19.vi.1976, G. Derra leg. (1918, RCGD)
    d) ♀, Italy, Lucania, Monticchio, Valle dell Ofanto, 300 m, ex l., 24.v.1967, F. Hartig leg. (2117, TLMF)
    e) ♂, France, without further data, detail of head and antenna, 10 × natural size (0177, NHMW)
    f) ♂, France, Toulon, 20.x.1962, R. Pinker leg., detail of head and antenna, 10 × natural size, inserted section

of antenna 30 × natural size (9832, NHMW)
    g) ♀, Italy, Sicily, Siracusa, 24.v.1906, A. Knitschke leg., detail of head and antenna, 10 × natural size (9833, NHMW)

50. *Agonopterix galicicensis* sp.n. –
    a) Holotype ♂, North Macedonia, Prespa pass, 1600m, viii.1977, F. Zürnbauer leg. (1641, TLMF)
    b) Paratype ♂, North Macedonia, Galicica NP, 25.vii.2015, Ig. Richter leg. (7260, RCIR)
    c) Paratype ♀, North Macedonia, Galicica NP, 25.vii.2015, Ig. Richter leg. (7259, RCIR)
    d) Paratype ♂, North Macedonia, Galicica NP, 25.vii.2015, Ig. Richter leg., detail of head and antenna, 10 × natural size (7261, RCIR)
    e) Paratype ♀, North Macedonia, Galicica NP, 25.vii.2015, Ig. Richter leg., detail of head and antenna, 10 × natural size (7259, RCIR)

51. *Agonopterix langmaidi* sp.n. –
    a) Holotype ♂, Italy, Sicily, Mistretta, 1100m, 10.ix.2015, T. Nupponen leg. (4721, RCTN)
    b) Paratype ♂, Italy, Sicily, Mistretta, 1100m, 12.ix.2015, T. Nupponen leg. (4720, RCTN)
    c) Paratype ♂, Greece, Crete, 5 km NE Hora Sfakia, 600m, 26.ix.2006, T. Nupponen leg. (4697, RCTN)
    d) Paratype ♂, Greece, Crete, Omalos Plateau, Chania, 3.x.2016, K. Larsen leg. (6714, RCKL)
    e) Details of head and antenna, 10 × natural size, same specimen as b)
    f) Details of head and labial palp, 10 × natural size, same specimen as c)

47a

47b

47c

47d

48a

48b

49a

49b

49c

49d

49e

49f

49g

50a

50b

50c

50d

50e

51a

51b

51c

51d

51e

51f

Colour-Plate 9

52. *Agonopterix lessini* Buchner, 2017 –
     a) Holotype ♀, Italy, Verona, Monte, 300 m, ex l., 24.vi.1986, K. Burmann leg. (1578, TLMF)
     b) Paratype ♂, Greece, Crete, Rethymnon, ex l., 24.iv.1996, R. Johansson leg. (2543, ZMUC)
     c) Paratype ♂, Italy, Interneppo, 14.vi.1968, J. Klimesch leg. (5155, ZSM)
     d) ♂, Turkey, Ercincan Province, Kolcekmez Dağı geçidi, 1200 m, 14.vii.1992, G. Baisch leg. (SMNK, 4379)
     e) ♀, Italy, Verona, Monte, 400 m, 15.ix.2001, A. Mayr leg. (1733, RCAM)
     f) Details of head and labial palp of holotype, 10 × natural size.

53. *Agonopterix selini* (Heinemann, 1870) –
     a) Neotype ♂, Germany, Oberlausitz, ex l., 6.vi.2011, F. Graf leg. (1862, TLMF)
     b) ♀, Italy, Friuli-Venezia Giulia, Venzone, Basovizza, ex l., 7.vi.2010, P. Sonderegger leg. (2139, NMBE)
     c) ♀, Slovenia, Kozina, 450 m, ex l., 23.v.2004, H. Deutsch leg. (2145, TLMF)
     d) ♂, Greece, Olympos, 800 m, 6.vii.1957, J. Klimesch leg. (2056, ZSM)
     e) ♀, Greece, Pindos mts., Dilofo, 30.vi.2012, Blochwitz leg. (1968, RCHB)

     f) ♂, Italy, Mt. Baldo, 1600 m, ex l., vi.1961, K. Burmann leg. (2103, TLMF). Details of head and labial palp, 10 × natural size

54. *Agonopterix ordubadensis* Hannemann, 1959 –
     a) ♂, Turkey, Kars Province, Karakurt, 1450 m, 12.ix.1993, M. Fibiger leg. (2367, ZMUC)
     b) ♀, Armenia, Ordubad, without date, N. Michailovitscha leg. (6601, ZIN)
     c) ♂, Turkey, Ankara Province, Kızılcahamam, 20.v.1971, J. Klimesch leg. (0875, ZSM)
     d) ♂, Armenia, Ordubad, without further data (6600, ZIN)
     e) ♂, Azerbeijan, 9 km SW Lerik, 1320 m, 29.vi.2016, J.-P. Kaitila leg. (6497, RCJK)
     f) ♂, Turkey, Gümüşhane Province, Kösedağı geçidi, 1850 m, 16.vii.2000, K. Larsen leg. (6497, RCKL)

55. *Agonopterix socerbi* Šumpich, 2012 –
     a) ♀, Italy, Friuli-Venezia Giulia, Trieste, Basovizza, ex l., 6.vi.2010, P. Sonderegger leg. (0534, NMBE)
     b) ♂, Slovenia, Mt. Nanos, 880 m, 22.viii.2015, L. Morin leg. (4205, RCLM)
     c) Details of head and labial palp, 10 × natural size, same specimen as a)

Colour-Plate 10

56. *Agonopterix angelicella* (Hübner, 1813) –
   a) ♀, Austria, Nordtirol, Innsbruck, ex l., 17.vii.1940, K. Burmann leg. (1619, TLMF)
   b) ♀, Nordtirol, Innsbruck, ex l., 17.vii.1940, K. Burmann leg. (1618, TLMF)
   c) ♀, Germany, Potsdam, 8.vii.1909, without further data (0161, NHMW)
   d) ♂, Nordtirol, Vennatal, 1500 m, ex l., 16.vii.1963, K. Burmann leg. (2110, TLMF)
   e) ♂, Italy, Friuli-Venezia Giulia, Alpi Carniche, Mt. Crostis, 1900 m, 2.viii.2013, L. Morin leg. (1555, RCLM)

57. *Agonopterix paraselini* Buchner, 2017 –
   a) Paratype ♀, Germany, Saarland, Perl, Hammelsberg, 15.vii.2016, A. Werno leg. (4773, RCAW)
   b) Paratype ♀, Austria, Niederösterreich, Neusiedel a.d. Zaya, Steinberg, ex l., 6.vi.2014, P. Buchner leg. (2121, RCPB)
   c) Paratype ♂, Austria, Niederösterreich, Eichkogel, 300 m, ex l., 24.vi.2011, P. Buchner leg. (1445, RCPB)
   d) ♀ Austria, Niederösterreich, Eichkogel, 300 m, 16.viii.2009, P. Buchner leg. (0313, RCPB)
   e) Paratype ♂, Austria, Niederösterreich, Eichkogel, 300 m, ex l., 11.vi.2008, P. Buchner leg., details of head and

labial palp, 10 × natural size (0572, RCPB)
   f) Paratype ♀, Switzerland, Neuenburg, Le Landeron, 730 m, ex l., 27.v.2005, P. Sonderegger leg., details of head and labial palp, 10 × natural size (0624, NMBE)

58. *Agonopterix parilella* (Treitschke, 1835) –
   a) ♂, Austria, Nordtirol, Telfs, ex l., 10.vii.1965, A. Hernegger leg. (1603, TLMF)
   b) ♀, Italy, Kaltern, viii.1957, K. Burmann leg. (1622, TLMF)
   c) ♂, Austria, Niederösterreich, Hundsheim, ex l., 8.vi.2008, P. Buchner leg. (0044, RCPB)
   d) ♀, Germany, Thalheim, ex l., 12.vii.1909, without further data (2026, ZSM)
   e) ♀, Austria, Niederösterreich, Hundsheim, ex l., 7.vii.2013, P. Buchner leg. (9826, RCPB)
   f) Details of head and labial palp, 10 × natural size, same specimen as e)

59. *Agonopterix ciliella* (Stainton, 1849) –
   a) ♀, Austria, Niederösterreich, Hochwokersdorf, ex l., 25.vi.2009, P. Buchner leg. (0360, RCPB)
   b) ♂, Germany, Oberlausitz, ex l., 17.vii.2003, F. Graf leg. (0358, RCFG)
   c) ♂, England, Norfolk, Horning, ex l., 5.viii.1979, B. Goater leg. (2476, ZMUC)

56a

56b

56c

56d

56e

57a

57b

57c

57d

57e

57f

58a

58b

58c

58d

58e

58f

59a

59b

59c

Colour-Plate 11

60. *Agonopterix orophilella* Rymarczyk, Dutheil & Nel, 2013 –
    a) ♀, France, Lanslebourg, Mont Cenis, 2070 m, ex l., 5.viii.2012, F. Rymarczyk & M. Dutheil leg. (7340, RCRD)
    b) ♂, France, Lanslebourg, Mont Cenis, 2070 m, ex l., 1.viii.2012, F. Rymarczyk & M. Dutheil leg. (7339, RCRD)
    c) ♂, France, Lanslebourg, Mont Cenis, Grand Croix, 1990 m, ex l., 5.viii.2012, F. Rymarczyk & M. Dutheil leg. (7328, RCRD)

61. *Agonopterix heracliana* (Linnaeus, 1758) –
    a) ♀, Portugal, Rio Assureira, Contim, Tras-os-Montes, ex l., 25.vii.2005, M. Corley leg. (0788, RCMC)
    b) ♂, Germany, Schleswig Holstein, Sylt, Morsum Kliff, ex l., 9.vi.2011, W. Wittland leg. (4454, RCWW)
    c) ♀, Germany, Oberlausitz, ex l., 23.vii.2002, F. Graf leg. (0353, RCFG)

62. *Agonopterix perezi* Walsingham, 1908 –
    a) ♂, Madeira, Ponto do São Lourenço, 0 m, ex l., 26.vi.1993, O. Karsholt leg. (3709, ZMUC)
    b) ♀, Canary Islands, Puerto de la Cruz, ex l., ix.1961, J. Klimesch leg. (0250, NHMW)
    c) ♀, Canary Islands, Puerto de la Cruz, ex l., 6.iv.1971, J. Klimesch leg. (0827, ZSM)

63. *Agonopterix putridella* ([Denis & Schiffermüller], 1775) –
    a) ♂, France, Southern France, 1887, A. Constant leg. (0741, NHMW)
    b) ♀, Austria, Niederösterreich, Eichkogel, 16.viii.2009, P. Buchner leg. (0012, RCPB)
    b) ♀, France, Var, Le Muy, 15.v.2011, P. Sonderegger leg. (0565, NMBE)

63.1. *Agonopterix putridella* ([Denis & Schiffermüller], 1775) ssp. *scandinaviensis* **subsp. n.** –
    a) ♀, Sweden, Öland, Hulterstad, ex l., 21.vii.1973, I. Svensson leg. (3811, ZMUC)
    b) ♀, Finland, Finström, Tjudö, ex l., iv.2006 larva leg., P. Välimäki & M. Mutanen leg. (3851, RCMM)

64. *Agonopterix quadripunctata* (Wocke, 1857) –
    a) ♂, Finland, Hanko, Tvaermine, 16.vii.1960, H. Krogerus leg. (4253, ZMUH)
    b) ♀, Poland, Breslau, without date, ex coll. C.S. Larsen (2497, ZMUC)
    c) ♂, Sweden, Öland, Hulterstad, ex l., 11.vi.2022, B.Å. Bengtsson leg. (9829, RCBB)

65. *Agonopterix hippomarathri* (Nickerl, 1864) –
    a) ♀, Italy, Bozen, 7.vii.1944, K. Burmann leg. (0410, TLMF)
    b) ♀, Slovenia, Nanos, Šembijska bajta, 3.ix.2005, C. Morandini leg. (1549, RCCM)
    c) ♂, Austria, Niederösterreich, Hundsheim, 27.vi.2004, P. Buchner leg. (0017, RCPB)
    d) ♀, Greece, Falakron mts, 1800 m, 31.viii.2008, W. Schmitz leg. (1935, RCWSc)

66. *Agonopterix astrantiae* (Heinemann, 1870) –
    a) ♀, Germany, Regensburg, ex l., 26.vii.1891, Hofmann leg. (0162, NHMW)
    b) ♀, Austria, Niederösterreich, Katzelsdorf/Leitha, ex l., 13.vi.2011, P. Buchner leg. (1525, RCPB)

67. *Agonopterix cnicella* (Treitschke, 1832) –
    a) ♂, Austria, Niederösterreich, Neunkirchen, ex l., 6.vi.2011, P. Buchner leg. (0651, RCPB)

60a 60b 60c

61a 61b 61c

62a 62b 62c

63a 63b 63c

63.1a 63.1b 64a

64b 64c 65a

65b 65c 65d

66a 66b 67a

Colour-Plate 12

67.   *Agonopterix cnicella* (Treitschke, 1832) –
      b) ♀, Greece, Crete, Lassithi, Makrygia-
          los, 19.vi.1988, R. Johansson leg. (3797,
          ZMUC)
      c) ♀, Italy, Friuli-Venezia Giulia, Carso
          Coriziano, Lago di Doberdo, 2.ix.1999,
          L. Morin leg. (1563, RCLM)
68.   *Agonopterix melancholica* (Rebel, 1917) –
      a) Syntype ♂, Russia, Orenburgskaya
          Oblast, 1892, Hansen leg. (0178,
          NHMW)
      b) ♀, Russia, Cheliabinskskaya Oblast,
          Moskovo, 16.vii.2007, K. Nupponen leg.
          (4607, RCTN)
69.   *Agonopterix alstromeriana* (Clerck, 1759) –
      a) ♂, Austria, Niederösterreich, Schwarzau
          am Steinfeld, ex l., 25.vii.2016, P. Buch-
          ner leg. (6381, RCPB)
70.   *Agonopterix capreolella* (Zeller, 1839) –
      a) ♂, Austria, Niederösterreich, Wasch-
          berg, 16.vi.2011, W. Stark leg. (0742,
          RCWS)
      b) ♀, Italy, Alto Adige, Bolzano, without
          date, leg. K. Burmann (1623, TLMF)
      c) ♂, Armenia, Vayots Dzor, Areni Nora-
          vank, 17.vii.2011, O. Karsholt leg. (2530,
          ZMUC)
71.   *Agonopterix yeatiana* (Fabricius, 1781) –
      a) ♀, Austria, Niederösterreich, Gum-
          poldskirchen, 28.ix.1930, A. Ortner leg.
          (0866, ZSM)
      b) ♀, Hungary, Simontornya, 1913, without
          further data (0106, NHMW)

      c) ♂, Italy, Sicily, Mistretta, 700 m, 20.vi.
          1952, J. Klimesch leg. (0865, ZSM)
72.   *Agonopterix silerella* (Stainton, 1865) –
      a) ♀, Austria, Nordirol, Schönwies, 800 m,
          ex l., viii.1963, K. Burmann leg. (0409,
          TLMF)
      b) ♂, Austria, Wien, Leopoldsberg, 30.vi.
          1946, A. Ortner leg. (0408, TLMF)
73.   *Agonopterix ligusticella* (Chrétien, 1908) –
      a) ♀, Greece, Peloponnes, Zachlorou, ex
          l., 15.vii.1958, J. Klimesch leg. (0856,
          ZSM)
      b) ♂, Italy, Lago di Garda, 250m, ex l.,
          25.vii.1963, K. Burmann leg. (0412,
          TLMF)
      c) ♂, Italy, Verona, Monte, 18.vi.1986,
          K. Burmann leg. (1580, TLMF)
74.   *Agonopterix medelichensis* Buchner, 2015 –
      a) Holotype ♀, Italy, Verona, Monte, 300
          m, ex l., 19.vii.1985, K. Burmann leg.
          (1642, TLMF)
      b) ♀, Italy, Alto Adige, Naturns, ex l.,
          6.viii.1935, J. Klimesch leg. (0867, ZSM)
75.   *Agonopterix irrorata* (Staudinger, 1871) –
      a) ♀, Greece, Crete, Litochron, 400 m,
          22.vi.1957, J. Klimesch leg. (0852, ZSM)
      b) ♂, Greece, Crete, Omalos, 1040 m,
          30.iv.2014, C. Hviid et al. leg. (2340,
          ZMUC)
      c) ♂, Bulgaria, Tuzlata, 100 m, 29.ix.2011,
          Z. Tokár leg. (2164, RCZT)
      d) ♂, Greece, Crete, Petres, Askifou, ex
          l., 30.iv.2012, P. Sonderegger leg. (1864,
          NMBE)

67b

67c

68a

68b

69a

70a

70b

70c

70d

71a

71b

71c

72a

72b

73a

73b

73c

74a

74b

75a

75b

75c

75d

Colour-Plate 13

76. *Agonopterix graecella* Hannemann, 1976 –
   a) ♂, Montenegro, Durmitor, 11.vii.2007, L. Srnka leg. (0968, RCLS)
   b) ♂, Italy, Mt. Baldo, 1550 m, 14.vii.1987, P. Huemer & G. Tarmann leg. (0383, TLMF)

77. *Agonopterix pseudoferulae* Buchner & Junnilainen, 2017 –
   a) Paratype ♀, Greece, Peloponnes, Chelmos, 2100 m, 10.vii.1963, J. Klimesch leg. (0873, ZSM)
   b) Paratype ♂, Italy, Puglia, Gargano, ex l., 4.iv.2016, P. Sonderegger leg. (4747, RCPB)

78. *Agonopterix subtakamukui* Lvovsky, 1998 –
   a) ♀, Austria, Vorarlberg, Koblach, 8.viii.1999, P. Huemer leg. (0377, TLMF, holotype of *A. cluniana*)
   b) ♂, Italy, Friuli-Venezia Giulia, Cialla di Prepotto, 21.vii.2004, L. Morin leg. (1552, RCLM)

79. *Agonopterix guanchella* **sp. n.** –
   a) Paratype ♂, Canary Islands, Gran Canaria, Fataga, 10.xii.2014, J. Junnilainen leg. (4558, MNCN)
   b) Holotype ♀, Canary Islands, Barranco de Guayadeque, 800 m, 9.–22.vi.2021, P. Falck leg. (9262, MNCN)

80. *Agonopterix furvella* (Treitschke, 1832) –
   a) ♀, Austria, Niederösterreich, Spitzerberg, ex l., 11.vi.2010, P. Buchner leg. (0019, RCPB)
   b) ♂, Switzerland, Valais, Leuk, Niedergampel, ex l., 13.vii.2017, R. Seliger leg. (5955, RCRS)

81. *Agonopterix pupillana* (Wocke, 1887) –
   a) ♂, Italy, Trento Sopra Sasso, ex l., viii.1945, J. Klimesch leg. (0845, ZSM)
   b) ♀, Italy, Trento Sopra Sasso, ex l., viii.1945, J. Klimesch leg. (0846, ZSM)

82. *Agonopterix rutana* (Fabricius, 1794) –
   a) ♀, Italy, Lago di Garda, Limone, ex l., 14.vii.1967, M. & W. Glaser leg. (0237, NHMW)
   b) ♀, Spain, Huesca, Candasnos, 30.v.2015, J. Viehmann leg. (3923, RCWSc)

Colour-Plate 14

83. *Agonopterix subumbellana* Hannemann, 1959 –
    a) ♀, Turkey, Nevşehir Province, Ürgüp, 1100 m, 20.vi.1995, G. Baisch leg. (4338, SMNK)
    b) ♂, Turkey, Niğde Province, Sivrihisar geçidi, 1700 m, 4.viii.1989, M. Fibiger & N. Esser leg. (3776, ZMUC)

84. *Agonopterix pallorella* (Zeller, 1839) –
    a) ♂, Austria, Niederösterreich, Mödling, 1888, W. Krone leg. (0134, NHMW)
    b) ♂, Austria, Burgenland, Zitzmannsdorfer Wiesen, ex l., 30.vii.1958, F. Kasy leg. (0240, NHMW)
    c) ♀, Spain, Sierra de la Nieves, La Yunquera, 13.vii.1991, A. Vives Moreno leg. (0768, MNCN)

85. *Agonopterix straminella* (Staudinger, 1859) –
    a) ♀, Syria, 25 km W Damaskus, 3.vi.1961, F. Kasy & E. Vartian leg. (0196, NHMW)
    b) ♂, Syria, 60 km NE Ladikije, 7.vi.1961, F. Kasy & E. Vartian leg. (0871, ZSM)

c) ♂, Lebanon, Bcharre, 1400 m, 10.vi.1931, H. Zerny leg. (0128, NHMW)
d) ♂, Portugal, Algarve, Carrapateira, ex l., 27.iv.2011, M. Corley leg. (0775, RCMC)
e) ♀, Spain, Teruel, Moscardon, 1500 m, 26.vi.2016, J. Viehmann leg. (4779, RCWSc)

86. *Agonopterix carduncelli* Corley, 2017 –
    a) ♀, Greece, Rhodos, Mt. Smith, 12.v.1975, J. Klimesch leg. (2046, ZSM)
    b) Paratype ♀, Portugal, Mexilhoeira Grande, Cruzinha, ex l., 15.v.2011, M. Corley leg. (0776, RCMC)
    c) Paratype ♂, Morocco, Ifrane, 30.vi.1972, F. Hahn leg. (1983, RCHB)
    d) Holotype ♂, Portugal, Algarve, Boliqueime, 24.xi.2010, M.J. Dale leg. (0758, NHMUK)

87. *Agonopterix bipunctosa* (Curtis, 1850) –
    a) ♂, Sweden, Ronneby, UTM33VWC 1921, 9.vii.1971, I. Svensson leg. (0401, TLMF)
    b) ♂, Sweden, Ronneby, UTM33VWC 1921, ex l., 4.vii.1971, I. Svensson leg. (2516, ZMUC)

Colour-Plate 15

88. *Agonopterix kyzyltashensis* Buchner & Šumpich 2020 –
    a) Paratype ♀, Russia, Altai mts, Chagan Uzun, Krasnaja Gorka Hill, 1870 m, 3.vii.2019, J. Šumpich leg. (7751, NMPC)
89. *Agonopterix squamosa* (Mann, 1864) –
    a) ♂, Turkey, Amasya Province, 1860, without further data. (0132, NHMW)
    b) ♀, Greece, Mitsikeli, Dikorifo, 1000 m, 8.vii.2005, J. Skyva leg. (6203, NMPC)
    c) ♂, Crimea, Karadagh, ex l., 7.v.1988, Y. Budashkin leg. (2162, RCZT)
90. *Agonopterix tschorbadjiewi* (Rebel, 1916) –
    a) Holotype ♀, Bulgaria, Burgas, 30.vii.1910, P. Tschorb leg. (0055, NHMW)
91. *Agonopterix volgensis* Lvovsky, 2018 –
    a) ♂, Russia, Orenburgskaya Oblast, Mt. Verbljushka, Donskoje, 19.vi.1999, K. & T. Nupponen leg. (4594, RCTN)
92. *Agonopterix kaekeritziana* (Linnaeus, 1767) –
    a) ♀, Austria, Niederösterreich, Perchtoldsdorf, 25.vi.2010, P. Buchner leg. (0325, RCPB)

    b) ♀, Austria, Niederösterreich, Fischa-wiesen, 22.vi.2011, W. Stark leg. (0725, RCWS)
    c) ♀, Spain, Sierra Nevada, Camino de la Veleta, 2250 m, 22.vii.1985, G. Baldizzone & E. Traugott Olsen leg. (2511, ZMUC)
    d) ♂, Danmark, Ellinge Lyng, ex l., 6.viii.1973, E. Palm leg. (5447, RCEP)
93. *Agonopterix broennoeensis* (Strand, 1920) –
    a) ♀, Norway, Kongsvold, 1000 m, ex l., 20.vii.1983, O. Karsholt leg. (3827, ZMUC)
    b) ♂, Norway, Rusånes, Saltdal Nai, ex l., 28.vi.1982, K. Larsen leg. (2514, ZMUC)
94. *Agonopterix uralensis* **sp.n.** –
    a) Paratype ♂, Russia, Orenburgskaya Oblast, Schibendy Valley 20 km S Prokrova, 29.vi.2003, K. Nupponen leg. (4600, RCTN)
    b) Paratype ♂, Russia, Orenburgskaya Oblast, Schibendy Valley 20 km S Prokrova, 29.vi.2003, K. Nupponen leg. (4602, RCTN)

Colour-Plate 16

95. *Agonopterix latipennella* (Zerny, 1934) –
   a) ♂, Turkey, Diyarbakır Province, Ergani, 22.xi.1939, without further data (3527, MHNG)
   b) Syntype ♂, North Lebanon, Becharré, 1400 m, 11.–20.vi.1931, H. Zerny leg. (3462, NHMW)

96. *Agonopterix invenustella* Hannemann, 1953 –
   a) ♀, Italy, Sicily, Caltanissetta, Butera, 29.viii.2020, F. Graf leg. (8979, RCFG)
   b) Paratype ♂, Algeria, Pont du Caid, 1892, "V. de B." leg., ex coll. Staudinger (6843, ZMHB)

97. *Agonopterix abditella* Hannemann, 1959 –
   a) ♂, Russia, Altai mts, Kosh Agach, Chagan Uzun, Krasnja Gorka, 1870 m, 4.vii.2014, M. Dvorak leg. (6229, NMPC)

98. *Agonopterix lidiae* Buchner, 2020 –
   a) Paratype ♂, Italy, Friuli-Venezia Giulia, Udine Province, Valle di Musi, Torrente Mea, 700 m, 15.viii.2007, H. Deutsch leg. (6412, MFSN)
   b) Paratype ♂, Italy, Friuli-Venezia Giulia, Udine Province, Venzone, Mt. Plauris southern part, 1300 m, 17.vii.2006, P. Huemer leg. (8822, TLMF)
   c) Holotype ♂, Italy, Udine, Tugliezzo, 500 m, 22.viii.2003, L. Morin leg. (4047, RCLM)

99. *Agonopterix subpropinquella* (Stainton, 1849) –
   a) ♀, Cyprus, Kyrenia, ex l., 12.v.1972, G. Deschka leg. (0860, ZSM)
   b) ♂, Italy, Sardinia, Mela Murgia, Quartucciu, iv.2014, L. Morin leg. (4207, RCLM)

   c) ♂, Greee, Crete, Zakros, 200 m, 26.v.1965, H. Reisser leg. (0853, ZSM)
   d) ♀, Croatia, Tromilja, ex l., 16.v.2011, Iv. Richter leg. (1326, NMPC)

100. *Agonopterix nanatella* (Stainton, 1849) –
   a) ♂, Switzerland, Valais, Bovernier, 1.vi.2006, P. Sonderegger leg. (0608, NMBE)
   b) ♂, Croatia, Split, 25.v.1908, A. Knitschke leg. (0138, NHMW)
   c) ♀, Austria, Niederösterreich, Oberweiden, 20.ix.2006, O. Rist leg. (0290, RCOR)

101. *Agonopterix kuznetzovi* Lvovsky, 1983 –
   a) ♂, Turkey, Kars Province, Karakurt, 1450 m, 12.ix.1993, M. Fibiger leg. (2380, ZMUC)
   b) ♀, Great Britain, Cornwall, Kynance Cove, ex l., 9.vii.1984, R.J. Heckford leg. (2472, ZMUC)
   a) ♂, Spain, Andalusia, Sierra de Gador, Barjali, 1650 m, 29.vii.2005, V. Červenka leg. (6215, NMPC)

102. *Agonopterix propinquella* (Treitschke, 1835) –
   a) ♂, Austria, Niederösterreich, Schwarzau am Steinfeld, 330 m, ex l., 3.ix.2010, P. Buchner leg. (0573, RCPB)
   b) ♀, Austria, Niederösterreich, Urschendorf, 360 m, 5.vii.2006, P. Buchner leg. (0024, RCPB)

103. *Agonopterix carduella* (Hübner, 1817) –
   a) ♂, Slovenia, Orešje, Kozja peč, Kozjansji park, 370 m, 26.vi.2003, S. Gomboc leg. (1892, RCSG)
   b) ♀, Austria, Niederösterreich, Schwarzau am Steinfeld, 330 m, ex l., 18.vi.2008, P. Buchner leg. (0023, RCPB)
   c) ♂, Austria, Tirol, Innsbruck, ex l., 15.viii.1940, K. Burmann leg. (2080, ZSM)

95a

95b

96a

96b

97a

98a

98b

98c

99a

99b

99c

99d

100a

100b

100c

101a

101b

101c

102a

102b

103a

103b

103c

Colour-Plate 17

104. *Agonopterix ivinskisi* Lvovsky, 1992 –
    a) ♀, Switzerland, Valais, Jeizinen, 1530 m, ex l., 9.ix.2010, P. Sonderegger leg. (0615, NMBE)
    b) ♂, Turkey, Ankara Province, 20 km W Kızılcahamam, ex l., 30.vi.1970, J. Klimesch leg. (0862, ZSM)
    c) ♂, Spain, Teruel, Moscardon, 1500 m, 14.ix.2007, J. Viehmann leg. (1930, RCWSc)

105. *Agonopterix ferocella* (Chrétien, 1910) –
    a) ♂, France, Douelle, Lot, ex l., 21.viii.1936, L. Lhomme leg. (0876, ZSM)
    b) ♂, Switzerland, Valais, Gampel, ex l., 13.vi.2010, P. Sonderegger leg. (1869, NMBE)
    c) ♂, Italy, Liguria, Testico, 5.iv.1981, Burgermeister leg. (0369, TLMF)

106. *Agonopterix laterella* ([Denis & Schiffermüller], 1775) –
    a) ♂, Turkey, Isparta Province, Sultan Dağları, 1800 m, 9.vii.2001, K. Larsen leg. (3135, RCKL)
    b) ♀, Greece, Peloponnes, Chelmos, 2000 m, 24.viii.2008, W. Schmitz leg. (1937, RCWSc)
    c) ♀, Italy, Udine, Bordano, ex l., 15.vi.2014, P. Sonderegger leg. (1990, NMBE)
    d) ♂, Spain, Teruel, Pozondon, 2.vii.2016, J. Viehmann leg. (0605, RCWSc)

107. *Agonopterix xeranthemella* Buchner, 2018 –
    a) Paratype ♀, North Macedonia, Galicica, 25.vii.2015, Ig. Richter leg. (7273, RCIR)

    b) Holotype ♀, Greece, Evro 35 km north Alexandropolis, 500 m, 8.vii.1986, M. Fibiger leg. (5485, ZMUC)
    c) Paratype ♂, France, La Deveze, ex l. 15.vii.1928, E. Dattin leg. (6934, MNHN)

108. *Agonopterix cinerariae* (Walsingham, 1908) –
    a) ♂, Canary Islands, Tenerife, Guimar, ex l. 15.iv.1965, J. Klimesch leg. (0882, ZSM)
    b) ♀, Canary Islands, Tenerife, Guimar, ex l. 15.iv.1965, J. Klimesch leg. (0878, ZSM)
    c) ♂, Canary Islands, Tenerife, Guimar, ex l. 16.iv.1965, J. Klimesch leg. (0879, ZSM)
    d) ♂, Canary Islands, Tenerife, Guimar, ex l. 15.iv.1965, J. Klimesch leg. (0877, ZSM)

109. *Agonopterix arenella* ([Denis & Schiffermüller], 1775) –
    a) ♀, Switzerland, Graubünden, Vicosoprano, ex l., 29.vii.2009, P. Sonderegger leg. (1990, NMBE)
    b) ♂, Austria, Niederösterreich, Urschendorf, 360 m, ex l., 20.vii.2008, P. Buchner leg. (0006, RCPB)
    c) Labial palp in lateral view, same specimen as b)
    d) ♀, Austria, Niederösterreich, Nöstach, ex l., 8.vii.2011, P. Buchner leg. (8496, RCPB)

110. *Agonopterix petasitis* (Standfuss, 1851) –
    a) ♂, Austria, Steiermark, Präbichl, 1100 m, ex l., 1.vi.2011, P. Buchner leg. (0650, RCPB)
    b) Labial palp in lateral view, same specimen as a)
    c) ♀, Austria, Steiermark, Lugauer, 3.viii.1910, H. Zerny leg. (0081, NHMW)

104a

104b

104c

105a

105b

105c

106a

106b

106c

106d

107a

107b

107c

108a

108b

108c

108d

109a

109b

109c (3x)

109d

110a

110b (3x)

110c

Colour-Plate 18

111. *Agonopterix cotoneastri* (Nickerl, 1864) [syn.: *A. senecionis* (Nickerl, 1864)] –
    a) ♂, Austria, Oberösterreich, Nationalpark Kalkalpen, ex l., 8.viii.2011, P. Buchner leg. (0652, RCPB)
    b) ♀, Italy, Abruzzo, Gran Sasso, Campo Imperatore, 2130 m, 13.viii.2012, T. Nupponen leg. (4727, RCTN)
    c) ♀, Poland, Śląsk [Silesia], ex coll. Staudinger, without further data (9261, NHMW)

112. *Agonopterix multiplicella* (Erschoff, 1877) –
    a) ♀, Italy, Alto Adige, Naturns, ex l., 7.viii.1935, J. Klimesch leg. (0863, ZSM)
    b) ♂, Slovakia, Strážske, 26.viii.1990, Z. Tokár leg. (2157, RCZT)

113. *Agonopterix doronicella* (Wocke, 1849) –
    a) ♂, Austria, Oberösterreich, Partenstein, ex l., 20.vi.1942, J. Klimesch leg. (0835, ZSM)
    b) ♂, Italy, Lucania, Mt. Vulture, 1000 m, ex l., 24.v.1966, J. Klimesch leg. (0833, ZSM)
    c) ♀, Austria, Oberösterreich, Partenstein, ex l., 20.vi.1942, J. Klimesch leg. (0834, ZSM)

114. *Agonopterix assimilella* (Treitschke, 1832) –
    a) ♀, Germany, Süd-Bayern, 25.v.1968, A. Speckmeier leg. (1605, TLMF)
    b) ♀, Austria, Niederösterreich, Scheiblingkirchen, ex l., 4.v.2008, P. Buchner leg. (0025, RCPB)

    c) ♂, Italy, Mt. Baldo, 1500 m, 25.vi.1970, K. Burmann leg. (0417, TLMF)
    d) ♂, Italy, Friuli-Venezia Giulia, Mt. San Simeone, ex l., 29.v.2010, P. Sonderegger leg. (0607, NMBE)
    e) ♂, Bulgaria, Gotse Delchew, Orelek, 1750 m, ex l., 25. v. 2010, O. Karsholt leg. (2319, ZMUC)

115. *Agonopterix atomella* ([Denis & Schiffermüller], 1775) –
    a) ♀, Switzerland, Vaud, Bonvillars, ex l., 28.vi.2005, P. Sonderegger leg. (0510, NMBE)
    b) ♂, Italy, Mt. Baldo, Graziani, Loc Canalette, 1620 m, 11.viii.2018, F. Theimer leg. (7871, RCFT)

116. *Agonopterix scopariella* (Heinemann, 1870) –
    a) ♂, Spain, Port Bon, 15.vi.1977, R. Pinker leg. (1927, TLMF)
    b) ♂, Balearics, Mallorca, Puig Mayor, 1000 m, ex l., 22.vi.1969, J. Klimesch leg. (2051, ZSM)
    c) ♀, Greece, Peloponnes, Mystras, ex l., 3.vi.2009, P. Sonderegger leg. (1867, NMBE)
    d) ♀, Switzerland, Bern, Cremines, 900 m, ex l., 25.vi.2005, P. Sonderegger leg. (0610, NMBE)
    e) ♀, Greece, Karpathos, Lefkos, 30 m, 26.v.1997, R. Sutter leg. (5629, SMNK)
    f) ♂, Greece, Karpathos, Lefkos, 30 m, 26.v.1997, R. Sutter leg. (5630, SMNK)

111a

111b

111c

112a

112b

113a

113b

113c

114a

114b

114c

114d

114e

115a

115b

116a

116b

116c

116d

116e

116f

Colour-Plate 19

117. *Agonopterix conciliatella* (Rebel, 1892) –
    a) ♂, Madeira, Encumeada, 1000 m, 15.vi.1993, ex l., O. Karsholt leg. (3723, ZMUC)
    b) ♀, Canary Islands, Tenerife, La Gomera, El Cedro, ex l., 21.vi.1965, J. Klimesch leg. (0893, ZSM)
    c) ♂, Madeira, Ilha, Quinta Grande, Fortinhas, 7.vii.1991, M. Meyer leg. (3971, ZfBS)
    d) ♀, Canary Isalnds, Tenerife, La Gomera, El Cedro, ex l., 8.vi.1965, J. Klimesch leg. (0836, ZSM)

118. *Agonopterix comitella* (Lederer, 1855) –
    a) ♂, Greece, Samos, Marathokambos, 320 m, 2.viii.2022, A. Blumberg leg. (9853, RCAM)
    b) ♂, Greece, Rhodos, Mt. Smith, ex l., 7.v.1983, J. Klimesch leg. (0885, ZSM)
    c) ♂, Greece, Karpathos, Pigadia, 20.vi.1935, O. Wettstein leg. (0069, NHMW)

119. *Agonopterix nervosa* (Haworth, 1811) –
    a) ♀, Germany, Kaiserstuhl, Vogtsburg, ex l., 10.vi.2010, P. Sonderegger leg. (0628, NMBE)
    b) ♀, Switzerland, Graubünden, Castaneda, ex l., 7.vii.2010, P. Sonderegger leg. (0627, NMBE)
    c) ♀, Great Britain, Cornwall, Porthcurno, ex l., 29.vi.2015, P. Buchner leg. (6392, RCPB)
    d) ♂, France, Corsica, Col de Sevi, 24.viii.1932, H. Reisser leg. (0054, NHMW)
    e) ♂, Great Britain, Scotland, Lochgelly, 24.viii.1932, S. Niemczyk leg. (3490, ZSM)
    f) ♂, France, Cantal, Albipierre, 1300 m, 15.viii.1994, J. Nel leg. (0587, RCJN)

120. *Agonopterix umbellana* (Fabricius, 1794) –
    a) ♂, Italy, Friuli-Venezia Giulia, Val Venzonassa, 5.x.2004, H. Deutsch leg. (1230, RCJN)
    b) ♀, Great Britain, Cornwall, Porthcurno, ex l., 5.viii.2016, P. Buchner leg. (6396, RCPB)
    c) ♂ "f. *lennigiella*", Germany, Rheingau, without date, ex coll. Staudinger (0058, NHMW)
    d) ♂, France, Cote d'Or, without date, ex coll. C.S. Larsen (2555, ZMUC)

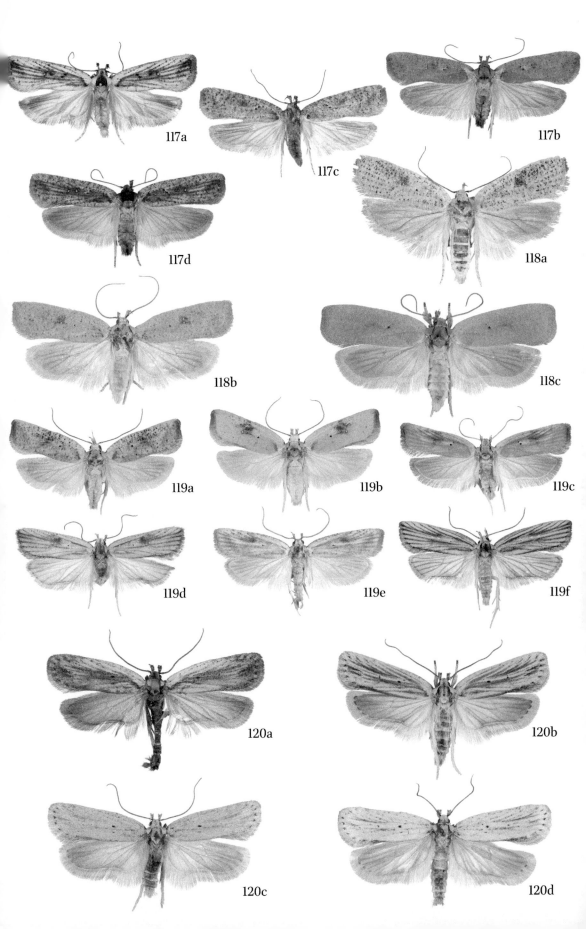

117a

117c

117b

117d

118a

118b

118c

119a

119b

119c

119d

119e

119f

120a

120b

120c

120d

Colour-Plate 20

121. *Agonopterix oinochroa* (Turati, 1879) –
    a) ♀, Spain, Aragon, Borau, 1190 m, 26.vii.2006, A. Mayr leg. (1738, RCAM)
    b) ♀, Austria, Niederösterreich, Mödling, ex l., 1888, without further data (0172, NHMW)
    c) ♂, Spain, Leon, Vilafeliz de Babia, Puerto Pinedo, 25.viii.2015, E. Palm leg. (3730, RCEP)

122. *Agonopterix aspersella* (Constant, 1888) –
    a) ♂, France, Cannes, ex coll. C.S. Larsen, without further data (2509, ZMUC)
    b) ♂, France, Gallia meridionalis, ex l., 1890, A. Constant leg. (0064, NHMW)
    c) ♂, ex coll. Staudinger, without further data (5725, ZMHB)

123. *Depressaria pulcherrimella* Stainton, 1849 –
    a) ♀, Finland, Somero, ex l., 2009, J. Junnilainen leg. (4491, RCJJ)
    b) ♀, Spain, Leon, Sena de Luna, 26.viii.2015, E. Palm leg. (3729, RCEP)

124. *Depressaria sordidatella* Tengström, 1848 –
    a) ♀, Sweden, Sk. Fjälkinge backe, 19.vii.1953, I. Svensson leg. (0923, ZSM)
    b) ♂, Finland, Kirkkonummi, Laehteelae, 20.viii.2002, J. Junnilainen leg. (4488, RCJJ)

    c) ♂, Sweden, Vb. Vännäs, Hällfors, 12.viii.1951, I. Svensson leg. (0922, ZSM)

125. *Depressaria floridella* Mann, 1864 –
    a) ♂, Austria, Niederösterreich, Bisamberg, 5.vii.2009, P. Buchner leg. (0038, RCPB)
    b) ♂, Italy, Gran Sasso, 1750 m, 15.vii.2010, A. Mayr leg. (1778, RCAM)
    c) ♂, Georgia, Rakisi River, 2015 m, 23.vii.2025, K. Nupponen leg. (4738, RCTN)

126. *Depressaria douglasella* Stainton, 1849 –
    a) ♀, Switzerland, Solothurn, Selzach, 670 m, ex l., 13.vi.2010, P. Sonderegger leg. (0641, NMBE)
    b) ♂, Bulgaria, Slavianka, 700 m, 14.vi.2014, J. Junnilainen leg. (4480, RCJJ)
    c) ♂, Cyprus, Kapoura, 1200 m, 3.ix.2003, Hentscholek leg. (9289, RCAM)

127. *Depressaria beckmanni* Heinemann, 1870 –
    a) ♂, Switzerland, Graubünden, Castaneda, 900 m, ex l., 13.iv.2008, P. Sonderegger leg. (3869, NMBE)
    b) ♂, Switzerland, Graubünden, Castaneda, 900 m, ex l., 28.vi.2006, P. Sonderegger leg. (7685, NMBE)
    c) ♂, Switzerland, Bern, Corcelles, 990 m, ex l., 6.iv.2008 larva leg., P. Sonderegger leg. (0636, NMBE)
    d) Neotype ♂, Austria, Styria, Bad Gleichenberg, Stradner Kogel, 480 m, 17.viii.1979, K. Rath leg. (8485, TLMF)

121a

121b

121c

122a

122b

122c

123a

123b

124a

124b

124c

125a

125b

125c

126a

126b

126c

127a

127b

127c

127d

Colour-Plate 21

128. *Depressaria incognitella* Hannemann, 1990 –
   a) ♂, Switzerland, Valais, Fully, 1240 m, ex l., 25.iv.2011 larva leg., P. Sonderegger leg. (0584, NMBE)
   b) ♂, Croatia, Ostarijska vrata, 9.vii.2010, Iv. Richter leg. (1288, RCIR)
   c) ♂, Switzerland, Valais, Leuk, 1560 m, ex l., 12.vii.2015, W. Wittland leg. (3932, RCWW)
   d) ♀, Switzerland, Valais, Fully, 1240 m, ex l., 1.v.2007, P. Sonderegger leg. (1521, NMBE)

129. *Depressaria nemolella* Svensson, 1982 –
   a) ♂, Austria, Kärnten, Heiligenblut, 9.viii.1954, G. de Lattin leg. (3950, ZfBS)
   b) ♀, Sweden, Up Östhammar, Raggarön Havsvik, RN 66793 16552, 7.–12.vii.2004, J. Björklund leg. (9246, RCJB)
   c) ♂, Sweden, Up Östhammar, Raggarön Havsvik, RN 66794 16551, 7.–12.vii.2004, J. Björklund leg. (4266, ECKU)

130. *Depressaria cinderella* Corley, 2002 –
   a) Paratype ♀, Portugal, Alto Alentejo, Portalegre, Minhota, 14.iv.1997, M. Corley leg. (0791, RCMC)
   b) Paratype ♂, Portugal, Alto Alentejo, Portalegre, São Mamede, ex l., 3.vi.2000, M. Corley leg. (0789, RCMC)

c) Paratype ♂, Portugal, Alto Alentejo, Portalegre, Minhota, 5.vi.1996, M. Corley leg. (0790, RCMC)

131. *Depressaria infernella* Corley & Buchner, 2019 –
   a) ♀, Portugal, Algarve, Serra de Monte Figo, 19.v.2002, M. Corley leg. (6676, RCMC).
   b) ♀, Portugal, Serra da Estrela, Poço do Inferno, 23.vi.2010, M. Corley leg. (6673, RCMC)
   c) Paratype ♂, Portugal, Serra da Estrela, Poço do Inferno, ex l., v.–18.vi.2003, M. Corley leg. (6675, RCMC)

132. *Depressaria indecorella* Rebel, 1917 –
   a) Holotype ♂, Russia, Orenburg, 10.vi.1892, Hansen leg. (0215, NHMW)
   b) ♂, Kazakhstan, 17 km NE Emba, 300 m, 20.v.2012, K. Nupponen leg. (4676, RCTN)

133. *Depressaria lacticapitella* Klimesch, 1942 –
   a) ♂, Austria, Ost-Tirol, Mittewald, 15.ix.2014, H. Deutsch leg. (1883, RCHD)
   b) Syntypus ♀, Austria, Oberösterreich, Gr. Pyhrgas, 1600 m, ex l., 21.viii.1940, J. Klimesch leg. (0190, NHMW)

134. *Depressaria hofmanni* Stainton, 1861 –
   a) ♂, Austria, Niederösterreich, Oberloiben, 18.vi.2010, W. Stark leg. (0271, RCWS)
   b) ♀, Switzerland, Neuenburg, Le Landeron, 730 m, ex l., 25.v.2005, P. Sonderegger leg. (0643, NMBE)

128a

128b

128c

128d

129a

129b

129c

130a

130b

130c

131a

131b

131c

132a

132b

133a

133b

134a

134b

Colour-Plate 22

135. *Depressaria albipunctella* ([Denis & Schiffermüller], 1775) –
     a) ♂, Switzerland, Neuenburg, Le Landeron, 620 m, ex l., 25.v.2005, P. Sonderegger leg. (0480, NMBE)
     b) ♂, Italy, Alto Adige, Vinschgau, Taufers, 23.vii.2000, A. Mayr leg. (1791, RCAM)
136. *Depressaria subalbipunctella* Lvovsky, 1981 –
     a) ♂, Crimea, Karadagh, 4.iii.2000, Y. Budashkin leg. (2165, RCZT)
     b) ♂, Turkey, İzmir Province, Bayındır, 600 m, 17.vii.1987, M. Fibiger leg. (5470, ZMUC)
137. *Depressaria krasnowodskella* Hannemann, 1953 –
     a) ♂, Morocco, Ht. Atlas, Tizi-n-Test, 1620 m, 3.v.2005, M. Dvořák leg. (6159, NMPC)
     b) ♂, Morocco, Ht. Atlas, Tachdirt, 2200 m, 25.vii.1933, H. Zerny leg. (0189, NHMW)

c) ♀, Portugal, Serra de Monte Figo, 24.v.2001, M. Corley leg. (0779, RCMC)
138. *Depressaria tenebricosa* Zeller, 1854 –
     a) ♂, North Macedonia, Treska gorge, 29.v.1955, J. Klimesch leg. (0927, ZSM)
     b) ♀, Bulgaria, Rupite, Kozhuh, 100 m, 7.vi.2009, J.-P. Kaitila & R. Haverinen leg. (4555, RCTN)
     c) ♀, Bulgaria, Pirin mts, Sandanski, Lilyanovo, 1000 m, 29.vi.1988, F. Eichler leg. (5841, ZMHB)
139. *Depressaria radiosquamella* Walsingham, 1898 –
     a) ♀, Morocco, Ifrane, 26.vi.1972, F. Hahn leg. (2014, RCHB)
     b) ♀, Spain, Valencia, Villargordo del Cabriel, Kikopark, 21.vi.2010, A. Stübner leg. (2001, RCAS)
140. *Depressaria emeritella* Stainton, 1849 –
     a) ♀, Austria, Niederösterreich, Schwarzau am Steinfeld, ex l., 14.vii.2009, P. Buchner leg. (0037, RCPB)
     b) ♀, Austria, Niederösterreich, Schwarzau am Steinfeld, ex l., 16.vii.2009, P. Buchner leg. (0036, RCPB)

135a

135b

136a

136b

137a

137b

137c

138a

138b

138c

139a

139b

140a

140b

Colour-Plate 23

141. *Depressaria olerella* Zeller, 1854 –
   a) ♂, Austria, Niederösterreich, Maiersdorf, ex l., 29.vi.2011, P. Buchner leg. (1480, RCPB)
   b) ♂, Germany, Brandenburg, Jaenischwalde, 19.viii.1997, A. Stübner leg. (1908, RCAS)
   c) same specimen as b), thorax with groups of whitish scales
   d) ♂, Austria, Niederösterreich, Gumpoldskirchen, ex l., 1.vii.2009, P. Buchner leg. (0033, RCPB)

142. *Depressaria leucocephala* Snellen, 1884 –
   b) ♀, Italy, Laas, 900 m, ex l., 25.vi.1976, K. Burmann leg. (2125, TLMF)
   b) ♂, Russia, Primorskaya Oblast, Slavyanki Ryasanovka, 7.viii.1983, A. Lvovsky leg. (7663, ZMHB)

143. *Depressaria depressana* (Fabricius, 1775) –
   a) ♀, Austria, Niederösterreich, Bisamberg, ex l., 15.viii.2008, P. Buchner leg. (0041, RCPB)
   b) ♀, Switzerland, Graubünden, Ftan, 1640 m, ex l., 28.viii.2005, P. Sonderegger leg. (0632, NMBE)

144. *Depressaria absynthiella* Herrich-Schäffer, 1865 –
   a) ♂, Germany, Bayern, 1912, without further data (0118, NHMW)
   b) ♀, Spain, Huesca, Ontinenja, 300 m, 28.vi.2015, J. Viehmann leg. (3907, RCWSc)

145. *Depressaria tenerifae* Walsingham, 1908, **stat. rev.** –
   a) ♀, Canary Islands, Tenerife, Guimar, ex l., 10.iii.1969, J. Viehmann leg. (3853, ZSM)

146. *Depressaria artemisiae* Nickerl, 1864 –
   a) ♂, Czech Republic, Praha, ex l., 2.vii.1932, V. Vlach leg. (0921, ZSM)
   b) ♂, Italy, Friuli-Venezia Giulia, Venzone, Basovizza, ex l., 22.vii.2010, P. Sonderegger leg. (0491, NMBE)
   c) ♀, Austria, Nordtirol, Zirl, ex l., 9.vi.1967, A. Hernegger leg. (0363, TLMF)

147. *Depressaria fuscovirgatella* Hannemann, 1967 –
   a) ♀, Russia, Orenburgskaya Oblast, Schibendy Valley 20 km S Prokrova, 22.vi.1999, K. & T. Nupponen leg. (4592, RCTN)
   b) ♂, Turkey, Gümüşhane Province, Kop geçidi, 14.ix.1993, M. Fibiger leg. (2362, ZMUC)
   c) ♂, Russia, Altai mts, Chulyshman Valley, 45 km N Ulagan, 600 m, 5.vii.2019, J. Šumpich leg. (7748, NMPC)

148. *Depressaria atrostrigella* Clarke, 1941 –
   a) ♀, Russia, Altai mts, Chulyshman Valley, 45 km N Ulagan, 600 m, 27.vi.2019, J. Šumpich leg. (7743, NMPC)
   b) ♂, Russia, Juldus, without further data (2552, ZMUC)
   c) ♂, Russia, Altai mts, Chulyshman Valley, 45 km N Ulagan, 600 m, 28.vi.2015, J. Šumpich leg. (7276, NMPC)

141a

141c (3x)

141b

141d

142a

142b

143a

143b

144a

144b

145a

146a

146b

146c

147a

147b

147c

148a

148b

148c

Colour-Plate 24

149. *Depressaria silesiaca* Heinemann, 1870 –
  a) ♂, Austria, Nordtirol, Ahrntal, 30.vi. 1966, A. Hernegger leg. (0393, TLMF)
  b) ♀, Italy, Pontresina, 15.viii.1954, K. Burmann leg. (0391, TLMF)
  c) ♂, Austria, Nordtirol, Serfaus, 11.viii. 1958, L. Süssner leg. (0392, TLMF)

150. *Depressaria zelleri* Staudinger, 1879 –
  a) ♂, Italy, Verona, Monte, ex l., 16.vii.1985, K. Burmann leg. (0422, TLMF)
  b) ♀, labial palp in lateral view, Bulgaria, Tri Ushi Hills, Aldomirovsko Marsh, 740 m, 10.viii.2020, S. Beshkov & A. Nahirnic leg. (9150, RCCP)
  c) ♀, Italy, Verona, Monte, ex l., 29.vii.1985, K. Burmann leg. (0421, TLMF)
  d) ♂, Greece, Ioannina, Katara Pass, 1600 m, 11.viii.1985, M. Fibiger leg. (2494, ZMUC)

151. *Depressaria pyrenaella* Šumpich, 2013 –
  a) Paratype ♂, France, Pyrénées Orientales, Vernet, viii.1924, K. Predota leg. (6178, NHMW)
  b) Holotype ♂, Spain, Jaca, 10.viii.1933, W. Fassnidge leg. (NHMUK10293057, NHMUK)

152. *Depressaria marcella* Rebel, 1901 –
  a) ♀, Spain, Cadiz, north of San Rogue, Pinar del Rey, 2.xi.2011, T. Nupponen leg. (4701, RCTN)
  b) ♀, Portugal, Herd. da Comenda, east of Elvas, ex l., 28.vii.2008, M. Corley leg. (0786, RCMC)
  c) ♂, Madeira, Porto Santo, 24.x.1994, O. Karsholt leg. (3722, ZMUC)
  d) ♀, Italy, Sicily, Caltanissetta, Babaurra, 27.vii.1957, Parvis leg. (0915, ZSM)
  e) ♀, France, Montpellier, 18.x.2001, T. Grünewald leg. (1906, RCTG)

153. *Depressaria peregrinella* Hannemann, 1967 –
  a) ♀, Greece, Peloponnes, Zachlorou, 11.vii.1958, ex l., J. Klimesch leg. (0893, ZSM)
  b) ♀, Turkey, Kars Province, 17 km SW Sarıkamış, 2200 m, 20.viii.1965, M. Achtelig & C. Naumann leg. (3955, ZfBS)

154. *Depressaria heydenii* Zeller, 1854 –
  a) ♀, Switzerland, Graubünden, Berninapass, ex l., 25.vii.2007, P. Sonderegger leg. (0547, NMBE)
  b) ♂, France, Col de Galilber, 2450 m, ex l., 18.viii.1996, J. Nel leg. (2227, TLMF)
  c) ♀, Italy, Trentino, Gruppo dell'Adamello, 2450 m, 27.vi.2014, A. Mayr leg. (1897, RCAM)

149a

149b

149c

150a

150b (3x)

150c

150d

151a

151b

152a

152b

152c

152d

152e

153a

153b

154a

154b

154c

Colour-Plate 25

155. *Depressaria fuscipedella* Chrétien, 1915 –
    a) ♀, Italy, Apulia, Gargano, Bosco Quarto, 12.viii.2020, F. Graf leg. (8978, RCAM)
156. *Depressaria ululana* Rössler, 1866 –
    a) ♂, Germany, Bayern, 1882, A.F. Rogenhofer leg. (0114, NHMW)
    b) ♀, Switzerland, Wallis, Fully, 1260 m, ex l., 25.vii.2007, P. Sonderegger leg. (0644, NMBE)
    c) ♂, Greece, Peloponnes, Exochori, ex l., 14.vi.2009, P. Sonderegger leg. (1865, NMBE)
    d) ♀, Greece, Kalavryta, ex l., 16.vi.2009, P. Sonderegger leg. (0645, NMBE)
157. *Depressaria manglisiella* Lvovsky, 1981 –
    a) ♀, Kosovo, Pristina, Grmiya, 700 m, 23.ix.1984, P. Jaksić leg. (8004, TLMF)
    b) ♂, Croatia, South Velebit, 26.viii.2011, Ig. Richter leg. (1245, RCIR)
    c) ♂, Croatia, South Velebit, 26.viii.2011, Ig. Richter leg. (1244, RCIR)

158. *Depressaria chaerophylli* Zeller, 1839 –
    a) ♀, Georgia, Tkibuli, 1165 m, 27.vii.2015, K. Nupponen & R. Haverinen leg. (4736, RCTN)
    b) ♀, Austria, Niederösterreich, Spitz an der Donau, 13.v.2015, W. Stark leg. (0274, RCWS)
    c) ♀, Slovakia, Chynorany, 16.ii.2003, Iv. Richter leg. (1256, RCIvR)
159. *Depressaria longipennella* Lvovsky, 1981 –
    a) ♂, Russia, Orenburgskaya Oblast, Schibendy Valley 20 km S Prokrova, 10.x.2007, K. Nupponen leg. (4496, RCTN)
    b) ♂, Tajikistan, Duschanbe, 5.xii.1961, Yu. Shchemkin leg. (5705, ZMHB)
160. *Depressaria daucella* ([Denis & Schiffermüller], 1775) –
    a) ♂, Great Britain, Cornwall, Porthcurno, ex l., 3.vii.2015, P. Buchner leg. (3403, RCPB)
    b) ♂, Bulgaria, Black Sea Coast, Veleka River, 0 m, 2.vii.2009, B. Zlatkov leg. (3519, RCBZ)

155

156a

156b

156c

156d

157a

157b

157c

158a

159a

158b

158c

159b

160a

160b

Colour-Plate 26

161. *Depressaria ultimella* Stainton, 1849 –
    a) ♀, Belgium, Mons, ex l., 20.viii.1936, A. Dufrane leg. (0917, ZSM)
    b) ♂, Belgium, Frameries, ex l., 24.viii.1935, A. Dufrane leg. (0918, ZSM)
    c) ♀, Moldova, Tighina, ix.1937, T. Hering leg. (5835, ZMHB)

162. *Depressaria halophilella* Chrétien, 1908 –
    a) ♀, Spain, Portbou, ex l., 17.iv.1967, J. Klimesch leg. (0819, ZSM)
    b) ♂, Spain, Portbou, ex l., 29.ix.1974, J. Klimesch leg. (0818, ZSM)
    c) ♀, France, Var, Saint-Cyr-sur-Mer, Port d'Alon, ex l., 31.iii.2014, P. Sonderegger leg. (1996, NMBE)

163. *Depressaria radiella* (Goeze, 1783) –
    a) ♀, Great Britain, Cornwall, Gorran Haven, ex l., 5.viii.2016, P. Buchner leg. (5148, RCPB)

    b) ♀, Austria, Niederösterreich, Hohe Wand, ex l., 21.viii.2009, P. Buchner leg. (0029, RCPB)
    c) ♂, Russia, Kabardno-Balkaria, Central Caucasus, Mt. Terscol, 2300 m, 29.viii.2013, L. Srnka leg. (2174, RCLS)

164. *Depressaria libanotidella* Schläger, 1849 –
    a) ♀, Switzerland, Jura, Soulce, 820 m, ex l., 1.ix.2008, P. Sonderegger leg. (0559, NMBE)
    b) ♀, Italy, Lago di Cavazzo, Mt. San Simeone, vii.1989, H. Deutsch leg. (1589, TLMF)
    c) ♂, Switzerland, Wallis, Leuk, Jeizinen, 2200 m, ex l., 25.viii.2017, R. Seliger leg. (5953, RCRS)

165. *Depressaria bantiella* Rocci, 1934 –
    a) ♂, France, Frejus, Route de Malpasset, 35 m, ex l., 18.vii.2013, F. Rymarczyk & M. Dutheil leg. (7327, RCRD)
    b) ♀, Greece, Crete, Troodos Mts, Olympus, 1950 m, ex l., 29.vi.1997, D. Nilsson et al. leg. (2416, ZMUC)

161a 161b 161c
162a 162b 162c
163a 163b
163c 164a
164b 164c
165a 165b

Colour-Plate 27

166. *Depressaria velox* Staudinger, 1859 –
    a) ♂, Bulgaria, Sandanski, Ploski, 200 m, ex l., 21.vi.2010, N. Savenkov & H. Roweck leg. (4281, ECKU)
    b) ♂, Portugal, Praia do Samouco, Estremadura, ex l., 14.vii.2015, J. Rosete leg. (3492, RCMC)
    c) ♂, Kazakhstan, Almaty, Shengeldy, 10 km SW Almatinskye, 480 m, 9.vi.2018, K. Larsen leg. (7400, RCKL)

167. *Depressaria pimpinellae* Zeller, 1839 –
    a) ♀, Austria, Nordtirol, Innsbruck, ex l., 19.viii.1940, K. Burmann leg. (1645, TLMF)
    a) ♂, Switzerland, Valais, Jeizinen, 1680 m, ex l., 6.ix.2008, P. Sonderegger leg. (0634, NMBE)
    c) ♀, Switzerland, Bern, Grindelwald, 1580 m, ex l., 8.viii.2006, P. Sonderegger leg. (0352, NMBE)

168. *Depressaria villosae* Corley & Buchner, 2018 –
    a) Holotype ♂, Portugal, Vilarinho, east of Parâmio, Trás-os-Montes, ex l., 20.vi.2007, M. Corley leg. (0799, NHMUK)

169. *Depressaria bupleurella* Heinemann, 1870 –
    a) ♀, Austria, Niederösterreich, Mannersdorf, 300 m, 31.iii.2016, W. Stark leg. (4962, RCWS)
    b) ♂, Germany, Pfalz, 1893, F. Eppelsheim leg. (0180, NHMW)
    c) ♂, Austria, Niederösterreich, Klosterneuburg, Buchberg, ex l., 11.ix.1922, H. Zerny leg. (0181, NHMW)

170. *Depressaria sarahae* Gastón & Vives, 2017 –
    a) ♂, Spain, Teruel, Albaracin. 1.x.2008, L. Srnka leg. (1005, RCLS)
    b) ♂, Spain, Granada, Sierra Nevada, 2430 m, 3.vii.2015, J. Tabell leg. (4025, RCJT)

170.1. *Depressaria sarahae* subsp. *tabelli* Buchner, 2017
    a) Holotype ♂, Canary Islands, Tenerife, Guimar, ex l., 16.iv.1907, T. Walsingham leg. (NHMUK010305296, NHMUK)

171. *Depressaria badiella* (Hübner, 1796) –
    a) ♂, North Macedonia, Ochrid, 11.vii.1939, J. Wolfschläger leg. (0925, ZSM)
    b) ♀, Spain, Valencia, Villargordo del Cabriel, 19.vi.2010, A. Stübner leg. (2005, RCAS)
    c) ♂, Turkey, Ankara Province, Kızılcahamam, 925 m, vi.1970, M. & W. Glaser leg. (5559, SMNK)

166a

166b

166c

167a

167b

167c

168a

169a

169b

169c

170a

170b

170.1a

171a

171b

171c

Colour-Plate 28

d) ♂, Greece, West Crete, Omalos Plateau, 1040 m, 20.vi.2014, C. Hviid, O. Karsholt & F. Vilhelmsen leg. (2342, ZMUK)

e) ♂, Austria, Niederösterreich, Rohrwald, Michelberg, 30.viii.1935, A. Ortner leg. (0924, ZSM)

172. *Depressaria pseudobadiella* Nel, 2011 –
a) ♂, Spain, Huesca, Candasnos, 30.v.2015, J. Viehmann leg. (3925, RCWSc)

173. *Depressaria subnervosa* Oberthür, 1888 –
a) ♀, Morocco, Ifrane, 26.vi.1972, F. Hahn leg. (2010, RCHB)
b) ♂, Morocco, Middle Atlas, 1900 m, 17.iv.2013, Z. Tokár leg. (2160, RCZT)

174. *Depressaria cervicella* Herrich-Schäffer, 1854
a) ♂, Hungary, Ofen, 1871, L. Anker leg. (0194, NHMW)

175. *Depressaria gallicella* Chrétien, 1908 –
a) ♂, France, Pyrenees, Val d'Ossoue, 1500 m, ex l., vii.1961, K. Burmann leg. (0399, TLMF)
b) ♀, Switzerland, Valais, Vouagnoz, 1500 m, ex l., 10.vii.2008, P. Sonderegger leg. (0647, TLMF)

176. *Depressaria altaica* Zeller, 1854 –
a) ♂, Russia, Orenburgskaya Oblast, Schibendy Valley 20 km S Prokrova, 28.ix.2005, K. Nupponen leg. (6465, RCTN)
b) Lectotype ♂, Kazakhstan, Altai, without further data (5913, ZMHB)

177. *Depressaria albarracinella* Corley, 2017 –
a) Paratype ♂, Spain, Huesca, Candasnos, 30.v.2015, J. Viehmann leg. (3919, RCWSc)
b) ♂, Greece, Central Greece, Archova, 1070 m, 9.vi.2013, P. Skou leg. (2326, ZMUC)

171d

171e

172a

173a

173b

174a

175a

175b

176a

176b

177a

177b

Colour-Plate 29

178. *Depressaria eryngiella* Millière, 1881 –
    a) ♀, Italy, Sardinia, Bacu Trotu, Ortuabis, 800 m, 23. viii 1978, G. Derra leg. (0647, RCGD)
    b) ♂, Croatia, South Velebit, 26.vii.2006, L. Srnka leg. (0974, RCLS)
    c) ♀, Turkey, Erzurum Province, Kop geçidi, 1750 m, 16. xi 1993, G. Derra leg. (0647, RCGD)

179. *Depressaria veneficella* Zeller, 1847 –
    a) ♀, Greece, Crete, Iraclion, Chersonisos, ex l., 9.v.1989, R. Johansson leg. (3789, ZMUC)
    b) ♀, Morocco, Ifrane, 2.vii.1972, F. Hahn leg. (2013, RCHB)

180. *Depressaria discipunctella* Herrich-Schäffer, 1854 –
    a) ♂, Spain, Andalusia, Sierra Nevada, Camino de la Veleta, 2050 m, 3.vii.1967, E. Traugott-Olsen leg. (2529, ZMUC)
    b) ♀, Croatia, Zengg, 6.vi.1916, Tobiasch leg. (0219, NHMW)
    c) ♂, Turkey, Ankara Province, Kızılcahamam, 1.vii.1968, J. Klimesch leg. (3855, ZSM)

181. *Depressaria hansjoachimi* **sp. n.** –
    a) Paratype ♂, Greece, Peloponnes, Achaia, Aroania, 2100 m, 1.vii.2001, T. Nupponen leg. (4683, RCTN)
    b) Paratype ♀, Greece, Peloponnes, Achaia, Aroania, 2200 m, 24.vi.1958, J. Klimesch leg. (6832, ZSM)

182. *Depressaria junnilaineni* Buchner, 2017 –
    a) Paratype ♂, Spain, Zaragossa, Bujaraloz, 300 m, 29.v.2015, J. Viehmann leg. (3906, MNCN)
    b) Paratype ♂, Spain, Val Kiko Park, Villargorda del Cabriel, 700 m, 22.vi.2010, F. Theimer leg. (1913, RCFT)
    c) Holotype ♂, Greece, Epirus, Ammoudia 10 km southeast Parga, 0 m, 23.viii.2008, W. Schmitz leg. (5997, TLMF)

183. *Depressaria pentheri* Rebel, 1904 –
    a) ♂, Croatia, Velebitski, 700 m, 5.vi.2008, J. Junnilainen leg. (4460, RCJJ)
    b) Holotype ♀, Bosnia and Herzegovina, Podasje, 1300 m, 22.vii.1901, A. Penther leg. (3471, TLMF)
    c) ♂, Croatia, Velebit Ostaria, vii.1911, M. Hilf leg. (0104, NHMW)

178a

178b

178c

179a

179b

180a

180b

180c

181a

181b

182a

182b

182c

183a

183b

183c

Colour-Plate 30

184. *Depressaria hannemanniana* Lvovsky, 1990 –
    a) ♀, Kazakhstan, Katutau mts, Konyro-len River, 1240 m, 2.x.2015, K. Nuppo-nen leg. (4703, RCTN)
    b) ♂, Russia, Orenburgskaya Oblast, Mt. Verbljushka, Donskoe, 260 m, 28.vi.2009, J. Šumpich leg. (6138, NMPC)
185. *Depressaria erzurumella* Lvovsky, 1996 –
    a) Holotype ♂, Turkey, 15 km south Erzu-rum, 3000 m, 20.vii.1989, M. Fibiger & Esser leg. (5420, ZMUC)
    b) ♂, Greece, Loutra Kilinis, south of Patras, 0 m, 28.vi.2007, J. Viehmann leg. (5999, TLMF)
186. *Depressaria dictamnella* (Treitschke, 1835) –
    a) ♂, Germany, Regensburg, ex l., 1891, no further data (0210, NHMW)
187. *Depressaria moranella* Chrétien, 1907 –
    a) ♂, North Macedonia, Ochrid, 19.ix.1936, J. Wolfschläger leg. (2068, ZSM)
    b) ♂, North Macedonia, Treska gorge, 25.vi.1967, R. Pinker leg. (6176, NHMW)

c) ♀, Lybia, Gharian, Wadi el Hira, 6.v.1983, U. Seneca leg. (2943, RCKL)
188. *Depressaria hystricella* Möschler, 1860 –
    a) ♂, Russia, Wolgogradskaya oblast, Sarepta (now Krasnoarmeysk), 1866, O. Staudinger leg. (0211, NHMW)
189. *Depressaria hirtipalpis* Zeller, 1854 –
    a) ♂, Turkey, Kayseri Province, Erciyes Dağı, 6.vii.1902, A. Penther leg. (0943, NHMW)
    b) ♂, Greece, Rhodos, Kamiros, ex l., 2.vi.1983, J. Klimesch leg. (0815, ZSM)
    c) ♂, Croatia, Krk, Stara Baska, 4.x.2012, May leg. (1667, RCRK)
190. *Depressaria erinaceella* Staudinger, 1870 –
    a) ♀, Italy, Sardinia, Sassari, ex l., 10.vi.1963, J. Klimesch leg. (0811, ZSM)
    b) ♂, Portugal, Algarve, Carrapateira, 6.x.1993, M. Corley leg. (0806, RCMC)
191. *Depressaria peniculatella* Turati, 1922 –
    a) ♀, Spain, Almeria, Tabernas, 400 m, 19.x.2009, J. Šumpich leg. (6137, NMPC)
    b) ♀, Portugal, Algarve, Sagres, 15.x.1996, A. Gardiner & P. Wallis leg. (0792, RCMC)
192. *Depressaria rungsiella* Hannemann, 1953 –
    a) ♂, Morocco, Middle Atlas, Ifrane, 1700 m, 15.x.1973, R. Pinker leg. (1672, ZSM).

184a

184b

185a

185b

186a

188a

187a

187b

187c

189a

189b

189c

190a

190b

191b

191a

192a

Male Genitalia Illustrations

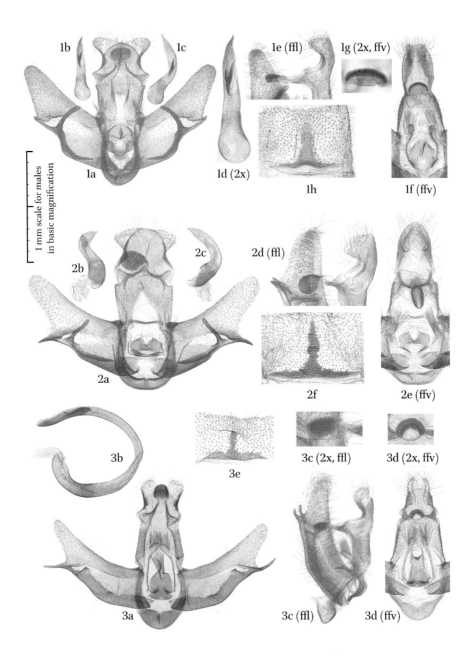

1. *Semioscopis avellanella* – a–e, g–h) Austria (8498, RCPB); f) Austria (8499, RCPB)
2. *Semioscopis oculella* – a–f) Austria (8502, NHMW)
3. *Semioscopis steinkellneriana* – a–e) Austria (8500, NHMW)

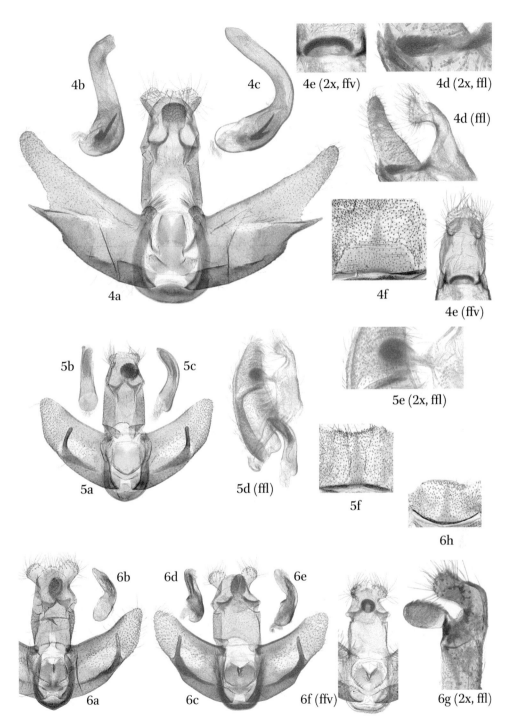

4.   *Semioscopis strigulana* – a–f) Austria (7304, NHMW)
5.   *Luquetia lobella* – a–f) Austria (8501, NHMW)
6.   *Luquetia orientella* – a–b) Greece (1924, RCHB); c–h) Albania (8176, RCCP)
7.   *Luquetia osthelderi* comb.nov. – males unknown

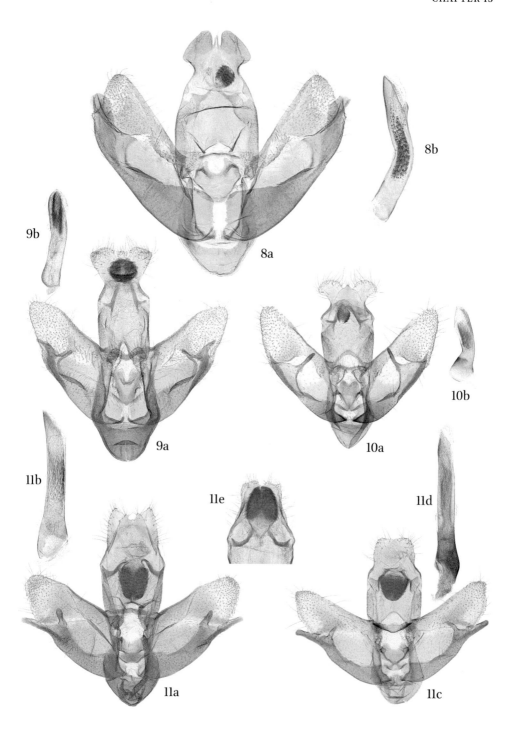

8.   *Exaeretia hepatariella* – a–b) Latvia (0707, NHMW)
9.   *Exaeretia allisella* – a–b) Denmark (2101, TLMF)
10.   *Exaeretia ciniflonella* – a–b) Russia (0703, NHMW)
11.   *Exaeretia preisseckeri* – a–b) Austria (0706, NHMW); c) Croatia (1240, RCIR);
      d–e) Hungary (6800, TLMF)

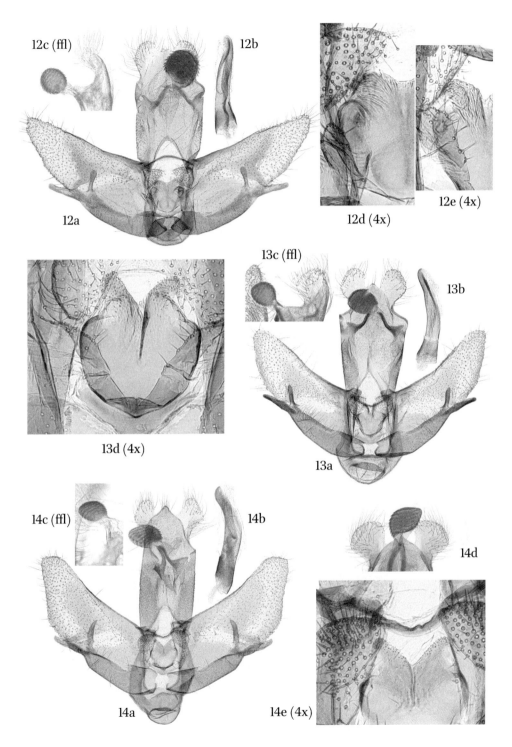

12. *Exaeretia lutosella* – a–b, d) Italy (1567, RCLM); c) Morocco (5839, ZMHB); e) Israel (6282, MGAB)

13. *Exaeretia thurneri* – a–b) North Macedonia (2258, NHMW); c–d) Romania (6342, MGAB)

14. *Exaeretia ledereri* – a–b, e) Croatia (6747, TLMF); c) Greece (6020, RCWSc); d) Tunisia (1909, RCHB)

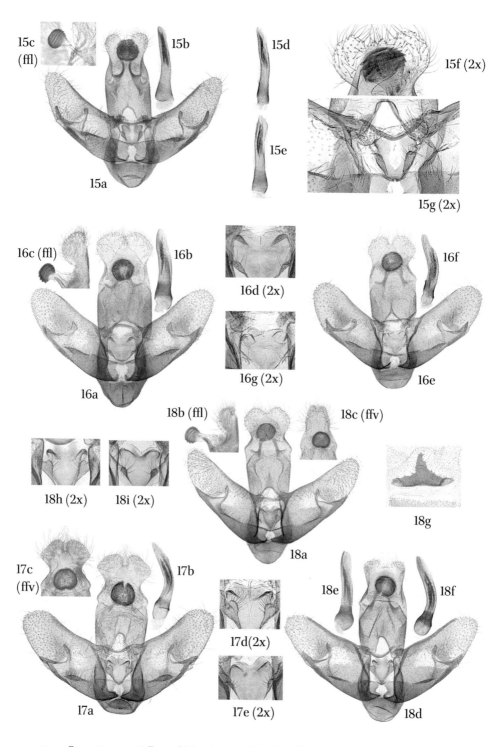

15. *Exaeretia praeustella* – a–b) Russia (1030, RCLS); c–g) Russia (6296, MGAB)
16. *Exaeretia nebulosella* – a–d) Russia (4599, RCTN); e–g) Russia (7131, ZIN)
17. *Exaeretia lepidella* – a–d) Russia (4581, RCTN); e) Russia (6172, NMPC)
18. *Exaeretia buvati* – a–c) France (8566, RCJJ); d–h) France (4571, RCJJ); i) France (8567, RCJJ)

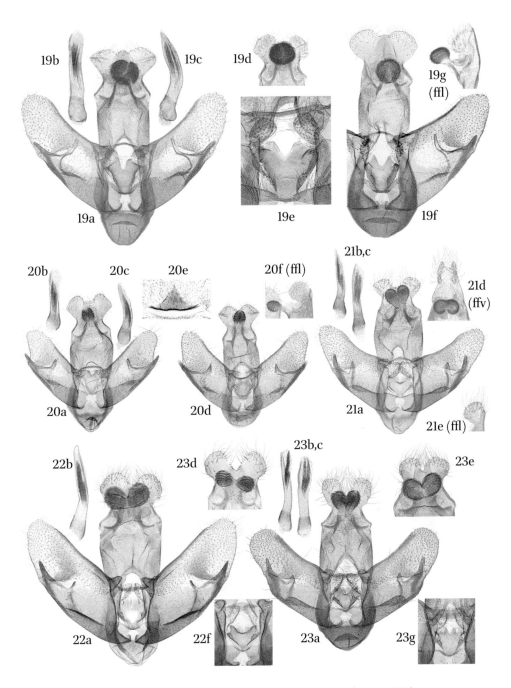

19. *Exaeretia mongolicella* – a–c) Mongolia (5888, HNHM); d–e) Russia (2004, RCHB); f) Russia (6299, MGAB); g) Mongolia (5887, HNHM)

20. *Exaeretia lvovskyi* – a–c) Russia (7264, NMPC); d–e) Russia (4611, RCTN); f) Russia (7266, NMPC)

21. *Exaeretia stramentella* – a–e) Germany (7305, NHMW)

22. *Exaeretia niviferella* – a–b) Kyrgyzstan (0953, NHMW)

23. *Exaeretia indubitatella* – a–c) Russia (4557, RCTN); d) Russia (1967, RCHB); e–f) Russia (6228, NMPC); g) Russia (4682, RCTN)

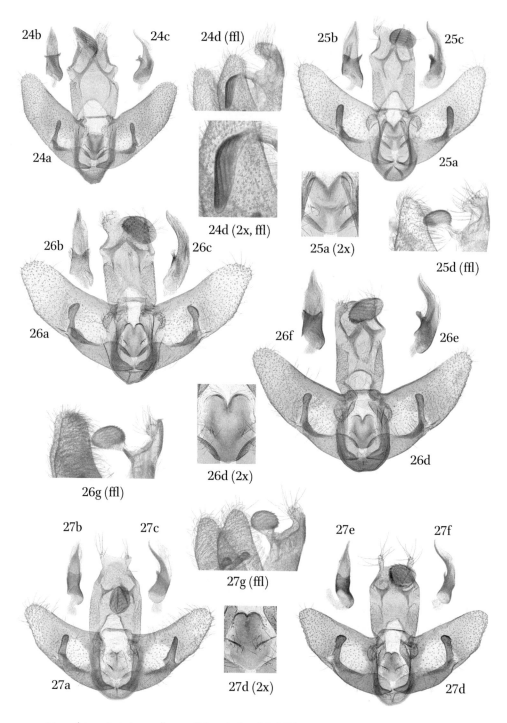

24. *Agonopterix impurella* – a–d) Austria (7296, RCPB)
25. *Agonopterix liturosa* – a–c) Russia (6468, RCTN); d) Austria (7236, RCWS)
26. *Agonopterix conterminella* – a–c) France (3741, ZMUC); d–f) Russia (6263, MGAB);
    g) Germany (6866, RCPB)
27. *Agonopterix arctica* – a–c) Norway (1914, RCTG). d–g) Sweden (0209, NHMW)

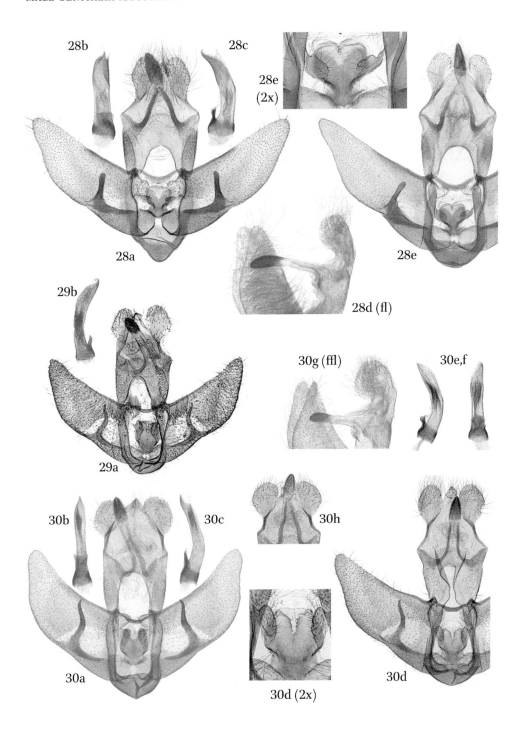

28. *Agonopterix ocellana* – a–c) Russia (4714, RCTN); d) Austria (0008, RCPB); e) Austria (0948, NHMW)

29. *Agonopterix fruticosella* – a–b) Spain (NHMUK1055274, J.F.G. Clarke slide 4991)

30. *Agonopterix rigidella* – a–c) Spain (RCWSc); d–g) Portugal (7321, RCMC); h) Spain (2464, ZMUC)

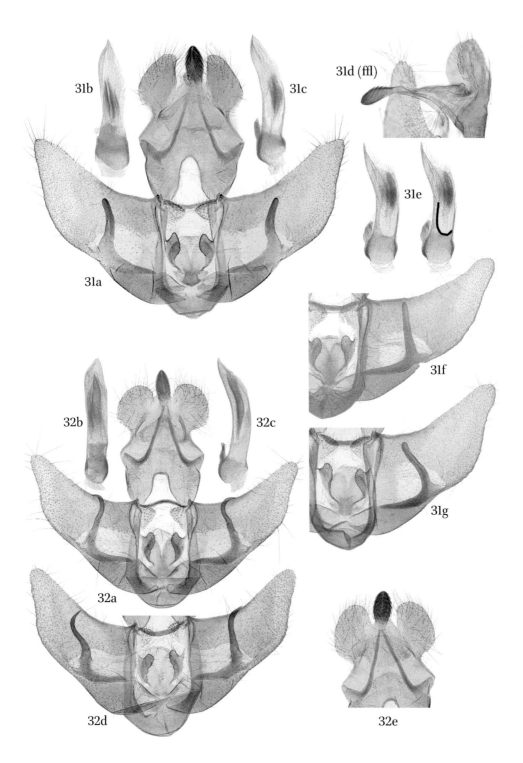

31.   *Agonopterix olusatri* – a) Sardinia (0700, NHMW); b–c, f) Morocco (6724, RCKL); d) Spain
      (4212, ZSM); e) Crete (3630, TLMF); g) Portugal (0805, RCMC)
32.   *Agonopterix leucadensis* – a–c) Greece (1797, RCTM); d–e) Greece (0486, NMBE)

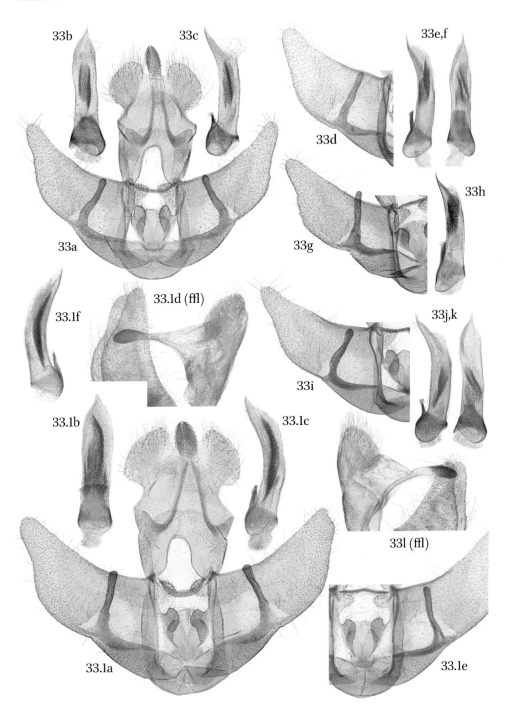

33. *Agonopterix adspersella* – a–c, l) France (1984, NMBE); d–f) Switzerland (7711, NMBE); g–h) Spain (1071, RCLS); i–k) Bulgaria (4544, RCTN)

33.1. *Agonopterix adspersella* ssp. *pavida* stat.rev. – a–c) Turkey (5389, ZMUC); d) Greece (5474, ZMUC); e–f) Turkey (2716, RCKL)

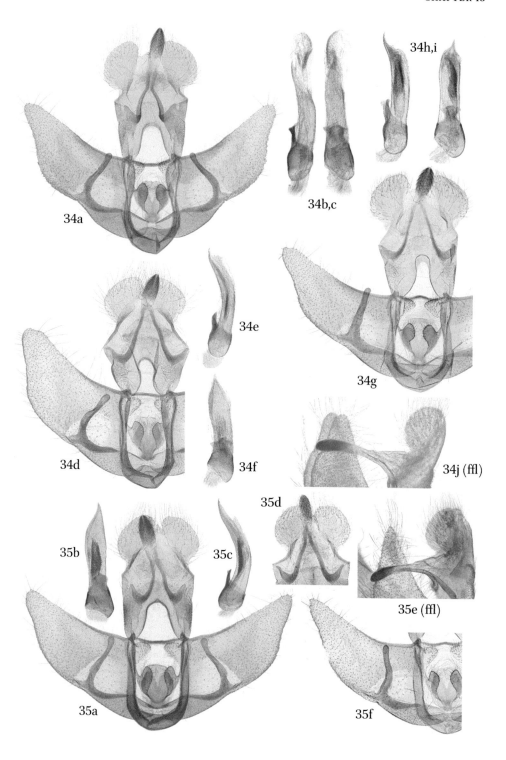

34.   *Agonopterix thapsiella* – a–c) Portugal (0803, RCMC); d–f) Sicily (6693, ZMUC);
      g–i) Portugal (0802, RCMC); j) Greece (4692, RCTN)
35.   *Agonopterix chironiella* – a–c) Italy (7696, NMBE); d) Italy (1635, TLMF); e–f) Italy (4213, ZSM)

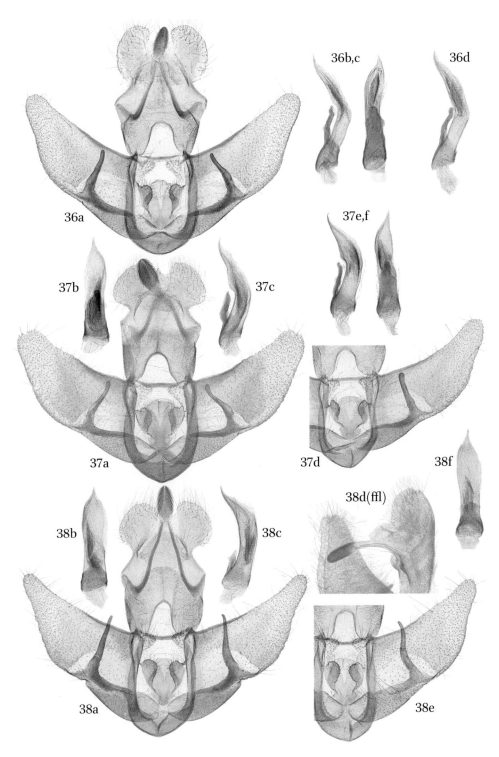

36. *Agonopterix cervariella* – a–c) France (3770, ZMUC); d) France (4174, NHMW)
37. *Agonopterix paracervariella* sp.n. – a–c) Italy (0894, ZSM); d–f) Slovenia (6358, NMPC)
38. *Agonopterix cadurciella* – a–c) Italy (4743, TLMF); d) Turkey (7310, NHMW); e–f) Italy (5717, ZMHB)

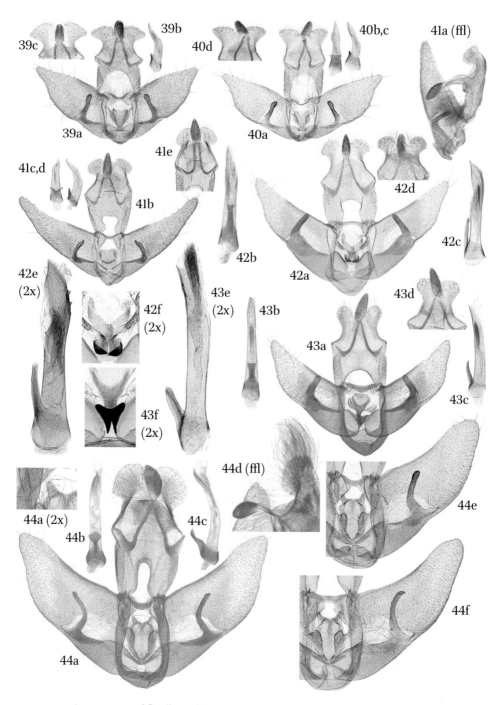

39. *Agonopterix nodiflorella* – a–b) France (1593, TLMF); c) Lebanon (0735, NHMW)
40. *Agonopterix rotundella* – a–c) Switzerland (6922, NMBE); d) Italy (1465, MFSN)
41. *Agonopterix purpurea* – a–d) Turkey (7308, NHMW); e) Sicily (0719, NHMW)
42. *Agonopterix curvipunctosa* – a–c) Austria (0718, RCWS); d–f) Austria (0109, RCWS)
43. *Agonopterix vendettella* – a, c) Sardinia (1566, RCLM); b) Canary Islands (3977, ZfBS); d) Madeira (3710, ZMUC); e–f) Canary Islands (2636, RCKL)
44. *Agonopterix alpigena* – a–d) Austria (6873, RCPB); e) Italy (1688, RCAM); f) Slovenia (1585, TLMF)

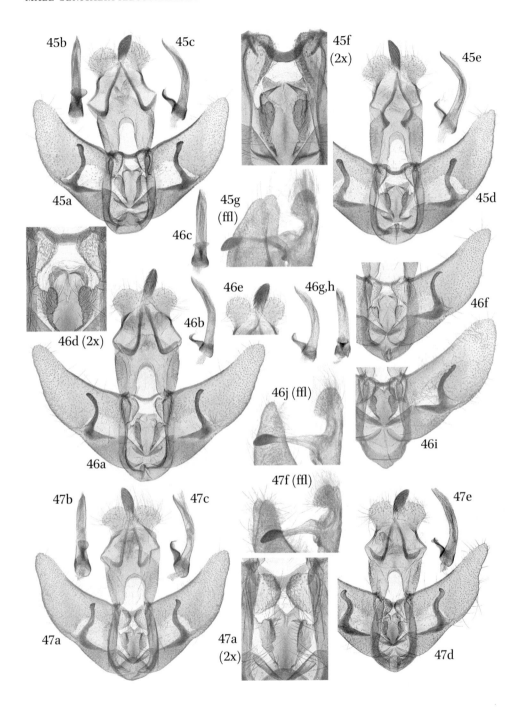

45. *Agonopterix richteri* sp.n. – a–c, g) Bulgaria (6484, RCJJ); d–e) Bulgaria (1143, RCIR); f) Bulgaria (1218, RCIR)

46. *Agonopterix coenosella* – a–c) Tajikistan (5695, ZMHB); d) Turkey (3116, RCKL); e) Russia (1045, RCLS); f–h) Kazakhstan (4598, RCTN); i) Romania (8145, RCZK); j) Turkey (5502, ZMUC)

47. *Agonopterix kayseriensis* – a–c) Turkey (4375, SMNK); d–e) Turkey (2369, ZMUC); f) Turkey (3117, RCKL)

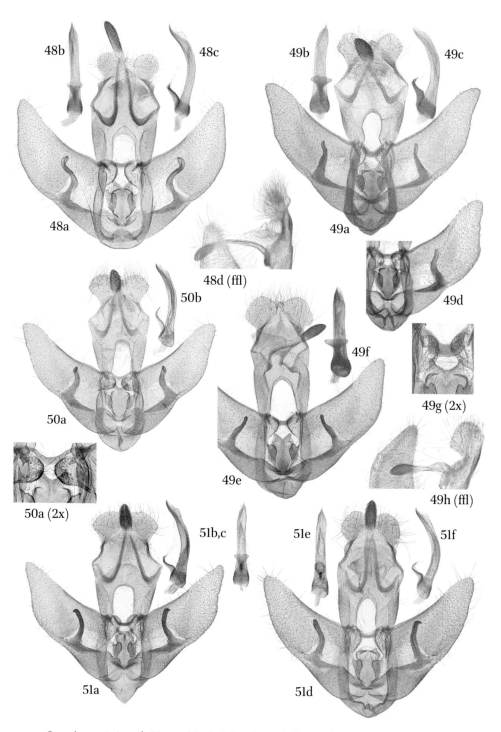

48. *Agonopterix cachritis* – a–c) Spain (5807, ZMHB); d) Spain (0903, ZSM)
49. *Agonopterix ferulae* – a–c) Corsica (7005, RCGF); d) Cyprus (2428, ZMUC); e–f, h) Sardinia (8529, RCGF); g) Portugal (0946, RCMC)
50. *Agonopterix galicicensis* sp.n. – a–b) North Macedonia (1641, TLMF)
51. *Agonopterix langmaidi* sp.n. – a–b) Crete(4697, RCTN); c) Crete (6714, RCKL); d–f) Sicily (4720, RCTN)

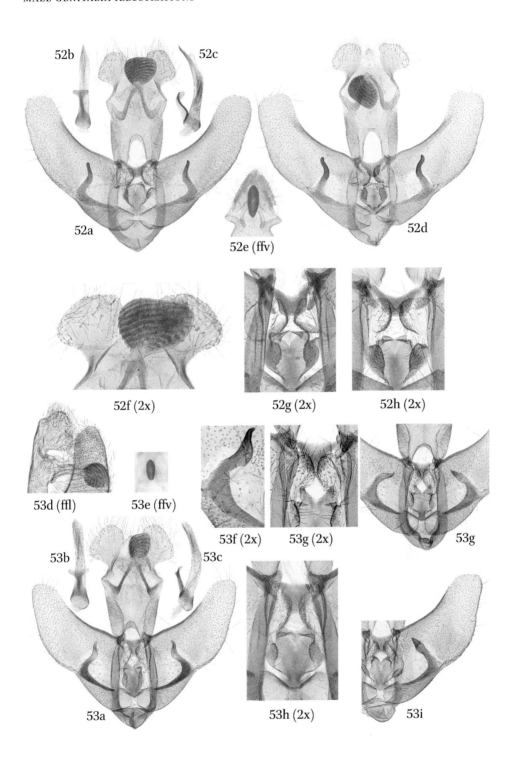

52e (ffv)

52f (2x)   52g (2x)   52h (2x)

53d (ffl)   53e (ffv)

53f (2x)   53g (2x)   53g

53h (2x)   53i

52. *Agonopterix lessini* – a–c) Italy (1558, RCLM); d) Italy (0674, RCRK); e) Italy (8576, RCAM);
f) Italy (1768, RCAM); g) Italy (1345, MFSN); h) Italy (1548, RCCM)

53. *Agonopterix selini* – a–e) Italy (1579, TLMF); f) Croatia (0964, RCLS); g) Italy (1436, MFSN);
h) Austria (1661, TLMF); i) Greece (1969, RCHB)

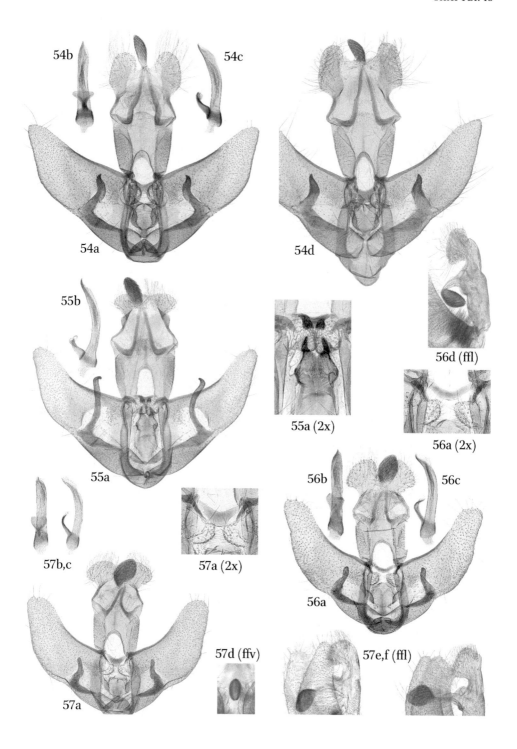

54. *Agonopterix ordubadensis* – a–c) Armenia (6600, ZIN); d) Turkey (3357, RCKL)
55. *Agonopterix socerbi* – a–b) Italy (2141, NMBE)
56. *Agonopterix angelicella* – a–c) Italy (2142, MFSN); d) Austria (7303, NHMW)
57. *Agonopterix paraselini* – a–c) Austria (5110, RCPB); d) Austria (1663, TLMF);
    e) Turkey (5129, NHMW); f) Austria (5112, RCPB)

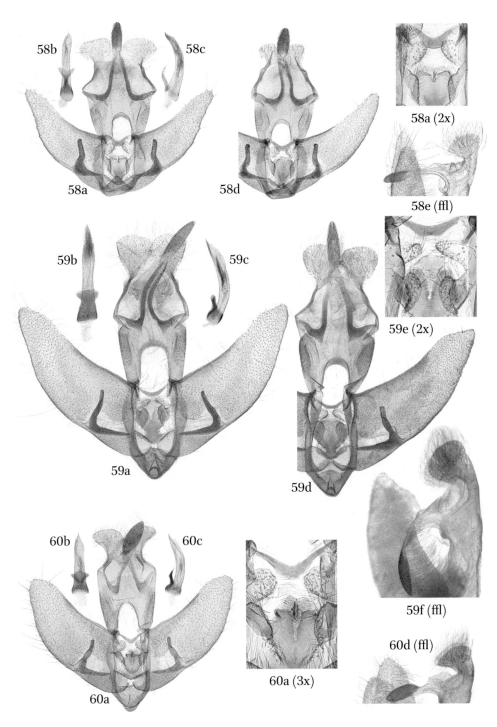

58. *Agonopterix parilella* – a, c) Slovakia (1014, RCLS); b, e) Austria (2127, TLMF); d) Austria (1603, TLMF)

59. *Agonopterix ciliella* – a, c) Austria (7297, RCPB); b, f) Austria (7291, RCPB); d) Italy (1385, MFSN); e) Russia (6301, MGAB)

60. *Agonopterix orophilella* – a–d) France (7339, RCRD)

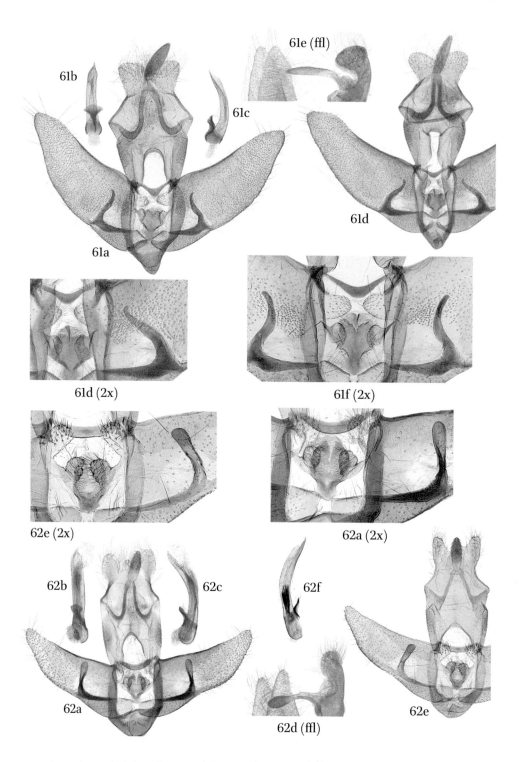

61.   *Agonopterix heracliana* – a, c) Germany (2563, RCPB); b) Germany (5443, RCPB);
      d) England (1863, RCPB); e) Austria (7225, RCWS); f) Italy (1193, MFSN)
62.   *Agonopterix perezi* – a) Canary Islands (0677, RCRK); b–d) Canary Islands (2063, ZSM);
      e–f) Madeira (3706, ZMUC)

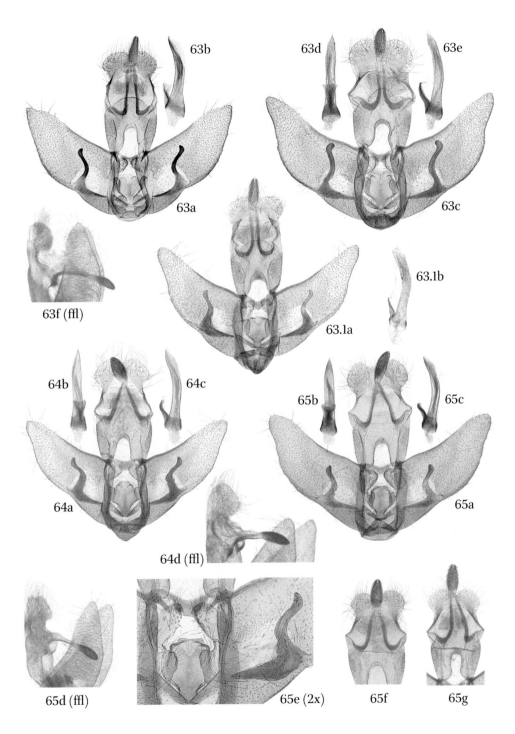

63b 63d 63e
63a 63c
63f (ffl)
63.1a 63.1b
64b 64c
64a
64d (ffl)
65b 65c
65a
65d (ffl) 65e (2x) 65f 65g

63. *Agonopterix putridella* – a–b) Austria (0929, RCPB); c–e) France (6297, MGAB); f) Russia (6615, ZIN)

63.1. *Agonopterix putridella* ssp. *scandinaviensis* ssp.n. – a–b) Sweden (2498, ZMUC)

64. *Agonopterix quadripunctata* – a) Poland (0714, NHMW); b–d) Estonia (4590, TUEE)

65. *Agonopterix hippomarathri* – a–c) Romania (8146, RCZK); d) Austria (0017, RCPB); e) Italy (4044, RCCM); f) Hungary (1625, TLMF); g) Croatia (1274, RCIR)

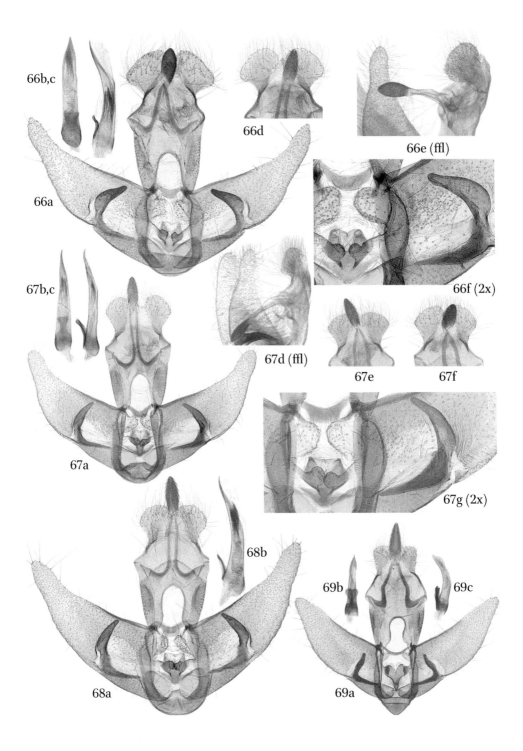

66b,c

66d

66e (ffl)

66a

66f (2x)

67b,c

67d (ffl)

67e   67f

67a

67g (2x)

68b

69b   69c

68a   69a

66.   *Agonopterix astrantiae* – a–c) Slovenia (1112, RCFG); d) Austria (0347, RCPB); e) Austria (7292, RCPB); f) Austria (1894, RCAM)

67.   *Agonopterix cnicella* – a–c) France (2890, RCKL); d) Austria (7294, RCPB); e) Slovenia (1669, RCRK); f–g) Greece (2757, RCKL)

68.   *Agonopterix melancholica* – a–b) Russia (3674, NHMW)

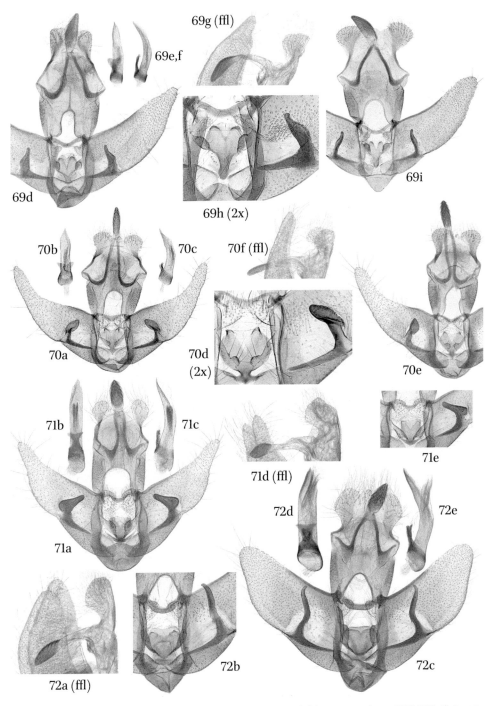

69. *Agonopterix alstromeriana* – d–g) Hungary (7312, NHMW); h) Germany (0734, NHMW); i) Austria (0349, RCPB)

70. *Agonopterix capreolella* – a–c) Turkey (4363, RCGB); d) Austria (0312, NHMW); e) Austria (1243, RCWS); f) Albania (8234, RCCP)

71. *Agonopterix yeatiana* – a–c) Greece (3653, TLMF); d) Spain (6378, NMPC); e) Italy (1354, MFSN)

72. *Agonopterix silerella* – a) Austria (7290, RCPB); b) Montenegro (1148, RCIR); c–e) Austria (0690, RCPB)

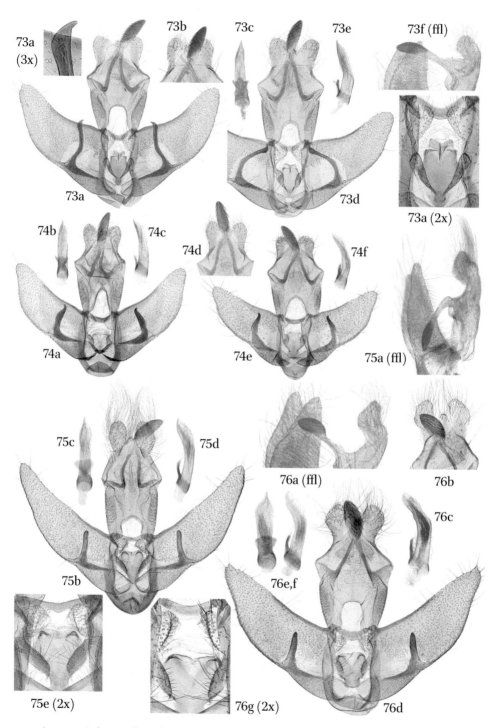

73. *Agonopterix ligusticella* – a) Italy (1556, RCLM); b) Croatia (0991, RCLS); c) Greece (1876, RCTG); d–e) Italy (0412, TLMF); f) Italy (8577, RCAM)
74. *Agonopterix medelichensis* – a–c) Italy (1697, RCAM); d, f) Greece (2736, RCKL); e) Turkey (2470, ZMUC)
75. *Agonopterix irrorata* – a) Cyprus (6700, ZMUC); b–d) Italy (1355, MFSN); e) Albania (1122, RCFG)
76. *Agonopterix graecella* – a–c) Italy (7914, TLMF); d–f) Greece (2816, RCKL); g) Montenegro (0968, RCLS)

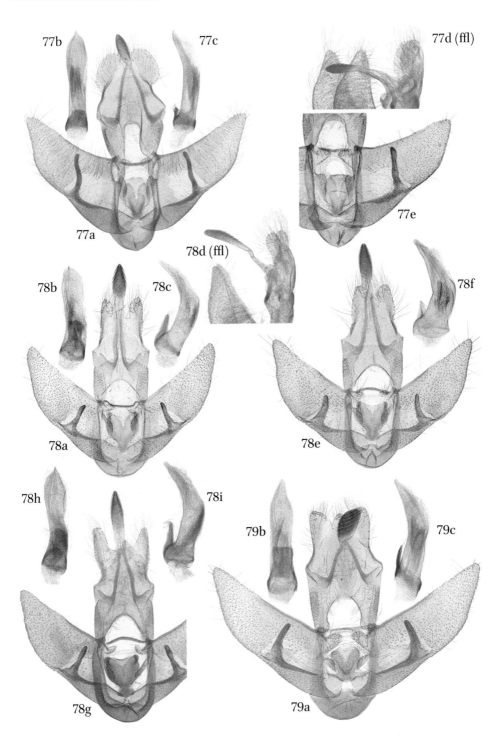

77.   *Agonopterix pseudoferulae* – a–d) Sardinia (6773, RCAM); e) Sicily (1928, RCHB)

78.   *Agonopterix subtakamukui* – a–c) Austria (1900, RCAM); e–f) Austria (1898, RCAM);
      d, g–i) Russia (7085, ZIN)

79.   *Agonopterix guanchella* – a–c) Canary Islands (4558, MNCN)

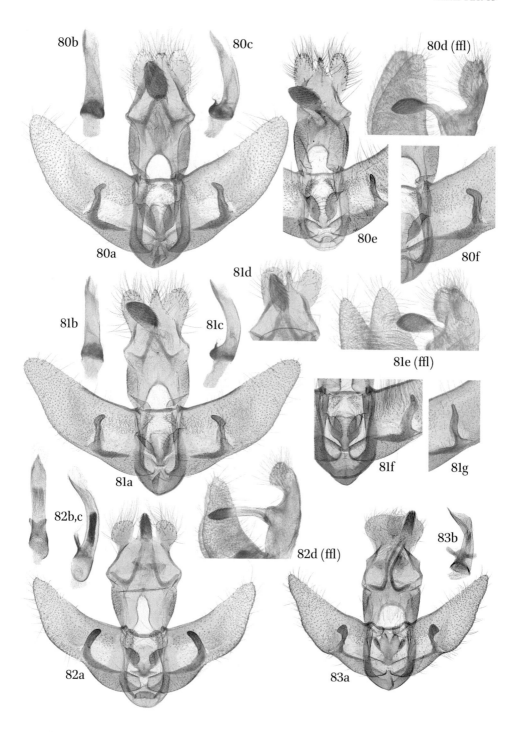

80. *Agonopterix furvella* – a–c) Germany (0738, NHMW); d) Switzerland (5956, RCRS); e) Austria (9466, RCPB); f) Austria (7298, RCPB)

81. *Agonopterix pupillana* – a) Italy (2061, ZSM); b–c, e, g) Italy (7933, TLMF); d) Italy (7986, TLMF); f) Albania (1338, RCFG)

82. *Agonopterix rutana* – a, c) Spain (1091, RCLS); b, d) Israel (6359, NMPC)

83. *Agonopterix subumbellana* – a–b) Turkey (3209, RCKL)

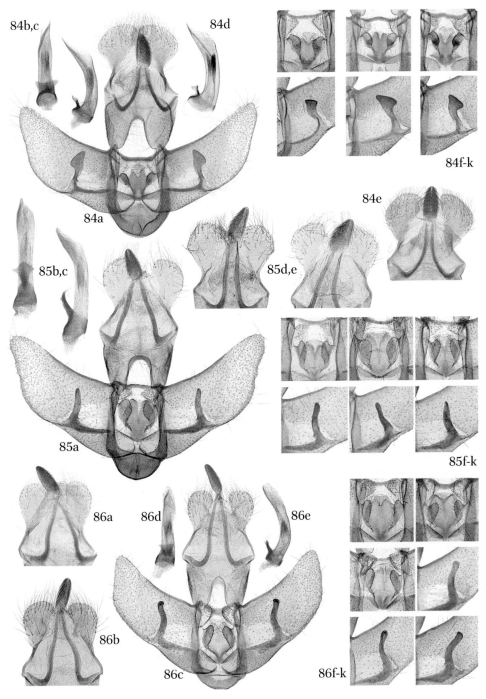

84. *Agonopterix pallorella* – a, d) Tunisia (0676, RCRK); b–c) Turkey (6430, RCJJ); e, j) Greece (4011, RCWSc); f, i) Tajikistan (5699, ZMHB); g) Spain (4185, NHMW); h, k) France (4013, RCWSc)

85. *Agonopterix straminella* – a–c) Tunisia (1981, RCRK); d, g) Morocco (1647, TLMF); e) Crete (2441, ZMUC); f, j) Spain (4776, RCWSc); h, k) Syria (0871, ZSM); i) Morocco (1950, RCAW)

86. *Agonopterix carduncelli* – a, i) Morocco (1983, RCRK); b, g) Greece (2047, ZSM); c–e) Spain (4700, RCTN); f) Portugal (0757, RCMC); h, k) Spain (2183, RCLS); j) Morocco (1677, ZSM)

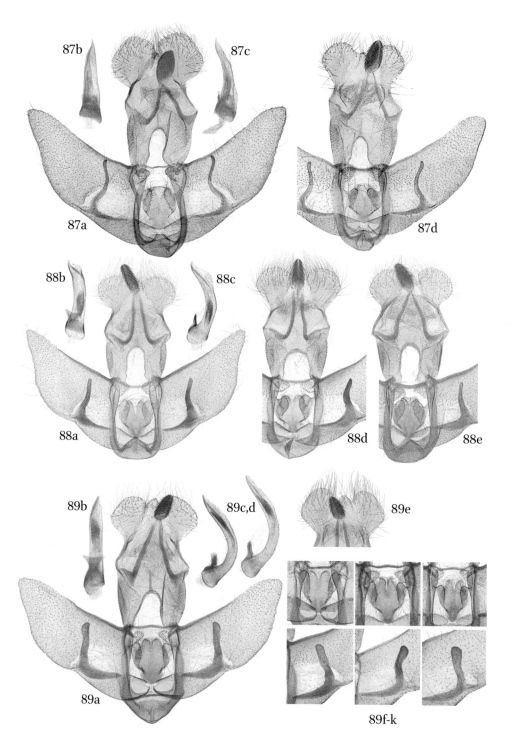

87.   *Agonopterix bipunctosa* – a–c) Russia (7274, NMPC); d) Sweden (2517, ZMUC)
88.   *Agonopterix kyzyltashensis* – a–c) Russia (4587, RCJK); d) Russia (1048, RCLS); e) Russia (4597, RCTN)
89.   *Agonopterix squamosa* – a) Turkey (0132, NHMW); b, f, j) Turkey (7646, ZMUC); c, e, h) Croatia (1188, RCIR); d, g) Turkey (2384, ZMUC); i) Croatia (1211, RCIR); k) Greece (4360, RCGB)

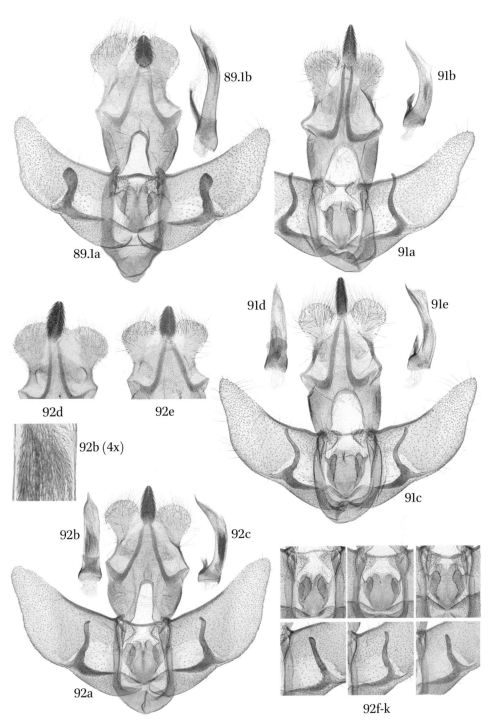

89.1.   *Agonopterix* near *squamosa* o₁ – a–b) Spain (2462, ZMUC)
90.   *Agonopterix tschorbadjiewi* – males unknown
91.   *Agonopterix volgensis* – a–b) Russia (6461, RCTN); c–e) Russia (4594, RCTN)
92.   *Agonopterix kaekeritziana* – a–c) Austria (4751, RCPB); d) Italy (1356, MFSN); e–f, i) Austria
        (7678, RCPB); g) Spain (5360, ZMUC); h) Greece (6047, RCWSc); j) Russia (4620, RCTN); k)
        Spain (7183, RCZT)

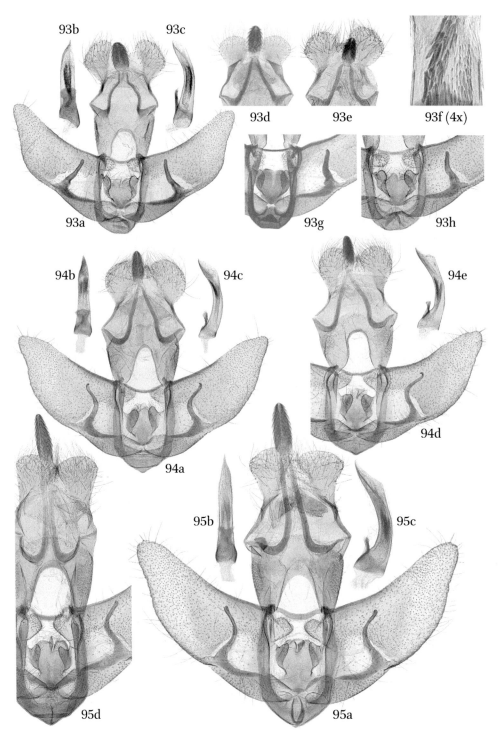

93.   *Agonopterix broennoeensis* – a–c) Finland (4579, RCJJ); d, f–g) Russia (7742, TLMF);
       e, h) Russia (2515, ZMUC)

94.   *Agonopterix uralensis* sp.n. – a–c) Russia (4495, RCTN); d) Russia (2187, RCLS); e) Russia (2200,
       RCLS)

95.   *Agonopterix latipennella* – a–c) Turkey (3527, MHNG); d) Cyprus (2454, ZMUC)

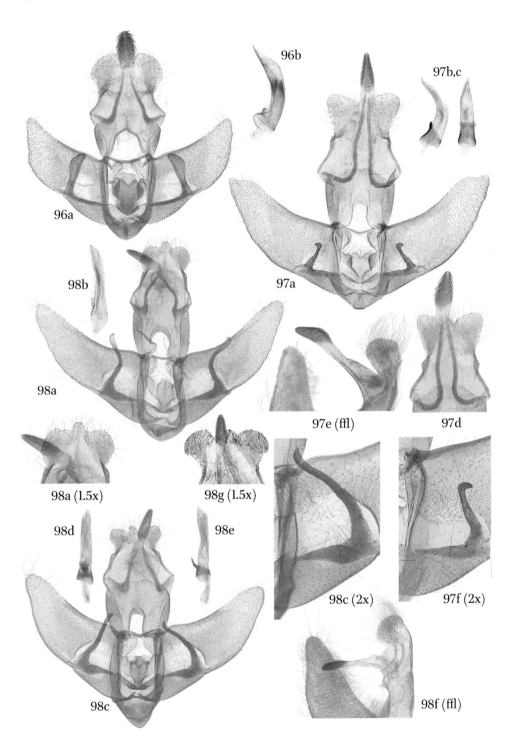

96. *Agonopterix invenustella* – a–b) Italy (3946, ZfBS)
97. *Agonopterix abditella* – a–c) Russia (4919, TLMF); d–f) Russia (3828, ZMUC)
98. *Agonopterix lidiae* – a–b) Italy (4047, RCLM); c–e) Italy (8803, TLMF); f) Italy (8748, TLMF); g) France (9842, RCFG, slide and photo Friedmar Graf)

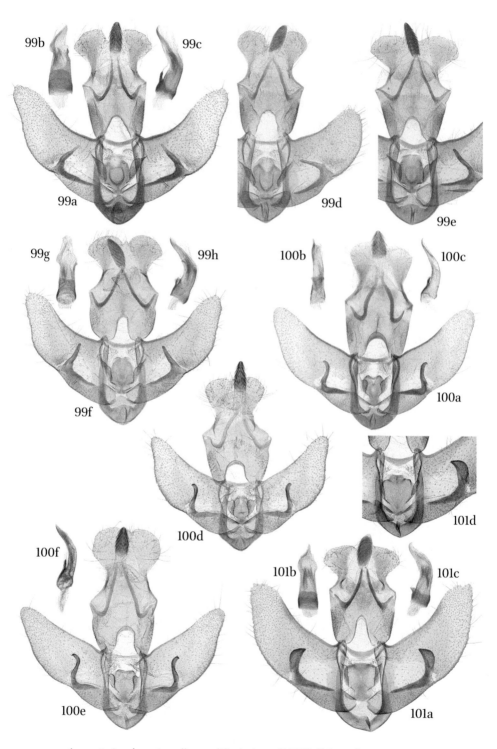

99.   *Agonopterix subpropinquella* – a–c) Spain (2704, RCKL); d) Spain (3895, RCWSc);
      e) Italy (4208, RCLM); f–h) Spain (4310, ECKU)

100. *Agonopterix nanatella* – a–c) Croatia (0138, NHMW); d) Spain (2202, RCLS); e–f) Italy (1353,
      MFSN)

101. *Agonopterix kuznetzovi* – a–c) Spain (6215, NMPC); d) Turkey (2380, ZMUC)

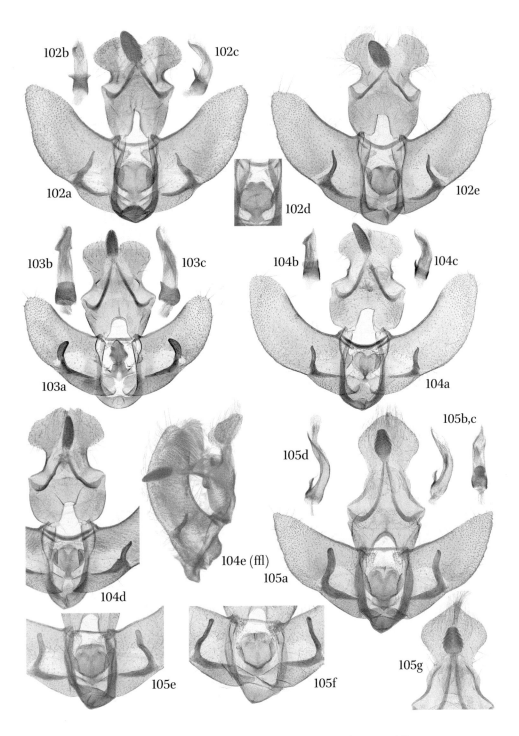

102. *Agonopterix propinquella* – a–c) Poland (0941, NHMW); d) Austria (8014, TLMF);
e) Russia (4669, RCTN)

103. *Agonopterix carduella* – a) Austria (0348, RCPB); b–c) Austria (7295, RCPB)

104. *Agonopterix ivinskisi* – a–c, e) Spain (6799, RCAM); d) Spain (1930, RCWSc)

105. *Agonopterix ferocella* – a–c, e) Bulgaria (4299, ECKU); d, f–g) Greece (4804, RCWSc);
e) Russia (2352, ZMUC)

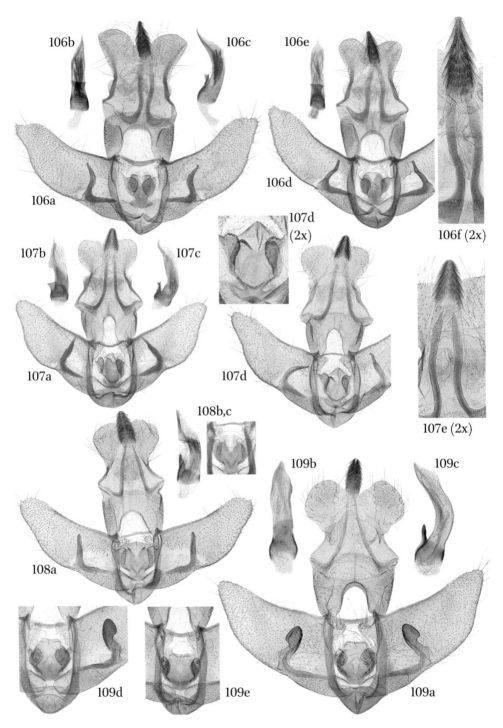

106. *Agonopterix laterella* – a–c) Romania (4091, RCZK); d–e) Spain (6300, MGAB); f) Italy (0708, NHMW)

107. *Agonopterix xeranthemella* – a–c) Turkey (3131, RCKL); d–e) Turkey (3136, RCKL);
        e) Turkey (3353, RCKL)

108. *Agonopterix cinerariae* – a–b) Canary Islands (2048, ZSM); c) Canary Islands (2096, TLMF)

109. *Agonopterix arenella* – a–c) Austria (0954, RCPB); d) Austria (4194, RCCM); e) Austria (7299, RCPB)

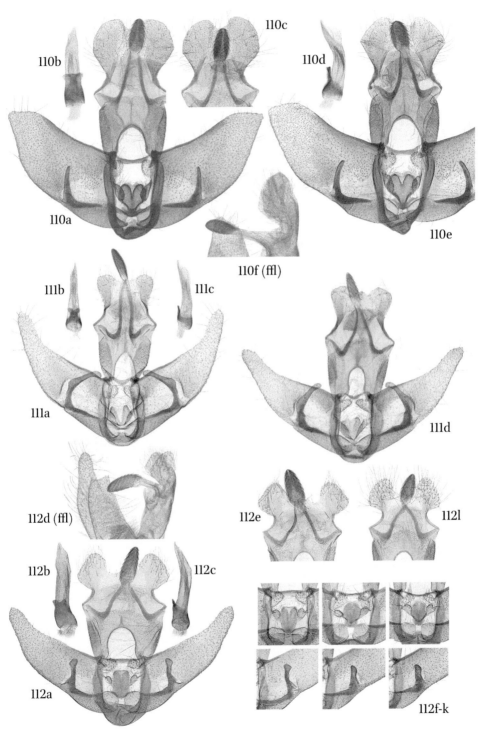

110. *Agonopterix petasitis* – a, f) Austria (0650, RCPB); b–d) Austria (7300, RCPB); e) Austria (0688, RCPB)
111. *Agonopterix cotoneastri* – a) Spain (2562, ZMUC); b–c) Spain (7306, NHMW); d) Austria (1725, RCAM)
112. *Agonopterix multiplicella* – a–d) unknown (6275, MGAB); e–f, i) India (NHMUK010891134, NHMUK); g, j, l) Slovakia (2157, RCZT); h, k) Estonia (2535, ZMUC)

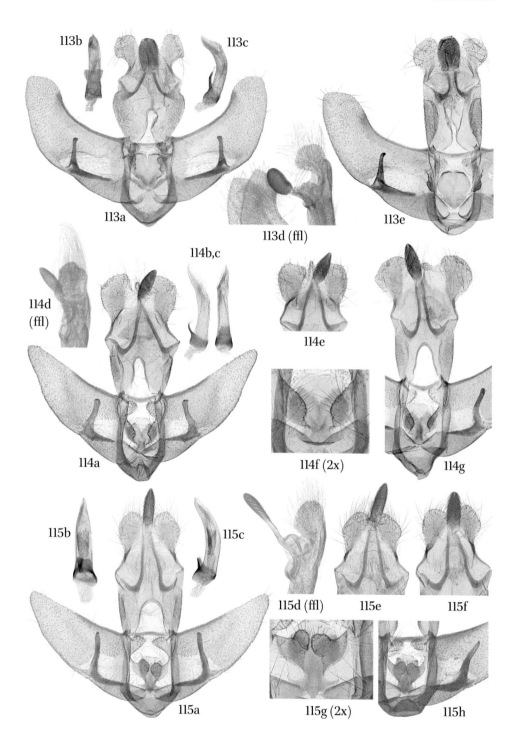

113.   *Agonopterix doronicella* – a–d) Greece (2774, RCKL); e) Austria (0135, NHMW)
114.   *Agonopterix assimilella* – a) Italy (1349, MFSN); b–c) Bulgaria (7302, NHMW);
        d) Turkey (6864, SMNK); e–f) Kazakhstan (6295, MGAB); g) Italy (1347, MFSN)
115.   *Agonopterix atomella* – a–c) Turkey (5812, ZMHB); d) Bulgaria (6862, SMNK);
        e, g) Turkey (3155, RCKL); f) Italy (1570, RCLM); h) Austria (1242, RCWS)

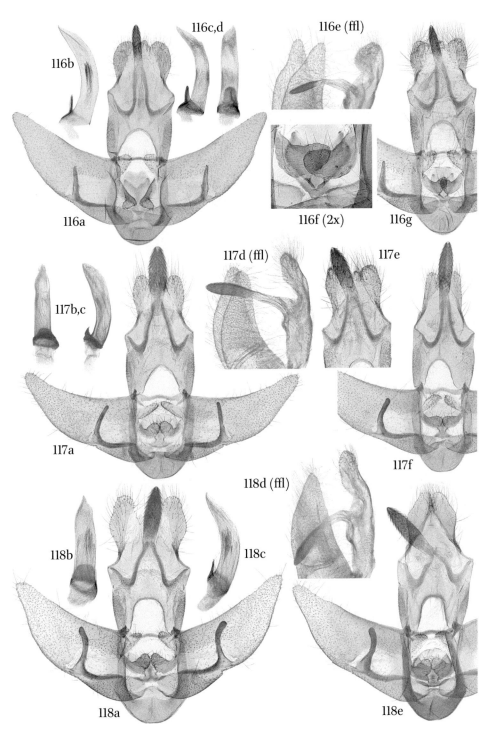

116. *Agonopterix scopariella* – a–b) Italy (1394, MFSN); c–d) Italy (0143, NHMW); e) Spain (8644, RCAM);
f) Canary Islands (2959, RCKL); g) Croatia (1270, RCLS)

117. *Agonopterix conciliatella* – a) Canary Islands (3892, RCWSc); b–c) Canary Islands (8591, RCAM);
d) Canary Islands (8590, RCAM); e) Canary Islands (2062, ZSM); f) Madeira (3971, ZfBS)

118. *Agonopterix comitella* – a) Crete (3769, ZMUC); b–d) Turkey (7315, NHMW); e) Greece (0069,
NHMW)

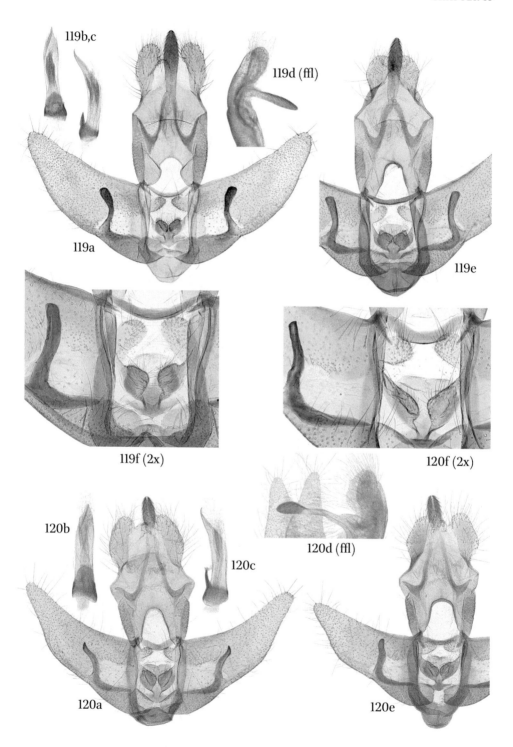

119b,c
119d (ffl)
119a
119e
119f (2x)
120f (2x)
120b
120c
120d (ffl)
120a
120e

119. *Agonopterix nervosa* – a–c) Austria (0329, RCPB); d) Spain (7059, RCWSc); e) Spain (4336, SMNK); f) Spain (2733, RCKL)
120. *Agonopterix umbellana* – a–d) France (7318, NHMW); e) Slovenia (2435, ZMUC); f) Germany (0058, NHMW)

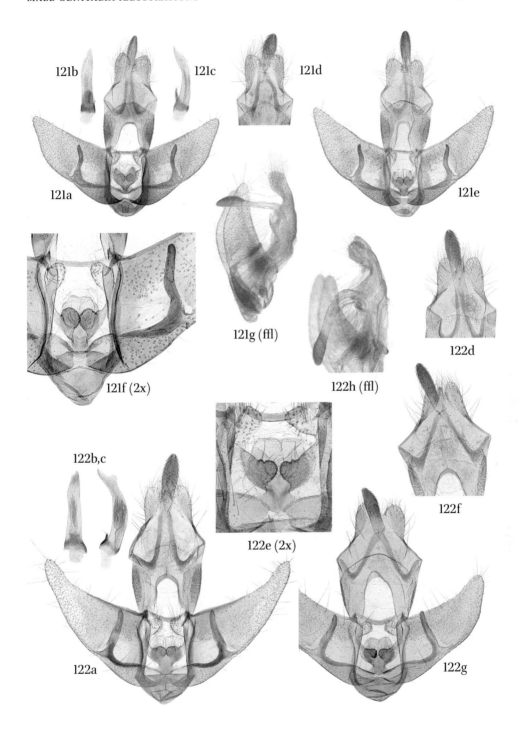

121. *Agonopterix oinochroa* – a) Spain (3730, RCEP); b–c, g) Serbia (7307, NHMW); d) Turkey (3208, RCKL); e) Greece (3877, RCWSc); f) Hungary (0709, NHMW)

122. *Agonopterix aspersella* – a) France (0127, NHMW); b–c, g–h) France (0064, NHMW); d) Spain (7378, RCKB); e–f) Portugal (2329, ZMUC)

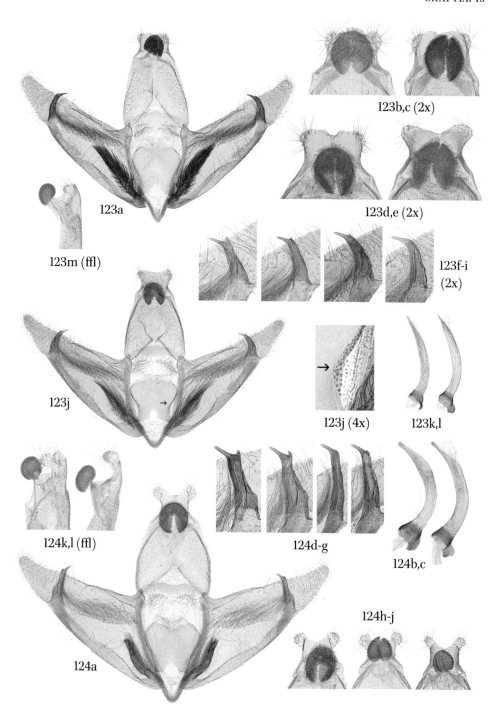

123a

123m (ffl)

123b,c (2x)

123d,e (2x)

123f-i (2x)

123j

123j (4x)

123k,l

124k,l (ffl)

124d-g

124b,c

124h-j

124a

123. *Depressaria pulcherrimella* – a) Slovakia (2837, RCKL); b) Spain (3733, ZMUC); c) Spain (3728, RCEP); d) Austria (8037, TLMF); e, l) Switzerland (7692, NMBE); f) Spain (3731, RCEP); g) Austria (1823, RCAM); h, j–k) Spain (9462, ZMUC); i, m) Liechtenstein (7661, NMSG)

124. *Depressaria sordidatella* – a–b) Russia (6457, RCTN); c) Germany (8076, TLMF); d) Slovakia (1308, RCIR); e, l) Finland (4488, RCJJ); f) Finland (4083, RCCS); g, h) France (6691, ZMUC); i, k) Germany (8077, TLMF); j) Romania (8143, RCZK)

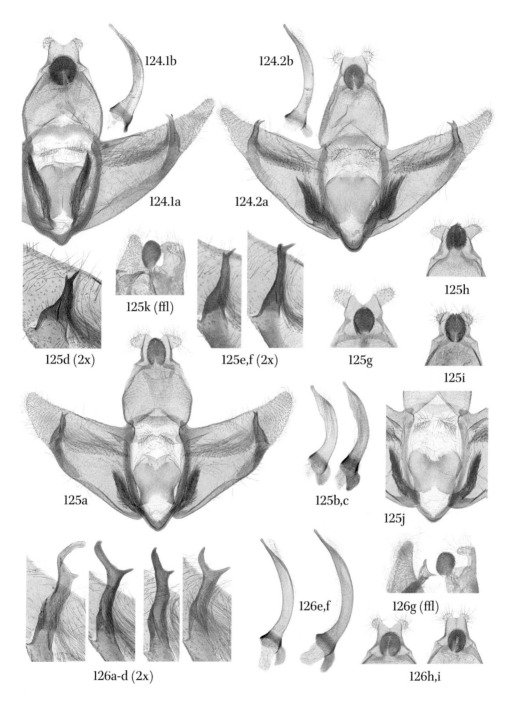

124.1. *Depressaria* near *sordidatella* 01 – a–b) Germany (3939, ZfBS)

124.2. *Depressaria* near *sordidatella* 02 – a–b) Russia (1027, RCLS)

125. *Depressaria floridella* – a–b) Greece (2422, ZMUC); c) Crimea (6661, ZIN); d) Greece (2401, ZMUC); e) Italy (1764, RCAM); f) Austria (0038, RCPB); g) Turkey (3309, RCKL); h) Bulgaria (4545, RCJK); i) Spain (5383, ZMUC); j) Spain (4782, RCWSc); k) Armenia (7899; ECKU)

126. *Depressaria douglasella* – a) Bulgaria (1163, RCIR); b) Italy (1405, MFSN); c) Bulgaria (1375, RCLS); d, f, h) Turkey (3183, RCKL); e) Spain (1912, RCFT); g) Bulgaria (4480, RCJJ); i) Slovakia (3014, RCKL)

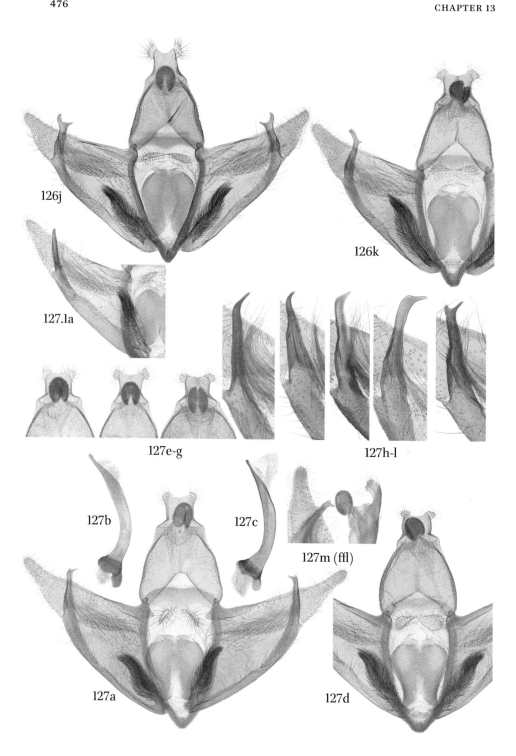

126. *Depressaria douglasella* – j) Hungary (0975, RCLS); k) Cyprus (7462, ZMUC)
127. *Depressaria beckmanni* – a–b, m) Austria (8485, TLMF); c–d) Austria (1853, TLMF);
e, i) Switzerland (2148, NMBE); f, h) Sardinia (6758, RCAM); g) Albania (8257, RCCP);
j) Latvia (4276, ECKU); k) Italy (4193, RCCM); l) Switzerland (3869, NMBE)
127.1. *Depressaria* near *beckmanni* ♂ – a) Sardinia (7018, RCGF)

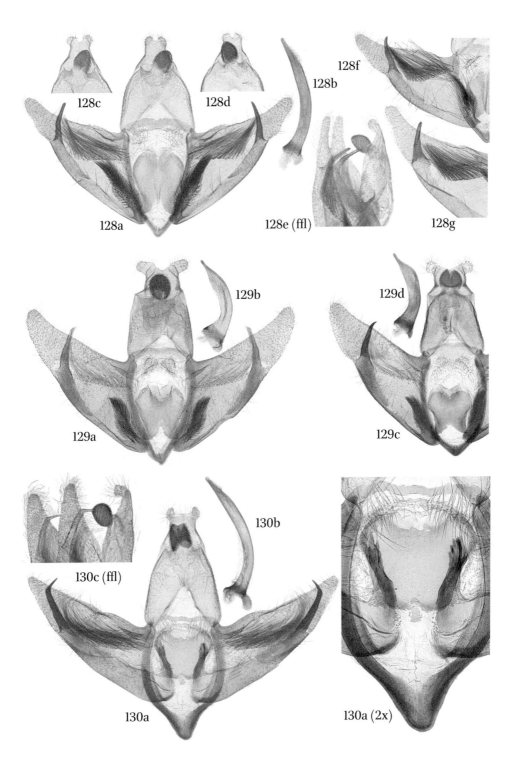

128. *Depressaria incognitella* – a–b) Switzerland (5434, RCEP); c) France (1839, RCAM); d) France (1940, RCWSc); e) France (8651, RCAM); f) Sardinia (4722, RCTN); g) Croatia (1161, RCIR)
129. *Depressaria nemolella* – a–b) Austria (3950, MfBS); c–d) Russia (6470, RCTN)
130. *Depressaria cinderella* – a–c) Portugal (0789, RCMC)

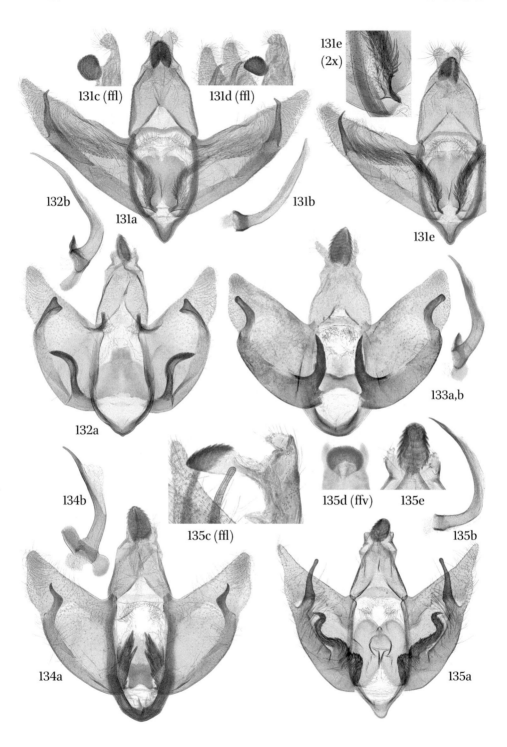

131c (ffl) 131d (ffl) 131e (2x)
131a 131b 131e
132b 132a
133a,b
134b 135c (ffl) 135d (ffv) 135e 135b
134a 135a

131.   *Depressaria infernella* – a–c) Spain (6848, ZMHB); d–e) Portugal (6672, RCMC)
132.   *Depressaria indecorella* – a–b) Kazakhstan (4676, RCTN)
133.   *Depressaria lacticapitella* – a–b) Austria (1883, RCHD)
134.   *Depressaria hofmanni* – a–b) Austria (2223, RCPB)
135.   *Depressaria albipunctella* – a–b) Morocco (1972, RCHB); c–e) Sardinia (8637, RCAM)

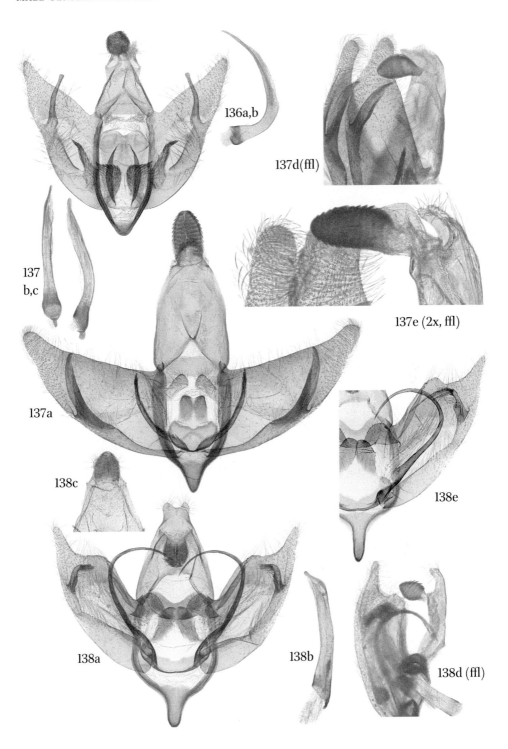

136.   *Depressaria subalbipunctella* – a) Crimea (2165, RCZT); b) Turkey (3269, RCKL)

137.   *Depressaria krasnowodskella* – a, d) Morocco (7673, NMPC); b) Morocco (0189, NHMW);
       c, e) Spain (6701, ZMUC)

138.   *Depressaria tenebricosa* – a–b) Bulgaria (1069, RCLS); c) Italy (2436, ZMUC); d) Cyprus (8652,
       RCAM); e) Turkey (0695, NHMW)

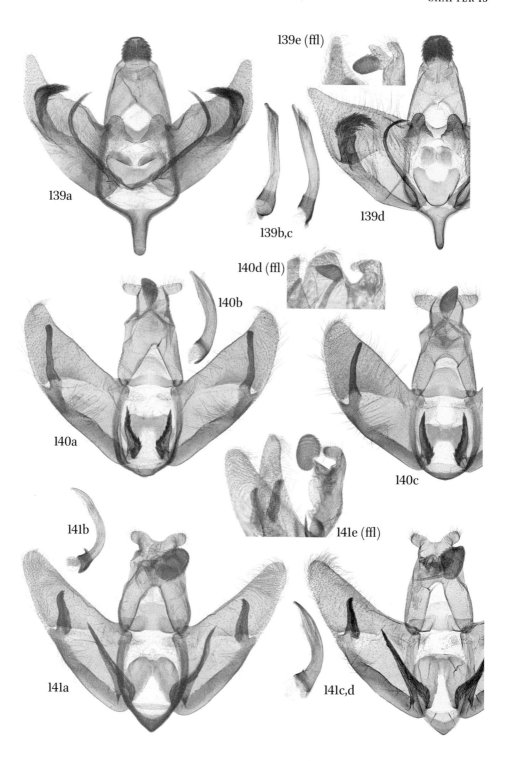

139e (ffl)

139a

139b,c

139d

140d (ffl)

140b

140a

140c

141e (ffl)

141b

141a

141c,d

139. *Depressaria radiosquamella* – a) Italy (1838, RCAM); b) Corsica (1652, TLMF); c–d) Morocco (1973, RCHB); e) Sardinia (8635, RCAM)
140. *Depressaria emeritella* – a) Germany (1907, RCAS); b–d) Kazakhstan (4649, RCTN)
141. *Depressaria olerella* – a–b, e) Germany (8075, TLMF); c–d) Austria (9463, RCPB)

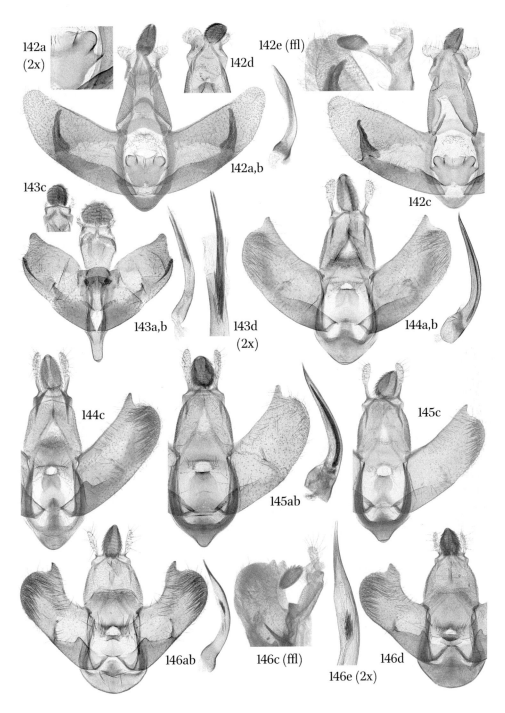

142a
(2x)

142d

142e (ffl)

142a,b

142c

143c

143a,b

143d
(2x)

144a,b

144c

145c

145ab

146ab

146c (ffl)

146d

146e (2x)

142. *Depressaria leucocephala* – a–b, e) Russia (6469, RCTN); c) Italy (0694, NHMW); d) Russia (6289, MGAB)

143. *Depressaria depressana* – a–b) Austria (0576, RCPB); c–d) Italy (1833, RCAM)

144. *Depressaria absynthiella* – a) Germany (0118, NHMW); b) Turkey (2392, ZMUC); c) Bulgaria (3506; RCBZ)

145. *Depressaria tenerifae* stat.rev. – a–b) Canary Islands (3942, ZfBS); c) Canary Islands (2640, RCKL)

146. *Depressaria artemisiae* – a–c) Austria (0721, RCWS); d) Russia (4632, RCTN); e) China (6919, MNHN)

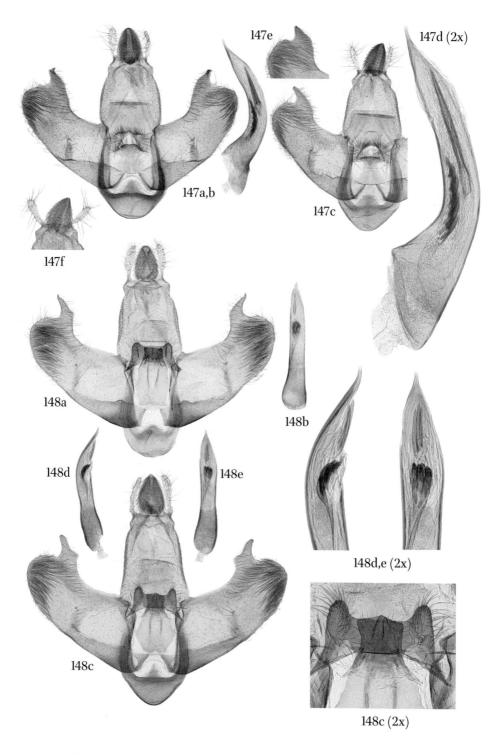

147.   *Depressaria fuscovirgatella* – a–b) Kazakhstan (4618, RCTN); c–d) Turkey (3385, RCKL);
       e–f) Turkey (2362, ZMUC)
148.   *Depressaria atrostrigella* – a–b) Russia (2552, ZMUC); c–e) China (6262, MGAB)

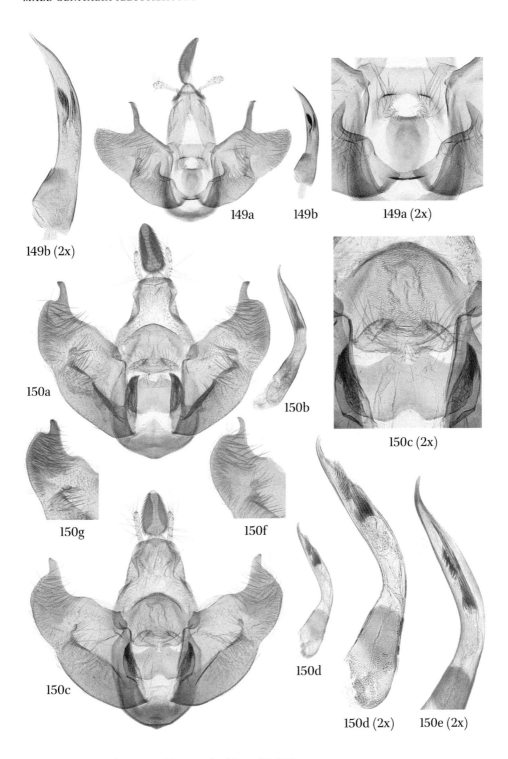

149. *Depressaria silesiaca* – a–b) Switzerland (0551, NMBE)
150. *Depressaria zelleri* – a–b) Greece (1939, RCWSc); c–d) Italy (2037, ZSM); e) North Macedonia (6836, ZSM); f) Greece (4684, RCTN); g) Russia (1051, RCLS)

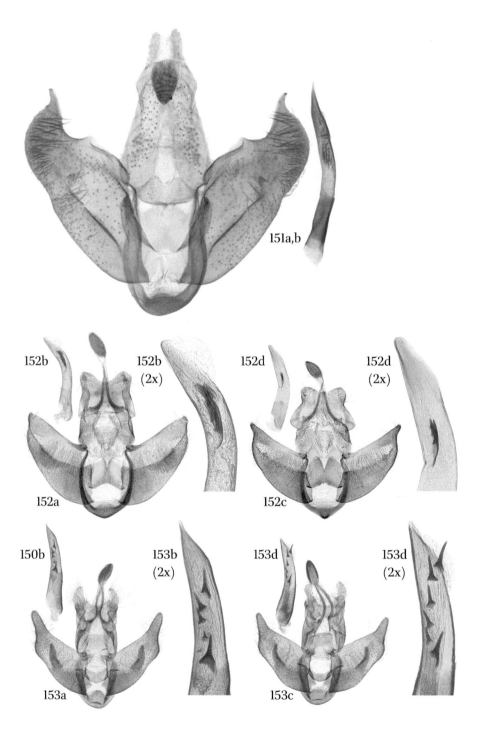

151.   *Depressaria pyrenaella* – a–b) Spain (NHMUK010293057, holotype, B.V. Ridout slide 19153, NHMUK)
152.   *Depressaria marcella* – a–b) Greece (1943, RCWSc). c–d) Albania (1121, RCFG)
153.   *Depressaria peregrinella* – a–b) Turkey (4362, SMNK); c–d) Greece (1943, RCTG)

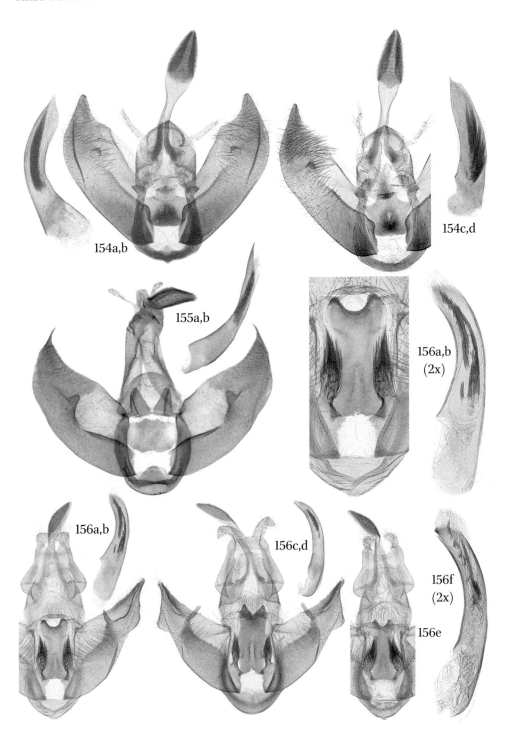

154. *Depressaria heydenii* – a–b) France (2225, TLMF); c–d) Austria (0928, RCPB)
155. *Depressaria fuscipedella* – a–b) Algeria ((6829, holotype, H.J. Hannemann slide 1002, photo Christian Gibeaux, MNHN)
156. *Depressaria ululana* – a–b) Germany (0935, NHMW); c–d) Turkey (4103, RCCS); e–f) Greece (4805, RCWSc)

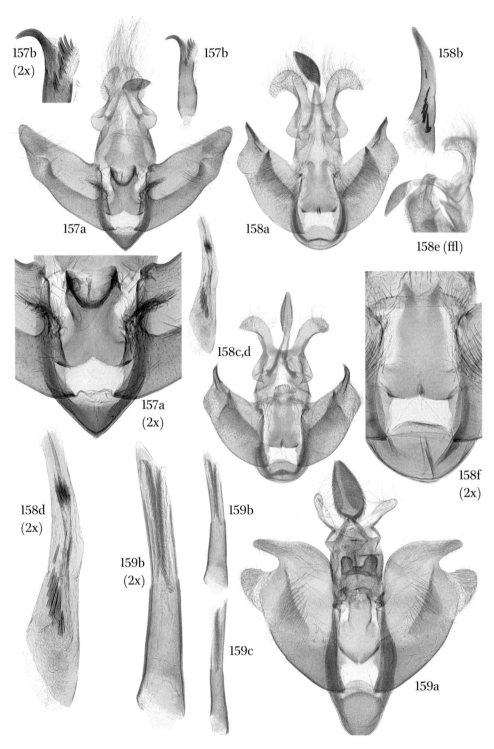

157.   *Depressaria manglisiella* – a–b) Croatia (1200, RCIR)
158.   *Depressaria chaerophylli* – a–b) Italy (1197, MFSN); c–e) Austria (0116, NHMW); f) Slovakia (2839, RCKL)
159.   *Depressaria longipennella* – a–c) Tajikistan (5705, ZMHB)

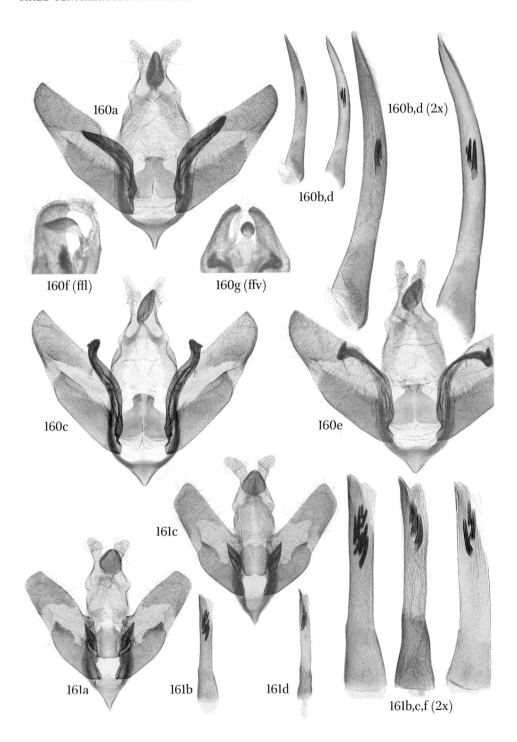

160a
160b,d (2x)
160b,d
160f (ffl)
160g (ffv)
160c
160e
161c
161a
161b
161d
161b,c,f (2x)

160. *Depressaria daucella* – a–b) Bulgaria (3515, RCBZ); c–d) Austria (0279, RCWS); e) Slovakia (2155, RCZT); f–g) Sardinia (8619, RCAM)
161. *Depressaria ultimella* – a–b) Austria (1659, TLMF); c–d) Denmark (5399, ZMUC); e) Romania (6315, MGAB); f) Austria (0122, NHMW)

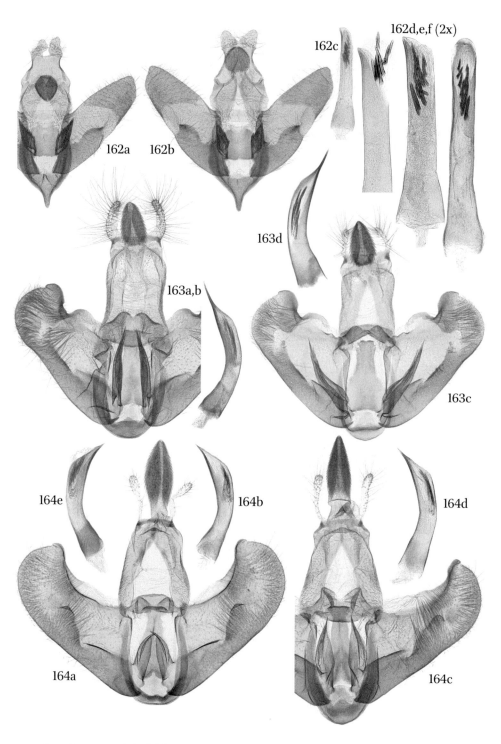

162. *Depressaria halophilella* – a, e) Sardinia (2419, ZMUC); b–c) France (1994, NMBE);
d) Greece (0681, RCRK); f) Madeira (3719, ZMUC)

163. *Depressaria radiella* – a–b) Russia (2174, RCLS); c–d) Austria (3439, RCPB)

164. *Depressaria libanotidella* – a–b) France (0185, NHMW); c–d) Italy (1851, TLMF); e) France
(6277, MGAB)

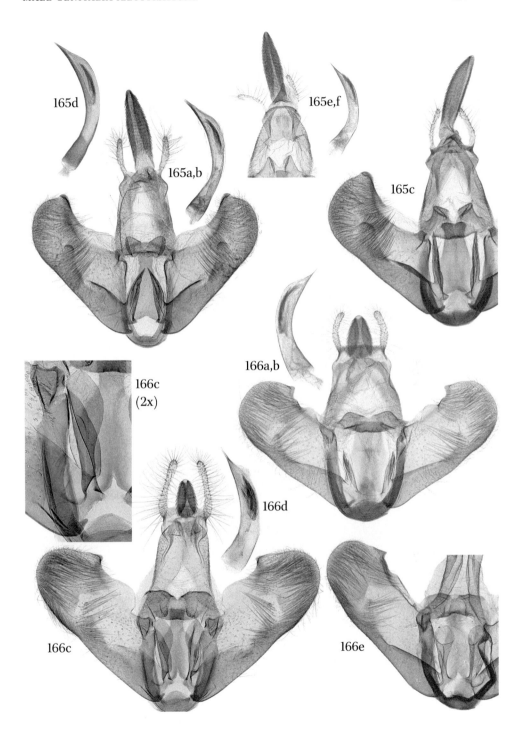

165. *Depressaria bantiella* – a–b) Crete (2411, ZMUC); c–d) Morocco (6160, NMPC); e–f) Turkey (3112, RCKL)

166. *Depressaria velox* – a–b) Slovenia (1550, RCCM); c–d) Iran (2339, ZMUC); e) Iran (5838, H.J. Hannemann slide 5156, ZMHB)

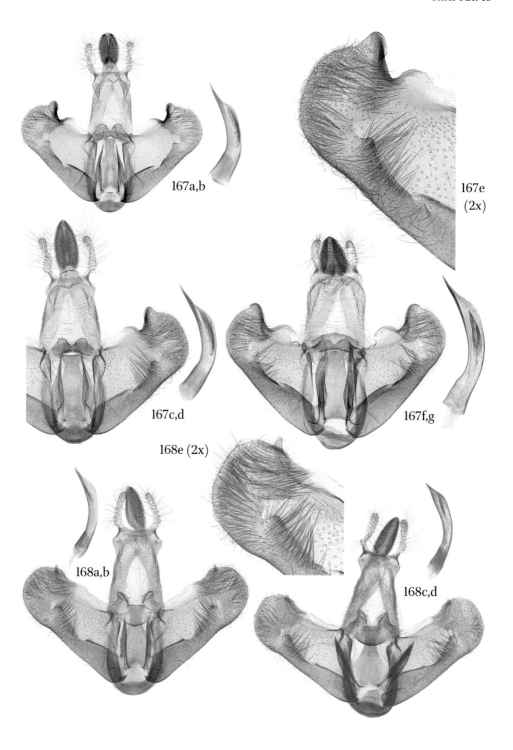

167a,b

167e (2x)

167c,d

167f,g

168e (2x)

168a,b

168c,d

167.   *Depressaria pimpinellae* – a–b) Austria (0282, RCWS); c–e) Turkey (4094, RCCS)
168.   *Depressaria villosae* – a–b) Greece (4368, SMNK); c–d) Spain (7545, ZMUC); e) Portugal (0799, holotype, NHMUK)

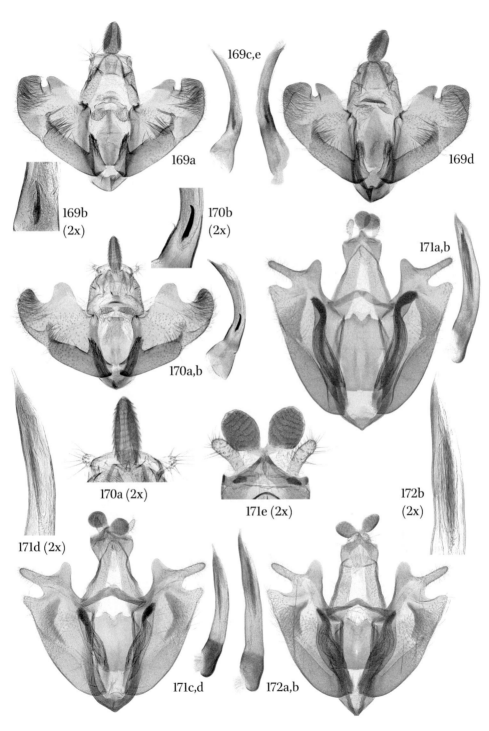

169. *Depressaria bupleurella* – a–b) Austria (0181, NHMW); c) Germany (0180, NHMW); d–e) Uzbekistan (5822, ZMHB)
170. *Depressaria sarahae* – a–b) Spain (1000, RCLS)
171. *Depressaria badiella* – a–b) Armenia (1660, RCRK); c–d) Austria (6291, MGAB); e) Corsica (6652, ZIN)
172. *Depressaria pseudobadiella* – a–b) Spain (3925, RCWSc)

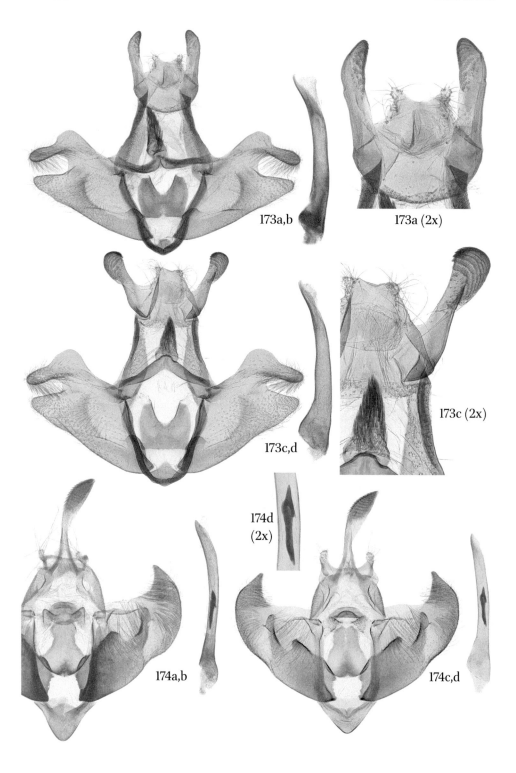

173a,b

173a (2x)

173c,d

173c (2x)

174d
(2x)

174a,b

174c,d

173. *Depressaria subnervosa* – a–b) Spain (2176, RCLS); c–d) Morocco (1574, NHMW)
174. *Depressaria cervicella* – a–b) Russia (6502, ZMUH); c–d) Austria (0732, NHMW)

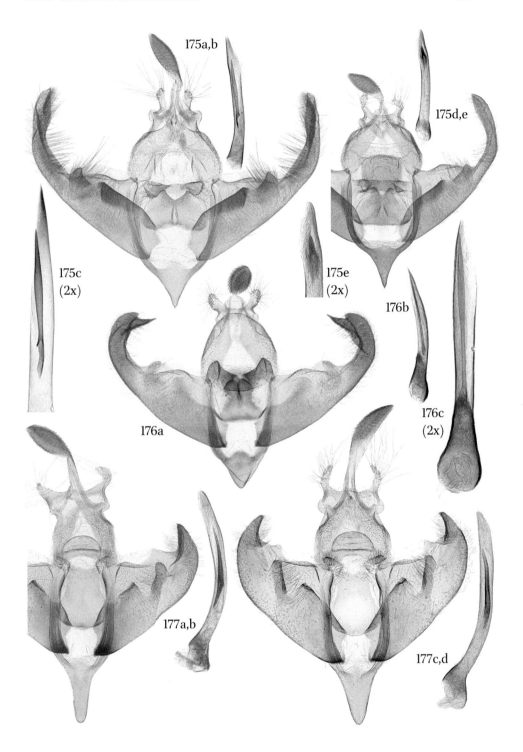

175. *Depressaria gallicella* – a–b) France (2075, ZSM); c) Switzerland (0483, NMBE);
     d–e) Kazakhstan (4653, RCTN)
176. *Depressaria altaica* – a–c) Russia (6465, RCTN)
177. *Depressaria albarracinella* – a–b) Greece (2326, ZMUC); c–d) Greece (5398, ZMUC)

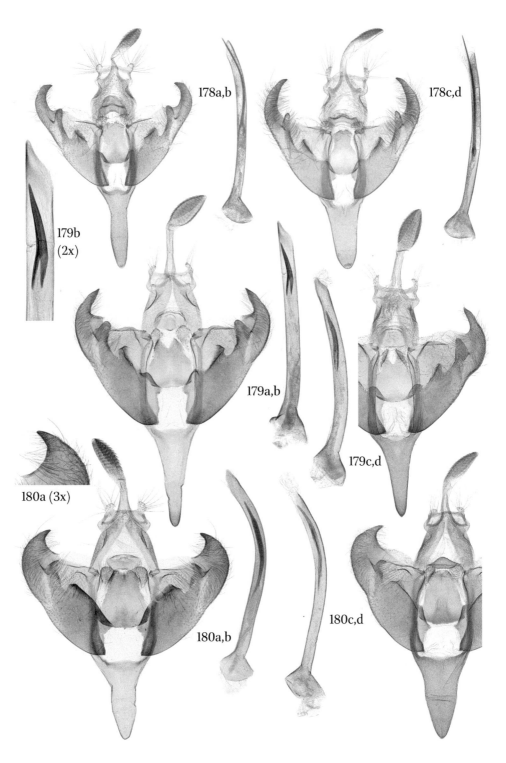

178. *Depressaria eryngiella* – a–b) Turkey (3105, RCKL); c) Croatia (0974, RCLS)
179. *Depressaria veneficella* – a–c) Sicily (0693, NHMW); c–d) Spain (2153, RCZT)
180. *Depressaria discipunctella* – a–b) Austria (1575, TLMF); c–d) Syria (2072, ZSM)

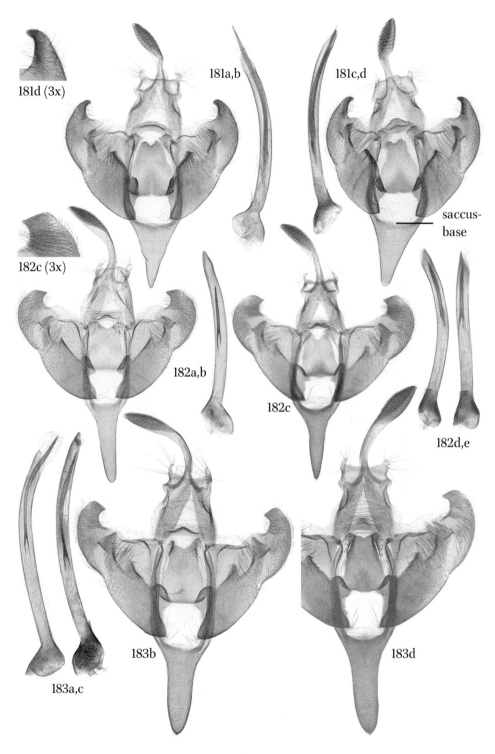

181d (3x)
181a,b
181c,d
saccus-base
182c (3x)
182a,b
182c
182d,e
183a,c
183b
183d

181. *Depressaria hansjoachimi* sp.n. – a–b) Greece (3090, RCKL); c–d) Croatia (2217, RCLS)
182. *Depressaria junnilaineni* – a–b) Spain (1913, RCFT); c–e) Greece (5997, holotype, TLMF)
183. *Depressaria pentheri* – a–b) Croatia (4460, RCJJ); c–d) Albania (8211, RCCP)

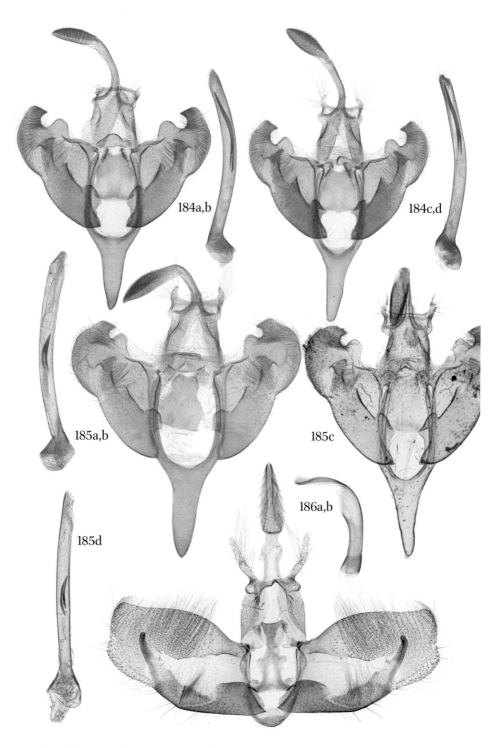

184. *Depressaria hannemanniana* – a–b) Russia (6138, NMPC); c–d) Kazakhstan (4718, RCTN)
185. *Depressaria erzurumella* – a–b) Greece (5999, TLMF); c–d) Turkey (5420, holotype, A. Lvovsky slide 3, ZMUC)
186. *Depressaria dictamnella* – a–b) Italy (0485, NMBE)

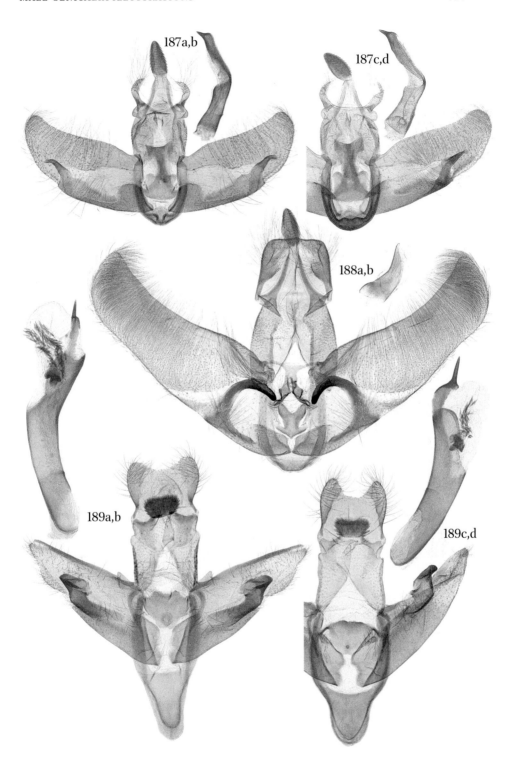

187. *Depressaria moranella* – a–b) North Macedonia (2081, ZSM); c–d) Libya (2940, RCKL)
188. *Depressaria hystricella* – a–b) Russia (0938, NHMW)
189. *Depressaria hirtipalpis* – a–b) Greece (1902, RCEV); c–d) Croatia (1670, RCRK)

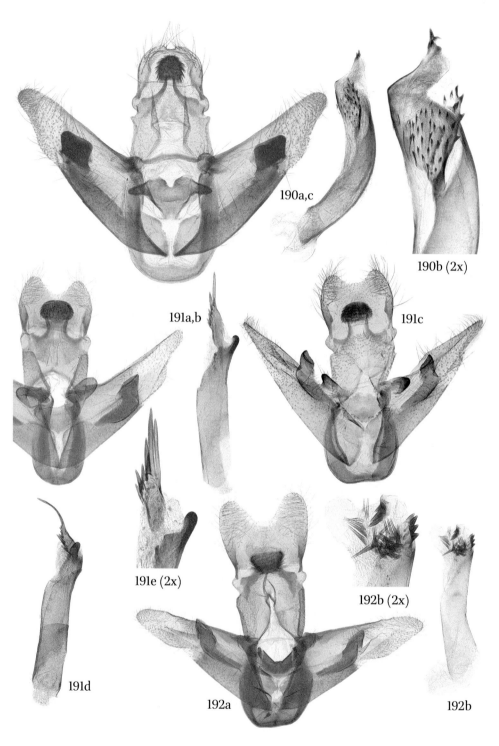

190a,c

190b (2x)

191a,b

191c

191e (2x)

192b (2x)

191d

192a

192b

190. *Depressaria erinaceella* – a–b) Spain (0697, NHMW); c) Spain (2182, RCLS)
191. *Depressaria peniculatella* – a–b) Spain (0945, NHMW); c–d) Crete (2539, ZMUC);
      e) Tunisia (2011, RCHB)
192. *Depressaria rungsiella* stat.rev. – a–b) Morocco (1672, ZSM)

Female Genitalia Illustrations

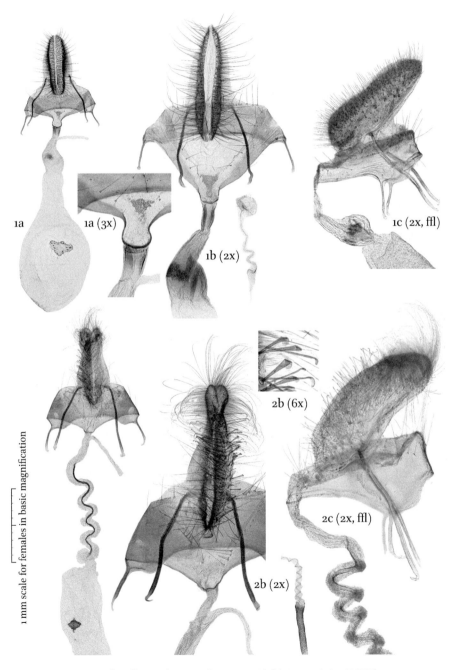

1 mm scale for females in basic magnification

1a

1a (3x)

1b (2x)

1c (2x, ffl)

2b (6x)

2b (2x)

2c (2x, ffl)

1. *Semioscopis avellanella* – a, c) Austria (0026, RCPB); b) Austria (7675, RCPB)
2. *Semioscopis oculella* – a, c) France (0494, NMBE); b) Austria (8886, TLMF)

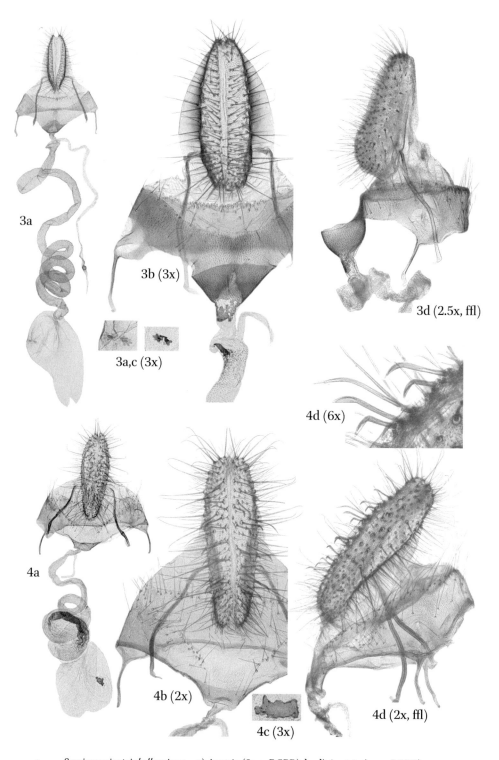

3.   *Semioscopis steinkellneriana* – a) Austria (8719, RCPB); b–d) Austria (1493, RCPB)
4.   *Semioscopis strigulana* – a) Switzerland (0431, NMBE); b–d) Austria (8884, TLMF)

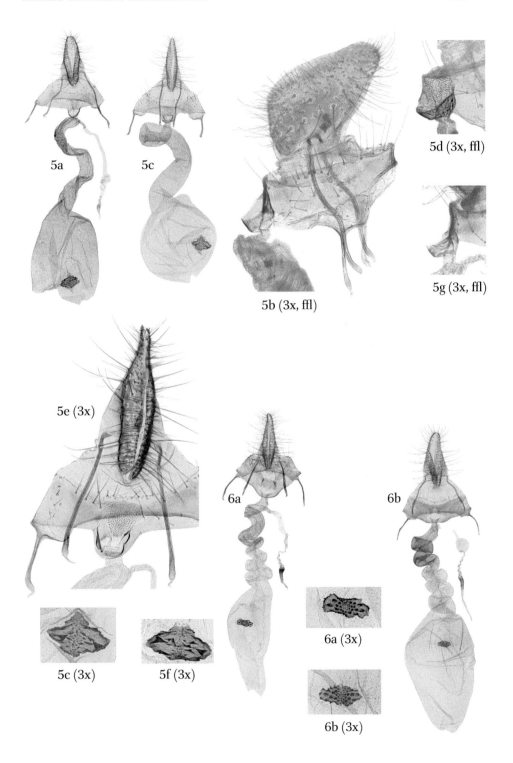

5. *Luquetia lobella* – a–b) Austria (9040, RCWS); c–d) Austria (9271, RCPB); e–g) (0028, RCPB)
6. *Luquetia orientella* – a) North Macedonia (1899, RCAM); b) Greece (8937, ZMUC)

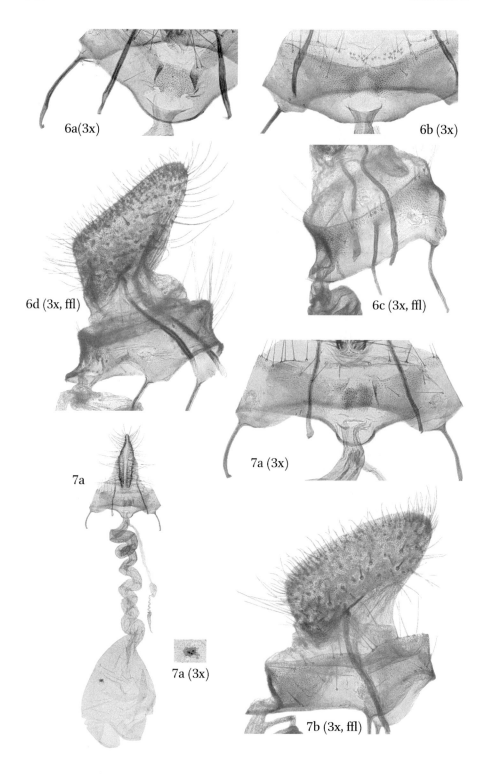

6a (3x) 6b (3x) 6d (3x, ffl) 6c (3x, ffl) 7a (3x) 7a 7a (3x) 7b (3x, ffl)

6.   *Luquetia orientella* – a, d) North Macedonia (1899, RCAM); b–c) Greece (8937, ZMUC)
7.   *Luquetia osthelderi* comb. nov. – a–b) Turkey (5107, NHMW)

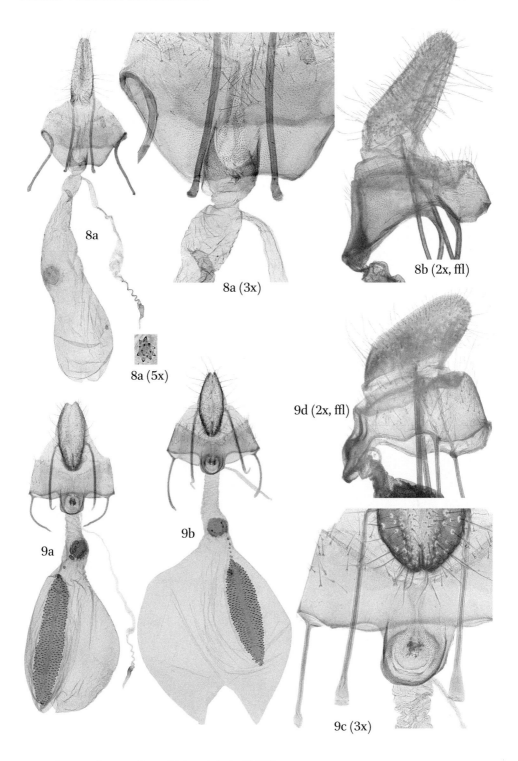

8a

8a (3x)

8a (5x)

8b (2x, ffl)

9d (2x, ffl)

9a

9b

9c (3x)

8. *Exaeretia hepatariella* – a–b) Austria (1483, TLMF)
9. *Exaeretia allisella* – a) Switzerland (7736, NMBE); b) Switzerland (0426, NMBE); c) Russia (7754, NMPC); d) Denmark (5441, RCEP)

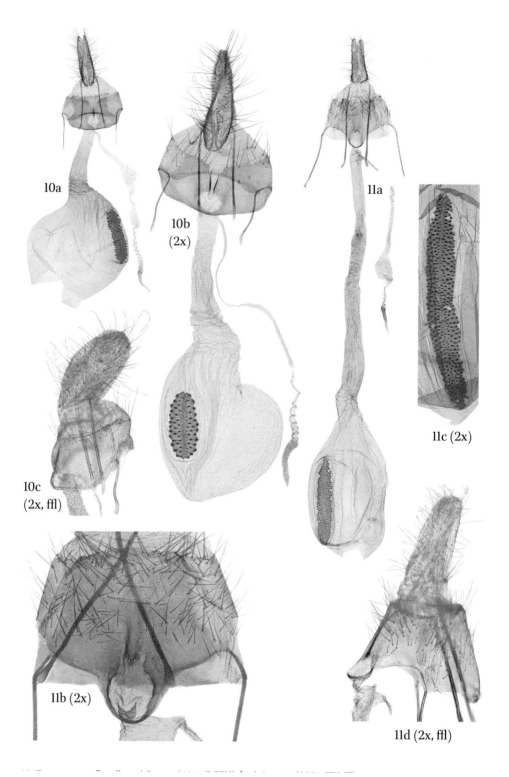

10.   *Exaeretia ciniflonella* – a) Russia (4674, RCTN); b–c) Austria (2097, TLMF)
11.   *Exaeretia preisseckeri* – a, d) Hungary (1229, RCAT); b–c) Austria (8721, RCPB)

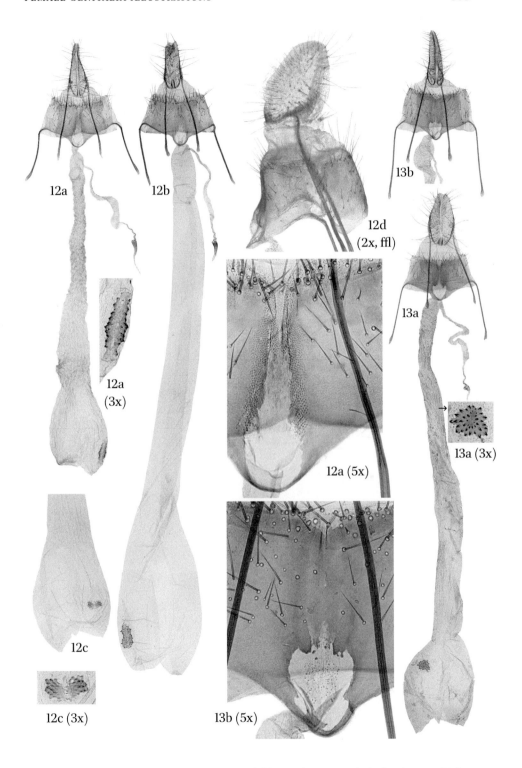

12. *Exaeretia lutosella* – a) France (3474, NHMW); b) Spain (1092, RCLS); c) Libya (2942, RCKL); d) Greece (6483, RCJJ)

13. *Exaeretia thurneri* – a) Greece (1878, RCTG); b) Romania (6332, MGAB)

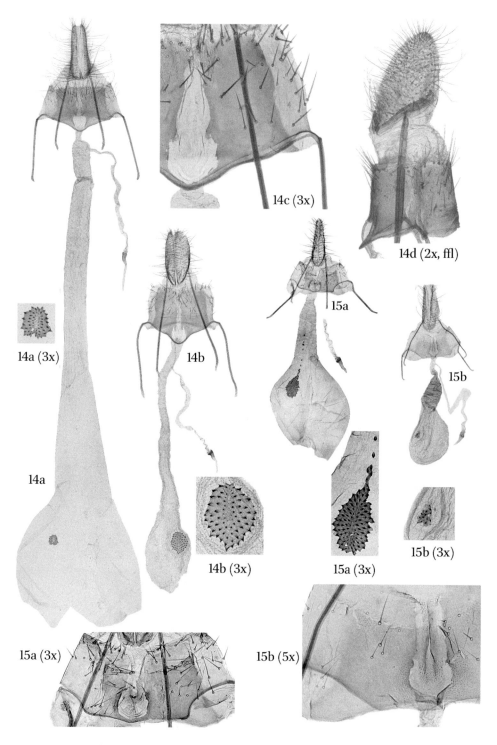

14a (3x)

14a

14b

14b (3x)

14c (3x)

14d (2x, ffl)

15a

15b

15b (3x)

15a (3x)

15a (3x)

15b (5x)

14.   *Exaeretia ledereri* – a) Turkey (4732, RCTN); b) Tajikistan (5715, ZMHB); c) Bulgaria (1060, RCLS); d) Turkey (HNHM, 4090)

15.   *Exaeretia praeustella* – a) Sweden (2488, ZMUC); b) Estonia (7130, ZIN)

16.   *Exaeretia nebulosella* – females unknown

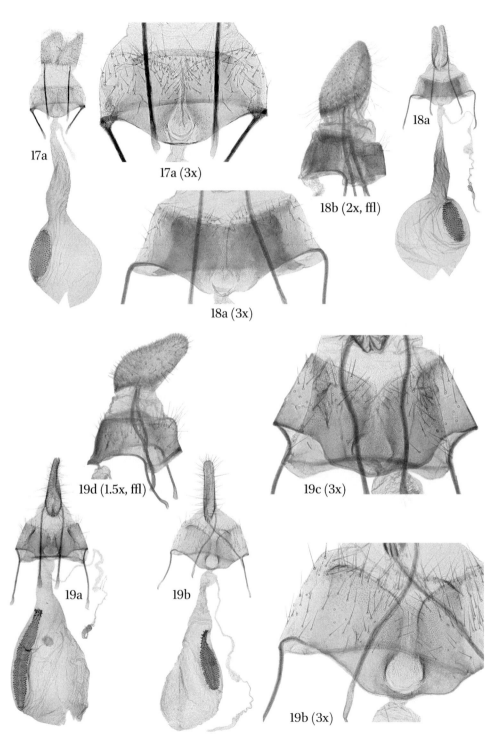

17. *Exaeretia lepidella* – a) Russia (7129, ZIN)
18. *Exaeretia buvati* – a–b) France (8570, RCJJ)
19. *Exaeretia mongolicella* – a) Mongolia (6904, MNHN); b) Russia (1922, RCIK); c–d) Russia (6298, MGAB)

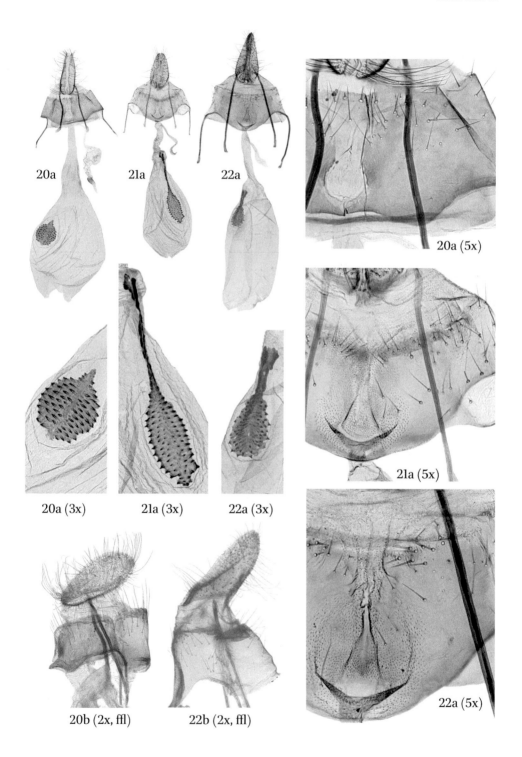

20a

21a

22a

20a (5x)

20a (3x)   21a (3x)   22a (3x)

21a (5x)

20b (2x, ffl)   22b (2x, ffl)

22a (5x)

20.   *Exaeretia lvovskyi* – a–b) Russia (7267, NMPC)
21.   *Exaeretia stramentella* – a) Poland (1537, NHMW)
22.   *Exaeretia niviferella* – a–b) China (1538, NHMW)

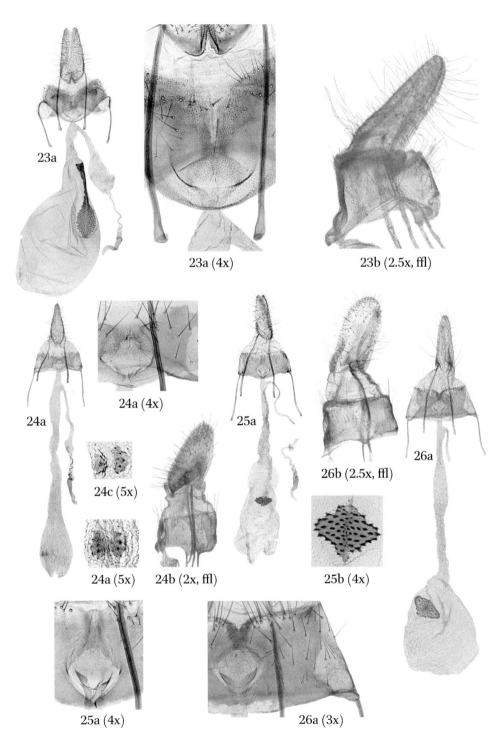

23a

23a (4x)

23b (2.5x, ffl)

24a

24a (4x)

25a

26b (2.5x, ffl)

26a

24c (5x)

24a (5x)   24b (2x, ffl)   25b (4x)

25a (4x)   26a (3x)

23. *Exaeretia indubitatella* – a–b) Russia (4924, TLMF)

24. *Agonopterix impurella* – a) Austria (4178, NHMW); b) Poland (3091, RCKL); c) Austria (1492, RCPB)

25. *Agonopterix liturosa* – a) Bulgaria (1088, RCLS); b) Turkey (2086, ZSM)

26. *Agonopterix conterminella* – a–b) Germany (1526, RCPB)

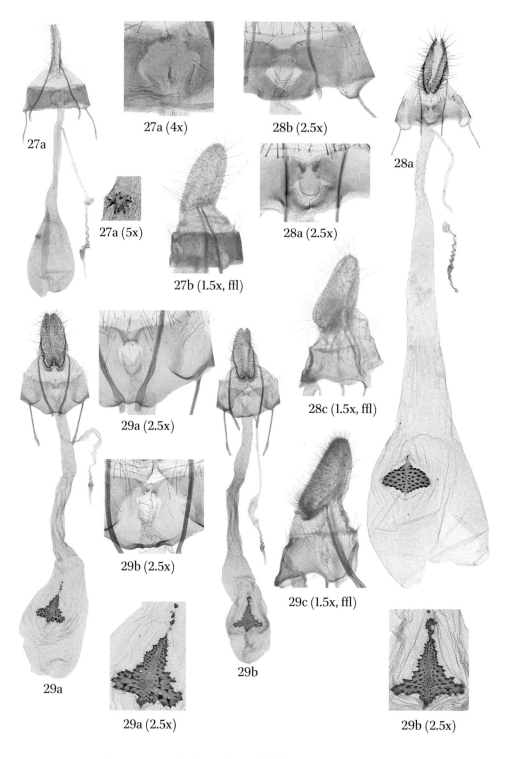

27a (4x)

28b (2.5x)

27a

28a

27a (5x)

28a (2.5x)

27b (1.5x, ffl)

28c (1.5x, ffl)

29a (2.5x)

29b (2.5x)

29a

29b

29c (1.5x, ffl)

29a (2.5x)

29b (2.5x)

27.   *Agonopterix arctica* – a–b) Finland (4548, RCJK)
28.   *Agonopterix ocellana* – a) Austria (1474, RCPB); b–c) Austria (8784, TLMF)
29.   *Agonopterix fruticosella* – a, c) Portugal (0784, RCMC); b) Portugal (0785, RCMC)

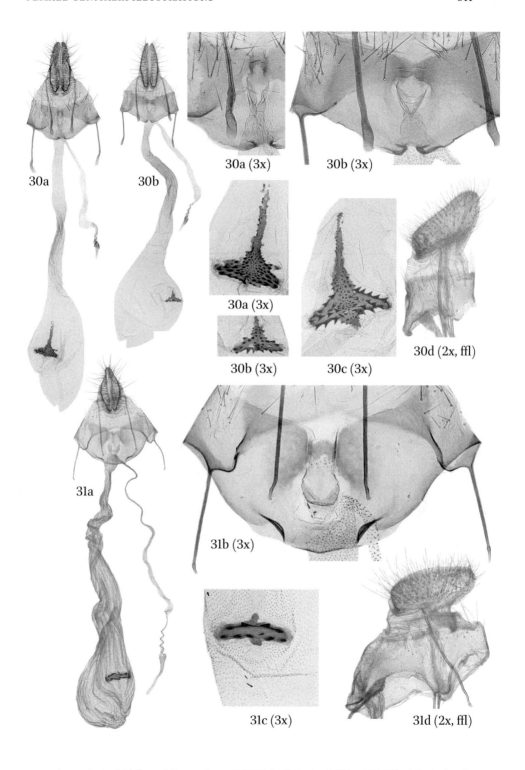

30a 30b

30a (3x) 30b (3x)

30a (3x)

30b (3x) 30c (3x) 30d (2x, ffl)

31a

31b (3x)

31c (3x) 31d (2x, ffl)

30. *Agonopterix rigidella* – a) France (7342, RCRD); b, d) Portugal (8891, RCMC); c) Spain (3998, RCWSc)

31. *Agonopterix olusatri* – a, d) Spain (0754, RCMC); b–c) Spain (2295, ZMUC)

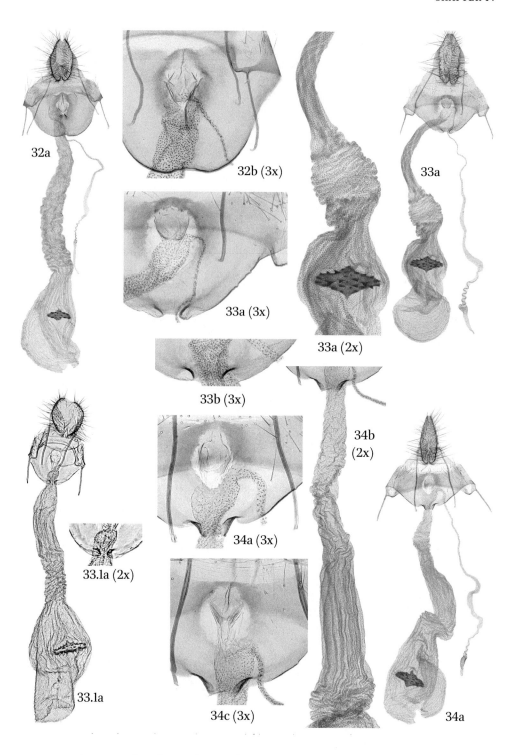

32a

32b (3x)

33a

33a (3x)

33a (2x)

33b (3x)

33.1a (2x)

33.1a

34a (3x)

34b (2x)

34c (3x)

34a

34a

32. *Agonopterix leucadensis* – a) Greece (0869, ZSM); b) Crete (2347, ZMUC)

33. *Agonopterix adspersella* – a) Austria (4250, RCPB); b) Croatia (0102, NHMW)

33.1. *Agonopterix adspersella* ssp. *pavida* stat.rev. – a) Turkey (NHMUK010293014, *A. pavida* holo-type, J.F.G. Clarke slide 9463, photo David Lees, NHMUK)

34. *Agonopterix thapsiella* – a) Greece (4246, NMBE); b–c) Crete (2348, NMBE)

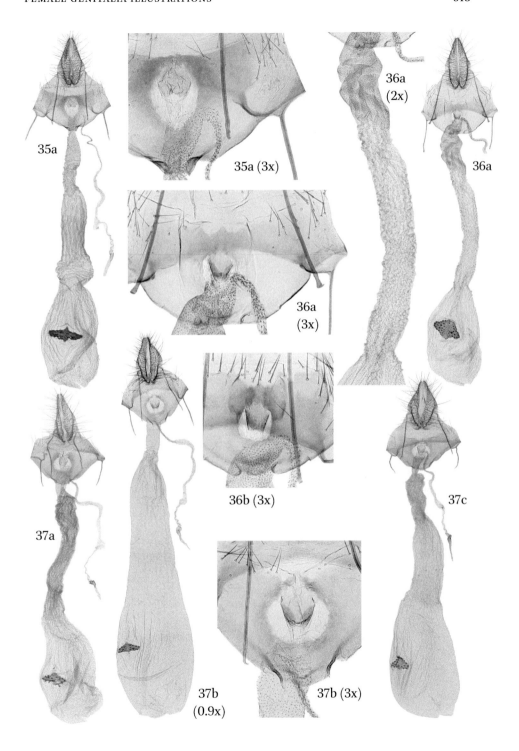

35. *Agonopterix chironiella* – a) Italy (1634, TLMF)
36. *Agonopterix cervariella* – a) Austria (0101, NHMW); b) Austria (3408, RCPB)
37. *Agonopterix paracervariella* sp.n. – a) Switzerland (7709, NMBE); b) Italy (5152, ZSM); c) Switzerland (7702, NMBE)

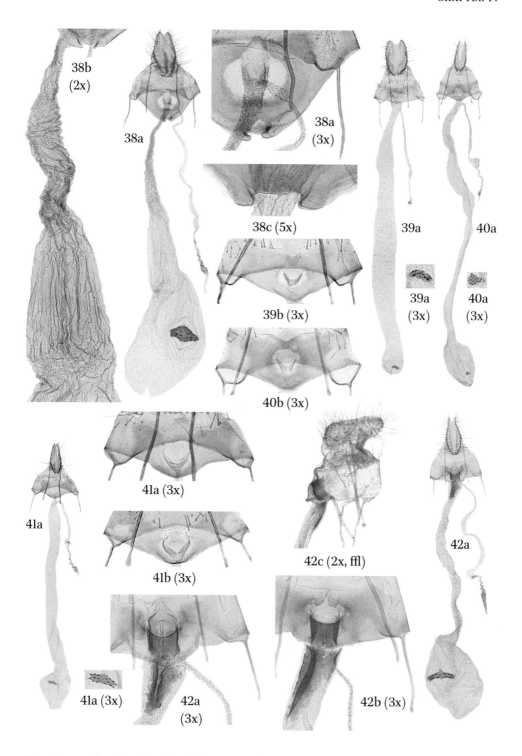

38. *Agonopterix cadurciella* – a) Italy (4457, TLMF); b) Croatia (4575, RCJJ); c) France (5166, TLMF)
39. *Agonopterix nodiflorella* – a) Turkey (3947, ZfBS); b) Sardinia (2113, TLMF)
40. *Agonopterix rotundella* – a) Italy (2112, TLMF)
41. *Agonopterix purpurea* – a) Italy (2241, MFSN); b) Turkey (3371, RCKL)
42. *Agonopterix curvipunctosa* – a) Crete (3052, RCKL); b) Spain (3896, RCWSc); c) Austria (9469, RCPB)

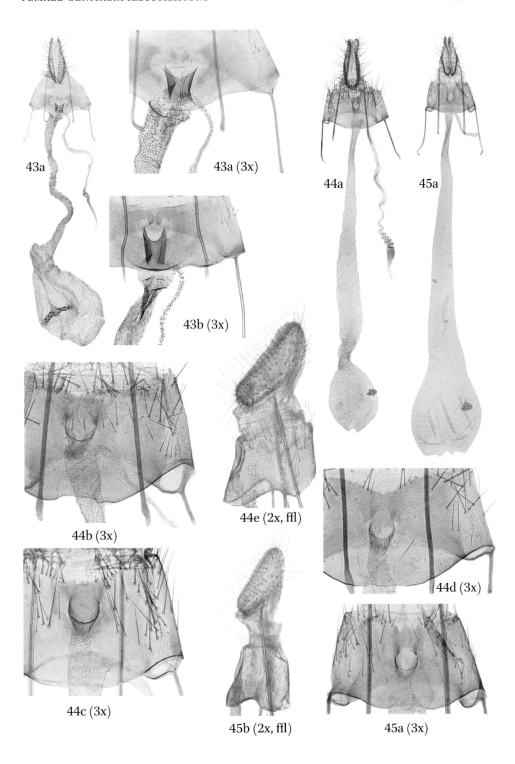

43a

43a (3x)

44a

45a

43b (3x)

44b (3x)

44e (2x, ffl)

44d (3x)

44c (3x)

45b (2x, ffl)

45a (3x)

43. *Agonopterix vendettella* – a) Madeira (3712, ZMUC); b) Italy (3549, RCFV)

44. *Agonopterix alpigena* – a) Slovenia (6205, NMPC); b) Slovenia (6214, NMPC); c) Croatia (1109, RCFG); d–e) Slovenia (5171, TLMF)

45. *Agonopterix richteri* sp.n. – a) Bulgaria (1031, RCLS); b) Bulgaria (9671, RCLS)

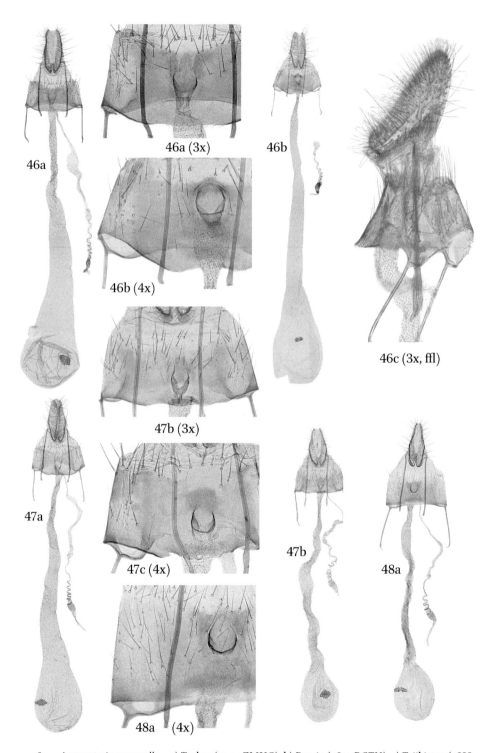

46a

46a (3x)

46b

46b (4x)

46c (3x, ffl)

47a

47b (3x)

47c (4x)

47b

48a

48a (4x)

46. *Agonopterix coenosella* – a) Turkey (5497, ZMUC); b) Russia (4631, RCTN); c) Tajikistan (5688, ZMHB)

47. *Agonopterix kayseriensis* – a) Turkey (4377, SMNK); b) Turkey (5504, ZMUC); c) Turkey (2381, ZMUC)

48. *Agonopterix cachritis* – a) Spain (2254, NHMW)

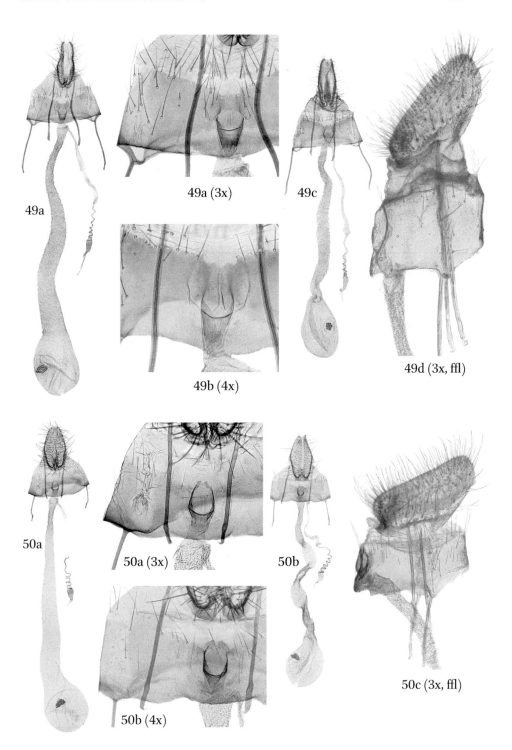

49a (3x)

49a

49c

49b (4x)

49d (3x, ffl)

50a

50a (3x)

50b

50b (4x)

50c (3x, ffl)

49. *Agonopterix ferulae* – a) Sardinia (1918, RCGD); b, d) Sardinia (1074, RCLS); c) France (2692, RCKL)

50. *Agonopterix galicicensis* sp.n. – a) North Macedonia (2343, ZMUC); b–c) North Macedonia (7259, RCIR)

51. *Agonopterix langmaidi* sp.n. – females unknown

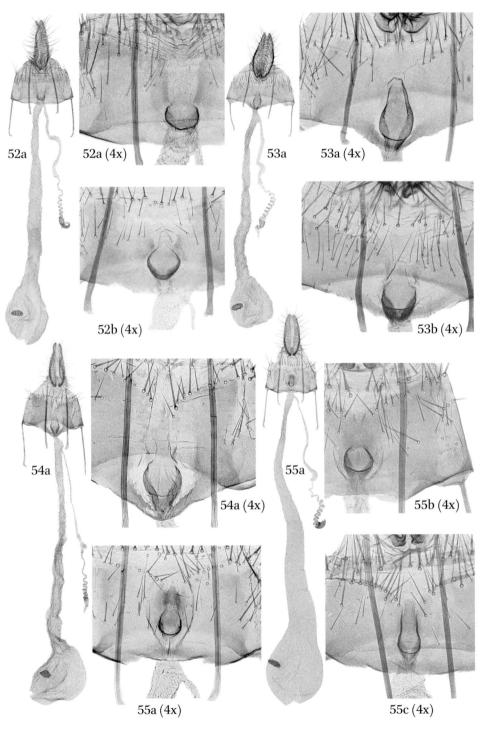

52.   *Agonopterix lessini* – a) Italy (1578, TLMF); b) Greece (2542, ZMUC)
53.   *Agonopterix selini* – a) Slovenia (2146, TLMF); b) Austria (6062, SMNK)
54.   *Agonopterix ordubadensis* – a) Armenia (6637, ZIN)
55.   *Agonopterix socerbi* – a, c) Italy (4269, NMBE); b) Slovenia (8316, TLMF)

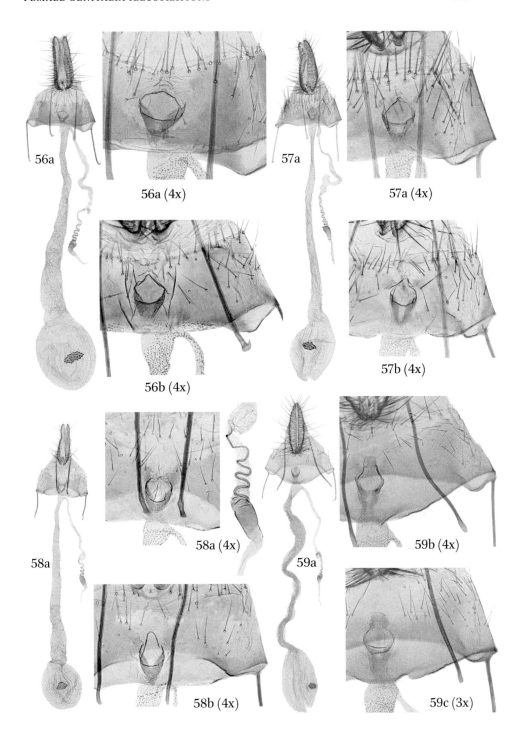

56a (4x)

56b (4x)

57a (4x)

57b (4x)

58a (4x)

58b (4x)

59b (4x)

59c (3x)

56. *Agonopterix angelicella* – a) Austria (1619, TLMF); b) Austria (0045, RCPB)
57. *Agonopterix paraselini* – a) Austria (2121, RCPB); b) Austria (1520, RCPB)
58. *Agonopterix parilella* – a) Slovenia (1111, RCFG); b) Germany (2029, ZSM)
59. *Agonopterix ciliella* – a) Russia (4625, RCTN); b) Austria (7996, TLMF); c) Austria (7946, TLMF)

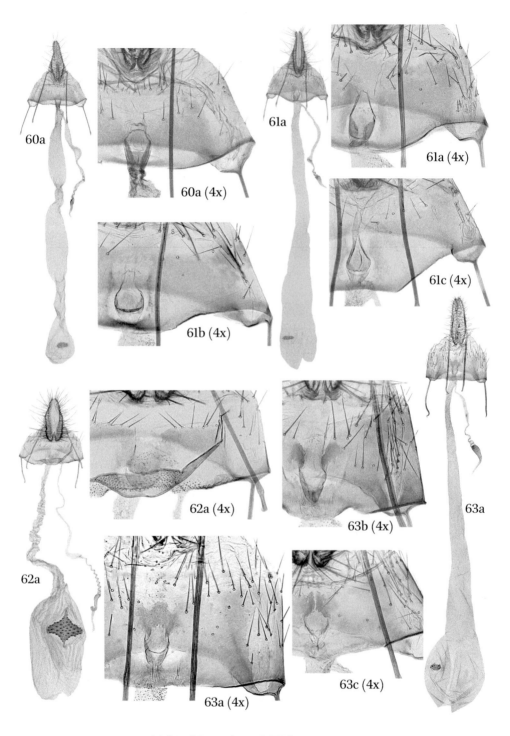

60a

60a (4x)

61a

61a (4x)

61b (4x)

61c (4x)

62a (4x)

63b (4x)

63a

62a

63a (4x)

63c (4x)

60.   *Agonopterix orophilella* – a) France (7340, RCRD)
61.   *Agonopterix heracliana* – a) Germany (5927, RCHB); b) Austria (5097, RCPB); c) Austria (8023, TLMF)
62.   *Agonopterix perezi* – a) Canary Islands (2064, ZSM)
63.   *Agonopterix putridella* – a) France (2429, ZMUC); b) France (2446, ZMUC); c) Austria (0012, RCPB)

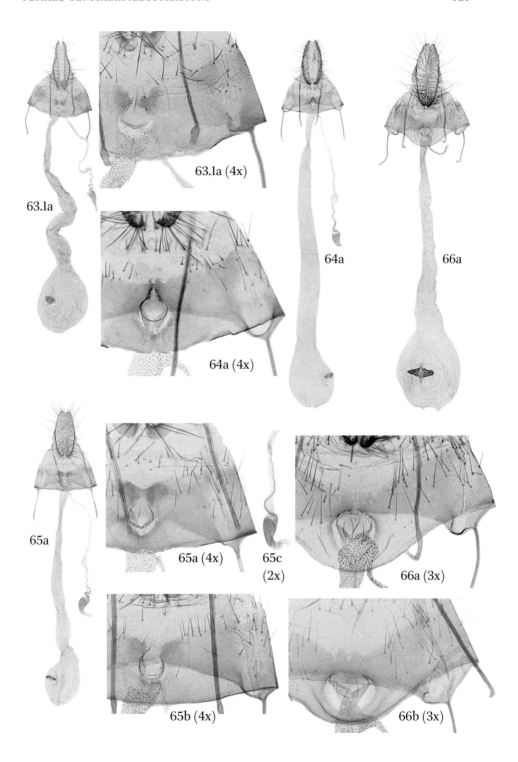

63.1a

63.1a (4x)

64a

64a (4x)

66a

65a

65a (4x)

65c (2x)

66a (3x)

65b (4x)

66b (3x)

63.1. *Agonopterix putridella* ssp. *scandinaviensis* – a) Finland (3851, RCMM)
64.   *Agonopterix quadripunctata* – a) Poland (3775, ZMUC)
65.   *Agonopterix hippomarathri* – a) Slovenia (2448, ZMUC); b–c) Slovenia (2444, ZMUC)
66.   *Agonopterix astrantiae* – a) Austria (1525, RCPB); b) Austria (8776, TLMF)

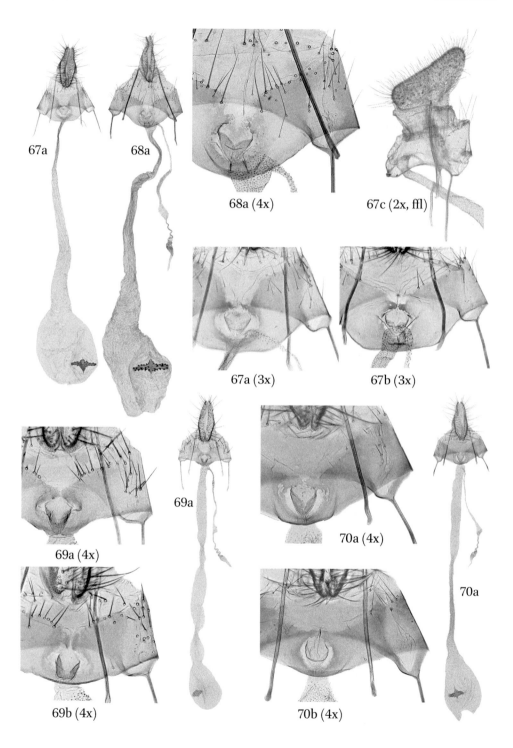

67a
68a
68a (4x)
67c (2x, ffl)
67a (3x)
67b (3x)
69a (4x)
69a
70a (4x)
70a
69b (4x)
70b (4x)

67.   *Agonopterix cnicella* – a) Austria (0020, RCPB); b–c) Italy (1487, RCRK)
68.   *Agonopterix melancholica* – a) unknown (5813, ZMHB)
69.   *Agonopterix alstromeriana* – a) Hungary (5114, NHMW); b) Austria (0002, RCPB)
70.   *Agonopterix capreolella* – a) Czechia (6221, NMPC); b) Hungary (4123, RCCS)

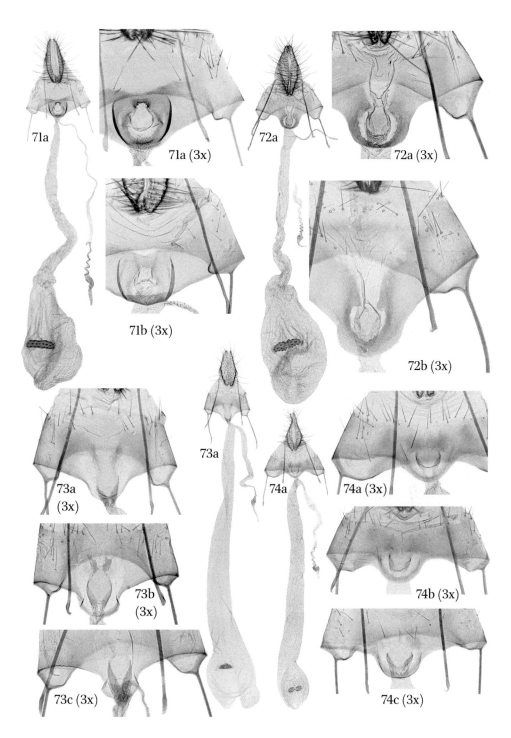

71. *Agonopterix yeatiana* – a) Greece (3657, TLMF); b) Netherlands (1531, NHMW)
72. *Agonopterix silerella* – a) Austria (1488, RCPB); b) Italy (4190, RCCM)
73. *Agonopterix ligusticella* – a) Croatia (1175, RCIR); b) Greece (1938, RCWSc); c) France (6948, MNHN)
74. *Agonopterix medelichensis* – a) Italy (1642, TLMF); b) Italy (1803, RCAM); c) Slovakia (2168, RCZT)

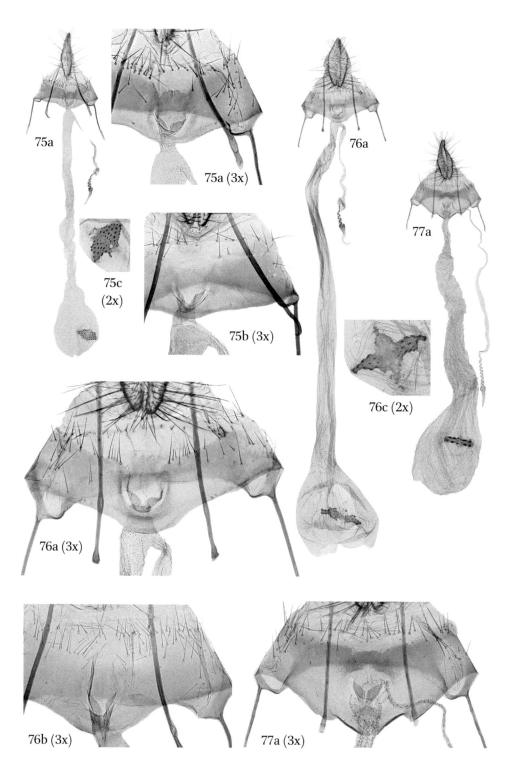

75a

75a (3x)

76a

77a

75c
(2x)

75b (3x)

76c (2x)

76a (3x)

76b (3x)

77a (3x)

75.   *Agonopterix irrorata* – a) Italy (1510, RCRK); b–c) Croatia (8881, TLMF)
76.   *Agonopterix graecella* – a) Italy (2474, ZMUC); b–c) Italy (8739, TLMF)
77.   *Agonopterix pseudoferulae* – a) Italy (1737, RCAM)

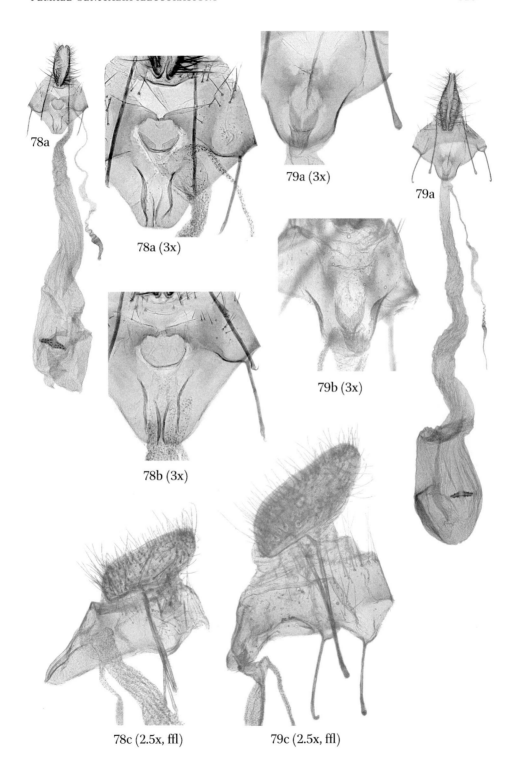

78a

78a (3x)

79a (3x)

79a

78b (3x)

79b (3x)

78c (2.5x, ffl)

79c (2.5x, ffl)

78. *Agonopterix subtakamukui* – a) Hungary (5906, HNHM); b–c) Austria (5188, TLMF)

79. *Agonopterix guanchella* – a–c) Canary Islands (9262, MNCN)

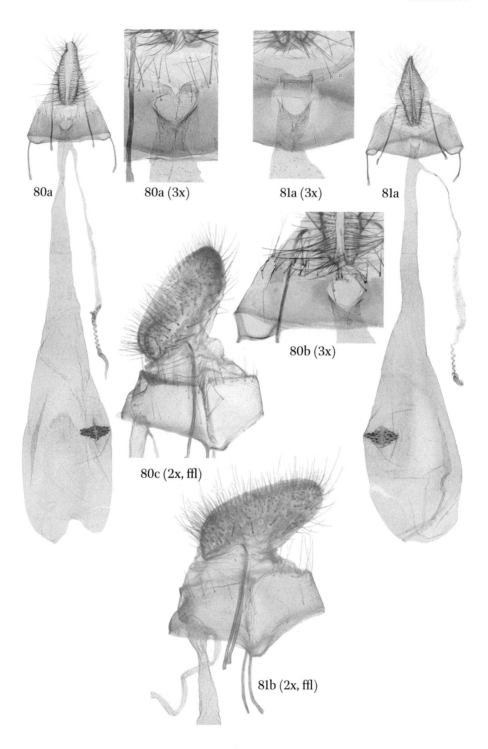

80a   80a (3x)   81a (3x)   81a

80b (3x)

80c (2x, ffl)

81b (2x, ffl)

80. *Agonopterix furvella* – a) Austria (9080, RCPB); b) Austria (9081, RCPB); c) Austria (1473, RCPB)
81. *Agonopterix pupillana* – a–b) Italy (8878, TLMF)

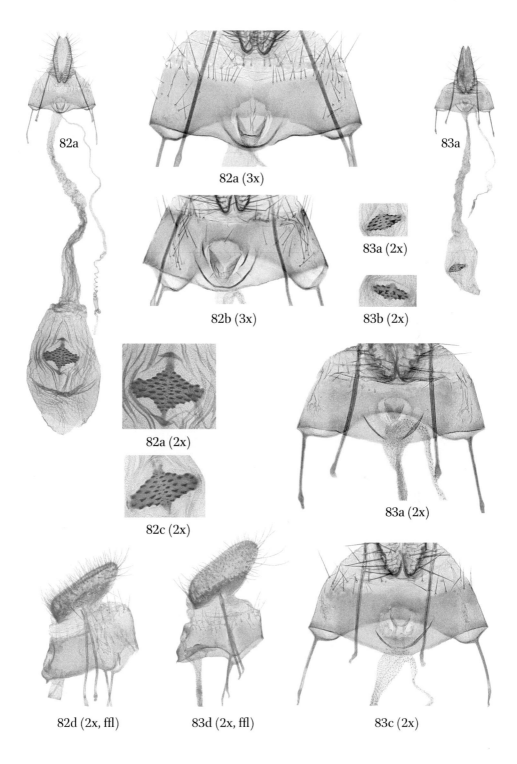

82a

82a (3x)

83a

82b (3x)

83a (2x)

83b (2x)

82a (2x)

82c (2x)

83a (2x)

82d (2x, ffl)

83d (2x, ffl)

83c (2x)

82. *Agonopterix rutana* – a) France (8872, TLMF); b–c) Spain (1096, RCLS); d) France (8539, RCGF)

83. *Agonopterix subumbellana* – a) Turkey (3374, RCKL); b–d) Iran (8427, TLMF)

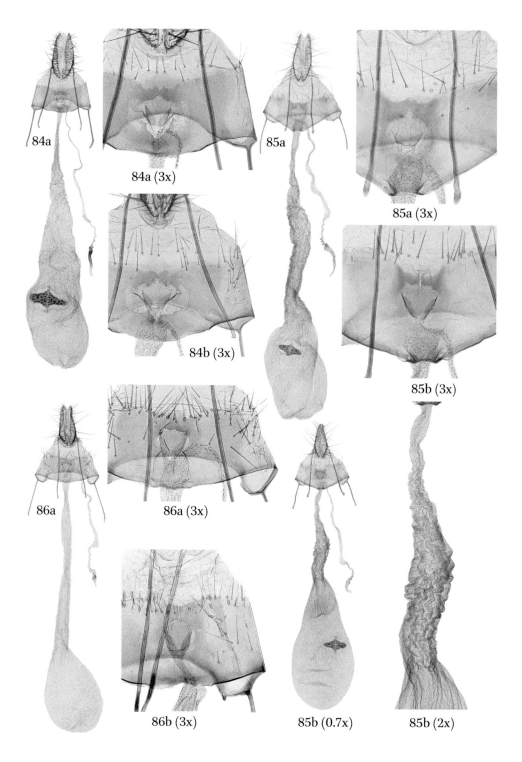

84a

84a (3x)

85a

85a (3x)

84b (3x)

85b (3x)

86a

86a (3x)

86b (3x)

85b (0.7x)

85b (2x)

84. *Agonopterix pallorella* – a) Spain (3727, RCEP); b) Turkey (6069, SMNK)

85. *Agonopterix straminella* – a) Tunisia (0129, NHMW); b) unknown (9470, NHMW)

86. *Agonopterix carduncelli* – a) Portugal (0776, RCMC); b) Morocco (1980, RCHB)

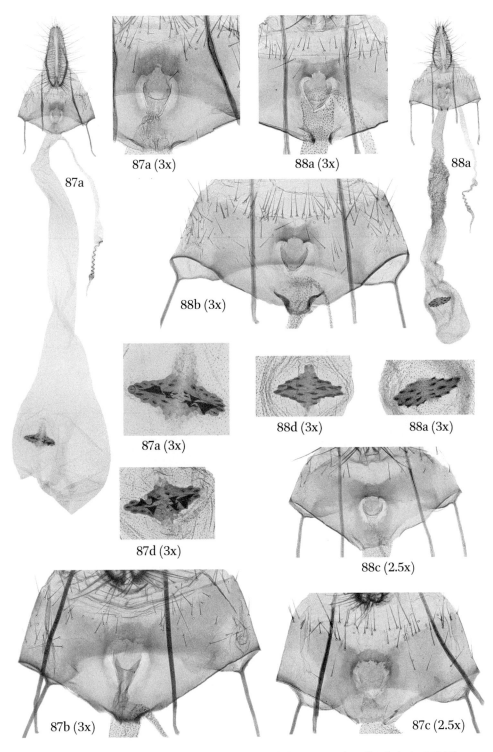

87a (3x)

88a (3x)

87a

88a

88b (3x)

88a

88d (3x)

88a (3x)

87a (3x)

87d (3x)

88c (2.5x)

87b (3x)

87c (2.5x)

87. *Agonopterix bipunctosa* – a) Italy (8373, TLMF); b) Italy (8386, TLMF); c–d) Italy (1560, RCLM)

88. *Agonopterix kyzyltashensis* – a) Russia (4296, ECKU); b) Russia (4596, RCTN); c–d) Russia (7751, NMPC)

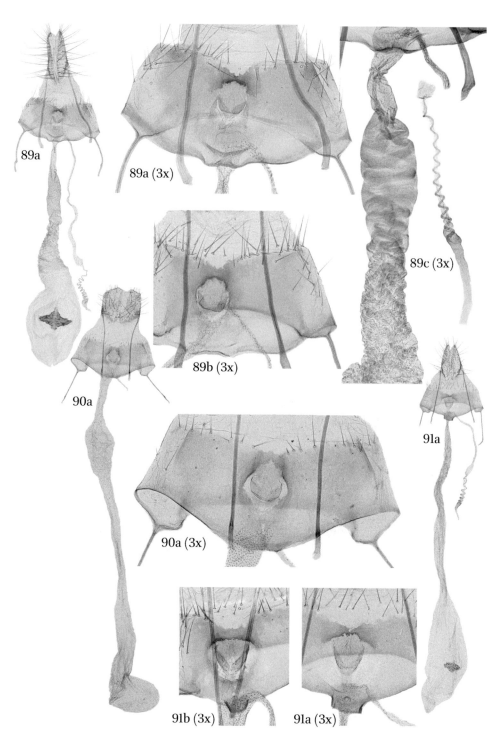

89.   *Agonopterix squamosa* – a) Croatia (8487, TLMF); b) Greece (2334, ZMUC); c) Croatia (0061, NHMW)

90.   *Agonopterix tschorbadjiewi* – a) Bulgaria (0055, holotype, H.J. Hannemannslide 795, museum id MV3132, NHMW)

91.   *Agonopterix volgensis* – a) Kazakhstan (4595, RCTN); b) Russia (2191, RCLS)

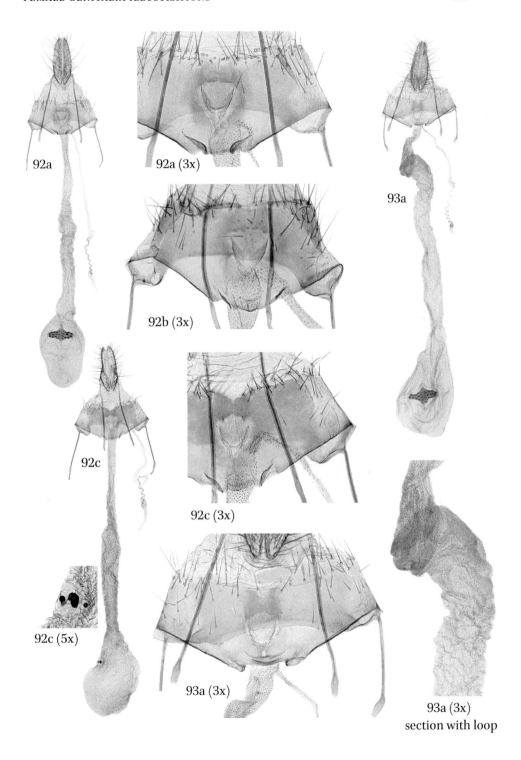

92a

92a (3x)

93a

92b (3x)

92c

92c (3x)

92c (5x)

93a (3x)

93a (3x)
section with loop

92. *Agonopterix kaekeritziana* – a) Italy (6408, MFSN); b) Austria (1524, RCPB) c) Spain (7184, RCZT)

93. *Agonopterix broennoeensis* – a) Norway (3827, ZMUC)

94. *Agonopterix uralensis* sp.n. – females unknown

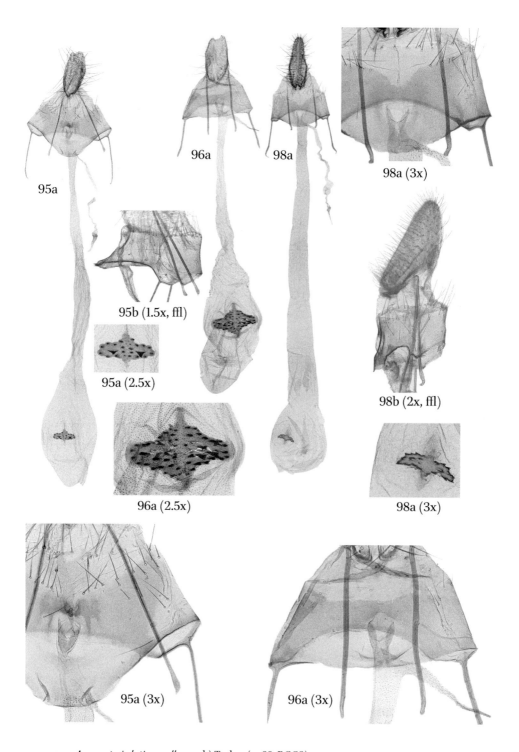

95a

96a

98a

98a (3x)

95b (1.5x, ffl)

98b (2x, ffl)

95a (2.5x)

96a (2.5x)

98a (3x)

95a (3x)

96a (3x)

95. *Agonopterix latipennella* – a–b) Turkey (4088, RCCS)
96. *Agonopterix invenustella* – a) Sicily (8979, TLMF)
97. *Agonopterix abditella* – no females available
98. *Agonopterix lidiae* – a–b) Italy (8804, TLMF)

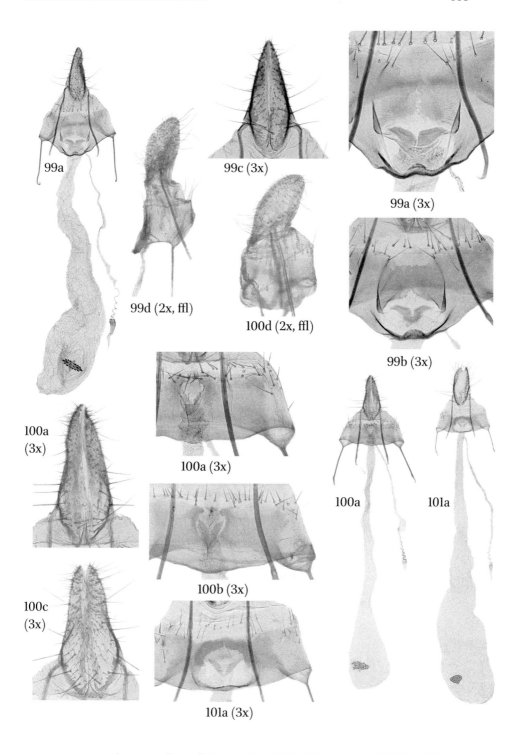

99a

99c (3x)

99d (2x, ffl)

99a (3x)

100d (2x, ffl)

99b (3x)

100a
(3x)

100a (3x)

100b (3x)

100a 101a

100c
(3x)

101a (3x)

99. *Agonopterix subpropinquella* – a, d) Cyprus (0860, ZSM); b) Turkey (3529, MHNG); c) Balearics (2055, ZSM)
100. *Agonopterix nanatella* – a) Austria (7962, TLMF); b–d) Greece (7377, RCKB)
101. *Agonopterix kuznetzovi* – a) England (2472, ZMUC)

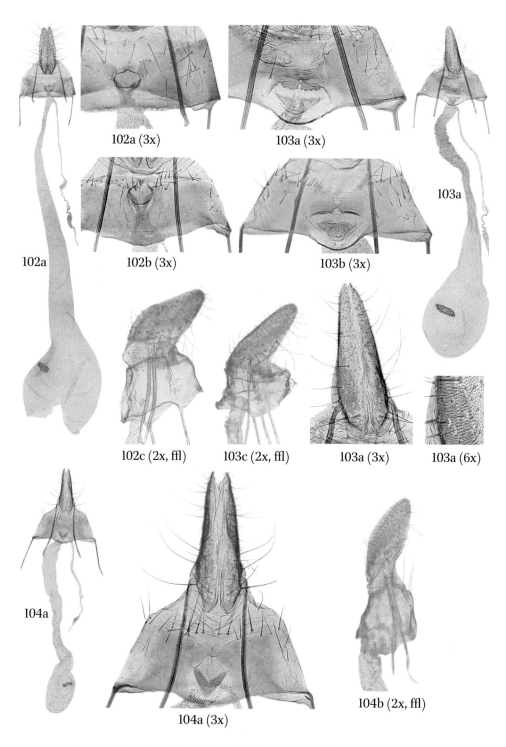

102a (3x)

103a (3x)

102a

102b (3x)

103b (3x)

103a

102c (2x, ffl)

103c (2x, ffl)

103a (3x)

103a (6x)

104a

104a (3x)

104b (2x, ffl)

102. *Agonopterix propinquella* – a) Switzerland (5693, ZMHB); b) France (6219, NMPC);
c) Austria (0024, RCPB)

103. *Agonopterix carduella* – a) Austria (4989, RCOR); b–c) Austria (0051, RCPB)

104. *Agonopterix ivinskisi* – a–b) Switzerland (0615, NMBE)

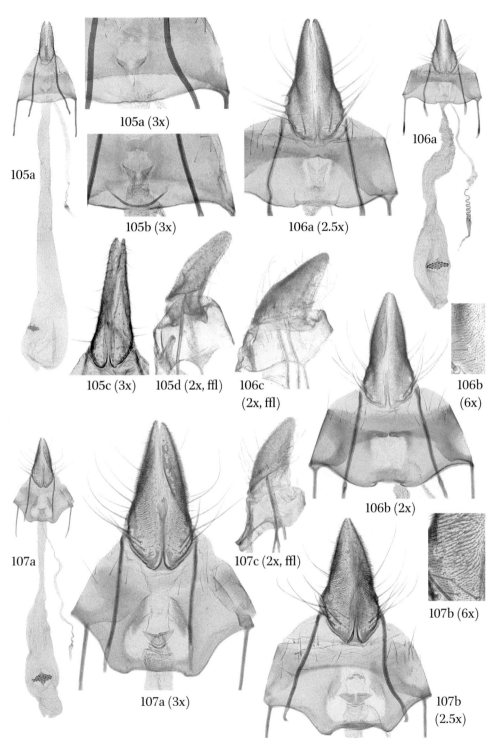

105a

105a (3x)

105b (3x)

106a

106a (2.5x)

105c (3x) 105d (2x, ffl) 106c (2x, ffl)

106b (6x)

106b (2x)

107a

107c (2x, ffl)

107b (6x)

107a (3x)

107b (2.5x)

105. *Agonopterix ferocella* – a) France (8879, TLMF); b) Slovenia (1545, MFSN); c–d) Austria (1539, NHMW)
106. *Agonopterix laterella* – a) Greece (1937, RCWSc); b) Germany (1965, RCHB); c) Austria (0088, NHMW)
107. *Agonopterix xeranthemella* – a, c) North Macedonia (7273, RCIR); b) Greece (5485, ZMUC)

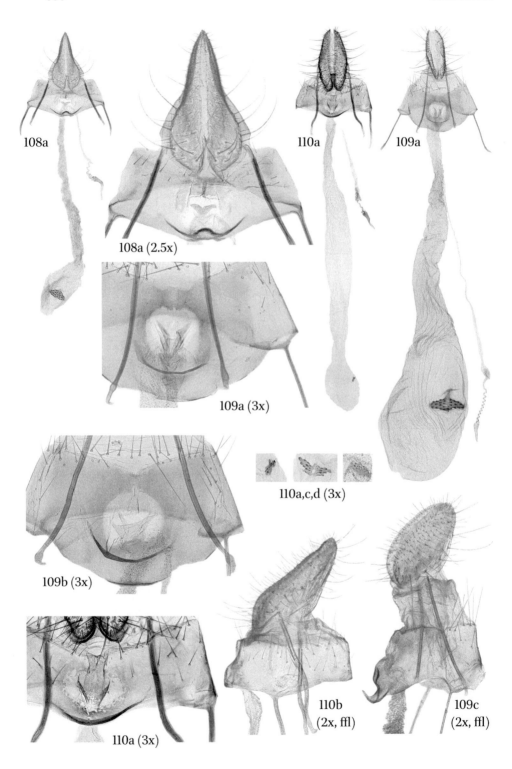

108a

108a (2.5x)

110a

109a

109a (3x)

109b (3x)

110a,c,d (3x)

110b
(2x, ffl)

109c
(2x, ffl)

110a (3x)

108.   *Agonopterix cinerariae* – a) Canary Islands (0878, ZSM)
109.   *Agonopterix arenella* – a) Austria (7777, RCPB); b) Austria (4929, RCWS); c) Austria (1471, RCPB)
110.   *Agonopterix petasitis* – a–b) Austria (1470, RCPB); c) Austria (9082, RCPB); d) Austria (9084, RCPB)

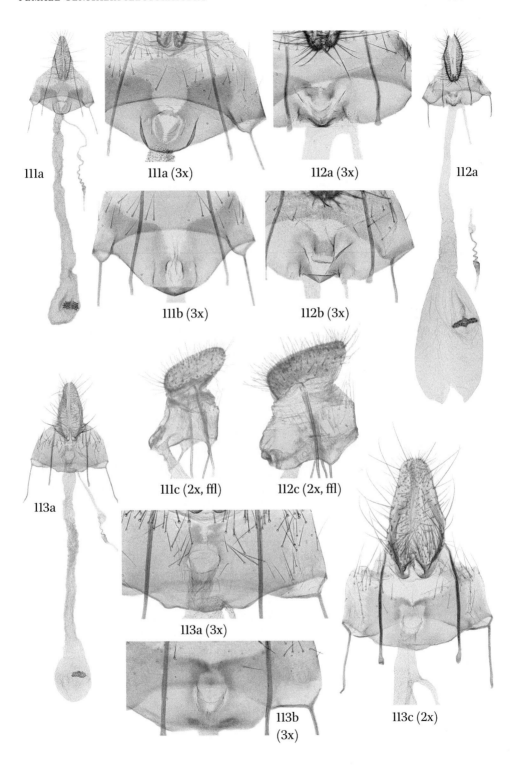

111a   111a (3x)   112a (3x)   112a

111b (3x)   112b (3x)

113a   111c (2x, ffl)   112c (2x, ffl)

113a (3x)

113b (3x)   113c (2x)

111. *Agonopterix cotoneastri* – a) Italy (6805, RCAM); b) Italy (4210, RCLM); c) Austria (1275, RCIR)
112. *Agonopterix multiplicella* – a) Italy (4233, TLMF); b) Russia (6285, MGAB); c) Finland (2337, ZMUC)
113. *Agonopterix doronicella* – a) Germany (8885, TLMF); b) Greece (2772, RCKL); c) Austria (1530, NHMW)

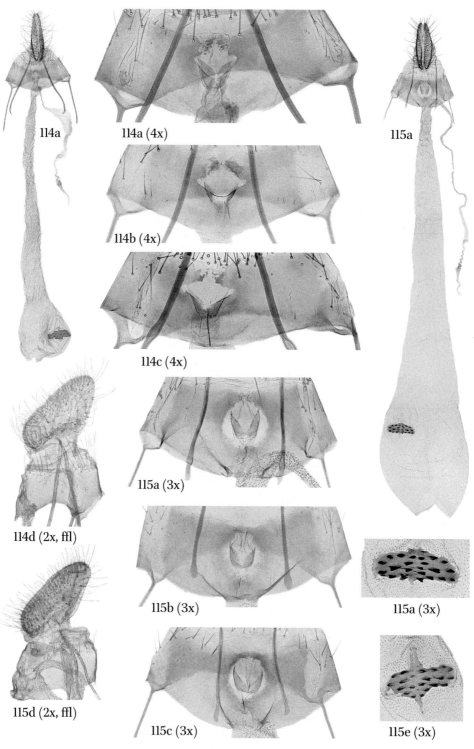

114a   114a (4x)   115a

114b (4x)

114c (4x)

115a (3x)

115b (3x)   115a (3x)

114d (2x, ffl)

115d (2x, ffl)   115c (3x)   115e (3x)

114.   *Agonopterix assimilella* – a, d) Austria (8720, KLM); b) Italy (4037, MFSN); c) France (2432, ZMUC)

115.   *Agonopterix atomella* – a, d) Austria (1491, RCPB); b) Austria (8293, TLMF); c, e) France (8271, TLMF)

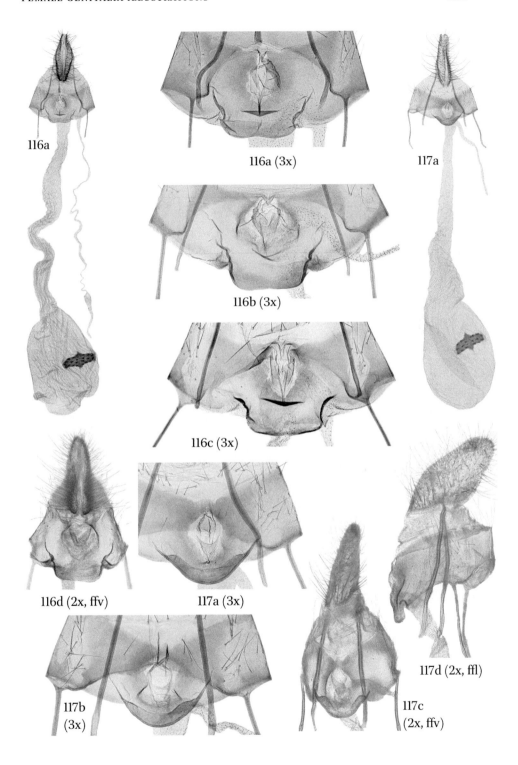

116a

116a (3x)

117a

116b (3x)

116c (3x)

116d (2x, ffv)

117a (3x)

117d (2x, ffl)

117b
(3x)

117c
(2x, ffv)

116. *Agonopterix scopariella* – a) Madeira (3705, ZMUC); b, d) Morocco (1946, ZfBS); c) France (1787, RCAM)

117. *Agonopterix conciliatella* – a) Canary Islands (1003, RCLS); b) Canary Islands (1496, RCRK); c–d) Canary Islands (8593, RCAM)

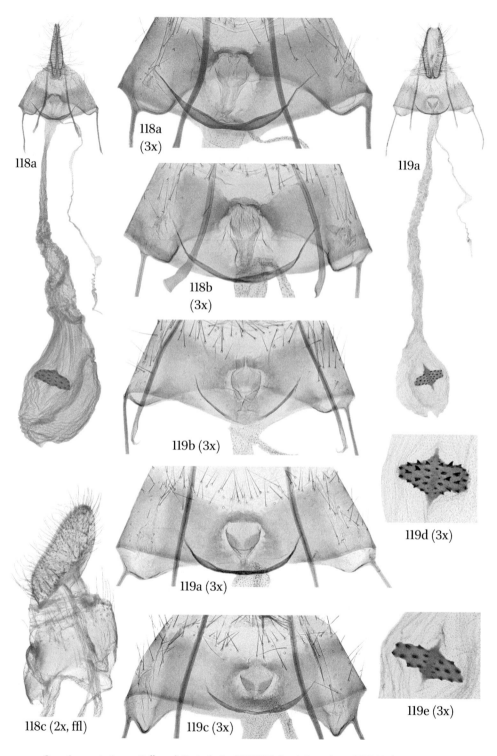

118. *Agonopterix comitella* – a) Crete (4169, NHMW); b–c) Crete (1541, NHMW)
119. *Agonopterix nervosa* – a) Hungary (4175, NHMW); b, d) Austria (1495, RCPB); c, e) Turkey (3395, RCKL)

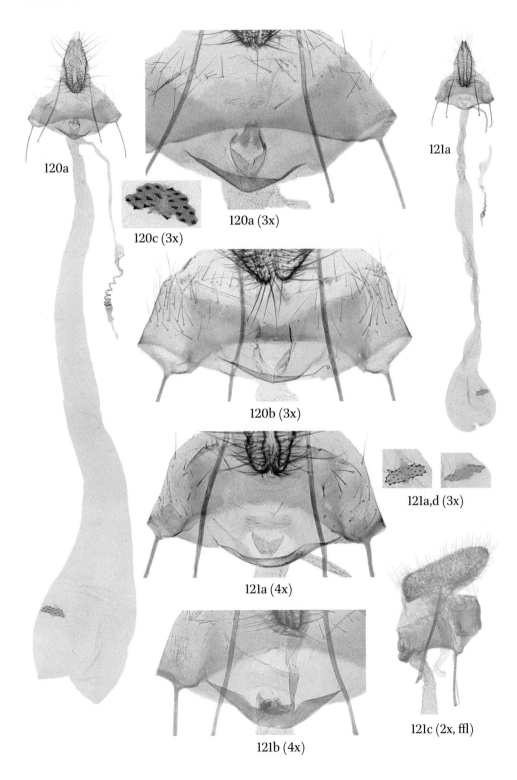

120a

120c (3x)

120a (3x)

120b (3x)

121a

121a,d (3x)

121a (4x)

121c (2x, ffl)

121b (4x)

120.   *Agonopterix umbellana* – a) Turkey (5401, ZMUC); b–c) Italy (1231, MFSN)

121.   *Agonopterix oinochroa* – a) Serbia (1533, NHMW); b–d) Austria (8297, TLMF)

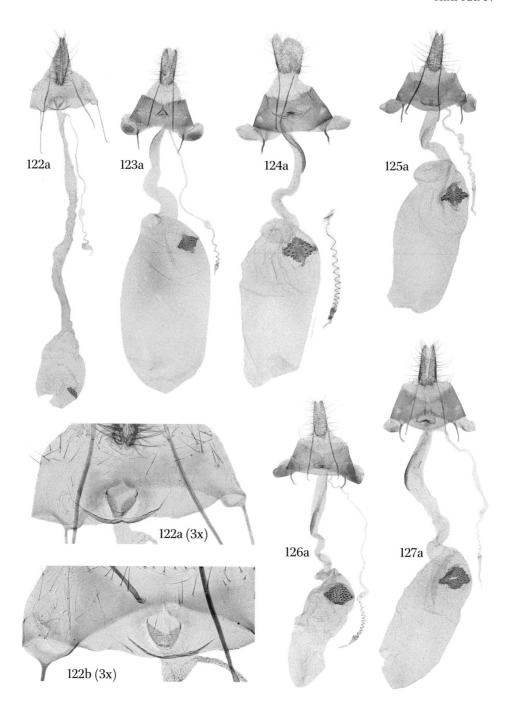

122a

123a

124a

125a

122a (3x)

126a

127a

122b (3x)

122. *Agonopterix aspersella* – a) Sicily (2510, ZMUC); b) France (1540, NHMW)
123. *Depressaria pulcherrimella* – a) Spain (3734, RCEP)
124. *Depressaria sordidatella* – a) Germany (1977, RCAS)
125. *Depressaria floridella* – a) Italy (1772, RCAM)
126. *Depressaria douglasella* – a) Switzerland (7729, NMBE)
127. *Depressaria beckmanni* – a) Switzerland (7688, NMBE)

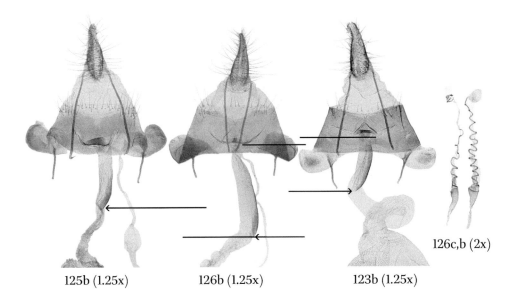

125b (1.25x)   126b (1.25x)   123b (1.25x)   126c,b (2x)

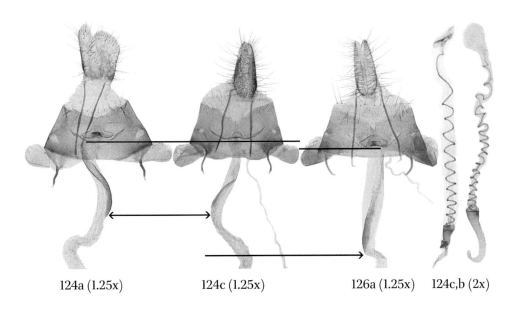

124a (1.25x)   124c (1.25x)   126a (1.25x)   124c,b (2x)

123. *Depressaria pulcherrimella* – b) Slovakia (1329, RCIR)
124. *Depressaria sordidatella* – a) Germany (1977, RCAS); b) Finland (4489, RCJJ); c) Austria (8818, TLMF)
125. *Depressaria floridella* – b) France (1866, NHBE)
126. *Depressaria douglasella* – a) Switzerland (7729, NMBE); b) Switzerland (0642, NMBE); c) Switzerland (7720, NMBE)

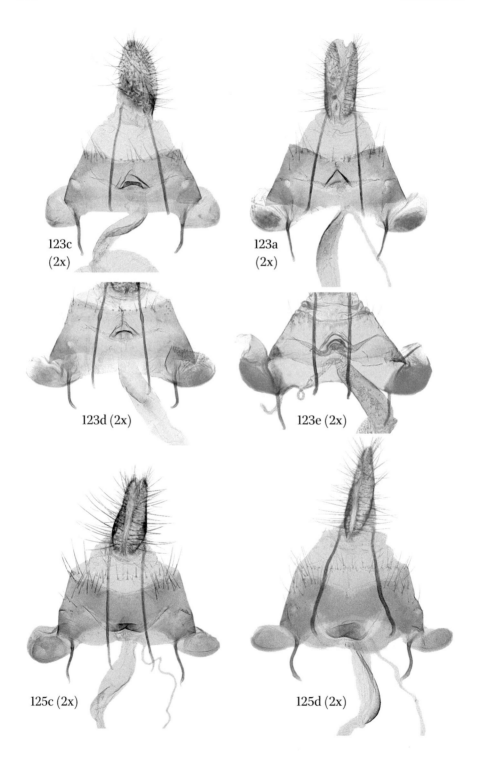

123c (2x)

123a (2x)

123d (2x)

123e (2x)

125c (2x)

125d (2x)

123. *Depressaria pulcherrimella* – a) Spain (3734, RCEP); c) Italy (1518, RCRK); d) Germany (7055, RCWSc); e) Austria (8052, TLMF)

125. *Depressaria floridella* – c) Turkey (3366, RCKL); d) France (3022, RCKL)

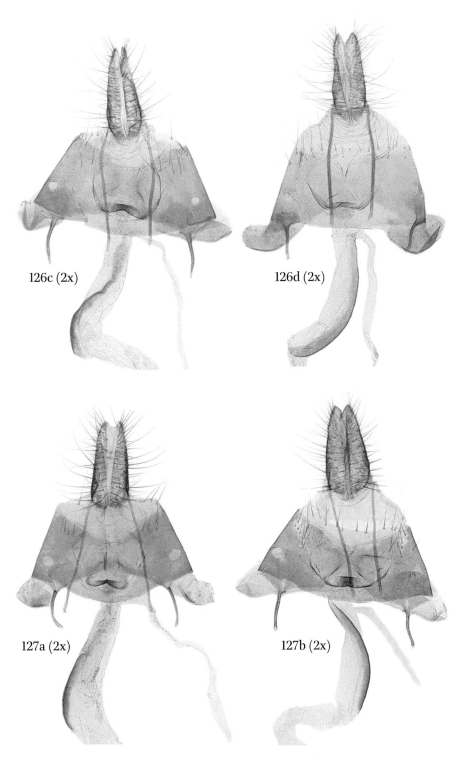

126c (2x)

126d (2x)

127a (2x)

127b (2x)

126. *Depressaria douglasella* – c) Switzerland (7720, NMBE); d) Greece (5634, SMNK)
127. *Depressaria beckmanni* – a) Switzerland (7688, NMBE); b) Switzerland (2152, NMBE)

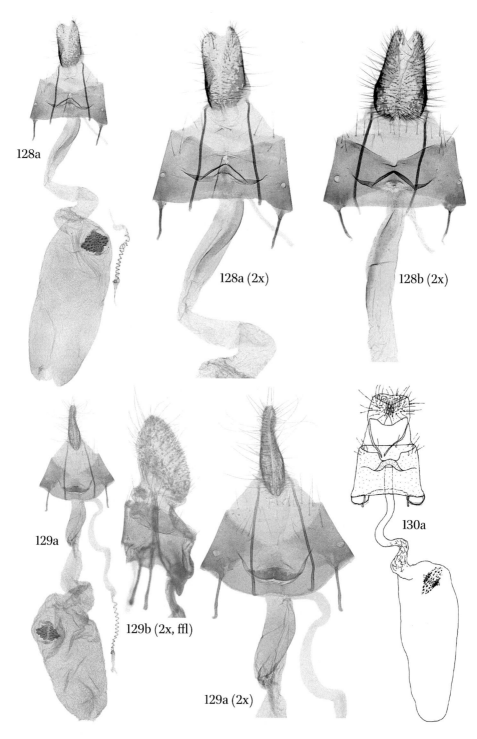

128. *Depressaria incognitella* – a) Italy (1775, RCAM); b) Switzerland (2150, NMBE)
129. *Depressaria nemolella* – a–b) Sweden (9246, RCJB)
130. *Depressaria cinderella* – a) Portugal (drawing from original description)

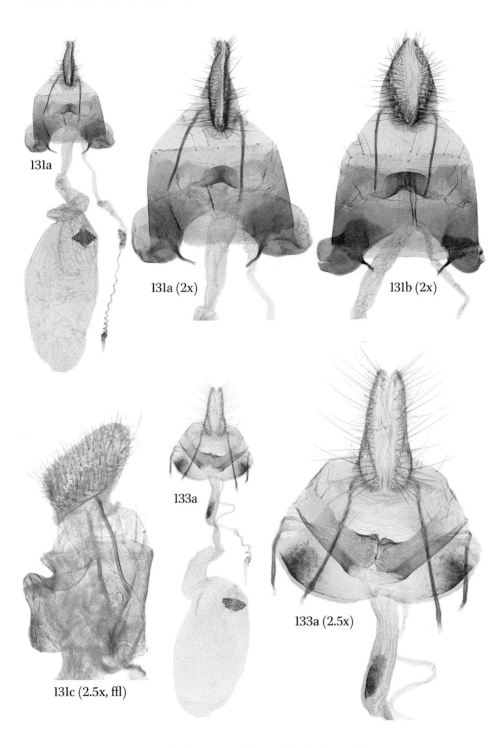

131a

131a (2x)

131b (2x)

131c (2.5x, ffl)

133a

133a (2.5x)

131. *Depressaria infernella* – a) Spain (6846, ZMHB); b–c) Spain (6847, ZMHB)
132. *Depressaria indecorella* – females unknown
133. *Depressaria lacticapitella* – a) Austria (3865, ZSM)

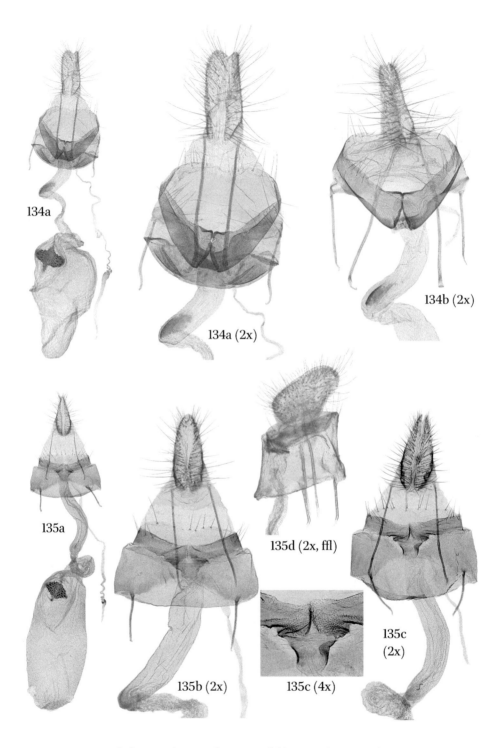

134.   *Depressaria hofmanni* – a) Austria (2222, RCPB); b) Austria (1479, RCPB)
135.   *Depressaria albipunctella* – a) Sardinia (1569, RCLM); b) Poland (2679, RCKL); c) Turkey (3307, RCKL); d) Austria (1476, RCPB)

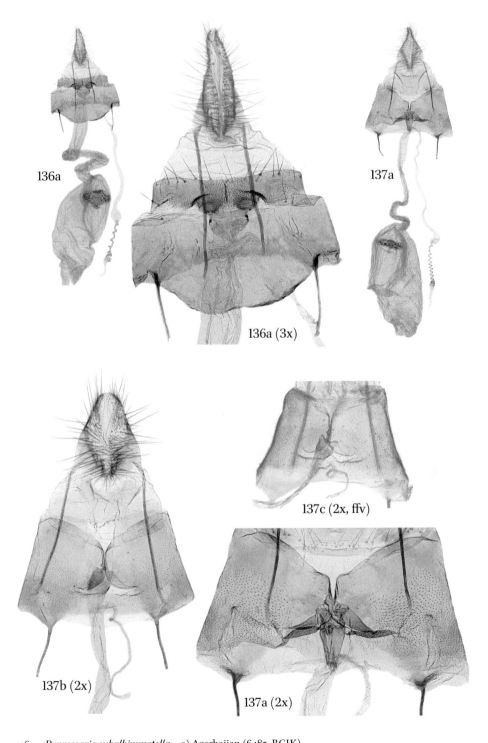

136a

136a (3x)

137a

137c (2x, ffv)

137b (2x)

137a (2x)

136. *Depressaria subalbipunctella* – a) Azerbaijan (6485, RCJK)
137. *Depressaria krasnowodskella* – a) Morocco (7190, RCZT); b–c) Spain (0765, ZMUC [c: uncompressed])

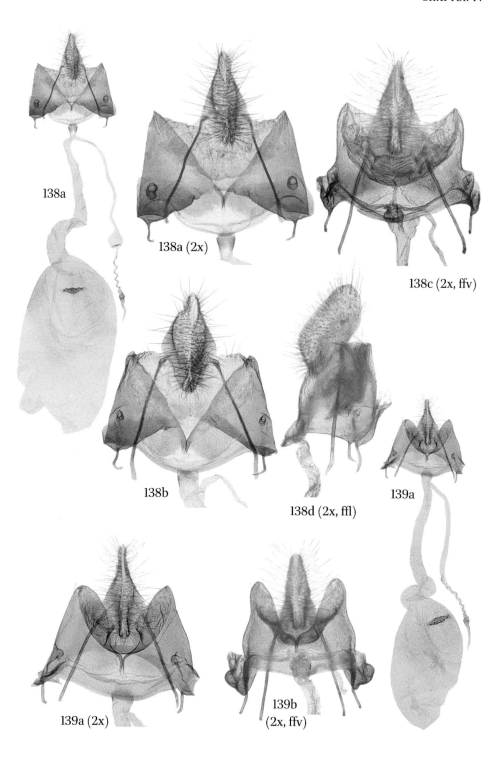

138a

138a (2x)

138c (2x, ffv)

138b

138d (2x, ffl)

139a

139a (2x)

139b
(2x, ffv)

138. *Depressaria tenebricosa* – a) Bulgaria (3521, RCBZ); b–c) Albania (8165, RCCP [c: uncompressed]); d) Croatia (1138, RCLS)

139. *Depressaria radiosquamella* – a–b) Sardinia (8631, RCAM [b: uncompressed])

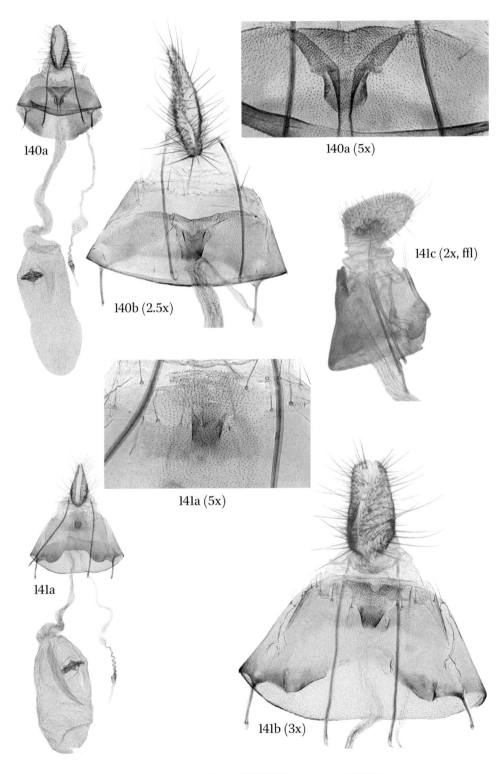

140a

140a (5x)

140b (2.5x)

141c (2x, ffl)

141a (5x)

141a

141b (3x)

140. *Depressaria emeritella* – a) Romania (6325, MGAB); b) Austria (0036, RCPB)

141. *Depressaria olerella* – a) Austria (8495, RCPB); b) Austria (1480, RCPB); c) Austria (8421, TLMF)

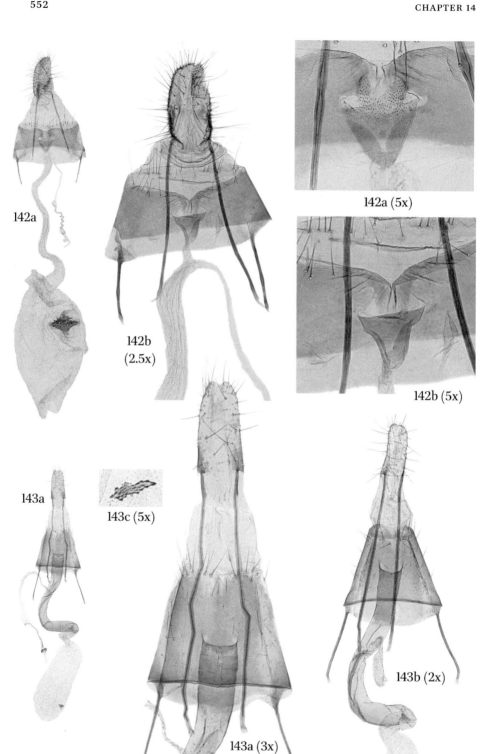

142a (5x)

142b (5x)

142a

142b
(2.5x)

143a

143c (5x)

143b (2x)

143a (3x)

142.   *Depressaria leucocephala* – a) Italy (0386, TLMF); b) Italy (8473, TLMF)
143.   *Depressaria depressana* – a) France (2693, RCKL); b–c) Austria (1481, RCPB)

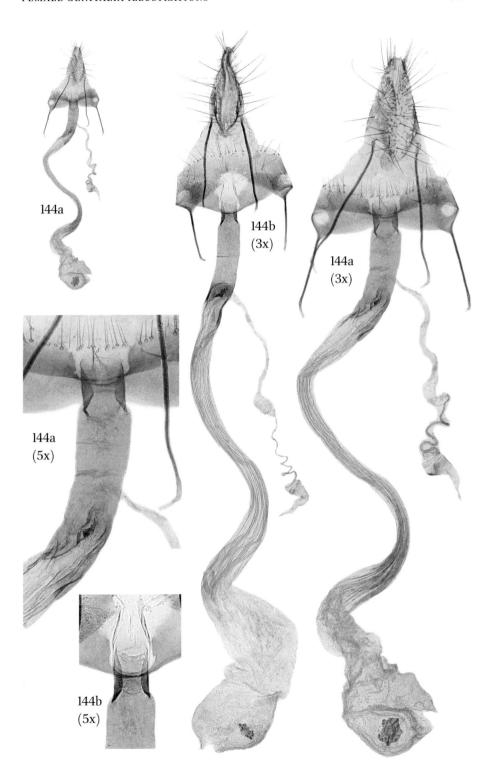

144. *Depressaria absynthiella* – a) Turkey (3130, RCKL); b) Spain (2524, ZMUC)

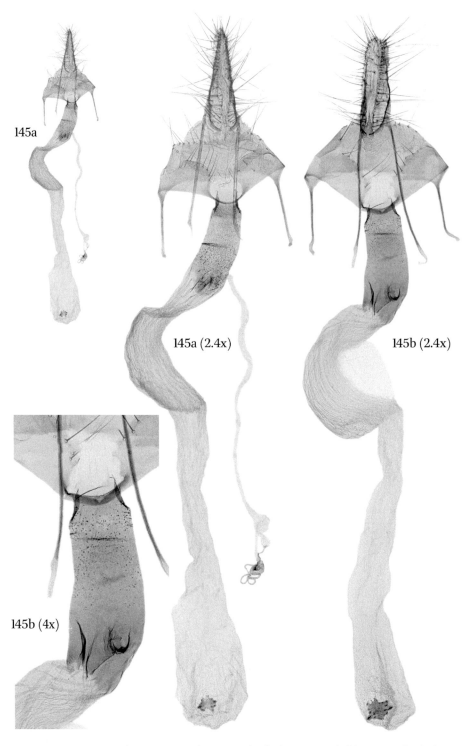

145a

145a (2.4x)

145b (2.4x)

145b (4x)

145.   *Depressaria tenerifae* stat. rev. – a) Canary Islands (2966, RCKL); b) Canary Islands (3854, ZMUC)

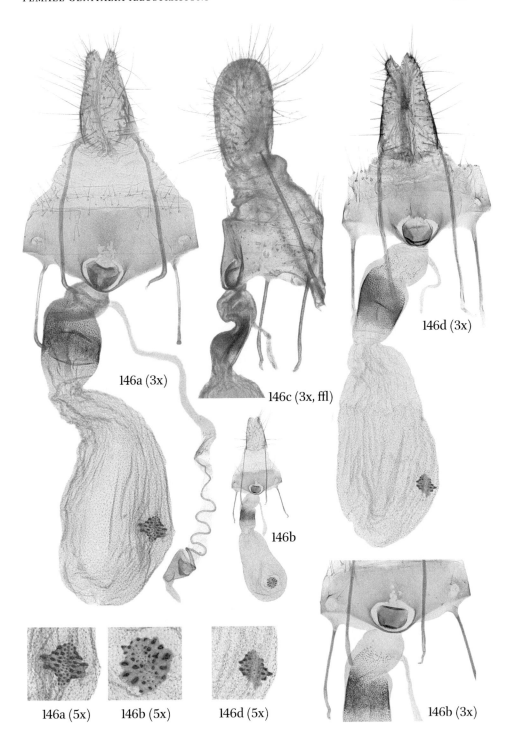

146a (3x)

146c (3x, ffl)

146d (3x)

146b

146a (5x) 146b (5x) 146d (5x) 146b (3x)

146. *Depressaria artemisiae* – a) Russia (8507, RCKL); b–c) Switzerland (0631, NMBE);
d) Hungary (1141, RCLS)

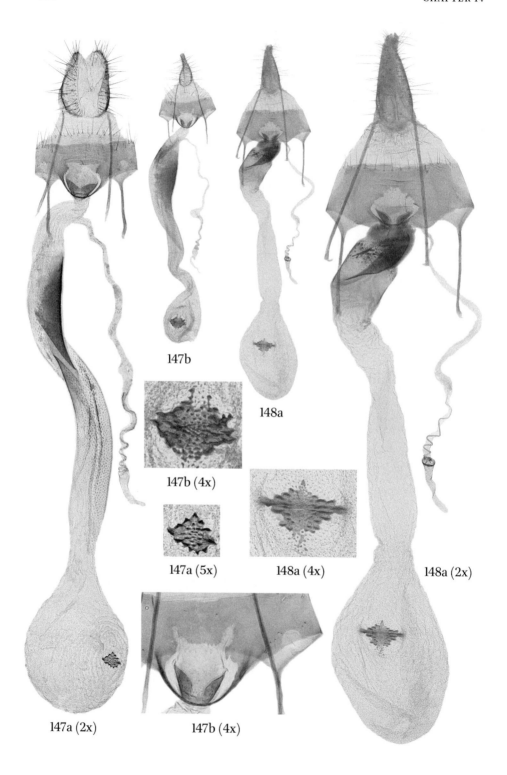

147b

147b (4x)

147a (5x)

148a

148a (4x)

148a (2x)

147a (2x)

147b (4x)

147. *Depressaria fuscovirgatella* – a) Kyrgyzstan (5090, KLM); b) Turkey (3306, RCKL)
148. *Depressaria atrostrigella* – a) Russia (7276, NMPC)

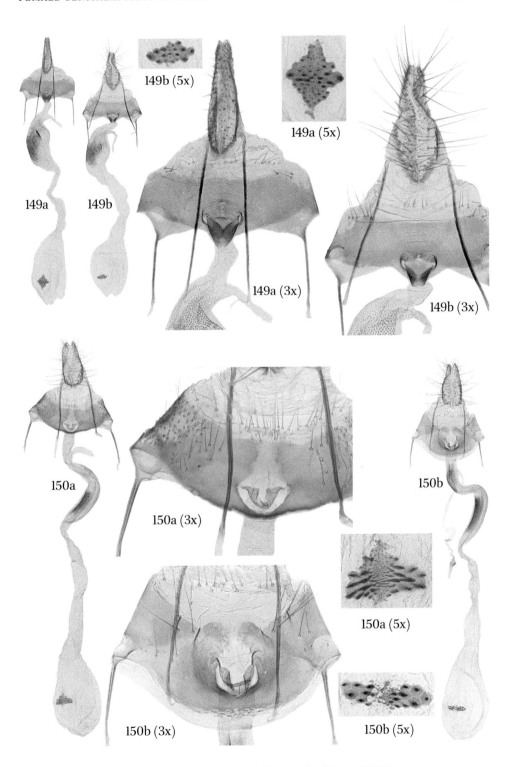

149b (5x)

149a (5x)

149a   149b

149a (3x)

149b (3x)

150a

150b

150a (3x)

150a (5x)

150b (3x)

150b (5x)

149.   *Depressaria silesiaca* – a) Austria (5190, TLMF); b) Switzerland (0633, NMBE)

150.   *Depressaria zelleri* – a) Greece (1936, RCWSc); b) Turkey (3213, RCKL)

151.   *Depressaria pyrenaella* – females unknown

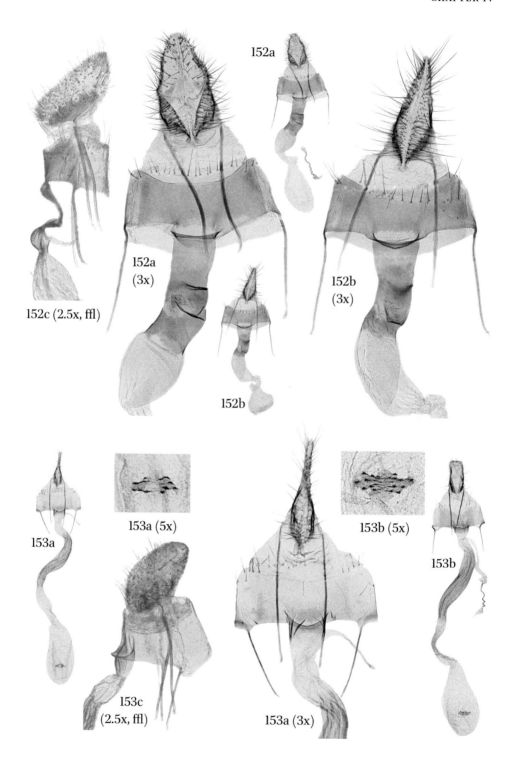

152. *Depressaria marcella* – a) Greece (2532, ZMUC); b–c) Spain (4701, RCTN)
153. *Depressaria peregrinella* – a) Greece (0893, ZSM); b–c) Turkey (3955, ZfBS)

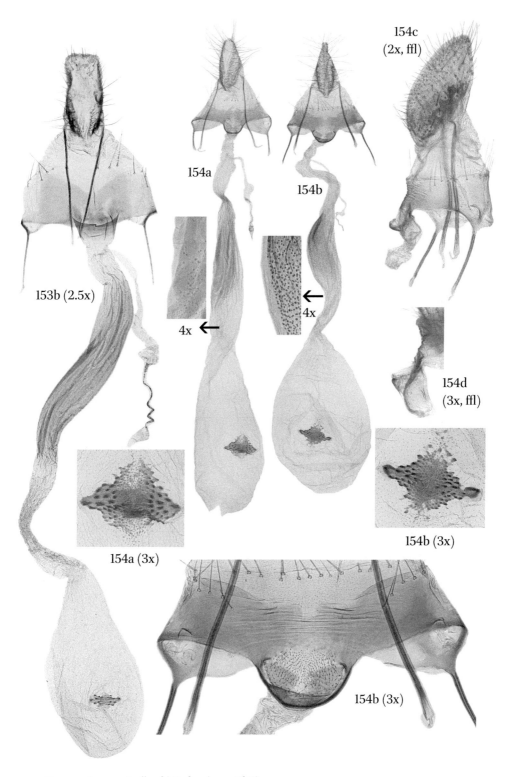

153b (2.5x)

154a

154b

154c
(2x, ffl)

4x

4x

154d
(3x, ffl)

154a (3x)

154b (3x)

154b (3x)

153. *Depressaria peregrinella* – b) Turkey (3955, ZfBS)

154. *Depressaria heydenii* – a) France (8863, TLMF); b–c) Romania (5904, HNHM): d) Austria (0316, RCPB)

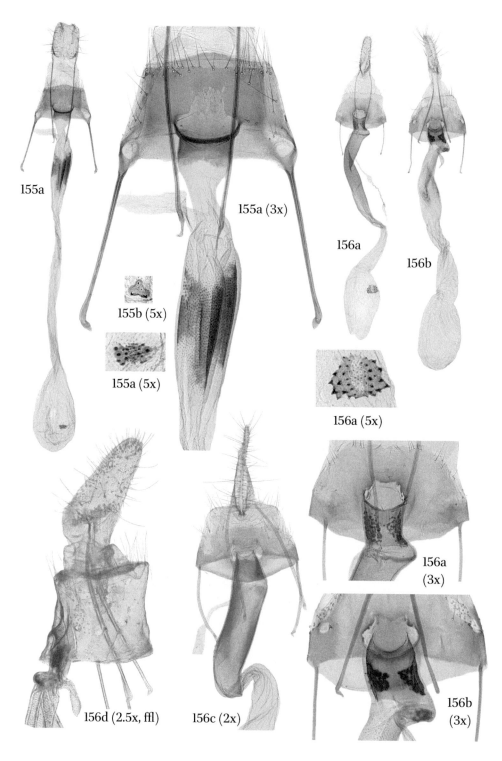

155a

155a (3x)

155b (5x)

155a (5x)

156a

156b

156a (5x)

156d (2.5x, ffl)

156c (2x)

156a (3x)

156b (3x)

155. *Depressaria fuscipedella* – a) Italy (8978, RCFG); b) Italy (9472, RCFG)

156. *Depressaria ululana* – a) Greece (4814, RCWSc); b, d) Switzerland (o644, NMBE);
 c) Armenia (8908, ECKU)

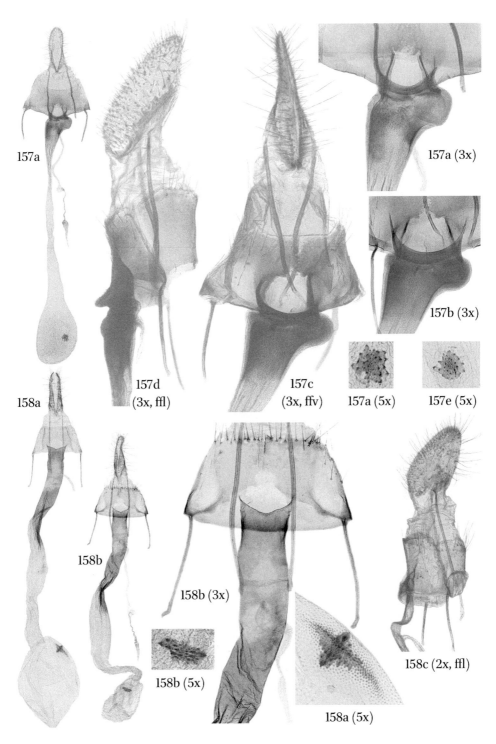

157a

157d
(3x, ffl)

157c
(3x, ffv)

157a (3x)

157b (3x)

157a (5x)

157e (5x)

158a

158b

158b (3x)

158b (5x)

158a (5x)

158c (2x, ffl)

157. *Depressaria manglisiella* – a) Serbia (8004, TLMF); b–e) Greece (6684, ZMUC [c: uncompressed])

158. *Depressaria chaerophylli* – a) Slovakia (1289, RCIR); b) Turkey (3956, ZfBS); c) Austria (0040, RCPB)

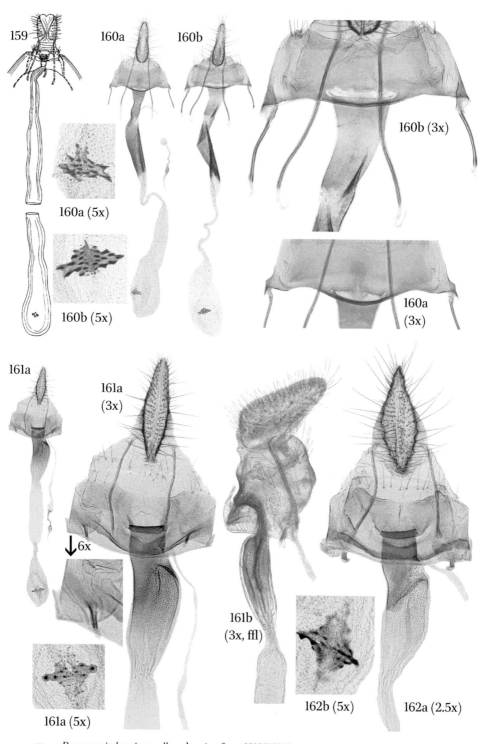

159. *Depressaria longipennella* – drawing from LVOVSKY, 2004
160. *Depressaria daucella* – a) Croatia (1137, RCIR); b) Austria (1477, RCPB)
161. *Depressaria ultimella* – a–b) Sicily (4494, RCTN)
162. *Depressaria halophilella* – a–b) France (0770, RCMC)

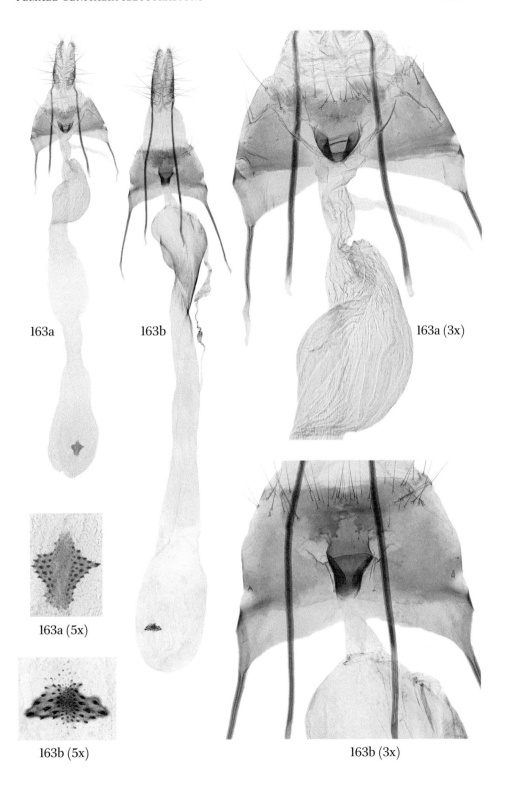

163a

163b

163a (3x)

163a (5x)

163b (5x)

163b (3x)

163. *Depressaria radiella* – a) Austria (0029, RCPB); b) Turkey (3212, RCKL)

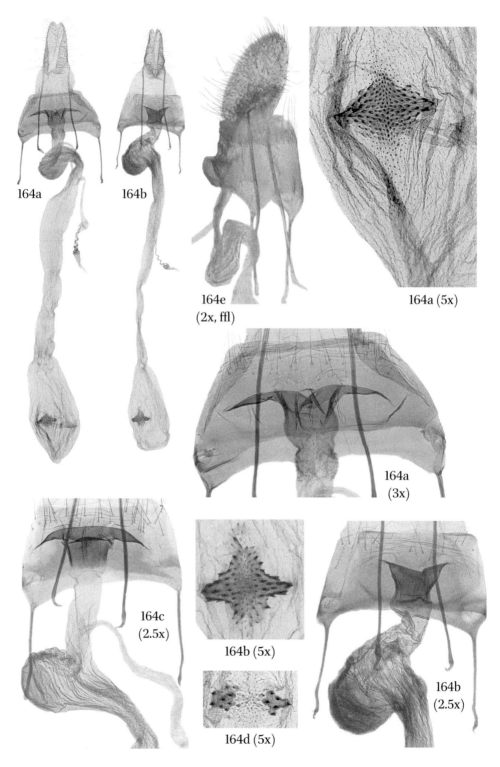

164a   164b

164e
(2x, ffl)

164a (5x)

164a
(3x)

164c
(2.5x)

164b (5x)

164d (5x)

164b
(2.5x)

164.   *Depressaria libanotidella* – a, e) Russia (7665, NMPC); b) Italy (6210, NMPC);
        c–d) Sweden (2503, ZMUC)

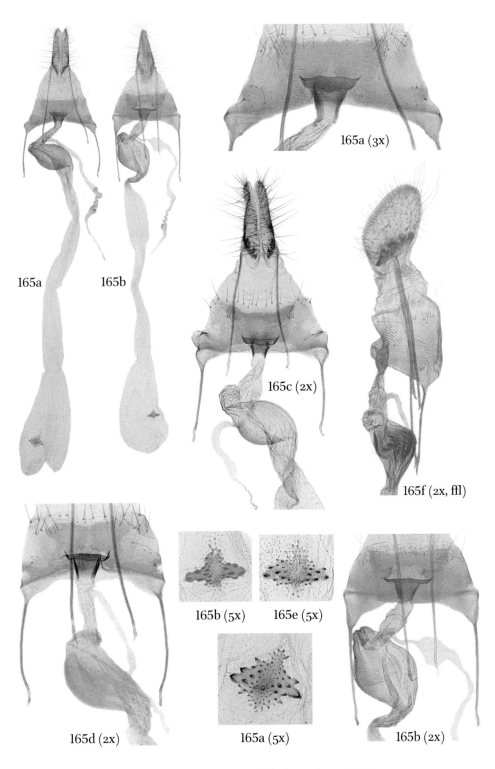

165a (3x)

165a 165b

165c (2x)

165f (2x, ffl)

165b (5x) 165e (5x)

165d (2x) 165a (5x) 165b (2x)

165. *Depressaria bantiella* – a) Greece (4807, RCWSc); b) Albania (8252, RCCP);
c) Montenegro (2137, RCFG); d–e) Cyprus (2416, ZMUC); f) Cyprus (7447, ZMUC)

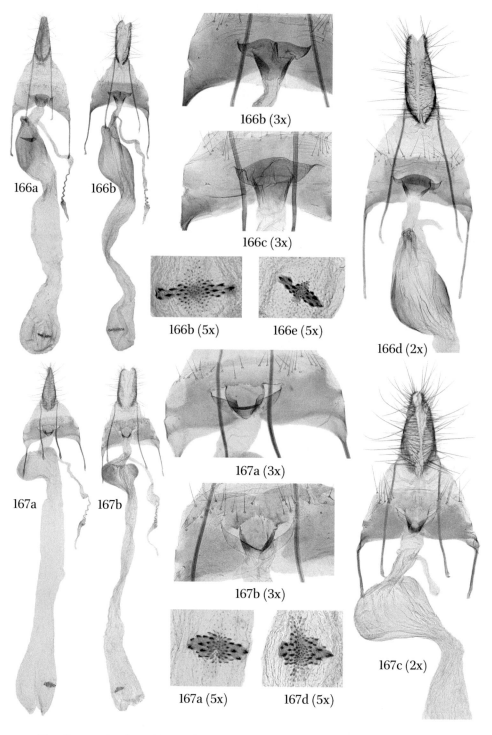

166b (3x)

166c (3x)

166b (5x)

166e (5x)

166a

166b

166d (2x)

167a (3x)

167b (3x)

167a (5x)

167d (5x)

167a

167b

167c (2x)

166. *Depressaria velox* – a) Greece (2413, ZMUC); b) Greece (4687, RCTN); c) Albania (1117, RCFG); d–e) Bulgaria (3507, RCBZ)

167. *Depressaria pimpinellae* – a) Slovakia (2840, RCKL); b) Russia (4624, RCTN); c–d) Austria (1478, RCPB)

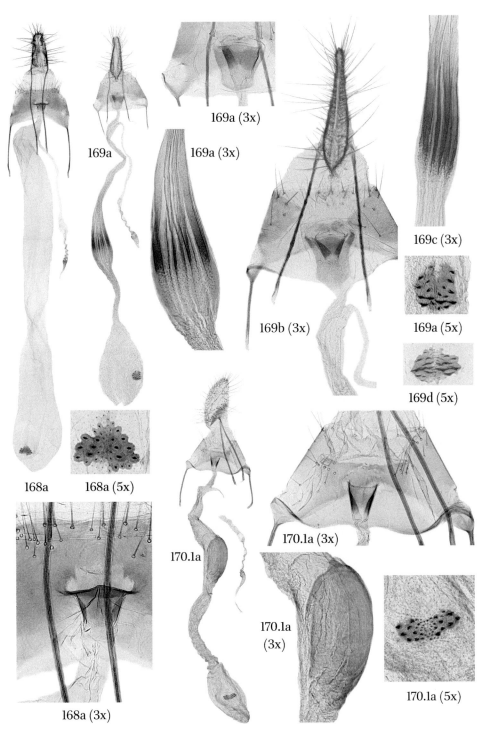

169a (3x)

169a

169a (3x)

169c (3x)

169b (3x)

169a (5x)

169d (5x)

168a

168a (5x)

170.1a (3x)

170.1a

170.1a
(3x)

170.1a (5x)

168a (3x)

168. *Depressaria villosae* – a) Sicily (2469, ZMUC)

169. *Depressaria bupleurella* – a) Italy (1646, TLMF); b–d) Austria (5325, RCWS)

170. *Depressaria sarahae* ssp. *sarahae*: females unknown

170.1. *Depressaria sarahae* ssp. *tabelli* – a) Canary Islands (2634, RCKL)

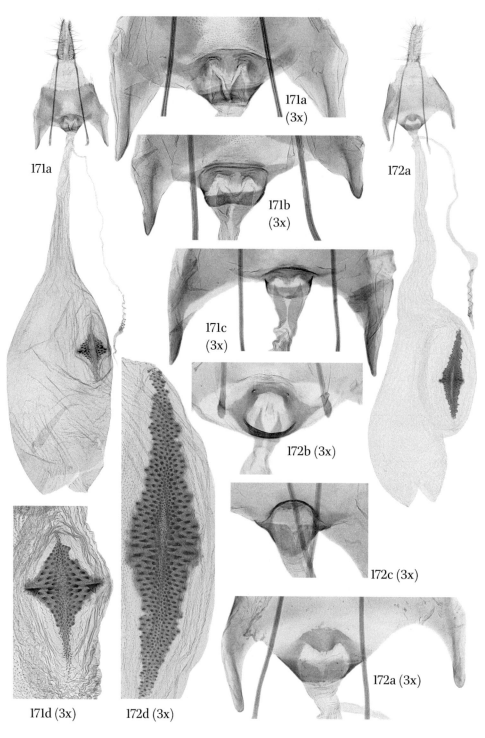

171a (3x)

171a

171b (3x)

172a

171c (3x)

172b (3x)

172c (3x)

172a (3x)

171d (3x)   172d (3x)

171. *Depressaria badiella* – a) Spain (6110, RCWSc); b) Romania (6316, MGAB); c–d) Italy (2439, ZMUC)

172. *Depressaria pseudobadiella* – a) Portugal (0752, RCMC); b) Turkey (3327, RCKL); c–d) France (3008, RCKL)

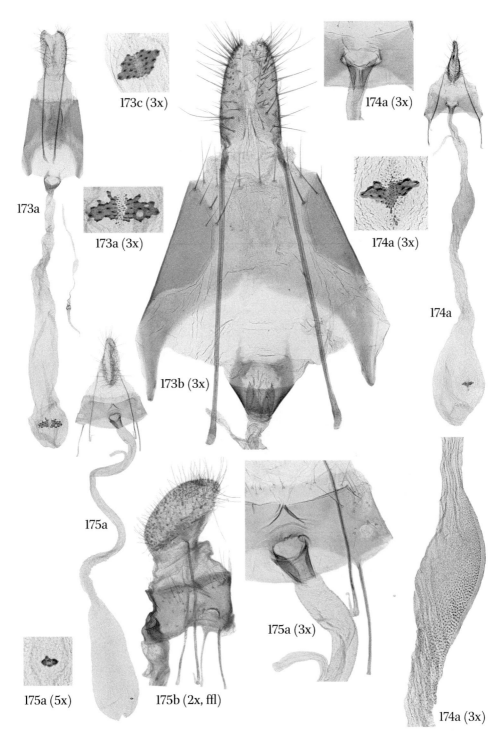

173c (3x)

174a (3x)

173a

173a (3x)

174a (3x)

174a

173b (3x)

175a

175a (3x)

174a (3x)

175a (5x)

175b (2x, ffl)

174a (3x)

173. *Depressaria subnervosa* – a) Morocco (2009, RCHB); b–c) Spain (2175, NHMW)
174. *Depressaria cervicella* – a) Hungary (1535, NHMW)
175. *Depressaria gallicella* – a–b) Switzerland (0647, NMBE)
176. *Depressaria altaica* – females unknown

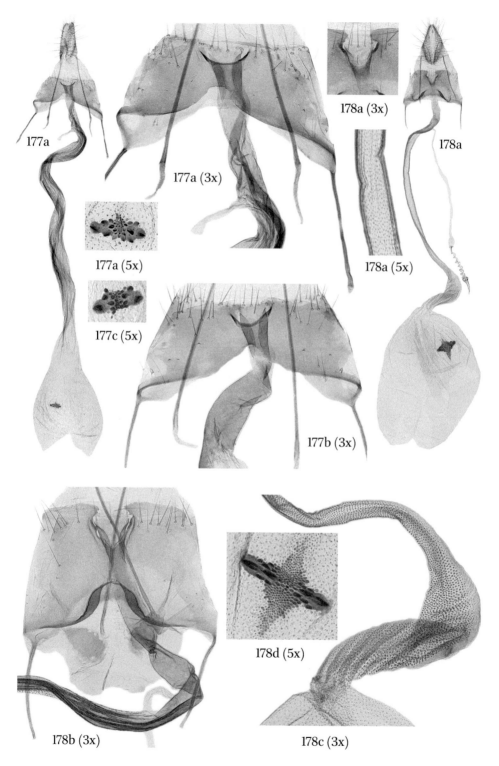

177a

177a (3x)

178a (3x)

178a

177a (5x)

178a (5x)

177c (5x)

177b (3x)

178d (5x)

178b (3x)

178c (3x)

177. *Depressaria albarracinella* – a) Spain (1786, TLMF); b–c) Spain (3904, RCWSc)
178. *Depressaria eryngiella* – a) Turkey (4471, RCJJ); b–d) Kyrgyzstan (5093, KLM)

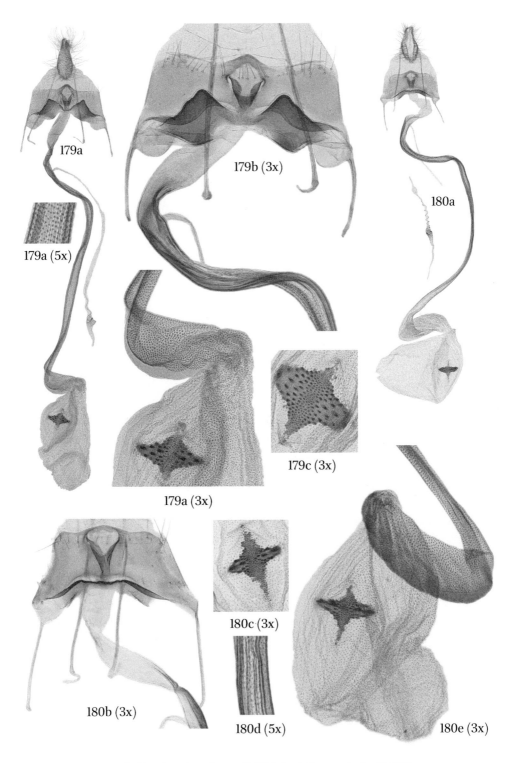

179a

179b (3x)

180a

179a (5x)

179c (3x)

179a (3x)

180c (3x)

180b (3x)

180d (5x)

180e (3x)

179.   *Depressaria veneficella* – a) Sardinia (5900, HNHM); b–c) Morocco (2360, ZMUC)
180.   *Depressaria discipunctella* – a) Greece (3086, RCKL); b–c) Turkey (2089, ZSM); d–e) Iran (5140, NHMW)

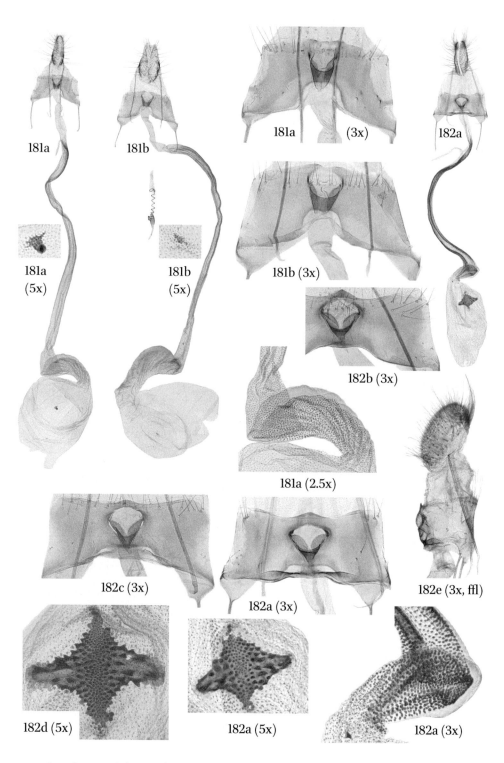

181a

181b

181a (3x)

182a

181a (5x)

181b (5x)

181b (3x)

182b (3x)

181a (2.5x)

182c (3x)

182a (3x)

182e (3x, ffl)

182d (5x)

182a (5x)

182a (3x)

181. *Depressaria hansjoachimi* sp.n. – a) Greece (1891, TLMF); b) Turkey (5404, ZMUC)

182. *Depressaria junnilaineni* – a, e) Greece (1483, RCRK); b) Spain (4792, RCWSc); c–d) Greece (5395, ZMUC)

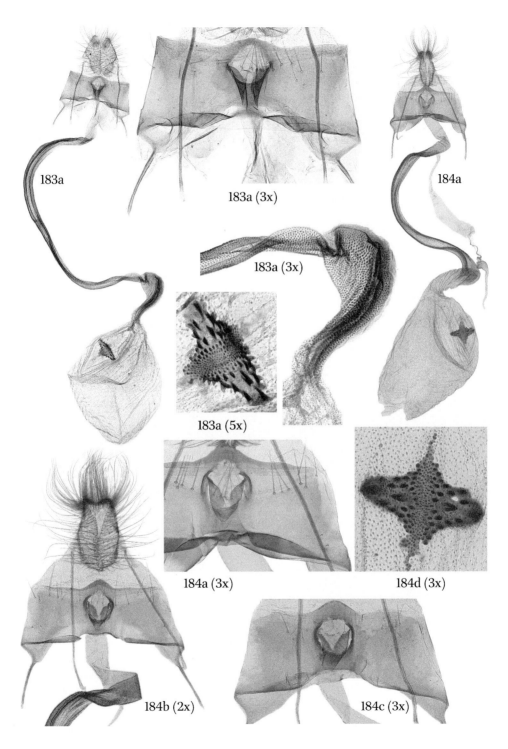

183a

183a (3x)

183a (3x)

183a (5x)

184a

184a (3x)

184d (3x)

184b (2x)

184c (3x)

183.   *Depressaria pentheri* – a) Bosnia and Herzegovina (3471, holotype, H.J. Hannemann slide 803,
       museum id MV3130, NHMW)

184.   *Depressaria hannemanniana* – a) Kazakhstan (4703, RCTN); b, d) Tajikistan (5690, ZMHB);
       c) Mongolia (5889, HNHM)

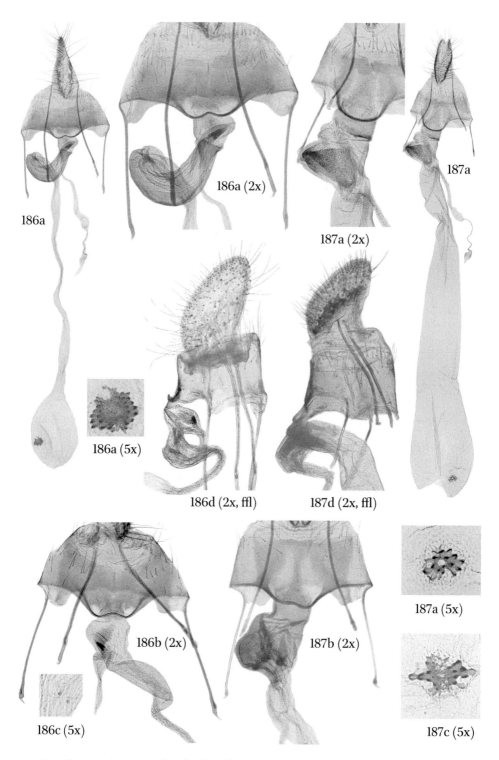

186a

186a (2x)

187a

187a (2x)

186a (5x)

186d (2x, ffl)   187d (2x, ffl)

186b (2x)   187b (2x)

187a (5x)

186c (5x)   187c (5x)

185. *Depressaria erzurumella* – females unknown
186. *Depressaria dictamnella* – a) Italy (8882, TLMF); b–d) Austria (1532, NHMW)
187. *Depressaria moranella* – a, d) Turkey (2085, ZSM); b–c) North Macedonia (6831, ZSM)

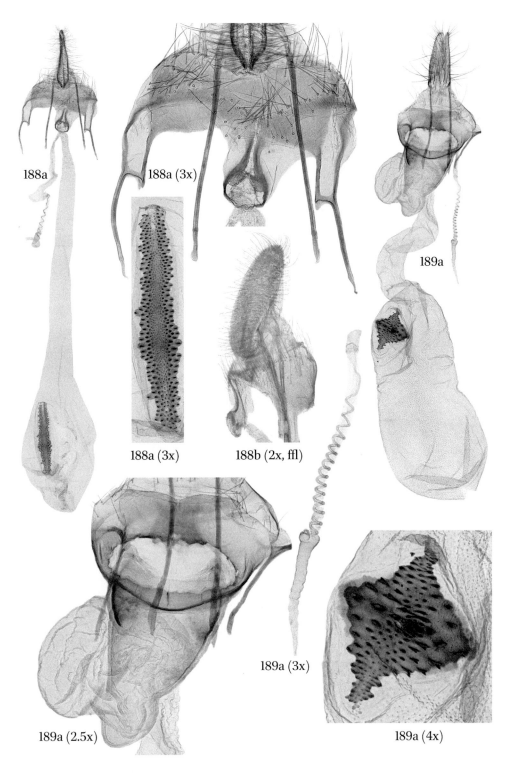

188a

188a (3x)

188a (3x)

188b (2x, ffl)

189a

189a (3x)

189a (2.5x)

189a (4x)

188. *Depressaria hystricella* – a–b) Russia (4470, RCTN)
189. *Depressaria hirtipalpis* – a) Croatia (0813, ZSM)

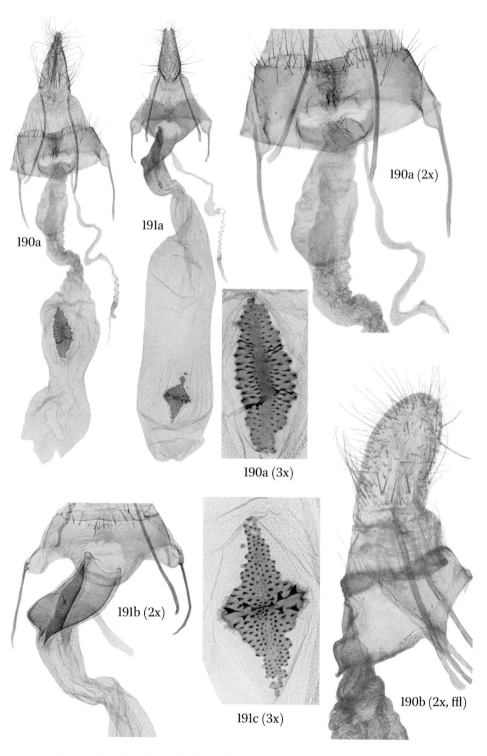

190a

191a

190a (2x)

190a (3x)

191b (2x)

191c (3x)

190b (2x, ffl)

190.   *Depressaria erinaceella* – a–b) Croatia (2073, ZSM)
191.   *Depressaria peniculatella* – a) Spain (4821, RCWSc); b–c) Algeria (1536, NHMW)
192.   *Depressaria rungsiella* – females unknown

References

Aarvik, L., Bengtsson, B.Å., Elven, H., Ivinskis, P., Jürivete, U., Karsholt, O., Mutanen, M. & Savenkov, N. 2017. Nordic-Baltic Checklist of Lepidoptera. *Norwegian Journal of Entomology, Supplement* **3**: 1–236.

Aarvik, L., Bengtsson, B.Å., Elven, H., Ivinskis, P., Jürivete, U., Karsholt, O., Mutanen, M. & Savenkov, N. 2021. Additions and corrections to the Nordic-Baltic Checklist of Lepidoptera. *Norwegian Journal of Entomology* **68**: 1–14.

Agenjo, R. 1954. Estudio de los tipos de las *Depressaria* (s.l.) de Chiclana, descritas por Staudinger en 1859, y de los de *Agonopterix subpallorella* (Stgr.), con algunos datos complementarios. *Eos* **30**: 59–75.

Aguiar, A.M.F. & Karsholt, O. 2006. Systematic catalogue of the Entomofauna of the Madeira Archipelago and Selvagens Islands. Lepidoptera. Vol 1. *Boletim do Museu Municipal do Funchal (História Natural)*. **Supl. No. 9**: 5–139.

Amsel, H.G. 1930. Die Microlepidopterenfauna der Mark Brandenburg nach dem heitigen Stande unserer Kenntnisse. *Deutsche entomologische Zeitschrift, Iris* **44**: 83–132.

Amsel, H.G. 1933. Die Lepidopteren Palästinas. *Zoogeographica* **2**: 1–146.

Amsel, H.G. 1935a. Neue palästinensische Lepidopteren. *Mitteilungen aus dem Zoologischen Museum in Berlin* **20**(2): 271–319, pls. 9–18.

Amsel, H.G. 1935b. Zur Kenntnis der Microlepidopterenfauna des südlichen Toten-Meer-Gebietes, nebst Beschreibungen neuer palästinensischer Macro- und Microlepidopteren. *Veröffentlichen aus dem Deutschen Kolonial- & Übersee-Museum, Bremen* **1**: 203–221.

Amsel, H.G. 1949. On the Microlepidoptera collected by E.P. Wiltshire in Irak and Iran in the years 1935 to 1938. *Bulletin de la Société Fouad 1er Entomologie* **33**: 217–351.

Amsel, H.G. 1958. Cyprische Kleinschmetterlinge (Aus den Landessammlungen für Naturkunde, Karlsruhe). *Zeitschrift der Wiener entomologischen Gesellschaft* **43**: 51–58, 69–75, 135.

Amsel, H.G. 1972. Depressarien aus Afghanistan, Iran, Irak und Arabien. Lepidoptera: Oecophoridae). *Beiträge für naturkundlichen Forschung in SüdwestDeutschland* **31**: 133–144.

Baez, M. 1998. *Mariposas de Canarias*. 216 pp. Madrid.

Balogh, I. 1951. A new Hungarian moth (Oecophoridae, Lep.). *Folia entomologica hungarica* (N.S.) **4**: 25–28.

Belshaw, R. 1993. Tachinid flies. Diptera: Tachinidae. *Handbooks for the Identification of British Insects* **10**(4a(i)): 1–169.

Benander, P. 1929. Zur Biologie einiger Kleinschmetterlinge III. *Entomologisk Tidskrift* **50**: 123–145.

Benander, P. 1955. *Agonopterix roseoflavella* n. sp. (Lep. Oecophoridae). *Opuscula Entomologica* **20**: 54–55.

Benander, P. 1964–1965. Notes on larvae of Swedish Microlepidoptera I–II. *Opuscula Entomologica* **29** (1964): 266–272; **30** (1965): 1–25.

Berthold, A.A. 1827. Natürliche Familien des Thierreichs, aus dem französischen, mit Anmerkungen und Zusätzen. *In* Latreille, P.A., *Natürliche Familien des Thierreichs*: 604 pp.

Bethune, C.J.S. 1870. Larva infesting the parsnip. *The Canadian Entomologist* 2: 1–4.

Biesenbaum, W., 2014. Familie: Depressariidae Meyrick, 1883. *Die Lepidopterenfauna der Rheinlande und Westfalens* 17: 1–193.

Billberg, G.J. 1820. *Enumeratio Insectorum*. 138 pp. Holmiae.

BOLD. 2007–2023. *The Barcode of Life Data System.* http://www.v3.boldsystems.org/index (Accessed December, 2023).

Bolt, D. & Schmid, J. 2024. Für die Schweiz neue Schmetterlingsnachweise von bis anhin zweifelhaften Schweizer Vorkommen (Lepidoptera). *Opuscula Lepidopterologica Alpina* 5: 1–16.

Bradley, J.D. 1966. Some changes in the nomenclature of British Lepidoptera. Part 4. *Entomologist's Gazette* 17: 213–235.

Brown, R. 1886. Trois lépidoptères nouveaux. *Actes de la Société linnéennes de Bordeaux* 40: lii–liii.

Bruand, T. 1851. Tineides. *In*: Catalogue systématique et synonymique des Microlépidoptères du department du Doubs. Tinéides. *Mémoires de la Société d'emulation du Doubs, Besançon* 3 (3): 1–102.

Buchner, P. 2015a. Two new species of *Agonopterix* (Depressariidae, Lepidoptera) from Europe. *Zootaxa* 3986 (1): 101–114.

Buchner, P. 2015b. Untersuchungen an „*Agonopterix thurneri*" und ihr überraschendes Ergebnis (Lepidoptera: Depressariidae: Depressariinae). *Quadrifina* 12: 1–15.

Buchner, P. 2017a. Redescription of *Agonopterix selini* (Heinemann, 1870) with description of *Agonopterix lessini* sp. n. and *Agonopterix paraselini* sp. n. (Lepidoptera, Gelechioidea). *Gortania* 38 (2016): 71–101.

Buchner, P. 2017b. *Depressaria junnilaineni*, a new species from the *veneficella* species-group (Depressariinae, Lepidoptera) from the West Palaearctic, with additional information on the rare species *D. pentheri* and *D. erzurumella*. *Centre for Entomological Studies, Miscellaneous Papers* 166: 1–19.

Buchner, P., 2017c. Faunistic records of Depressariidae (Lepidoptera, Gelechioidea) from Turkey – a result of studies for "Microlepidoptera of Europe: Depressariinae". *Cesa News* 134: 1–34.

Buchner, P. 2018. *Agonopterix xeranthemella*, a new species of Depressariidae (Lepidoptera) from Europe and Turkey. *Centre for Entomological Studies, Miscellaneous Papers* 177: 1–9.

Buchner, P. 2019. Vier neue Schmetterlingsarten für Österreich, darunter *Depressaria nemolella* Svensson, 1982, neu für Frankreich (Lepidoptera). *Beiträge zur Entomofaunistik* 20: 41–46.

Buchner, P. 2020a. *Agonopterix kayseriensis*, a new species of the *Agonopterix alpigena* group (Lepidoptera: Depressariidae) from Turkey and Romania. *Centre for Entomological Studies, Miscellaneous Papers* 213: 1–18.

Buchner, P. 2020b. Three new species of Depressariinae (Lepidoptera) from Europe and Western Asia and establishment of three new synonyms. *Centre for Entomological Studies, Miscellaneous Papers* 217: 1–32.

Buchner, P. 2022. *Agonopterix guanchella* Buchner, sp. n., a new species of Depressariidae from Canary Islands (Spain) (Lepidoptera: Depressariidae). *SHILAP Revista de lepidopterología* 50(199): 395–404.

Buchner, P., Corley, M. & Junnilainen, J. 2017. Three new species and a new subspecies of Depressariinae (Lepidoptera) from Europe. *ZooKeys* 684: 119–154.

Buchner, P. & Corley, M. 2019. *Agonopterix olusatri*, a new species of Depressariidae (Lepidoptera) from the West Palaearctic region. *Centre for Entomological Studies, Miscellaneous Papers* 196: 1–13.

Buchner, P., Junnilainen, J. & Nupponen, K. 2019. *Agonopterix sideensis* from Turkey and *Exaeretia lvovskyi* from Russia, two new species of Depressariidae (Lepidoptera) from the Palaearctic region, and the transfer of *Exaeretia montuosella* (Hannemann, 1976) into the genus *Agonopterix* Hübner, [1825]. *Centre for Entomological Studies, Miscellaneous Papers* 184: 1–25.

Buchner, P. & Karsholt, O. 2019. Depressariinae of Madeira and the Azores Islands (Lepidoptera: Depressariidae). *Beiträge zur Entomologie* 69: 331–353.

Buchner, P. & Šumpich, J. 2018. Faunistic records of *Agonopterix* and *Depressaria* from continental Spain, and updated checklist (Lepidoptera: Depressariidae). *SHILAP Revista de lepidopterología* 46(184): 681–694.

Buchner, P. & Šumpich, J. 2020. Depressariidae (Lepidoptera) of the Russian Altai Mountains: new species, new records and updated checklist. *Acta Entomologica Musei Nationalis Pragae* 60: 201–244.

Buhl, O., Falck, P., Karsholt, O., Larsen, K. & Schnack, K. 1989. Fund af småsommerfugle fra Danmark i 1987 (Lepidoptera). *Entomologiske Meddelelser* 57: 123–135.

Busck, A. 1922. Microlepidoptera from British Columbia. *The Canadian Entomologist* 53: 276–280.

Căpuşe, I. & Kovács, A. 1987. *Catalogue de la collection de lépidopteres "László Diószeghy" du Musée Départamental Covasna, Sfîntu Gheorghe.* Institut de Spéologie "Émile Racovitza", Bucharest. 397 pp.

Caradja, A. 1920. Beitrag zur Kenntnis der geographischen Verbreitung der Mikrolepidopteren des palaearktischen Faunengebietes nebst Beschreibung neuer Formen. 3.Theil. *Deutsche entomologische Zeitschrift, Iris* 34: 75–179.

Carvalho, J. Passos de & Corley, M.F.V. 1995. Additions to the Lepidoptera of Algarve, Portugal. *SHILAP Revista de lepidopterología* 23(91): 191–230.

Celli, G. 1970. Studies on a lepidopterous pest, *Depressaria marcella* Rebel (Lepidoptera: Oecophoridae), injurious to the seed carrot crops (*Daucus carota* L.) and research for a rational control. *Bollettino dell'Istituto di Entomologia della Università di Bologna* (1969) 29:1–44.

Chrétien, P. 1896. Description de Microlépidoptères nouveaux de France et d'Algerie. *Le Naturaliste* (2) 10: 104–105.

Chrétien, P. 1907a. Les chenilles des Buplèvres. *Le Naturaliste* 29: 89–91.

Chrétien, P. 1907b. Description de deux espèces nouvelles de *Depressaria* d'Algérie. *Bulletin de la Société Entomologique de France* 1907: 276–279.

Chrétien, P. 1908a. Microlépidoptères nouveaux pour la Faune française. *Le Naturaliste* 30: 59–60, 126–128, 186–188, 245–246.

Chrétien, P. 1908b. Nouveaux Microlépidoptères de France et de Corse. *Le Naturaliste* 30: 258–261.

Chrétien, P. *In* Spuler, A. 1910. *Die Schmetterlinge Europas, Kleinschmetterlinge*. 2. 523 pp. Stuttgart.

Chrétien, P. 1914. Description d'une espèce nouvelle de *Depressaria*. *Bulletin de la Société Entomologique de France* 1914: 159.

Chrétien, P. 1915. Contribution á la connaissance des Lépidoptères du Nord de l'Afrique. *Annales de la Société Entomologique de France* 84(3): 289–374.

Chrétien, P. *In* Oberthür, C. 1922. Les Lépidoptères du Maroc. *Études de Lépidoptérologie Comparées* 19: 13–402.

Chrétien, P. 1925. La légende de *Graellsia isabellae*. Appendice. *L'Amateur de Papillons* 2: 241–247, 257–263.

Chrétien, P. 1929. *In* Lhomme, L., Les *Depressaria*. *L'Amateur de Papillons* 4: 193–200.

Christoph, H. 1872. Neue Lepidoptera des europaeischen Faunengebietes. *Horae Societatis Entomologicae Rossicae* 9(1): 9–39.

Christoph, H. 1877. Sammelergebnisse aus Nordpersien, Krasnowodsk in Turkmenien und dem Daghestan. *Horae Societatis Entomologicae Rossicae* 12(3): 197–299.

Christoph, H. 1882. Neue Lepidoptera des Amurgebietes. *Bulletin de la Société Impériale des Naturalistes de Moscou* 57: 5–47.

Clarke, J.F.G. 1941. Revision of the North American moths of the family Oecophoridae, with descriptions of new genera and species. *Proceedings of the United States National Museum* 90: 33–286, i–viii, pls 1–48.

Clarke, J.F.G. 1947. Notes on Oecophoridae. *Journal of the Washington Academy of Sciences* 37: 1–18.

Clerck, C. 1759. *Icones Insectorum rariorum, cum nominibus eorum trivialibus, locisque e C.Linnaei*. [xii] + [iii] pp., 55 pls. Holmiae.

Constant, A. 1884. Notes sur quelques Lépidoptères nouveaux. *Annales de la Société entomologique de France* 4: 201–216.

Constant, A. 1888. Descriptions de Lépidoptères nouveaux ou peu connus. *Annales de la Société entomologique de France* 6: 161–172.

Constant, A. 1893–1894. Descriptions d'espèces nouvelles de Microlépidoptères. *Annales de la Société entomologique de France* 62(3): 392–400, (4): 401–404. Pl. 11.

Corley, M.F.V. 2002. Two new species of Depressariidae (Lepidoptera) from Portugal. *Nota lepidopterologica* 24: 25–33.

Corley, M.F.V. 2005. Further additions to the Lepidoptera of Algarve, Portugal. II. (Insecta: Lepidoptera). *SHILAP Revista de lepidopterología* 33(131): 347–364.

Corley, M.F.V. 2015. *Lepidoptera of Continental Portugal. A fully revised list*. 288 pp. Martin Corley, Faringdon.

Corley, M.F.V., Marabuto, E., Maravalhas, E., Pires, P. & Cardoso, J.P. 2009. New and interesting Portuguese Lepidoptera records from 2008 (Insecta: Lepidoptera). *SHILAP Revista de lepidopterología* 37(148): 463–484.

Corley, M.F.V., Marabuto, E., Maravalhas, E., Pires, P. & Cardoso, J.P. 2011. New and interesting Portuguese Lepidoptera records from 2009 (Insecta: Lepidoptera). *SHILAP Revista de lepidopterología* **39**(153): 15–35.

Corley, M.F.V., Nunes, J., Rosete J., Terry, R. & Ferrerira, S. 2020. New and interesting Portuguese Lepidoptera records from 2019 (Insecta: Lepidoptera). *SHILAP Revista de lepidopterología* **48**(192): 609–641.

Corley, M. & Buchner, P. 2018. *Depressaria villosae* sp. nov., a new species from Portugal, Spain and Greece (Depressariidae). *Entomologist's Record and Journal of Variation* **130**: 105–111.

Corley, M.F.V., Buchner, P. & Ferreira, S. 2019. *Depressaria infernella* Corley & Buchner, a new Iberian species of the *Depressaria douglasella* group (Lepidoptera, Depressariidae). *SHILAP Revista de lepidopterología* **47**(186): 293–300.

Corley, M.F.V., Buchner, P., Rymarczyk, F. & Dutheil, M. 2020. *Agonopterix rigidella* (Chrétien, 1907) resurrected from synonymy (Lepidoptera, Depressariidae). *Entomologist's Record and Journal of Variation* **132**: 281–296.

Curtis, J. 1828. British Entomology, being illustrations and descriptions of the genera of Insecta found in Great Britain and Ireland: containing coloured figures from nature of the most rare and beautiful species and in many instances of the plants upon which they are found. *British Entomology* **5**: pl. 195–241.

Curtis, J. 1850. Notes upon the smaller British Moths, with descriptions of some nondescript or imperfectly characterized species. *Annals and Magazine of Natural History* (2) **5**: 110–121.

Denis, J.N.C.M. & Schiffermüller, I. 1775. *Ankündung eines systematischen Werkes von den Schmetterlingen der Wienergegend.* 323 pp., 3 pls. Wien.

De Prins, W. & Steeman, C. 2003–2023. *Catalogue of the Lepidoptera of Belgium.* Belgian Biodiversity Platform. http://www.projects.biodiversity.be/Lepidoptera (Accessed December, 2023).

Derra, G. 1989. Bemerkenswerte Kleinschmetterlinge (Microlepidoptera). Entomofauna. *Zeitschrift für Entomologie* **10**(30): 465–471.

deWaard, J.R., Ivanova, N.V., Hajibabaei, M. & Hebert, P.D.N. 2008. Assembling DNA Barcodes: Analytical Protocols. *In*: Cristofre, M. (Ed.) *Methods in Molecular Biology: Environmental Genetics.* Humana Press Inc., Ottowa, 275–293.

Donovan, E. 1806. *The Natural History of British Insects.* **11**: 100 pp. London.

Douglas, J.W. 1846. Descriptions of ten new British moths. *Zoologist* **4**: 1266–1270.

Duponchel, P.A.J. 1838. Nocturnes, 8. *In* Godart, J.-B. *Histoire naturelle des Lépidoptères ou Papillons de France* **11**: 1–720, pls 287–314.

Erschoff, N. 1874. Travels in Turkestan. 2. Zoogeographical Investigations. Lepidoptera. *In* Fedschenko, A.P., *Travels in Turkestan* **2**(5): 1–128. Pls 1–6.

Erschoff, N. 1877. Diagnosen neuer Lepidopteren aus den verschiedenen Provinzen des Russischen Reiches. *Horae Societatis Entomologicae Rossicae* **12**: 336–348.

Euro+Med 2006 + [continuously updated]: Euro+Med PlantBase – the information resource for Euro-Mediterranean plant diversity. – Published at http://www.europlusmed.org [accessed November 2023].

Eversmann, E.F. von. 1844. *Fauna Lepidopterologica Volgo-Uralensis exhibens Lepidopterorum species, quas per viginti quinque annos in provinciis Volgam fluvium inter et montes Uralenses sitis observavit et descripsit.* xiv + 633 pp. Casani.

Fabricius, J.C. 1775. *Systema Entomologiae, sistens Insectorum Classes, Ordines, Genera, Species, Adiectis Synonymis, Locis, Descriptionibus, Observationibus.* 1–832.

Fabricius, J.C. 1777. *Genera insectorum eorumque characteres naturales secundum numerum, figuram, situm et proportionem omnium partum oris adiecta mantissa specierum nuper detectarum.* Chilonii, Carolus Ernesto Bohnii. viii + 1–310.

Fabricius, J.C. 1781. *Species Insectorum Exhibentes Eorum Differentia Specifica, Synonymia Auctorum, Loca Natalia, Metamorphosin Adiectis, Observationibus, Decriptionibus.* 2: 1–517.

Fabricius, J.C. 1787. *Mantissa Insectorum sistens species nuper detectas adiectis synonymis, observationibus, descriptionibus, emendationibus. Mantissa Insectorum* 2: 1–382.

Fabricius, J.C. 1794. *Entomologica systematica emendata et aucta. Secundum classes, ordines, genera, species adiectis, synonimis, locis, observationibus. descriptionibus.* 3(2): 349 pp. Hafniae.

Fabricius, J.C. 1798. *Supplementum Entomologiae Systematicae.* [4] + 572 pp. Hafniae.

Fazekas, I. & Schreurs, A. 2013. *Depressaria ululana* Rössler, 1866, new species in Hungary (Lepidoptera: Elachistidae). *Microlepidoptera hungarica* 6: 3–6.

Fetz, R. 1994. Larvalmorphologische Beiträge zum phylogenetischen System der ehemaligen Oecophoridae (Lepidoptera, Gelechioidea). *Neue entomologische Nachrichten* 33: 3–273.

Fletcher, T.B. 1929. *Memoirs of the Department of Agriculture in India* (Entomological series) 11: 77.

Frey, H. 1868. Die schweizerischen Microlepidopteren. *Mitteilungen der Schweizerischen Entomologischen Gesellschaft* 2: 376–380.

Frey, H. 1870. Ein Beitrag zur Kenntnis der Microlepidopteren. *Mitteilungen der Schweizerischen Entomologischen Gesellschaft* 3: 244–256, 277–296.

Frey, H. 1880. *Die Lepidopteren der Schweiz.* 1–454.

Freyer, C.F. 1835. *Neuere Beiträge zur Schmetterlingskunde mit Abbildungen nach der Natur* 2: 83–114.

Fuchs, A. 1903. Alte und neue Kleinfaltern der Europäischen Fauna. *Entomologische Zeitung, Stettin* 64: 227–247.

Fujisawa, K. 1985. On eight species of the genus *Agonopterix* Hübner (Lepidoptera: Oecophoridae) from Japan, with description of six new species. *Tinea* 12: 33–40.

Gastón, J. & Vives Moreno, A. 2017. Descripción de una especie nueva del género *Depressaria* Haworth, 1811 (Lepidoptera: Depressariidae). *Arquivos Entomolóxicos* 17: 351–354.

Geoffroy, E.L. 1785. *In* Fourcroy, A.F. de, *Entomologia Parisiensis, sive catalogus Insectorum quae in agro parisiensi reperiuntus.* Paris 2: 1–544.

Georgesco, M. 1965. Contribution à l'étude des microlépidoptères (Lep. Oecophoridae – Gen. *Agonopteryx*) des grottes de Roumanie. *Revue roumaine de Biologie (Série de Zoologie)* 10: 111–115.

Glitz, C.T. 1863. Erster Nachtrag zu dem Verzeichnisse der bei Hannover vorkommenden Schmetterlinge. *Jahresbericht der Naturhistorisches Gesellschaft zu Hannover* **14**: 39–41.

Goeze, J.A.E. 1783. *Entomologische Beyträge zu des Ritter Linné zwölften Ausgabe des Natursystems* 3(4). Leipzig, Weidmanns Erben & Reich. xx + 178 pp.

Grange, J.-C., Grange, D. & Nel, J. 2011. *Depressaria zelleri* Staudinger, 1879, espèce nouvelle pour France (Lep. Depressariidae) *Oreina* **16**: 10.

Gravenhorst, J.L.C. 1849. 2. Bericht über die Arbeiten der entomologische Sektion im Jahre 1849. *Jahresbericht der Schlesischen Gesellschaft für Vaterlandische Kultur. Naturwissenschaftlich-Medizinische Reihe. Breslau* **1849**: 65–73.

Hannemann, H.J. 1953. Natürlichen Gruppierung der europäischen Arten der Gattung *Depressaria* s.l. (Lep. Oecoph.). *Mitteilungen aus dem zoologischen Museum in Berlin* **29**: 269–373.

Hannemann, H.J. 1954. Anhang zur Natürlichen Gruppierung der europäischen Arten der Gattung *Depressaria* s.l. (Lep. Oecoph.). *Mitteilungen aus dem zoologischen Museum in Berlin* **30**: 35–37.

Hannemann, H.J. 1957. Über die weiblichen Genitalapparate der Gattung *Martyrhilda* Clarke, 1941 (Lep. Oecoph.). *Deutsche entomologische Zeitschrift, Neue Folge* **4**: 103–111.

Hannemann, H.J. 1958a. Die Gruppierung weiterer Depressarien nach dem Bau ihrer Kopulationsorgane, Teil 1 (Lep. Oecophoridae). *Mitteilungen aus dem zoologischen Museum in Berlin* **34**: 3–47.

Hannemann, H.J. 1958b. Die Eingruppung weiterer Depressarien nach dem Bau ihrer Kopulationsorgane, Teil 2 (Lep. Oecophoridae). *Deutsche entomologische Zeitschrift, Neue Folge* **5**: 456–465.

Hannemann, H.J. 1959. Neue Depressarien aus der Sammlung S. Toll. *Deutsche entomologische Zeitschrift, Neue Folge* **6**: 34–43.

Hannemann, H.J. 1967. Die Microlepidopteren der Brandtschen Iran-Ausbeute. Teil Depressariini. *Entomologisk Tidskrift* **88**: 164–169.

Hannemann, H.J. 1971. Ergebnisse der zoologischen Forschungen von Dr. Z. Kaszab in der Mongolei. Lepidoptera, Oecophoridae, Depressariini. *Acta Zoologica Academiae Scientarum Hungaricae* **17**: 261–270.

Hannemann, H.J. 1976a. Depressarien-Studien. *Deutsche entomologische Zeitschrift, Neue Folge* **23**: 233–252.

Hannemann, H.J. 1976b. Depressarien aus der Sammlung E. Turati. *Deutsche entomologische Zeitschrift, Neue Folge* **23**: 207–211.

Hannemann, H.J. 1982. Beitrag zur Kenntnis von *Depressaria silesiaca* Heinemann, 1870. *Deutsche entomologische Zeitschrift, Neue Folge* **29**: 483–487.

Hannemann, H.J. 1983. Neue Synonyme bei den Depressarien. *Deutsche entomologische Zeitschrift, Neue Folge* **30**: 373–376.

Hannemann, H.J. 1989. Studien an Depressarien (Lep., Oecophoridae). *Deutsche entomologische Zeitschrift, Neue Folge* **36** (4–5): 389–399.

Hannemann, H.J. 1990. Neue Depressarien (Lep. Oecophoridae). *Deutsche entomologische Zeitschrift, Neue Folge* **37**: 137–144.

Hannemann, H.J. 1995. Kleinschmertterlinge oder Microlepidoptera IV. Flachleibmotten (Depressariidae). *Die Tierwelt Deutschlands* **69**: 7–192.

Hannemann, H.-J. 1996. Depressariidae *In* Karsholt, O. & Razowski, J. (Eds.) *The Lepidoptera of Europe. A distributional checklist.* Apollo Books, Stenstrup. 380 pp.

Harper, M.W., Langmaid, J.R. & Emmet, A.M. 2002. Oecophoridae. *In* Emmet, A.M. & Langmaid, J.R. (Eds) *The Moths and Butterflies of Great Britain and Ireland* **4**(1). Harley Books, Colchester. 326 pp.

Haworth, A.H. 1811. *Lepidoptera Britannica; sistens digestionem novam insectorum lepidopterorum quae in Magna Britannia reperiuntur, larvarum pabulo, temporeque pascendi; expansione alarum; mensibusque volandi; synonymis atque locis observationibusque variis.* Part III (1811): 377–512. Londini.

Heckford, R.J. 1983. *Apium nodiflorum*: a previously unrecognized foodplant of *Depressaria ultimella* Stainton. *Entomologist's Record and Journal of Variation* **95**: 229–231.

Heckford, R.J. 2004. A note on the history and larvae of *Levipalpus hepatariella* (Lienig & Zeller, 1846) (Lepidoptera: Oecophoridae) and *Dichomeris juniperella* (Linnaeus, 1761) (Lepidoptera: Gelechiidae) in the British Isles. *Entomologist's Gazette* **55**: 1–13.

Heikkilä, M., Mutanen, M., Kekkonen, M. & Kaila, L. 2014. Morphology reinforces proposed molecular phylogenetic affinities: a revised classification for Gelechioidea (Lepidoptera). *Cladistics* **30**: 563–589.

Heinemann, H. von. 1870. *Die Schmetterlinge Deutschlands und der Schweiz. Kleinschmetterlinge* **2**(1). 388 pp. Braunschweig.

Hering, E.M. 1924. Beitrag zur Kenntnis der Microlepidopteren-Fauna Finlands. *Notulae Entomologicae* **4**: 75–84.

Hering, M. 1936. Blattminen von Spanien. *Eos* **11**: 331–384.

Herrich-Schäffer, G.A.W. 1853 ["1853–1855"] b. *Systematische Bearbeitung der Schmetterlinge von Europa, zugleich als Text, Revision und Supplement zu Jakob Hübner's Sammlung europäischer Schmetterlinge.* **5**. Die Schaben und Federmotten. pp. [1] 2–394 + [1] 2–52, pls. Regensburg.

Herrich-Schäffer, G.A.W. 1865. Ein Ausflug ins Ober-Engadin. *Correspondenz-blatt des Zoologisch-mineralischen Vereines in Regensburg* **19**(8): 109–113, 115–117.

Hodges, R.W. *In* Dominick, R.B., Ferguson, D.C., Franclemont, J.G., Hodges, R.W. & Munroe, E.G. 1974. *The Moths of America North of Mexico* Fasc. **6**(2). *Gelechioidea: Oecophoridae* (in part). i–x, 1–142, pls 1–7.

Hodges, R.W., Dominick, T., Davis, D.R., Ferguson, D.C., Franclemont, J.G., Munroe, E.G. & Powell, J.A. 1983. *Check List of the Lepidoptera of America North of Mexico.* E.W. Classey, London. i–xxiv, 1–284.

Hübner, J. 1793. *Sammlung auserlesener Vögel und Schmetterlinge, mit ihren Namen herausgegeben auf hundert nach der Natur ausgemalten Kupfern.* pp. [1]–[5] 6–16, 100 pls. Augsburg.

Hübner, J. 1796–1836 ["1796"] b. *Sammlung europäischer Schmetterlinge.* **8**. Tineae–Schaben. 1–78 pp. (1796), 71 pls 1–71 (1796–[1836]). Augsburg.

Hübner, J. 1816–1826. ["1825"]. *Verzeichniß bekannter Schmettlinge* [sic]. (1): [1–3], 4–16 (1816); (2): 17–32 (1819); (3): 33–48 (1819); (4): 49–64 (1819); (5): 65–80 (1819); (6): 81–96 (1819); (7): 97–112 (1819); (8): 113–128 (1819); (9): 129–144 (1819): (10): 145–160 (1819); (11): 161–176 (1819); (12): 177–192 (1820); (13): 193–208 (1820); (14): 209–224 (1821); (15): 225–240 (1821); (16): 241–256 (1821); (17): 257–272 (1823); (18): 273–288 (1823); (19): 289–304 (1823); (20): 305–320 (1825); (21): 321–336 (1825); (22): 337–352 (1825); (23–27): 353–431 ([1825]). Augsburg.

Huemer, P. & Lvovsky, A. 2000. *Agonopterix cluniana* sp.n., a surprising discovery from the northern Alps (Lepidoptera: Depressariidae). *Nachrichten des entomologischen Vereins Apollo, N.F.* **21**: 135–142.

Huisman, K.J. 2012. The Micro moth genus *Agonopterix* in the Netherlands (Lepidoptera: Elachistidae: Depressariinae). *Nederlandse faunistische Mededelingen* **37**: 45–104.

Huisman, K.J. & Sauter, W. 2002. Redescription of the female and distribution of *Depressaria incognitella* Hannemann, 1990 (Depressariidae). *Nota lepidopterologica* **24**: 35–41.

Iglesias, C., Sinobas, J. & Varés, L. 2002. *Hasenfussia erinaceella* (Staudinger, 1870) (Lepidoptera, Depressaridae), un taladro del cardo. *Boletin de Sanidad Vegetal Plagas* **28**:103–106.

Jacobs, S.N.A. 1954. The British Oecophoridae. Part III. *Proceedings and Transactions of the South London entomological and natural History Society* **1954–1955**: 54–76.

Jakšić, P. 2016. Tentative check list of Serbian microlepidoptera. *Ecologica Montenegrina* **7**: 33–258.

Japan Moths. 2002–2023. http://www.jpmoth.org/Depressariidae (Accessed December, 2023).

Kaila, L., Mutanen, M. & Nyman, T. 2011. Phylogeny of the mega-diverse Gelechioidea (Lepidoptera): Adaptations and determinants of success. *Molecular Phylogenetics and Evolution* **61**: 801–809.

Karsholt, O., Lvovsky, A.L. & Nielsen, C. 2006. A new species of *Agonopterix* feeding on giant hogweed (*Heracleum mantegazzianum*) in the Caucasus with a discussion of the nomenclature of *A. heracliana* (Linnaeus) (Depressariidae). *Nota lepidopterologica* **28**: 177–192.

Karsholt, O. & Razowski, J. 1996. *The Lepidoptera of Europe. A Distributional Checklist.* 380 pp. Apollo Books, Stenstrup.

Klemensiewicz, S. 1898. O nowych i malo znanych gatunkach motyli fauny Galicyjskiej. *Sprawozdanie Komisyi Fizyograficznej, Kraków* **33**: 113–190.

Klimesch, J. 1953. Die Raupe von *Depressaria* (*Schistodepressaria*) *cervicella* H.-S. (Lep. Oecophoridae). *Zeitschrift der Wiener entomologischen Gesellschaft* **38**: 22–25.

Klimesch, J. 1985. Beiträge zur Kenntnis der Microlepidopteren-Fauna des Kanarischen Archipels. 7. Oecophoridae, Symmocidae, Holcopogonidae. *Vieraea* **14**: 131–151.

Koçak, A.Ö. & Kemal, M. 2009. Revised Checklist of the Lepidoptera of Turkey. *Priamus* Suppl. **17**: 1–253.

Kollar, V. 1832. Systematisches Verzeichniss der Schmetterlinge im Erzhergzogthume Oesterreich. *Beiträge zur Landeskunde Oesterreich's unter der Enns* **8**(2): 1–101.

Kovács, Z. & Kovács, S. 2020. Contributions to the knowledge of the Depressariidae, Peleopodidae, Ethmiidae and Fuchsiini (Lepidoptera, Gelechioidea) of Romania, with an annotated checklist. *Travaux du Muséum National d'Histoire Naturelle "Grigore Antipa"* **63**: 203–254.

Kristensen, N.P. 2003. Skeleton and muscles: adults. Pp. 39–131. – *In* N.P. Kristensen (ed.): *Lepidoptera, Moths and Butterflies. Volume 2: Morphology, Physiology and Development – Handbook of Zoology IV Arthropoda: Insecta.* Part 36. Berlin and New York.

Krulikovsky, L.K. 1903. Petites notices lépidoptérologiques VII. *Revue Russe d'Entomologie* **3**: 177–182 (in Russian).

Kumar, S., Stecher, G., Li, M., Knyaz, C. & Tamura, K. 2018. MEGA X: Molecular Evolutionary Genetics Analysis across computing platforms. *Molecular Biology and Evolution* **35**: 1547–1549.

Langmaid, J.R. & Pelham-Clinton, E.C. 1984. *Agonopterix kuznetzovi* Lvovsky (Lepidoptera, Oecophoridae), a species new to the British Isles. *Entomologist's Gazette* **35**: 67–72.

Laštůvka, Z. & Liška, J. 2011. *Komentovaný seznam motýlů České republiky. Annotated checklist of moths and butterflies of the Czech Republic (Insecta: Lepidoptera)*. Biocont Laboratory spol. s.r.o., Brno: 1–146.

Latreille, P.A. 1829. Suite et fin des Insectes. – *In* Cuvier, G. 1829. *Le Règne Animal distribué d'après son organisation, pour servir de base à l'Histoire Naturelle des animaux et d'introduction à l'anatomie comparée.* xxiv + 556 pp. Paris.

Lederer, J. 1855. Beitrag zur Schmetterlings-Fauna von Cypern, Beirut und einem Theile Klein-Asiens. *Verhandlungen des zoologisch-botanischen Vereins in Wien* **5**: 177–254.

Lepiforum e.V. 2006–2023. https://www.lepiforum.org/wiki/page (Accessed December, 2023).

Leraut, P. 1991. Contribution à l'étude des Oecophoridae (s. l.) 2. Deux nouveaux noms de genres. *Alexanor* **17**: 232.

Leraut, P. 2023. *Moths of Europe. 8. Microlepidoptera 2, Epipyropidae to Pterophoridae.* 655 pp. N.A.P. Editions, Verrières-le-Buisson.

Lhomme, L. 1929. Les *Depressaria*. *L'Amateur de Papillons* **4**: 193–200.

Lhomme, L. 1945. Oecophoridae, pp. 702–783. In Lhomme, L. 1935–[1963]. *Catalogue des lépidoptères de France et de Belgique* **2**: 1–1253. Le Carriol.

Lienig, F. & Zeller, P.C. 1846. Lepidopterologische Fauna von Lievland und Curland, Bearbeitet von Friederike Lienig, geb. Berg, mit Anmerkungen von P.C. Zeller. *Isis von Oken, Leipzig* **1846** (3–4): 175–302.

Linnaeus, C. 1758. *Systema naturae per regna tria naturae, secundum classes, ordines, genera, species, cum characteribus, differentiis, synonymis, locis.* **1** [edn. 10]: 824 pp. Holmiae.

Linnaeus, C. 1767. *Systema Naturae,* **1** (2) (Edn. 12). 1327 pp. Holmiae.

Liu, S. & Wang, S. 2010. One new species and three newly recorded species of the genus *Exaeretia* Stainton, 1849 (Lepidoptera: Elachistidae: Depressariinae) from China. *Zootaxa* **2444**: 45–50.

Lucas, D. 1940. Contribution a l'étude des Lépidoptères de l'Afrique du Nord. *Bulletin de la Société entomologique de France* **44**: 226–229.

Lucas, D. [1950] 1951. Contribution a l'étude des Lépidoptères Nord-Africains. *Bulletin de la Société entomologique de France* **55**: 141–144.

Lvovsky, A.L. 1981a. Oecophoridae. *In* Medvedev, G.S. *Key to insects in the European part of the USSR. 4: Lepidoptera* part 2: 560–638. Leningrad.

Lvovsky, A.L. 1981b. New species of the broad-winged moths of the genus *Depressaria* Hw. (Lepidoptera, Oecophoridae) of the fauna of the USSR. *Trudy Zoologicheskogo Instituta* **103**: 73–83.

Lvovsky, A.L. 1983. A new species of broad-winged moth from the genus *Agonopterix* Hbn. (Lepidoptera, Oecophoridae). *Entomologicheskoe Obozrenie* **62**(3): 594–595. Translated in: *Entomological Review, Washington* **62**: 136–137.

Lvovsky, A.L. 1990. New and little known species of the Microlepidoptera (Lepidoptera: Oecophoridae, Xyloryctidae, Tortricidae) of the fauna of the USSR and neighbouring countries. *Entomologicheskoe Obozrenie* **69**: 638–655.

Lvovsky, A.L. 1996. New and little known species of *Depressaria* Haworth, 1811 (Lepidoptera, Depressariidae). *Atalanta* **27**: 421–425.

Lvovsky, A.L. 1998a. New and little known species of flat moths (Lepidoptera, Depressariidae) from the fauna of Russia and neighbouring countries. *Entomological Review, Washington* **78**: 466–474.

Lvovsky, A.L. 1998b. On the little-known species *Depressaria caucasica* Christoph, 1877 (Lepidoptera: Depressariidae). *Zoosystematica Rossica* **7**: 311–312.

Lvovsky, A.L. 2001. A review of the flat moths of the genus *Depressaria* Haworth, 1811 (Lepidoptera, Depressariidae) of the fauna of Russia and neighbouring Countries: I. *Entomologicheskoe Obozrenie* **80**: 680–705.

Lvovsky, A.L. 2004. A review of the flat moths of the genus *Depressaria* Haworth, 1811 (Lepidoptera, Depressariidae) of the fauna of Russia and neighbouring Countries: II. *Entomologicheskoe Obozrenie* **83**: 190–213.

Lvovsky, A.L. 2006. Check-list of the broad-winged and flat moths (Lepidoptera: Oecophoridae, Chimabachidae, Amphisbatidae, Depressariidae) of the fauna of Russia and adjacent countries. *Proceedings of the Zoological Institute, St. Petersburg* **307**: 1–118. [In Russian].

Lvovsky, A.L. 2013a. A review of the flat moths of the genus *Exaeretia* Stainton, 1849 (Lepidoptera, Depressariidae) of the fauna of Russia and neighbouring countries. *Entomologicheskoe Obozrenie* **92**: 780–801.

Lvovsky, A.L. 2013b. *Agonopterix* (*Subagonopterix*) *vietnamella* subgen. nov. et spec. nov., of flat moths from South-Eastern Asia (Lepidoptera, Depressariidae). *Atalanta* **44**: 131–132.

Lvovsky, A.L. 2014. *Agonopterix comitella* (Lederer, 1855) (Lepidoptera: Depressariidae), a new species to the fauna of Russia. *Entomologicheskie i Parazitologicheskie Issledovaniya v Povolzhje* [2014] **11**: 135–137.[In Russian with English summary].

Lvovsky, A.L. 2018. New systematic and distribution data of flat-body moth genus *Agonopterix* Hübner, [1825] (Lepidoptera, Depressariidae) of the fauna of Russia. *Entomologicheskoe Obozrenie* **97**: 317–324. [In Russian].

Lvovsky, A.L. 2019. Depressariidae. *In*: S.Yu. Sinev (ed.). *Catalogue of the Lepidoptera of Russia*. Edition 2. St. Petersburg: Zoological Institute RAS: 53–57. [In Russian].

Lvovsky, A.L. & Anikin, V.V. 2009. "Exaeretia nebulosella (Lepidoptera, Depressariidae), a New Species to the Fauna of Russia," *Entomologicheskie i Parazitologicheskie Issledovaniya v Povolzhje* [2008] **7**: 39–41 (2009). [In Russian].

Lvovsky, A.L. & Jalava, J. 1993. *Depressaria sordidatella* Tengström is the valid name for *Depressaria weirella* Stainton (Lepidoptera, Oecophoridae). *Atalanta* 24: 299–300.

Lvovsky, A.L. & Knyazev, S.A. 2013. *Agonopterix rotundella* (Lepidoptera, Depressariidae) – a new species to the fauna of Russia. *Amurian zoological Journal* 5 (2): 151–152. [In Russian with English summary]. Available from: http://omflies.narod.ru/Publications /Lvovsky_Knyazev_2013.pdf.

Lvovsky, A.L. & Schernijasova, R.M. 1992. The Fauna of Lepidoptera, Oecophoridae of Tajikistan. [In Russian]. *Journal of the Academy of Sciences of the Republic of Tajikistan, Department of Biological Sciences* 2(126): 3–7.

Lvovsky, A.L., Sinev, S. Yu., Kravchenko, V.D., & Müller, G.C. 2016. A contribution to the Israeli fauna of Microlepidoptera: Oecophoridae, Autostichidae, Depressariidae, Cryptolechiidae and Lecithoceridae with ecological and zoogeographical remarks (Lepidoptera: Gelechioidea). *SHILAP Revista de lepidopterología* 44(173): 97–113.

Lvovsky, A.L. & Stanescu, M. 2019. Taxonomic notes on five species of the family Depressariidae (Lepidoptera: Gelechioidea), described by Aristide Caradja from the Russian Far East. *Zoosystematica rossica* 28: 251–257.

Mann, J. 1855. Die Lepidopteren gesammelt auf einer entomologischen Reise in Corsika im Jahre 1855. *Verhandlungen der kaiserlich-königlichen zoologisch-botanischen Gesellschaft in Wien* 5: 529–572.

Mann, J. 1861. Zur Lepidopteren-fauna von Amasia. *Wiener Entomologische Monatschrift* 5: 155–162, 183–193.

Mann, J. 1864. Nachtrag zur Schmetterling-Fauna von Brussa. *Wiener Entomologische Monatschrift* 8: 173–190.

Mann, J. 1869. Lepidopteren gesammelt während dreier Reisen nach Dalmatein in den Jahren 1850, 1862 und 1868. *Verhandlungen der kaiserlich-königlichen zoologisch-botanischen Gesellschaft in Wien* 19: 371–388.

McKenna, D. & Berenbaum, M. 2003. A field investigation of *Depressaria* (Elachistidae) host plants and ecology in the western United States. *Journal of the Lepidopterists' Society* 57 (1): 36–42.

Meert, R., in prep.: First observation of *Depressaria ultimella* (Lepidoptera, Depressariidae) in Spain.

Meyrick, E. 1906. Descriptions of Australian Tineina. *Transactions of the Royal Society of South Australia* 30: 33–66.

Meyrick, E. 1910. Descriptions of Indian micro-lepidoptera. The *Journal of the Bombay Natural History Society* 20: 143–168, 435–462.

Meyrick, E. 1913. *Exotic Microlepidoptera* 1(1–5): 1–160. Marlborough.

Meyrick, E. 1920. *Exotic Microlepidoptera* 2(10): 289–320. Marlborough.

Meyrick, E. 1921a. *Exotic Microlepidoptera* 2(13): 385–416. Marlborough.

Meyrick, E. 1921b. *Depressaria autocnista* sp. n. *Entomologist* 54: 76.

Meyrick, E. 1923a. *Exotic Microlepidoptera* 2(20): 609–640. Marlborough.

Meyrick, E. 1923b. Three new micro-lepidoptera from Cyprus. *Entomologist* 56: 277–278.

Meyrick, E. 1927. *A revised handbook of British Lepidoptera.* vi, 914 pp. London.

Meyrick, E. 1928. Oecophoridae. *Exotic Microlepidoptera* 3(15): 467–478. Marlborough.

Meyrick, E. 1936a. *Exotic Microlepidoptera* 4(20): 609–642. Marlborough.

Meyrick, E. 1936b. *Exotic Microlepidoptera* 5(1–3): 1–96. Marlborough.

Millière, P. 1866. *Iconographie et Description de Chenilles et Lépidoptères inédits* 2: 1–506.

Millière, P. 1881. *Lépidoptérologie* 7: 1–19, pl 8, f. 8–9.

Minet, J. 1985. Ébauche d'une classification moderne de l'ordre des Lépidoptères. *Alexanor* 14(7): 291–313.

Minet, J. 1991. Remaniement partiel de la classification des Gelechioidea, essentiellement en fonction de caractères pré-imaginaux (Lepidoptera Ditrysia). *Alexanor* 1989, 16(4): 239–255.

Morris, F.O. 1870. *A Natural History of British Moths* 4: 1–321. London.

Möschler, H.B. 1860. Vier neue südrussische Schmetterlinge. *Wiener Entomologische Monatschrift* 4: 273–276.

Müller-Rutz, J. 1922. Die Schmetterlinge der Schweiz. (4. Nachtrag, Kleinschmetterlinge). *Mitteilungen der Schweizerischen Entomologische Gesellschaft* 13: 217–259.

Mutanen, M., Hausmann, A., Hebert, P.D.N., Landry, J.-F., Waard, J.R. De & Huemer, P. 2012. Allopatry as a Gordian knot for taxonomists: Patterns of DNA barcode divergence in Arctic-Alpine Lepidoptera. *PLoS ONE* 7(10)(e47214): 1–9.

Nel, J. 2011. *Depressaria pseudobadiella* n. sp. décrite du sud de la France (Lep.Depressariidae). *Oreina* 16: 4–5.

Nel, J., Doux, Y., Taurand, L., Thibault, M. & Varenne, T. 2022. Quelques Lépidoptères peu cités ou nouveaux pour la faune de France (Lepidoptera, Meessiidae, Lyonetiidae, Depressariidae, Batrachedridae, Coleophoridae, Urodidae, Crambidae, Geometridae, Noctuidae). *Revue de l'Association Roussillonnaise d'Entomologie* 31(2): 136–142.

Nel, J. & Grange, J.-C. 2014. Description d'*Exaeretia buvati* sp. n. des Pyrénées, espèce voisine d'*E. lepidella* (Christoph, 1872) d'Asie Centrale (Lepidoptera, Elachistidae, Depressariinae). *Revue de l'Association Roussillonnaise d'Entomologie* 23: 52–55.

Nickerl, F.A. 1864. Neue Microlepidopteren. *Wiener Entomologische Monatschrift* 8 (1): 1–8.

Nowicki, M.S. 1860. *Enumeratio Lepidopterorum Haliciae orientalis.* 269 pp.

Nunes, J., Buchner, P. & Corley, M. 2024. *Agonopterix cachritis* (Staudinger, 1859) (Lepidoptera, Depressariidae) new to Portugal, rediscovery of a lost species. *Boletin de la SAE* 34: 000–000.

Oberthür, C. 1888. Lépidoptères d'Europe et d'Algerie. *Etudes d'Entomologie* 12: 21–44.

Palm, E. 1989. Nordeuropas Prydvinger. *Danmarks Dyreliv* 4: 1–247.

Palm, N.-B. 1943. Two new species of Swedish Tineina. *Opuscula Entomologica* 8: 25–28.

Pastorális, G. 2022. Zoznam motýľov (Lepidoptera) zistených na Slovensku. Checklist of Lepidoptera recorded in Slovakia. *Entomofauna carpathica*, **34** (Supplementum 2): 1–181.

Pastorális, G. & Buschman, F. 2018. A Magyarországon előforduló molylepkefajok névjegyzéke, 2018. A checklist of Hungarian micro-moths (Lepidoptera). *Microlepidoptera.hu* 5: 77–258.

Patočka, J. & Turčáni, M. 2005. *Lepidoptera pupae. Central European species.* 1: 1–542, 2: 1–321. Apollo Books, Stenstrup.

Pfaffenzeller, F. 1870. Neuer Tineinen. *Entomologische Zeitung, Stettin* 31: 320–324.

Pinzari, M. & Pinzari, M. 2013. Two interesting species of elachistid moth: *Depressaria eryngiella*, new to Italy, and *Depressaria halophylella* (Lepidoptera, Elachistidae). *Bollettino dell'Associazione Romana di Entomologia* 67 (2012): 69–74.

Predota, K. 1934. Neue Macro- und Microlepidopteren aus den Ostpyrenäen, Spanien und Algerien. *Zeitschrift des Verereins der Naturbeobachter und Samler, Wien* 9: 1–2.

Prinz, J. 1917. Versammlung am 3. November 1916. *Verhandlungen der kaiserlich-königlichen zoologisch-botanischen Gesellschaft in Wien* 67: 15–27.

Prosser, S.W.J., de Waard, J.R., Miller, S.E. & Hebert, P.D.N. 2016. DNA barcodes from century old specimens using next-generation sequencing. *Molecular Ecology Resources* 16 (2): 487–497.

Ragonot, E.L. 1874. *Annales de la Société entomologique de France* 4: 585.

Ragonot, E.L. 1889. *In* Laboulbène, A. [Prés.] 1889. Séance du 22 mai 1889. *Bulletin des séances et bulletin bibliographique de la Société entomologique de France* 1889: xcvii–cxii.

Ragonot, E.L. 1895. Microlépidoptères de la Haute-Syrie recoltés par M. Ch. Delagrande, et descriptions des espèces nouvelles. *Bulletin de la Société entomologique de France* 1895 (4): 94–109.

Ratnasingham, S. & Hebert, P.D.N. 2007. The Barcode of Life Data System. *Molecular Ecology Notes* 7: 355–364.

Rebel, H. 1889. Beiträge zur Microlepidopteren-Fauna Oesterreich-Ungarns. *Verhandlungen der kaiserlich-königlichen zoologisch-botanischen Gesellschaft in Wien* 39: 293–326.

Rebel, H. 1891. Beitrag zur Microlepidopteren-Fauna Dalmatiens. *Verhandlungen der kaiserlich-königlichen zoologisch-botanischen Gesellschaft in Wien* 41: 610–639.

Rebel, H. 1892. Beitrag zur Microlepidopterenfauna des canarischen Archipels. *Annalen des kaiserlich-königlichen naturhistorischen Hofmuseums* 7(3): 241–284.

Rebel, H. 1893. Neue oder wenig gekannte Microlepidoptera des palaearktischen Faunengebietes. *Entomologische Zeitung, Stettin* 54: 37–59.

Rebel, H. 1901. *Catalog der Lepidopteren des palaearctischen Faunengebietes. II Theil. Famil. Pyralidae – Micropterygidae.* Berlin, R. Friedlander & Sohn. 1–368.

Rebel, H. 1904. Studien über die Lepidopterenfauna der Balkanländer. II. Teil. Bosnien und Herzegowina. *Annalen des Naturhistorisches Museums in Wien* 19: 97–377.

Rebel, H. 1916a. Beitrag zur Lepidopterenfauna Bulgariens. *Verhandlungen der kaiserlich-königlichen zoologisch-botanischen Gesellschaft in Wien* 66: 36–46.

Rebel, H. 1916b. Die Lepidopterenfauna Kretas. *Annalen des Naturhistorisches Museums in Wien* 30: 66–172, pl. 4, fig. 4.

Rebel, H. 1917a. Beschreibung von 7 neuen paläarktischen Arten der Gattung *Depressaria* Hw. *Verhandlungen der kaiserlich-königlichen zoologisch-botanischen Gesellschaft in Wien* 67: 18–27.

Rebel, H. 1917b. Ueber eine Mikrolepidopterenausbeute aus dem östlichen Tannuola-Gebiet. *Deutsche entomologische Zeitschrift, Iris* 30(2): 186–195.

Rebel, H. 1927. Versammlung am 3 Dezember 1926. *Verhandlungen der kaiserlich-königlichen zoologisch-botanischen Gesellschaft in Wien* 77: 1–8.

Rebel, H. 1929. Versammlung am 1 März 1929. *Verhandlungen der kaiserlich-königlichen zoologisch-botanischen Gesellschaft in Wien* 79: 41–48.

Rebel, H. 1932. Griechische Lepidopteren. *Zeitschrift des Österreichischen Entomologen Vereins, Wien* 17: 53–56.

Rebel, H. 1936a. Neue Mikrolepidopteren von Sardinien. *Deutsche entomologische Zeitschrift, Iris* 50: 36, 92–100.

Rebel, H. 1936b. In Osthelder, Lepidoptera-Fauna von Marasch in türkisch Nordsyrien. *Mitteilungen der Münchner entomologische Gesellschaft* 25: 67–90.

Rebel, H. 1937. Zwei neue Gelechioidea. *Zeitschrift des Österreichischen Entomologen Vereins, Wien* 22: 13–16.

Retzius, A.I. 1783. *Caroli Lib. Bar. de Geer ... Genera et species insectorum e generosissimi auctoris scriptis extraxit, digessit, latine quoad partem reddidit, et terminologiam insectorum Linneanam addidit.* 220 + 32 pp. (index). Lipsiae.

Roberti, D. 1968. La difesa del carciofo dai parassiti animali. *Entomologica – Annali di Entomologia generali ed applicata, Bari* 5:127–165.

Robinson, G.S., Ackery, P.R., Kitching, I.J., Beccaloni, G.W. & Hernández, L.M. 2010. *HOSTS – A Database of the World's Lepidopteran Hostplants.* Natural History Museum, London. http://www.nhm.ac.uk/hosts [Accessed November 2023].

Rocci, U. 1934. La "*Depressaria* dell'Anice" in Italia. (Lep. – Gelechiidae). *Bolletino della Societa Entomologica Italiana* 66: 221–230.

Rössler, A. 1866. Verzeichnis der Schmetterlinge des Herzogthums Nassau. *Jahrebücher des nassauischen Vereins für Naturkunde* 19–20: 99–442.

Rymarczyk, F., Dutheil, M. & Nel, J. 2012. *Agonopterix dictamnephaga* n. sp., espèce nouvelle découverte dans les Alpes-Maritimes (France) (Lep. Elachistidae Depressariinae). *Oreina* 20: 14–16.

Rymarczyk, F., Dutheil, M. & Nel, J. 2013. *Agonopterix feruliphila* (Millière, 1866), stat. rest. *Agonopterix silerella* (Stainton, 1865) en France et description de deux nouvelles espèces, *Agonopterix orophilella* sp. nov. et *Agonopterix centaureivora* sp. nov. 2ᵉ contribution à la connaissance des Depressariinae de France (Lep. Elachistidae Depressariinae). *Oreina* 21: 13–24.

Rymarczyk, F., Dutheil, M. & Nel, J. 2015a. *Depressaria bantiella* (Rocci, 1934), stat. rev., bona species. *Agonopterix seraphimella* (Chrétien, 1929) synonyme junior d'*Agonopterix alpigena* (Frey, 1870). 4ᵉ contribution à la connaissance des Depressariinae de France (Lepidoptera, Elachistidae Depressariinae). *Revue de l'Association Roussillonnaise d'Entomologie* 24: 6–13.

Rymarczyk, F., Dutheil, M. & Nel, J. 2015b. *Depressaria millefoliella* Chrétien, 1908, synonyme junior de *D. silesiaca* Heinemann, 1870, espèce authentifiée en France. 6ᵉ contribution à la connaissance des Depressariinae de France (Lepidoptera, Elachistidae Depressariinae). *Revue de l'Association Roussillonnaise d'Entomologie* **24**: 19–23.

Rymarczyk, F., Dutheil, M. & Nel, J. (*in litt.*). *Unpublished list of host-plant records for Depressariidae from France.*

Sammut, P. 1984. A systematic and synonymic list of the Lepidoptera of the Maltese Islands. *Neue Entomologische Nachrichten* **13**: 1–124.

Savchuk, V.V. & Kajgorodova, N.S. 2015. New records of Lepidoptera in Crimea. *Caucasian entomological Bulletin* **11**: 175–182.

Savchuk, V.V. & Kajgorodova, N.S. 2017. New data on fauna and biology of Lepidoptera of Crimea. *Caucasian entomological Bulletin* **13**: 111–124.

Savchuk, V.V. & Kajgorodova, N.S. 2020. New data on the fauna and bionomics of Lepidoptera of Crimea. Part II. *Caucasian entomological Bulletin* **16** (2): 255–264.

Schläger, F. 1849. Ueber verschienen Microlepidoptern. *Bericht des lepidopterologischen Tauschvereins über das Jahre 1849*: 38–48.

Schmid, J. 2019. *Kleinschmetterlinge der Alpen: Verbreitung, Lebensraum, Biologie.* 800 pp. Bern.

Sinev, S. Yu., Baryshnikova, S.V., Lvovsky, A.L., Anikin, V.V. & Zolotuhin, V.V. 2017. Volga – Ural Microlepidoptera described by E. Eversmann, pp. 384–379. *In*: Anikin, V.V., Sachkov, S.A. & Zolotuhin, V.V. "Fauna lepidopterologica Volgo-Uralensis" from P. Pallas to present days. *Proceedings of the Museum Witt, Munich* **7**: 1–696.

Snellen, P.C.T. 1884. Nieuwe of weinig bekende Microlepidoptera van Noord-Azie. 2. Tineina en Pterophorina. *Tijdschrift voor Entomologie* **27**: 151–196, pl. 8–10.

Sohn, J.-C., Regier, J.C., Mitter, C., Adamski, D., Landry, J.-F., Heikkilä, M., Park, K.-T., Harrison, T., Mitter, K., Zwick, A., Kawahara, A.Y., Cho, S., Cummings, M.P. & Schmitz, P. 2016. Phylogeny and feeding trait evolution of the mega-diverse Gelechioidea (Lepidoptera: Obtectomera): new insight from 19 nuclear genes. *Systematic Entomology* **41**: 112–132.

Sonderegger, P. 2013. *Agonopterix flurii* sp. nov. aus dem Wallis, Schweiz (Lepidoptera, Depressariidae). *Contributions to Natural History* **21**: 1–14.

Sonderegger, P. 2013. *Agonopterix ferocella* (Chrétien, 1910) (Lepidoptera, Depressariidae) neu für die Schweiz. *Entomo Helvetica* **6**: 123–127.

Sonderegger, P. (*in litt.*). *Biologie Depressariidae Europa.* Unpublished list of species of Depressariidae and their biology.

Spuler, A. 1910. *Die Schmetterlinge Europas, Kleinschmetterlinge.* 2. 523 pp. Stuttgart.

Stainton, H.T. 1849. On the species of *Depressaria*, a Genus of Tineidae, and the allied Genera *Orthotaelia* and *Exaeretia. The Transactions of the entomological Society of London* **5**: 151–173.

Stainton, H.T. 1854. *Insecta Britannica. Lepidoptera: Tineina.* viii + 313 pp., 10 pls. London.

Stainton, H.T. 1861. *The Natural History of the Tinea* 6: ix + 283 pp., 10 pls. London.

Stainton, H.T. 1865. Notice of an undescribed species of the genus *Depressaria. Entomolologist's Monthly Magazine* **1**: 221–222.

Standfuss, M. 1851. Lepidopterologische Beiträge zur Kenntniss der Seefelder bei Heinerz und ihrer Umgebung. *Zeitschrift für Entomologie, Breslau* **16**: 49–58.

Staudinger, O. 1859. Diagnosen nebst kurze Beschreibung neuer andalusischer Lepidopteren. *Entomologische Zeitung, Stettin* **20**(7–9): 211–259.

Staudinger, O. 1870–1871a. Beschreibung neuer Lepidopteren des europäischen Faunengebiets. *Berliner entomologische Zeitschrift* **14**: 97–132, 193–208 (1870); 273–330 (1871).

Staudinger, O. (1870) 1871b: Beitrag zur Lepidopterenfauna Griechenlands. *Horae societatis entomologicae rossicae* **7**: 3–304.

Staudinger, O. 1879–1880. Lepidopteren-Fauna Kleinasien's (Fortsetzung). *Horae societatis entomologicae rossicae* **15**: 159–368 (1879); 369–435 (1880).

Stephens, J.F. 1829. *The nomenclature of British insects, being a compendious list of such species as are contained in the systematic catalogue of British insects, and forming a guide to their classification*. 68 pp.

Stephens, J.F. 1834. *Illustrations on British Entomology. Insecta Haustellata* **4**: 436 pp., 23–40 pls. London.

Strand, E. 1902. *Depressaria arctica* Strand n. sp. *Archiv for Mathematik og Naturvidenskab, Christiania* B **24**(7): 3–4.

Strand, E. 1920. Beiträge zur Lepidopterenfauna Norwegens und Deutschlands. *Archiv für Naturgeschichte* (1919) **85**A (4): 1–82.

Strand, E. 1927. Neubenennungen palaearktischer Lepidoptera und Apidae. *Archiv für Naturgeschichte* (Abt. A) **91 (12)**: 281–283.

Šumpich, J. 2013. *Depressaria pyrenaella* sp. n. – A confused species from south-western Europe (Lep.: Depressariidae). *Entomologist's Record and Journal of Variation* **125**: 114–118.

Šumpich, J. & Liška, J. 2018. New records of butterflies and moths from the Czech Republic, and update the Czech Lepidoptera checklist since 2011. *Journal of the National Museum (Prague), Natural History Series* **187**: 47–64.

Šumpich, J. & Skyva, J. 2012. New faunistic records for a number of Microlepidoptera, including description of three new taxa from Agonoxenidae, Depressariidae and Gelechiidae (Gelechioidea). *Nota lepidopterologica* **35**: 161–179.

Svensson, L., 1980. Anmärkningsvärde fynd av Microlepidoptera i Sverige 1979. *Entomologisk Tidskrift* **101**: 75–86.

Svensson, I. 1982. Four new species of Microlepidoptera from northern Europe. *Entomologica scandinavica* **13**: 293–300.

Swederus, N.S. 1787. Et nytt genus och femtio nya species of insekter beskriven. *Kongliga Svenska Vetenskaps Academiens nya Handlingar* **8**: 276–290.

Szent-Ivány, J. 1943. Depressarien-Angaben (Lepidopt.) aus der Sammlung des Ungarischen National-Museums, 2. *Fragmenta faunistica hungarica* **6**: 98–101.

Tengström, J.M.J. [1848]. Bidrag till Finlands Fjäril-Fauna. *Notiser ur Sällskapets pro Fauna et Flora fennica Förhandlingar* **1**: 69–164.

Thunberg, C.P. 1794. *Dissertatio Entomologica sistens Insecta svecica*. **7**: 83–98, 1 pl. Upsaliae.

Treitschke, F. 1832. *Die Schmetterlinge von Europa. (Fortsetzung des Ochsenheimer'schen Werkes)*. **9**(1), viii + 272 pp. Leipzig.

Treitschke, F. 1833. *Die Schmetterlinge von Europa (Forsetzung des Ochsenheimer'schen Werkes)*. **9**(2), 294 pp. Leipzig.

Treitschke, F. 1835. *Die Schmetterlinge von Europa (Fortsetzung des Ochsenheimer'schen Werkes)*. **10**(3), 302 pp. Leipzig.

Turati, E. 1879. Contribuizione alla fauna lepidotterologica Lombarda. *Bollettino della Società entomologica italiana* **11**: 153–208.

Turati, E. 1921. Nuove forme di lepidotteri. IV. Correzione e note critiche. *Naturalista siciliano* (1919) **23**: 203–351.

Turati, E. 1924. Spedizione lepidotterologica in Cirenaica 1921–1922. *Atti della Società italiana di Scienze naturali e del Museo Civico di Storia Naturali in Milano* **63**: 17–191.

Turati, E. 1927. Novita di Lepidotterologia in Cirenaica. 2. *Atti della Società italiana di Scienze naturali e del Museo Civico di Storia Naturali in Milano* **66**: 313–344.

Turati, E. & Zanoni, V. 1922. Materiali per una faunula lepidotterologia di Cirenaica. *Atti della Società italiana di Scienze naturali e del Museo Civico di Storia Naturali in Milano* **61**: 138–178.

Tutt, J.W. 1893. *Depressaria aurantiella*, n. sp? *Entomologist's Record and Journal of Variation* **4**: 241.

Van Laar, W. 1961. Female genitalia of the species of *Depressaria* Hw. s.l. (Lepidoptera, Oecophoridae) occurring in the Netherlands. *Zoologische Mededelingen* **38**: 15–40.

Van Laar, W. 1964. Male genitalia of the species of *Depressaria* Haworth s.l. (Lepidoptera, Oecophoridae) occurring in the Netherlands. *Zoologische Mededelingen* **39**: 391–408.

Van Nieukerken, E.J., Karsholt, O., Brown, R.L.,Heikkilä, M., Huemer, P., Kaila, L., Landry, J.-F., Li, H., Ponomarenko, M.G. & Sinev, S. 2022. Case 3841 – Epigraphiidae Guenée, 1845 (Lepidoptera, Gelechioidea): proposed suppression to conserve the widely used family-group name Depressariidae Meyrick, 1883. *The Bulletin of Zoological Nomenclature* **79**(1): 18–30.

Vives Moreno, A. 2014. Catálogo sistemático y sinonímico de los Lepidoptera de la Península Ibérica, de Ceuta, de Melilla y de las Islas Azores, Baleares, Canarias, Madeira y Salvajes. (Insecta: Lepidoptera.). *SHILAP Revista de Lepidopterología Suppl.*: 1–1184.

Vives Moreno, A. & Gastón, J. 2017. Contribución al conocimiento de los Microlepidoptera de España, con la descripción de una especie nueva (Insecta: Lepidoptera). *SHILAP Revista de lepidopterología* **45**(178): 317–342.

Wallengren, H.D.J. 1881. Genera nova Tinearum. *Entomologisk Tidskrift* **2**: 94–97.

Walsingham, Lord. 1881. On some North-American Tineidae. *Proceedings of the zoological Society of London* **1881**: 301–325.

Walsingham, Lord. 1898. New Corsican Microlepidoptera. *Entomologist's monthly Magazine* **34**: 131–134, 166–172.

Walsingham, Lord. 1903. Spanish and Moorish Micro-Lepidoptera. *Entomologist's monthly Magazine* **39**: 179–187, 209–214, 262–268, 292–293.

Walsingham, Lord. 1907a. Algerian Microlepidoptera. *Entomologist's monthly Magazine* (2) **18**(43): 147–154.

Walsingham, Lord. 1908. Microlepidoptera of Tenerife. *Proceedings of the zoological Society of London* **1907**: 911–1034.

Wang, Q.-Y. & Li, H.-H. 2020. Phylogeny of the superfamily Gelechioidea (Lepidoptera: Obtectomera), with an exploratory application on geometric morphometrics. *Zoologica Scripta* **49**(10): 307–328.

Weber, P. 1945. Die Schmetterlinge der Schweiz. 7. Nachtrag Mikrolepidoptera. *Mitteilungen der Schweizerischen Entomologischen Gesellschaft* 19: 347–407.

Wocke, M.F. 1849. 2. Bericht über die Arbeiten der entomologische Sektion im Jahre 1849. *Jahresbericht der Schlesischen Gesellschaft für Vaterlandische Kultur. Naturwissenschaftlich-Medizinische Reihe. Breslau* 1849: 65–73.

Wocke, M.F. 1857. Neue schlesische Falter. *Jahresbericht der Schlesischen Gesellschaft für Vaterlandische Kultur. Naturwissenschaftlich-Medizinische Reihe. Breslau* 1857: 116–119.

Wocke, M.F. 1887. Zwei neu Gelechiden. *Zeitschrift für Entomologie, Breslau (N.F.)* 12: 62–64.

Ylla, J., Requena, E. & Macià, R. 2020. Noves dades faunístiques d'interès dels gèneres *Agonopterix* Hübner, [1825] 1816, i *Depressaria* Haworth, 1811, per a la península Ibèrica i Balears (Lepidoptera: Depressariidae, Depressariinae). *Butlletí de la Societat Catalana de Lepidopterologia* 111: 11–30.

Zeller, P.C. 1839. Versuch einer naturgemaessen Eintheilung der Schaben. *Isis von Oken, Leipzig* 1839 (3): 167–220.

Zeller, P.C. 1847. Bemerkungen über die auf einer Reise nach Italien und Sicilien beobachteten Schmetterlingsarten. *Isis von Oken, Leipzig* 1847(2): 121–159; (3): 213–233; 284–308; (6): 401–457; (7) 483–522; (8): 561–594; (9): 641–673; (10) 721–771; (11): 800–859; (12): 881–914.

Zeller, P.C. 1850. Verzeichniss der von Herrn Jos. Mann beobachteten Toscanischen Microlepidoptera. *Entomologische Zeitung, Stettin* 11(2): 59–64; (4): 134–136; (5): 139–162; (6): 195–212.

Zeller, P.C. *In:* Schmidt, F. 1851. Lepidopterologische Beobachtungen. *Entomologische Zeitung, Stettin* 12: 74–83.

Zeller, P.C. 1854. Die Depressarien und einige ihnen nahe stehende Gattungen. *Linnaea Entomologica* 9: 189–403.

Zeller, P.C. 1868. Beiträge zur Naturgeschichte der Lepidoptern. *Stettiner entomologische Zeitung* 29: 401–429.

Zerny, H. 1934. Lepidopteren aus dem nördlichen Libanon. *Deutsche entomologische Zeitschrift, Iris* 48: 1–28.

Zerny, H. 1940. Mikrolepidopteren aus dem Elburs-Gebirge in Nord-Iran. *Zeitschrift des Wiener Entomologen-Vereins* 25: 20–24.

Zetterstedt, J.W. 1838–1840. *Insecta lapponica.* 1140 pp. Lipsiae.

Zhu, X., Zhang, L. & Wang, S. 2023. New species and newly recorded species of the genus *Agonopterix* Hübner, [1825] (Lepidoptera: Depressariidae) from China. *Zootaxa* 5258 (4): 379–404.

Index to Host-Plants

The numbers refer to species numbers. Synonyms are not listed.
Questionable host-plant records are not included.

Achillea millefolium   109, 141, 149
   setacea   149
Adenocarpus complicatus   116
   hispanicus   120
   sp.   117
Aegopodium podagraria   56, 59, 61, 124
Amelanchier ovalis   3
Anagyris foetida   118
Anchusa officinalis   92
Anethum sp.   143
Angelica sp.   124
   archangelica   42, 59, 71
   sylvestris   56, 59, 61, 143
Antennaria dioica   8
Anthriscus caucalis   42
   sylvestris   41, 42, 59, 60, 61, 75, 124, 126, 135, 158
Apium graveolens   71
Arctium sp.   103
   lappa   99, 102, 105, 109
   minus   102, 109
Artemisia absinthium   144
   borealis   146
   campestris   9, 15, 146
   canariensis   145
   chamaemelifolia   144
   dracunculus   146
   frigida   144
   laciniata   15
   vallesiaca   144
   vulgaris   9, 19, 112, 142, 149
Astrantia major   59, 66
Athamanta cretensis   33, 37, 133
   turbith   65

Berula erecta   161
Betula sp.   1, 2, 10
Bunium bulbocastanum   128, 156
Bupleurum angulosum   169
   falcatum   30, 33, 169
   frutescens   33
   fruticosum   29
   petraeum   33, 169
   ranunculoides   169

   rigidum   30
   rotundifolium   169
   salicifolium   170
   stellatum   33, 169
Cachrys libanotis   48
Calicotome spinosa   116
Carduus sp.   102, 103
   crispus   99
   defloratus   99
   litigiosus   99, 105
   personata   103, 109
   pycnocephalus   99
   tenuifolius   99, 100
Carlina acanthifolia   105
   corymbosa   100
   vulgaris   100, 109
Carpinus betulus   1
Carthamus caeruleus   86
Carum sp.   59, 71, 126
   carvi   61, 143, 154, 160
   verticillatum   156, 160
Centaurea sp.   91, 102
   aspera   84, 99
   jacea   84, 92, 99, 103, 109
   leucophaea   99, 104
   nigra   84, 92, 99, 103, 109
   nigrescens   92, 109
   paniculata   84, 89, 92, 99
   pectinata   92, 99
   scabiosa   84, 87, 92, 99, 104, 109
   sphaerocephala   85, 99
   sterilis   89
   stoebe   84, 99
   uniflora   92, 103, 104, 106
   vallesiaca   84, 99
Chaerophyllum aureum   59, 61, 158
   bulbosum   135, 158
   hirsutum   154
   temulum   41, 42, 61, 71, 124, 135, 158
   villarsii   59, 61, 154
Cicuta douglasii   160
   virosa   59, 71, 160, 161

Cirsium sp. 92
  acaulon 99, 103
  arvense 99, 102, 103, 109
  eriophorum 99, 102, 103, 109
  ferox 99, 105
  heterophyllum 103
  monspessulanum 109
  tuberosum 105, 109
  vulgare 99, 102, 103, 105, 109
Cnidium dubium 123
Conium maculatum 43, 61, 69, 124, 135
Conopodium sp. 130
  majus 61, 123, 131, 135, 156
Coriandrum sp. 143
Corylus avellana 1
Cotoneaster sp. 3
Crataegus sp. 3, 5
Crithmum maritimum 71, 143, 162
Cyanus montanus 92, 104, 106
  segetum 99, 103, 106
  triumfettii 106
Cynara cardunculus 99
  scolymus 190
Cytisophyllum sessilifolium 115, 119
Cytisus decumbens 120
  grandiflorus 116
  hirsutus 119, 122
  oromediterraneus 116, 119
  scoparius 114, 116, 119
  striatus 116
  triflorus 122
  villosus 114, 122

Daucus carota 33, 40, 41, 59, 61, 70, 71, 75, 123,
    126, 135, 143, 152, 164
  crinita 126
  muricatus 41, 152
Dictamnus albus 80, 81, 186
Distichoselinum tenuifolium 40
Doronicum austriacum 113
  carpetanum 113
  pardalianches 113
Drusa oppositifolia 62

Echinops ritro 105
  sphaerocephalus 105
  spinosissimus 85
Elaeoselinum asclepium 77
Elwendia persica 156

Eryngium campestre 67, 178
  maritimum 67

Falcaria vulgaris 70, 143
Ferula communis 31, 39, 49, 143
  glauca 39, 49
  linkii 179
  litwinowiana 184
  tingitana 31
Ferulago campestris 52, 73
Foeniculum vulgaris 31, 33, 39, 143

Galactites tomentosus 99
Genista anglica 115, 119, 120
  cinerea 114, 119, 120
  florida 114
  germanica 119
  monspessulana 116, 122
  pilosa 114, 115, 116, 119, 120, 121
  radiata 119
  sagittalis 115, 119
  scorpius 116, 120
  tinctoria 115, 116, 119, 121

Haplophyllum suaveolens 14
  tuberculatum 187
Helosciadium nodiflorum 160, 161, 163
Hieraceum sp. 171
Heracleum sp. 56
  austriacum 154
  juranum 154
  sphondylium 59, 61, 124, 143, 154, 163
Hypericum hyssopifolium 25
  maculatum 25
  perforatum 24, 25
  pulchrum 24
  tetrapterum 24
  undulatum 25
Hypochaeris radicata 171

Inula montana 109

Knautia arvensis 92, 109

Laburnum alpinum 116
  anagyroides 116, 119
Laser trilobum 72
Laserpitium gallicum 44, 58, 143, 175
  halleri 59, 143, 154, 164

Laserpitium (*cont.*)
 krapfii 56, 154
 latifolium 44, 56, 59, 61, 154, 164
 nestleri 44
 peucedanoides 37, 65
 prutenicum 175
 siler 33, 44, 59, 72, 98, 154, 175
Lembotropis nigricans 119
Leontodon sp. 171
Ligusticum ferulaceum 59, 164
 lucidum 53, 55, 59, 73, 164, 166
 mutellina 154
 scoticum 61
Lupinus arboreus 116, 119

Mantisalca salmantica 99
Meum athamanticum 33, 59, 61, 123, 154
Myrrhis odorata 61

Oenanthe sp. 61
 aquatica 160, 161
 crocata 61, 71, 160, 161
 fistulosa 160
 pimpinelloides 71, 75, 160
 pteridifolia 43
 sarmentosa 160
 silaifolia 59
Onopordum acanthium 99, 109
Opopanax chironium 35, 42
Orlaya daucoides 33
 grandiflora 40, 143

Pastinaca sativa 33, 56, 59, 61, 75, 124, 143, 163
Pericallis appendiculata 108
 tussilaginis 108
Petasites albus 110
 hybridus 110
 paradoxus 110
Peucedanum sp. 61, 124
 alsaticum 59, 63
 cervaria 36, 57, 58, 70
 officinale 63, 89
 oreoselinum 53, 58, 135, 143
 ostruthium 56, 59, 154
 palustre 53, 56, 59, 71, 143
 venetum 59, 154
Pimpinella sp. 135
 anisum 165
 major 59, 127, 154, 167
 nigra 59, 167
 peregrina 71, 75, 165

 saxifraga 33, 56, 59, 70, 123, 143, 154, 167
 villosa 34, 168
Populus sp. 10
 tremula 4, 28
Prangos trifida 33
Prunus spinosa 3, 5
Ptychotis heterophylla 135
 saxifraga 156

Rhaponticum coniferum 99
 scariosum 109
Ruta angustifolia 12, 82
 chalepensis 82
 graveolens 12, 82
 montana 82
 pinnata 62

Salix sp. 10, 27
 alba 26, 28
 aurita 26, 28
 caprea 26, 28
 cinerea 26, 28
 daphnoides 26
 eleagnos 28
 fragilis 26, 28
 myrsinites 27, 28
 myrtilloides 27
 purpurea 26, 28
 repens 26, 28
 triandra 26
 viminalis 26, 28
Salvia fruticosa 189
 lavandulifolia 189
 officinalis 189
Sanicula europaea 61, 66
Saussurea alpina 93
Scabiosa columbaria 92
Scaligeria napiformis 156
Scandix australis 33
 pecten-veneris 32, 33
Selinum carvifolia 53, 58, 59
 dubium 64
 silaifolia 42, 53
Senecio cacaliaster 111
 doria 111
 doronicum 111
 nemorensis 111
 ovatus 111
 pyrenaicus 111
 sarracenicus 111
Serratula tinctoria 84, 87, 101, 102, 109

Seseli sp. 125, 126, 135
  annuum 125
  arenarium 125
  austriacum 65
  gummiferum 33
  hippomarathrum 65, 74, 125
  leucospermum 38
  libanotis 33, 42, 58, 59, 61, 63, 123, 129, 134,
    143, 164
  longifolium 65
  montanum 38, 65
  pallasii 38
  tortuosum 33, 38, 67, 143, 166
Silaum silaus 41, 59, 61, 70, 71, 143
Silybum marianum 99
Sison amomum 61, 160
Sium latifolium 59, 70, 71, 143, 160, 161
Smyrnium olusatrum 31, 43, 61, 75
Sonchus sp. 109
  arvensis 171
Sorbus sp. 3
Sorbus aucuparia 5

Spartium junceum 120
Spiraea media 188

Tanacetum corymbosum 21, 141
  vulgare 140, 141, 149
Taraxacum sp. 171
Thapsia garganica 34, 179
  villosa 34, 179
Tilia cordata 1
  platyphyllos 1
Tordylium maximum 41
Torilis arvensis 41
  japonica 41, 61, 126, 135
Trinia glauca 33, 63, 65, 74, 125, 150
Tussilago farfara 110

Ulex sp. 119
  europaeus 120
  gallii 120
  minor 120

Xeranthemum inapertum 107

Index to Entomological Genus Names

The numbers refer to the page numbers.
Synonyms are in italics.

Agonopterix   82
Agonopteryx   83

Ctenioxena   83

Depressaria   251
Depressariodes   61

Enicostoma   57
Epeleustia   82
Epigraphia   50
Exaeretia   61

Haemylis   82
Hasenfussia   252
Horridopalpus   252

Levipalpus   61
Luquetia   57

Martyrhilda   61

Piesta   251
Pinaris   82

Schistodepressaria   251
Semioscopis   50
Siganorosis   251
Subagonopterix   83
Syllochitis   83

Tichonia   82

Volucra   251
Volucrum   251

Index to Entomological Species-Group Names

The numbers refer to the species number.
Synonyms are in italics.
Species mentioned in text with page number are extralimital or excluded.

abchasiella   6
abditella   97
absinthivora   144
absynthiella   144
adspersella   33
adspersella sensu Rymarczyk *et al.*   38
adustatella   139
aegopodiella   135
agyrella   p. 151, 361
albarracinella   177
albidana   71
albidella   69
albiocellata   138
albipuncta   135
albipunctella   135
alienella   2
allisella   9
almatinka   p. 332
alpigena   44
alstromeriana   69
altaica   176
amanthicella   33
amasiella   143
amasina   42
amblyopa   138
amilcarella   99
amurella   19
anchusella   144
anella   2
angelicella   56
annexella   59
anthriscella   75
apiella   160
applana   61
arabica   187
archangelicella   p. 151
arctica   27
arenella   109
aridella   100
artemisiae   146
aspersella   122
assalella   p. 276
assimilella Treitschke   114
assimilella Zeller   100

astrantiae   66
atomella   115
atomella   4
atricornella   71
atrostrigella   148
aurantiella   171
autocnista   122
avellanella   1

badiella   171
banatica   41
bantiella   165
beckmanni   127
bipunctifera   p. 178
bipunctosa   87
blackmori   119
bluntii   143
boicella   119
broennoeensis   93
brunneella   171
budashkini   89
bupleurella   169
buryatica   p. 151
buvati   18

cachritis   48
cadurciella   38
calycotomella   116
campestrella   178
caprella   70
capreolella   70
carduella   103
carduncelli   86
caucasica   163
centaureivora   104
cerefolii   61
cervariella   36
cervicella   174
chaerophylli   158
characterella   3
characterella   28
characterosa   3
chironiella   35
chneouriella   152

cicutella 61
ciliella 59
cinderella 130
cinerariae 108
ciniflonella 10
cluniana 78
cnicella 67
coenosella 46
colarella 143
comitella 118
conciliatella 117
consimilella 4
conterminella 26
corichroella 92
corticinella 171
costosa 119
cotoneastri 111
crassiventrella 33
crispella 99
cruenta 152
cryptipsila 85
culcitella 21
cuprinella 152
curvipunctosa 42
cyrniella 116

daghestanica 104
daucella 160
daucivorella 164
deliciosella 178
delphinias 139
depressana 143
depressella 143
depunctella 119
deverrella   p. 330, 334, 335, 343
dictamnella 186
dictamnephaga 81
discipunctella 180
divergella 84
doronicella 113
douglasella 126
dracunculi 146
dryadoxena 119
dumitrescui p. 361
duplicatella 139

echinopella 85
emeritella 140
epicachritis 48
erinaceella 190
eryngiella 178

erzurumella 185
exquisitella 19

ferocella 105
ferulae 49
feruliphila 33
flavella 92
floridella 125
flurii 104
freyi 149
frigidella 171
frustratella 171
fruticosella 29
fumicella 1
funebrella 68
furvella 80
fuscipedella 155
fusconigerella 173
fuscovenella 71
fuscovirgatella 147

galicicensis 50
gallicella 175
genistella 116
geoffrella p. 57
gilvella 109
gozmanyi 11
graecella 76
granulosella 42
guanchella 79
gudmanni 124

halophilella 162
hamriella p. 211
hannemanniana 184
hansjoachimi 181
hepatariella 8
heraclei auct. 163
heraclella 106
heracliana 61
heracliana auct. 163
heydenii 154
himmighofenella 99
hippomarathri 65
hirtipalpis 189
hofmanni 134
homochroella 14
huebneri 25
humerella 58
hungarica 167
hypericella 25

hypericella auct. 24
hystricella 188

iliensis 43
immaculana 109
impurella 24
incarnatella 106
incognitella 128
indecorella 132
indubitatella 23
infernella 131
inoxiella 39
intermediella 99
invenustella 96
irrorata 75
irrorella 114
isa 10
isabellina 66
ivinskisi, Agonopterix 104
ivinskisi, Depressaria p. 332, 335, 339

junnilaineni 182

kaekeritziana 92
kailai p. 332, 334
karmeliella 33
kayseriensis 47
keltella 99
klamathiana 10
klimeschi 112
knitschkei 120
kotalella 42
krasnowodskella 137
kuznetzovi 101
kyzyltashensis 88

lacticapitella 133
laetella 113
langmaidi 51
larseniana 124
laserpitii 164
laterella 106
latipennella 95
lechriosema 9
lederi 14
lennigiella 120
lepidella 17
lessini 52
leucadensis 32
leucocephala 142
leucostictella 19

leviella 14
libanotidella 164
lidiae 98
ligusticella 73
linolotella 34
liodryas 85
liturella 25
liturella 92
liturosa 25
liupanshana p.
lobella 5
longipennella 159
lugubrella 5
lutosella 12
lvovskyi 20

manglisiella 157
marcella 152
marmotella 111
medelichensis 74
melancholica 68
mendesi 85
mesopotamica 164
mikomoensis 84
millefoliella 149
miserella 126
mongolicella 19
monilella 69
moranella 187
multiplicella 112
mutatella 117

nanatella 100
nebulosella 16
nemolella 129
nervosa Haworth 119
nervosa Stephens 160
nigromaculata 13
niviferella 22
nodiflorella 39
nordlandica 27
novaspersella 122

obolucha 178
obscurana 119
occaecata 99
ocellana 28
oculella 2
oglatella 71
oinochroa 121
olerella 141

olusatri 31
ontariella 163
ordubadensis 54
orientella 6
orophilella 60
osthelderi 7

pagmanella 147
pallorella 84
paracervariella 37
paraselini 57
parilella 58
pastinacella Duponchel 163
pastinacella Stainton 180
pavida 33.1
peloritanella 40
peniculatella 191
pentheri 183
peregrinella 153
perezi 62
perpallorella 87
perstrigella 119
petasitis 110
pimpinellae 167
praeustella 15
prangosella 143
preisseckeri 11
propinquella 102
prostratella 120
pseudobadiella 172
pseudoferulae 77
puella 69
pulcherrimella 123
pulverella 115
pulverella 167
punctata 61
pupillana 81
purpurea 41
putrida 71
putridella 63
pyrenaella 151

quadripunctata 64
quintana 175

radiata 163
radiella 163
radiosquamella 139
ragonoti 44
rebeli 30
remota 99

respersella 115
retiferella 82
rhodochlora 143
rhodochrella 99
riadella p. 343
richteri 45
rigidella 30
rimulella p. 361
rjabovi p. 332, 339
roseoflavella 93
rotundella 40
rubescens 116
rubricella 160
rubripunctella 33
rungsiella Hannemann, 1953 192
rungsiella sensu Hannemann, 1976 191
rutana 82
ruticola p. 284

sabulatella 33
salevensis 44
sarahae 170
sardoniella 190
sarracenella 111
scandinaviensis 63.1
schaidurovi p. 339
schmidtella 113
scopariella 116
selini 53
semenovi 123
senecionis 111
seneciovora 111
septicella p. 151
seraphimella 44
signella 28
signosa 28
silerella 72
sileris 44
silesiaca 149
smolandiae 10
socerbi 55
sordidatella 124
sparrmanniana 92
sphondiliella 163
squamosa 89
steinkellneriana 3
stramentella 21
straminella 85
strigulana 4
subalbipunctella 136
sublutella 99

subnervosa 173
subpallorella 84
subpropinquella 99
subtakamukui 78
subtenebricosa 139
subumbellana 83

tabelli 170.1
takamukui p. 172
tenebricosa 138
tenerifae 145
thapsiella 34
thomanniella 142
thoracica 99
thunbergiana 5
thurneri 13
tortuosella 166
tripunctaria 73
tschorbadjiewi 90
turbulentella 42

uhrykella 171
ulicetella 120
ultimella 161
ululana 156

umbellana 120
uralensis 94

vaccinella 41
variabilis 99
velox 166
vendettella 43
veneficella 179
ventosella 71
venustella 157
villosae 168
volgensis 91

weirella 124

xeranthemella 107
xyleuta 14

yeatiana 71
yomogiella p. 230

zapryagaevi p. 306
zelleri 150
zephyrella 42

Printed in the United States
by Baker & Taylor Publisher Services